Assisted Reproductive Technologies in Animals Volume 2

Juan Carlos Gardón • Katy Satué Ambrojo
Editors

Assisted Reproductive Technologies in Animals Volume 2

Future Developments and Considerations

Springer

Editors
Juan Carlos Gardón
Faculty of Veterinary and Experimental Sciences
Department of Animal Medicine and Surgery, Catholic University of Valencia-San Vicente Mártir
Valencia, Spain

Katy Satué Ambrojo
Medicine and Animal Surgery Department
CEU-Cardenal Herrera University
Alfara del Patriarca, Valencia, Spain

ISBN 978-3-031-87197-9 ISBN 978-3-031-87198-6 (eBook)
https://doi.org/10.1007/978-3-031-87198-6

This Springer imprint is published by the registered company Springer Nature Switzerland AG
The registered company address is: Gewerbestrasse 11, 6330 Cham, Switzerland

Foreword

The continuous advancement of science and technology has led to a profound reshaping of our understanding of the biological world. This has resulted in a series of remarkable breakthroughs that were previously thought to be beyond reach. Among these remarkable strides, assisted reproductive technologies (ART) have revolutionized not only human medicine but also the field of animal reproduction.

Assisted Reproductive Technologies in Animals: Future Developments and Considerations is a testament to the profound impact these technologies have had on the management, conservation, and improvement of animal species across the globe. The journey of ART in animals began with a simple yet revolutionary idea: to overcome the natural barriers of reproduction and enhance the genetic potential of animals. Today, this field encompasses a wide array of techniques, including artificial insemination, in vitro fertilization, embryo transfer, and cloning as well as basic concepts about physiological reproductive events. These methods are not only critical for agricultural productivity but also play a pivotal role in wildlife conservation and the preservation of endangered species.

This book provides an extensive exploration of the current trends and practices in animal ART, offering invaluable insights for veterinarians, researchers, and students. It highlights the latest advancements, from the refinement of existing techniques to the development of novel approaches that push the boundaries of what is possible. The contributors to this volume are leading experts in their respective fields, each bringing a wealth of knowledge and experience to the table. This ensures that readers are equipped with the most up-to-date information and practical guidance.

One of the most significant contributions of this book is its emphasis on the ethical and practical implications of ART. As we endeavor to reconcile technological expertise with ethical considerations, it is imperative to comprehend the influence of these technologies on animal welfare, genetic diversity, and ecosystem stability. These technologies must recognize the responsibility that comes with such powerful tools, and we have to use them ethically and prudently.

Furthermore, the current trends in reproductive management also emphasize the collaborative nature of this field. The integration of knowledge from various disciplines, including genetics, endocrinology, veterinary medicine, and beyond, demonstrates that progress is often the result of interdisciplinary efforts. Such collaboration not only accelerates innovation but also ensures that the benefits of ART are maximized across different sectors and species.

As we stand on the threshold of discoveries and innovations, this book serves as both a reflection on the progress that has been made and a beacon guiding us toward future possibilities. This book is an essential resource for anyone involved in animal reproduction, providing a comprehensive foundation for understanding and applying ART in various contexts. In conclusion, we would like to express our profound gratitude to the authors and contributors who have dedicated their expertise and passion to this volume. Their dedication to the advancement of the field of animal reproductive technologies is evident in each chapter, and their work will undoubtedly serve as an inspiration to future generations of scientists and practitioners.

It is our hope that this book will serve as a cornerstone for continued progress and a source of inspiration for all those who seek to harness the power of ART for the betterment of animal health, conservation, and productivity.

Valencia, Spain — Juan Carlos Gardón
Valencia, Spain — Katy Satué Ambrojo

Preface

The field of assisted reproductive technologies (ART) in animals has undergone a profound evolution over the past few decades, progressing from experimental concepts to indispensable tools in modern animal husbandry, conservation biology, and biomedical research. This book, *Assisted Reproductive Technologies in Animals: Future Developments and Considerations*, aims to provide a comprehensive overview of the latest advancements, methodologies, and practical applications of ART in various animal species.

The impetus for compiling this book stems from the recognition of the profound impact that ART has on animal production, genetic improvement, and biodiversity conservation. As global demand for food increases, efficient and sustainable livestock production becomes more critical. ART plays a crucial role in meeting these demands by enhancing reproductive efficiency, accelerating genetic gain, and mitigating the impacts of disease and environmental stressors. Moreover, the application of ART in wildlife conservation offers hope for endangered species, providing tools to preserve genetic diversity and prevent extinction.

This book is structured to guide readers through the multifaceted world of animal ART. We begin with foundational principles, exploring the biology of reproduction and the historical context of ART development. Subsequent chapters delve into specific technologies such as artificial insemination, in vitro fertilization, embryo transfer, and cloning. Each chapter is crafted to not only explain the technical aspects but also discuss the practical implications, challenges, and prospects of these technologies.

A notable aspect of this volume is its interdisciplinary approach. The volume is enriched by contributions from leading experts in veterinary medicine, reproductive biology, genetics, and conservation science, which collectively provide a well-rounded perspective. The inclusion of case studies and real-world applications provides readers with practical insights and demonstrates the versatility of ART across different contexts and species.

The text incorporates ethical considerations, reflecting the growing importance of responsible and humane use of reproductive technologies. As we endeavor to expand the boundaries of what is feasible, it is of the utmost importance to consider the well-being of the animals involved, the ecological implications, and the ethical dimensions of genetic manipulation. The objective of this book is to encourage a balanced view of the advancements in ART while promoting ethical stewardship.

Furthermore, the text addresses the latest research trends and emerging technologies that promise to shape the future of animal reproduction. The field of assisted reproductive technologies (ART) is undergoing rapid expansion, encompassing genomic selection and gene editing, advanced cryopreservation techniques, and biobanking. These innovations have the potential to revolutionize animal breeding, conservation efforts, and biomedical research, opening new avenues for exploration and application.

In preparing this book, we have been privileged to work with an outstanding group of contributors who have generously shared their knowledge and expertise. Their dedication to advancing the science and practice of animal ART is evident in every chapter, and their contributions are what make this book an authoritative resource.

We hope that *Assisted Reproductive Technologies in Animals: Future Developments and Considerations* will serve as a valuable reference for researchers, practitioners, and students alike. Whether you are engaged in livestock production, wildlife conservation, or academic research, we believe that this book will provide the information and inspiration needed to advance the field of animal reproduction.

We would like to express our gratitude to all those who have embarked on this journey with us. We look forward to the discoveries and innovations that lie ahead, and we are confident that this book will play a part in shaping the future of reproductive management in animals.

Valencia, Spain — Juan Carlos Gardón
Valencia, Spain — Katy Satué Ambrojo

Contents

Abbreviations

ADS	Abnormal development syndrome
AFP	Antral follicle population
Ag	Silver
AI	Artificial insemination
AIRN	Antisense of IGF2R non-protein coding RNA
Al	Aluminum
ALH	Lateral head displacement
AMH	Anti-mullerian hormone
Anim	Animal
AOS	Abnormal offspring syndrome
ARCA	Official Catalog of Livestock Breeds of Spain
ART(s)	Assisted reproductive technolog(ies)
Assoc	Association
Au	Gold
AUFI	Absolute uterine factor infertility
AV	Artificial vagina
AVAMUR	Murciano-Levantina Cow Breeders Association
BBM	Behavior-based management
BCF	Beat cross frequency
BEGAIN	Brain-enriched guanylate kinase-associated protein
B-mode or 2D mode	Two-dimensional mode
BPD	Biparietal diameter
BSA	Bovine serum albumin
BTS	Beltsville thawing solution
BWS	Beckwith–Wiedemann syndrome
CAD	Canadian dollar
CASA	Computer-assisted sperm analysis
CAT	Catalase
CB	Carbon black
CCTV	Closed-circuit television
CDKN1C	Cyclin-dependent kinase inhibitor 1C
CeO_2	Cerium oxide
CL	Corpus luteum
CLBP	Corpus luteum blood perfusion

COREMUR	Association for the conservation and recovery of the Murciano-Levantina cow
CPU	Central processing unit
CTCF	1-zinc finger protein CCCTC-binding factor
Cu	Copper
dECM	Decellularized extracellular matrix
Develop	Development
dH_2O	Distilled water
DIRAS3	DIRAS family GTPase 3
DMR	Differentially methylated region
DMSO	Dimethyl sulfoxide
DNA	Deoxyribonucleic acid
Dom.	Domestic
DRA	Disease risk analysis
ECM	Extracellular matrix
Ed	Editorial
EDTA	Ethylenediaminetetraacetic acid
EGTA	Ethylene glycol tetraacetic acid
E-N	Eosin-nigrosin stain
EPHA5	EPH receptor A5
ERK	Extracellular signal-regulated kinase
ET	Embryo transfer
EU	European Union
Fe_3O_4	Iron oxide
Fert.	Fertility
FSH	Follicle-stimulating hormone
GA	Grade A
GAGs	Glycosaminoglycans
GB	Grade B
GC	Grade C
GSH	Glutathione
GSH	Glutathione peroxidase
GT	Genital tubercle
GV	Germinal vesicle
H19	H19 imprinted maternally expressed transcript
H3K27me3	Trimethylation on histone 3 lysine 27
H3K4me3	Trimethylation on histone 3 lysine 4
H3K9me3	Trimethylation on histone 3 lysine 9
HA	Hydroxyapatite
HiC	High-throughput chromosome conformation capture technique
HIREC	Human-induced rapid environmental change
HOX	Homeobox (homeotic) genes
HPG	Hypothalamic-pituitary-gonadal
HSP	Heat shock proteins

HVJ	Hemagglutinating virus of Japan
ICR	Imprinting control region
ICSI	Intracytoplasmic sperm injection
IETS	International Embryo Transfer Society
IGF2	Insulin-like growth factor 2
IGF2R	Insulin-like growth factor 2 receptor
IM	Individual motility
IMIDA	Murcian Institute for Agricultural and Environmental Research and Development
iPSCs	Induced pluripotent stem cells
IUCN	International Union for Conservation of Nature
IVC	In vitro culture
IVEP	In vitro embryo production
IVF	In vitro fertilization
IVM	In vitro maturation
IVP	In vitro production
Jpn	Japan
KCNQ1OT1	Potassium voltage-gated channel subfamily Q member 1 overlapping transcript 1
KRP	Krebs-Ringer phosphate
LH	Luteinizing hormone
LIN	Linearity index
lncRNA	Long noncoding RNA
LOC535280	Uncharacterized bovine transcript
LOS	Large offspring syndrome
LPO	Lipoperoxidación
MAD	Mean angular displacement of the head
MAGEL2	MAGE family member L2
MAPA	Ministry of Agriculture, Fisheries and Food
MDA	Malondialdehyde
MG	Murciano-Granadina
MII	Metaphase II of meiosis II
min	Minute
miRNA	microRNA
ml	Milliliter
ML	Murciano-Levantino
MMP	Mitochondrial membrane potential
MMP2	Matrix metallopeptidase 2
MMPs	Matrix metalloproteinase
MNP	Magnetic nanoparticles
MOET	Multiple ovulation and embryo transfer
Mol	Molecular
MoO_3	Molybdenum trioxide
MOT (%)	Percentage of motile spermatozoa
MPF	Maturation promoting factor

mRNA	Messenger ribonucleic acid
mSOF	Modified synthetic oviduct fluid
MTT	3-[4,5-dimethylthiazol-2-yl]-2,5-diphenyltetrazolium bromide
LN_2	Liquid nitrogen
NaCl	Sodium chloride
NAP1L5	Nucleosome assembly protein 1-like 5
ncRNA	Noncoding RNA
NGOs	Non-governmental organizations
Ni	Nickel
NNAT	Neuronatin
NP	Nanoparticles
OMIM	Online Mendelian Inheritance in Man
OPU	Ovum pick up
P4	Progesterone
PB1	First polar body
PBS	Phosphate-buffered saline
PEG10	Peternally espressed gene 10
PEG3	Peternally espressed gene 3
pers. obs	Personal observations
PHLDA2	Pleckstrin homology-like domain family A member 2
PI	Pulsatility index
PI	Propidium iodide
PLAGL1	PLAG1-like zinc finger 1
PLGA	Polylactic-co-glycolic acid
PMN	Polymorphonuclear cells
PNA	Peanut agglutinin
PolII	RNA polymerase II
PRF	Pulse repetition frequency
Prod	Production
Rec	Record
Reprod	Reproduction
RFM	Retained fetal membranes
RGS17	Regulator of G protein signaling 17
ROS	Reactive oxygen species
RT	Reaction time
RT	Room temperature
SAOS	Spontaneous abnormal offspring syndrome
Sci.	Science
SCNT	Somatic cell nuclear transfer
SDC	Sodium deoxycholate
SDS	Sodium dodecyl sulfate
Se	Selenium
SLOS	Spontaneous large offspring syndrome
SMGT	Sperm-mediated gene transfer

SNRPN	Small nuclear ribonucleoprotein polypeptide N
SOD	Superoxide dismutase
SPION	Superparamagnetic iron oxide nanoparticle
STR	Straightness index
TAD	Topologically associating domain
T-AOC	Total antioxidant capacity
TCM-199	Tissue culture medium 199
TiO2	Titanium dioxide
tRfs	tRNA-derived fragments
tRNA	Transfer RNA
TRNP	Tetramethylrhodamine
TSA	Trichostatin A
UBF	Uterine blood flow
USD	United States dollar
UV	Ultraviolet light
VAP	Average path velocity
VCL	Curvilinear velocity
VEGF	Vascular endothelial growth factor
Vet	Veterinary
VSL	Straight-line velocity
WGA	Wheat germ agglutinin
WOB	Oscillation index
WRC	Wildlife rehabilitation centers
ZnO	Zinc oxide
ZP	Zona pellucida
%	Percentage
°C	Celsius degrees
3D	Three-dimensional
4D	Four-dimensional
6-DMAP	6-dimethylamino purine

Part I

Current Trends for Reproductive Management

The Use of Ultrasonography in Female Bovine Reproduction

Juan Carlos Gardón, Begoña Álvarez Acero, and Salvador Ruiz López

Abstract

Ultrasonography is an indispensable instrument in the realm of bovine female reproduction, offering substantial advantages for both diagnosis and monitoring of reproductive health. It facilitates early and precise pregnancy detection, often as early as 3 weeks post-insemination, thereby enabling informed decision-making for breeding management. Additionally, ultrasonography facilitates a comprehensive evaluation of ovarian structures, encompassing follicular development, corpus luteum dynamics, and the identification of abnormalities such as ovarian cysts. Furthermore, ultrasonography is indispensable in the diagnosis of uterine pathologies, including infections, which can adversely affect fertility. Advances in Doppler ultrasonography have further enhanced reproductive evaluations by providing insights into blood flow, which is crucial for assessing organ viability and functionality. Bovine ovum pick-up (OPU) is a technique that facilitates the retrieval of oocytes from live donor cows for in vitro fertilization (IVF)

J. C. Gardón (✉)
Faculty of Veterinary and Experimental Sciences, Department of Animal Medicine and Surgery, Catholic University of Valencia-San Vicente Mártir, Valencia, Spain
e-mail: jc.gardon@ucv.es

B. Á. Acero
Faculty of Veterinary and Experimental Sciences, Catholic University of Valencia-San Vicente Mártir, Valencia, Spain
e-mail: balvarez@mail.ucv.es

S. R. López
Faculty of Veterinary Medicine, Department of Physiology, Espinardo Campus, University of Murcia, Campus Mare Nostrum (CMN), Murcia, Spain

Murcian Institute for Biosanitary Research Pascual Parrilla (IMIB), Campus of Health Sciences, Murcia, Spain
e-mail: sruiz@um.es

J. C. Gardón, K. Satué Ambrojo (eds.), *Assisted Reproductive Technologies in Animals Volume 2*, https://doi.org/10.1007/978-3-031-87198-6_1

and embryo production. The quality and yield of retrieved oocytes are influenced by various factors, including donor selection, hormonal stimulation protocols, and operator expertise. Continuous advancements in equipment, protocols, and understanding of follicular dynamics promise to further enhance the efficiency and success rates of OPU in bovine reproductive management. The noninvasive nature and precision of ultrasonography make it a cornerstone of reproductive management strategies, contributing to improved fertility rates and overall herd productivity. Ongoing technological advancements in ultrasonography are poised to expand its role in bovine reproduction, offering novel opportunities for enhancing reproductive performance.

Keywords

Bovine reproduction · Pregnancy diagnosis · Ovum pick-up · Doppler ultrasonography · Breeding programs

1 Introduction

1.1 Basics of Ultrasonography

The use of ultrasound can be attributed to the discovery of the piezoelectric effect by the Curie brothers in the mid-nineteenth century. The piezoelectric effect demonstrates that the passage of an electric current through a crystal results in a vibration that generates a beam of ultrasound (Varshney and Chaudhary 2022a).

Ultrasonography is one of the most commonly utilized diagnostic imaging modalities in veterinary medicine and field clinics. A substantial body of clinical tests for reproductive diagnosis at the field level has demonstrated the superiority of ultrasound in diagnosing a range of conditions, including reproductive organ pathology, soft tissue alterations, and fluid accumulation, among others. Moreover, ultrasound can be employed in nondiagnostic applications, including needle guidance for cystocentesis and cytology, as well as the safe collection of tissue samples (Ligtvoet et al. 1989; Varshney and Chaudhary 2022a).

An ultrasound scanner is comprised of three principal components: a transducer or probe, a central processing unit (CPU), and a display (Ligtvoet et al. 1989; Mattoon and Berry 2021).

The transducers utilize piezoelectric crystals that emit ultrasound beams in a pulsed manner. In a subsequent phase, the crystals capture reflected ultrasound (echoes) and relay them to the CPU, which generates an image on the screen (Ligtvoet et al. 1989).

The tonality of the echographic image is dependent upon the amplitude of the wave (echo) received by the transducer, which is then converted into a gray scale. The greater the echogenicity or brightness of the object, the greater the number of echoes that are returned to the transducer (e.g., bone), or the lesser the echogenicity (anechoic) or brightness (e.g., liquid), the fewer echoes that are returned to the transducer (Ligtvoet et al. 1989).

The following section presents a series of frequently asked questions that provide an introduction to the fundamental principles of ultrasound and its applications in the training of new veterinarians.

1.1.1 How Do Ultrasound Devices Work?

As previously stated, ultrasonic systems are equipped with a CPU, which serves as the primary processing and integral control unit of the system. The CPU transmits an electrical current to the transducer or probe, which contains an array of piezoelectric crystals. Upon exposure to an electric current, the crystals undergo a rapid change in shape and vibration, resulting in the generation of sound waves that propagate within the animal's body (Ribadu and Nakao 1999; Varshney and Chaudhary 2022b).

These sound waves interact with the tissues and encounter boundaries between them, such as those between bone and soft tissue, or fluid and soft tissue. Some sound waves are reflected back to the transducer, while others continue until they reach another boundary and are reflected. Upon reaching the transducer, the reflected sound waves are absorbed by the piezoelectric crystals, which then emit electrical currents. These are processed by the CPU to form an image. The CPU performs calculations to determine the intensity of the echoes, which allows for the assessment of image brightness or darkness. Additionally, it calculates the distance between boundaries, thereby providing depth information. Ultimately, the aforementioned calculations are displayed on the screen in the form of a two-dimensional image (Ligtvoet et al. 1989). The emission of sound by the transducer is not constant; rather, it occurs at a rate of approximately once every 10^{-6} s, with the subsequent reception occurring at a rate of approximately once every 10^{-3} s (Fig. 1). It is also important to note that the CPU controls the emission and reception of sound, and by determining the transmission time of sound (from emission to reception), the processing unit can calculate the depth of different tissues (Varshney and Chaudhary 2022b).

It is similarly crucial to have the requisite software to enhance diagnostic capabilities, thereby improving image quality and offering a variety of analytical tools. It is of great importance to ensure that the ultrasound software is compatible with existing systems and to consider the significance of software support.

1.1.2 What Are Transducers or Probes?

Transducers, also referred to as probes, are an essential component of ultrasound technology. They are available in a wide range of sizes and shapes, designed for use in specific examinations. The most prevalent categories are as follows:

Linear probes are distinguished by an array of crystals arranged in a linear configuration along the length of the probe, which presents an image with homogeneous density and exhibits excellent near gain. These probes are employed for imaging superficial structures, vascular examinations, and small anatomical components (Rantanen and Ewing 1981).

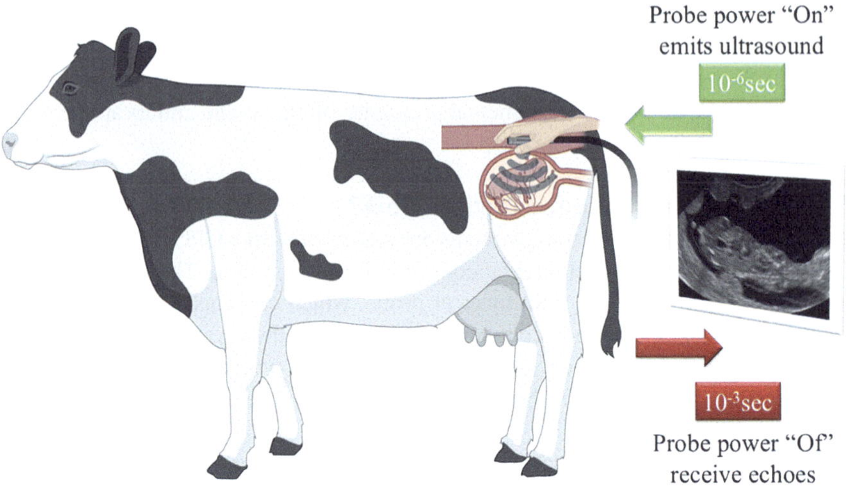

Fig. 1 The central processing unit (CPU) governs the emission and reception of sound. Through the determination of the transmission time of sound (from emission to reception), it becomes evident that the emission of sound by the transducer is not constant. Rather, it occurs at a rate of approximately once every 10^{-6} s, with the subsequent reception occurring at a rate of approximately once every 10^{-3} s

Curvilinear or convex probes feature a convex configuration of crystals, moderate sound propagation, and a favorable far image, rendering them well-suited for transabdominal imaging (Rantanen and Ewing 1981).

These two types of probes are included in the category of endocavitary probes, which are used for endorectal or endovaginal examinations. Finally, three-dimensional (3D) and four-dimensional (4D) probes provide detailed static or live video images, while phased array probes are used for cardiac examinations (Varshney and Chaudhary 2022b).

1.1.3 Why Are There Different Types of Transducers (Probes)?

Transducers, or probes, are available in a wide range of sizes and shapes, with specifications tailored to specific applications. The depth of penetration and image detail or resolution are dependent upon the different fields of view and acoustic wave frequencies of the transducers (Rantanen and Ewing 1981; Rantanen 1986; Varshney and Chaudhary 2022b).

The majority of probes are designed in a manner that allows them to pass through the surface of the animal's body, with shapes and configurations optimized for specific applications or organs. The two most prevalent types of ultrasound probes utilized in veterinary medicine are microconvex and linear array transducers. Moreover, there are specialized transducers that are capable of entering body openings, such as the rectum, to achieve a closer proximity to specific organs. Such transducers are frequently employed in the context of large animal reproductive ultrasounds (Mattoon and Berry 2021).

The emission frequency of an ultrasound scanner is a principal factor influencing the depth of penetration. Frequencies in the range of 10–16 MHz are optimal for imaging at shallow or superficial depths, while frequencies in the range of 1–2 MHz are more suitable for imaging at deeper depths, such as in abdominal examinations of large animals (Varshney and Chaudhary 2022b).

1.1.4 What Are the Different Imaging Modes?

The imaging mode is the parameter that defines how the ultrasound data is displayed to the operator and the format in which the ultrasound data is presented visually on the screen. The most commonly utilized ultrasound scanners employ the following imaging modes:

Two-dimensional mode (B-mode or 2D-mode) is defined as follows:

This mode entails the generation of a two-dimensional image derived from the signals acquired by the probe. The image is representative of the plane formed by the ultrasound beams emitted by each of the piezoelectric crystals aligned along the surface of the probe. The resulting image is displayed in gray scale, wherein the varying shades indicate the attenuation of the ultrasound by the underlying tissue. Consequently, regions of the tissue exhibiting greater ultrasound attenuation appear lighter in the image, due to the higher reflection coefficient, whereas regions with lower attenuation appear darker, indicating enhanced ultrasound transmission. The ultrasound scanner generates the image by emitting ultrasound pulses from the crystals of the probe, thereby creating a two-dimensional image (Fig. 2a). The succession of these images in time allows the generation of dynamic images, analogous to a video, as opposed to static images (Mattoon and Berry 2021).

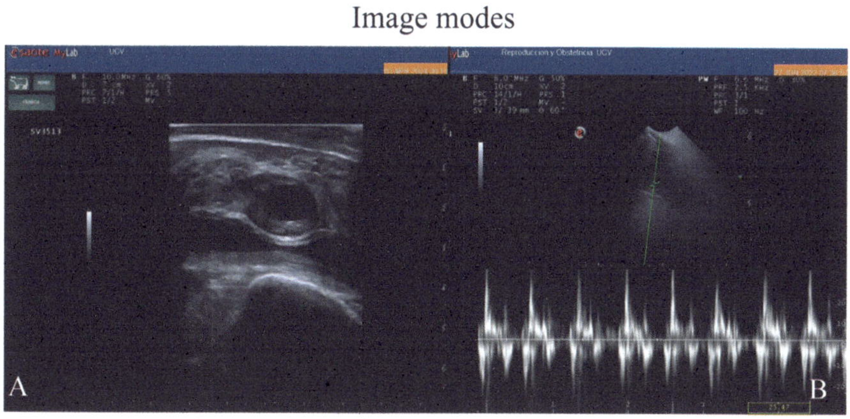

Fig. 2 The following image modes are employed for the most common ultrasound machines. (**a**) B-mode or 2D-mode. The ultrasound scanner generates the image by emitting ultrasound pulses from the crystals of the probe, thereby creating a two dimensional image. (**b**) M-mode. This mode is typically presented with the abscissa representing time and the ordinate indicating the distance from the transducer. This distance is derived from the time delay between the emission, reflection, and detection of the ultrasound pulses

M-mode represents an advancement over B-mode technology. This mode generates a series of images over time, thereby facilitating the observation and measurement of motion and its amplitude. An M-mode ultrasound is typically presented with the abscissa representing time and the ordinate indicating the distance from the transducer (Fig. 2b). This distance is derived from the time delay between the emission, reflection, and detection of the ultrasound pulses (Mattoon and Berry 2021).

Color Doppler imaging is a medical imaging technique that employs the Doppler effect to generate a color-coded representation of the direction and velocity of flowing blood. This imaging mode is employed for the visualization and analysis of organ-specific flow, direction, and vascularization, as well as for the assessment of fetal viability and luteal functionality, among other applications (Mattoon and Berry 2021).

The generation of images in the various modes of Doppler ultrasound is based on changes in the frequency of ultrasound emitted by the transducer in comparison to those reflected by the tissue, particularly when the latter is reflected by small moving particles (e.g., blood figurative elements). If ultrasound is reflected from a static structure, the frequency of the reflected beam is identical to that of the emitted beam. However, when ultrasound is reflected from a moving structure, its frequency changes, increasing or decreasing depending on whether the structure is approaching or moving away from the transducer. The frequency changes are proportional to the velocity of the moving particle, thereby enabling the scanner to calculate the velocity (Mattoon and Berry 2021).

The utilization of the Doppler effect in ultrasound enables the acquisition of diverse imaging modes (Mattoon and Berry 2021).

- Color Doppler imaging: This assigns a color from a predefined scale to the observed flow in a two-dimensional image, according to its direction and speed. This allows for a qualitative evaluation of the movement (Fig. 3a).
- Power color Doppler: Qualitative detection of motion is achieved through the use of a single color. It does not assess the direction or velocity of the flow but is capable of detecting flows with lower velocities than those observed with color Doppler (Fig. 3b).
- Pulsed Doppler: This generates a graph representing the flow velocity at a specific point on the two-dimensional image versus time. It is of the utmost importance to correct the insonation angle at the specified point to achieve optimal alignment with the direction of the flow to be measured, with an ideal angle of 15° or less.
- Continuous Doppler: This mode is utilized for the measurement of significantly higher velocities than those accessible through pulsed Doppler. In this mode, the transducer transmits and receives ultrasound continuously, measuring flow velocity along the ultrasound beam rather than at a single point.

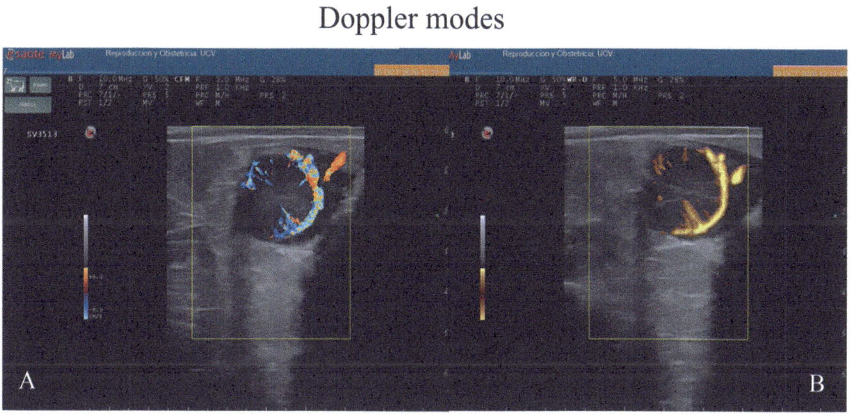

Fig. 3 Color Doppler imaging is a medical imaging technique that employs the Doppler effect to generate a color-coded representation of the direction and velocity of flowing blood. (**a**) Color Doppler of a corpus luteum. This mode assigns a color from a predefined scale to the observed flow in a two-dimensional image, according to its direction and speed. This allows for a qualitative evaluation of the movement. (**b**) Power color Doppler of a corpus luteum. This mode does not assess the direction or velocity of the flow but is capable of detecting flows with lower velocities than those observed with color Doppler

1.1.5 Would It Be Beneficial to Take into Account the Dimensions of the Ultrasound Scanners?

Ultrasound scanners are available in a range of sizes, with the specific dimensions dependent on the user's requirements. In the present era, a significant proportion of high-quality ultrasound scanners are portable, rendering them an optimal selection for individuals who necessitate mobility. Some portable models have a weight of less than 500 g, which allows them to be stored in a bag with minimal effort. These devices can be wireless or connected to a smartphone or tablet. The compact and portable nature of these ultrasound scanners makes them especially advantageous for use on farms. As previously stated, some portable ultrasound scanners weigh less than a bag of sugar and can therefore be easily stored. It is generally accepted that larger devices are bulkier, although this is not necessarily problematic if they remain stationary. Conversely, for a machine that will be moved frequently, a smaller, lighter option is preferable (Mattoon and Berry 2021; Varshney and Chaudhary 2022b).

1.1.6 What Is the Level of Image Quality That They Can Provide?

In the context of ultrasound scanners, image quality and screen dimensions are of paramount importance. In the case of portable devices, it is of the utmost importance to ensure that the screen is sized to meet the user's needs, given that the screens of portable ultrasound scanners are typically smaller.

Although many practitioners are seeking the optimal portable ultrasound system, console-based systems frequently offer superior images with enhanced resolution and sophisticated features. Portable systems are capable of displaying up to

250 gray scale and can incorporate color displays; however, these capabilities are not universal across all models (Ribadu and Nakao 1999; Varshney and Chaudhary 2022b).

Furthermore, brightness is another factor that affects image quality. For instance, veterinarians frequently utilize ultrasound scanners in outdoor settings, wherein the brightness can be calibrated without compromising image quality or readability.

1.1.7 Is Portability a Crucial Factor When Choosing an Ultrasound Scanner?

The evolution of ultrasound technology over the past five decades has resulted in a notable transformation from large, bulky systems to compact, portable equipment. These portable ultrasound scanners afford clinicians the flexibility and mobility to accompany them and facilitate real-time imaging of patients in a variety of settings. This mobility permits practitioners to obtain high-quality images without relying on the availability of a single machine. These compact systems require less space and can be relocated expeditiously from one room to another, thereby reducing waiting periods, enhancing operational efficiency, and improving the patient experience. Nevertheless, it is necessary to ascertain whether a portable ultrasound machine is a more appropriate choice than a fixed console. If the ultrasound machine is required to be relocated frequently, a portable option is to be preferred. Fixed equipment is unwieldy and susceptible to damage during transportation. However, portable ultrasound machines may have limitations in terms of image quality, probe options, software features, and the size and functionality of their keyboards and monitors (Varshney and Chaudhary 2022b).

In the context of on-farm consultations, the key considerations are portability, system size, uptime, and reliability. Ultrasound systems must provide outstanding image quality in B-mode, M-mode, power color Doppler, and color Doppler to facilitate a comprehensive range of applications and examination types. Such applications include obstetrics and gynecology. In the context of daily on-farm work, cost-effective and complete solutions that can adapt and grow with the user's needs are often sought. Compact systems are optimal for environments that require fast-paced work and advanced features in a compact and versatile format. Some portable systems are equipped with advanced imaging and artificial intelligence components, enabling image capture and facilitating the diagnostic process (Fricke 2002; Lüttgenau and Bollwein 2014; Fontes and Oosthuizen 2022).

The battery life of portable ultrasound scanners is of paramount importance. It is recommended that a battery with a life of several hours be selected, with the option of a second battery if necessary. This is of particular importance in regions where access to reliable electrical power is limited.

It is essential to consider the level of training that will be required for the intended users of the ultrasound system. It is crucial to consider the level of expertise of the technicians who will be utilizing the new ultrasound machine, as additional training may be necessary. It would be prudent to ascertain whether the ultrasound equipment supplier offers online or on-site training, bearing in mind that the latter may entail additional costs.

It is recommended that ultrasound scanners be selected from reputable brands with a long history of exemplary customer service and the offering of state-of-the-art models. Furthermore, it is crucial to ascertain the warranty conditions, the availability of service contracts, and the scope of technical assistance.

It would also be prudent to consider additional features such as image storage capacity, ease of image transfer, and integration and connectivity with other peripheral devices.

1.1.8 Interpretation of Ultrasound Images

Interpreting ultrasound images of the reproductive system requires a solid understanding of image composition and the potential artifacts that can lead to misdiagnosis. As sound waves pass through tissues, they may be altered in several ways. When sound waves encounter different tissue interfaces, the returning echoes vary in strength, resulting in diverse image patterns (Ribadu and Nakao 1999).

The ability of a tissue to reflect sound waves determines its ultrasonic properties. Liquids, which do not reflect sound waves (anechoic), appear black on the screen, as seen in structures like ovarian follicles and embryonic vesicles. Dense tissues, such as bone, reflect a large amount of sound waves (echogenic) and are displayed as light gray or white. Various tissues and reproductive tract structures appear in different shades of gray depending on their echogenicity (Rantanen and Ewing 1981).

Common artifacts include distant enhancement (when the sound beam strikes the far wall of a fluid-filled structure, such as a follicle or embryonic vesicle), refraction artifacts, specular reflection, reverberation artifacts, and acoustic shadowing (Peter et al. 1992; Kirberger 1995).

2 Ultrasonography in Bovine Reproduction

It is a common misconception among veterinarians and breeders that ultrasound is only used for the early diagnosis of pregnancy in bovine animals. However, ultrasonography is a fundamental tool in the comprehensive evaluation of the bovine reproductive tract, as it offers a noninvasive, real-time approach to assessing reproductive health. Consequently, this technique has gained increasing popularity in veterinary practices due to its capacity to facilitate the early and precise identification of a range of reproductive conditions, including pregnancy (Barrón Bravo et al. 2023).

One of the most significant advantages of utilizing ultrasound in bovine reproductive management is the enhancement of reproductive efficiency across the entire herd. By accurately diagnosing the reproductive status of individual cows, decisions regarding breeding, culling, and treatment can be made in a more efficient manner, which in turn leads to higher conception rates and shorter calving intervals. This efficiency is of paramount importance for the economic success of both dairy and beef cattle farms (Dubuc et al. 2011; Quintela et al. 2012).

2.1 Uterine Evaluation

The use of ultrasonography in veterinary medicine has become a standard practice, offering detailed insights into the health and reproductive status of the bovine uterus. This noninvasive technique enables the real-time visualization of the uterus, which is of paramount importance for the early diagnosis of pregnancy, the evaluation of uterine conditions, and the monitoring of postpartum recovery. By enabling the early detection of pregnancy, typically 26–30 days after breeding, ultrasonography markedly enhances reproductive cycle management and overall herd fertility (Pierson and Ginther 1984; Fricke and Lamb 2002, 2005).

One of the primary advantages of uterine ultrasound is its capacity to assess the uterine environment both before and following breeding. For instance, ultrasound can discern the presence of fluid within the uterus, which may signify infection or inflammation (Quintela et al. 2012). Furthermore, postpartum ultrasound examinations are crucial for ensuring that the uterus has undergone involution correctly and that there are no pathological changes, such as retained fetal membranes or metritis. These evaluations are of critical importance in determining the optimal timing for rebreeding and in reducing the incidence of infertility (LeBlanc et al. 2011; Elmetwally 2018).

Moreover, ultrasound is a valuable tool in the treatment of reproductive disorders related to the uterus. It enables precise evaluation of uterine cysts, adhesions, and other anomalies that may impact fertility (Binelli et al. 2023). An early and accurate diagnosis by ultrasound allows for prompt intervention, which may include medical treatment or surgical procedures, thus improving reproductive outcomes and increasing herd productivity (Okano and Tomizuka 1987; Kamimura et al. 1993; LeBlanc et al. 2011; Vallejo-Timaran et al. 2020).

2.1.1 The Physiological Uterus

Ultrasonography is a valuable tool for the evaluation of the bovine uterus, providing detailed, real-time visualization of its structure and function. It is of the utmost importance to be able to distinguish between healthy and pathological conditions in the bovine uterus through the use of ultrasound. This is particularly crucial in the context of reproductive management (Suthar et al. 2018).

The procedure is relatively straightforward, but it is essential to have a comprehensive understanding of bovine anatomy, to adhere to the appropriate technical procedures, and to comply with the relevant safety protocols. The following steps delineate the methodology for an efficacious bovine uterine ultrasound. Before commencing the ultrasound examination, it is of paramount importance to restrain the cow in a manner that minimizes movement and ensures the safety of both the animal and the operator. It is essential to restrain the cow with a headlock or sleeve and to tie or restrain the tail to prevent it from obstructing the procedure. It is of significant importance to ensure thorough cleansing of the rectal region to minimize the risk of contamination and enhance the quality of the image (Fricke and Lamb 2002).

The ultrasound procedure commences with the insertion of the transducer into the rectum, which facilitates access to the uterus, situated close to the rectal wall. The operator is required to don a lubricated rectal glove and then insert the hand, with the transducer held in place, into the rectum. The operator must apply gentle and steady pressure to advance the transducer until it is positioned above the uterus. The transducer must be correctly oriented to capture sagittal and transverse images, which will provide a comprehensive view of the uterine horns, body, and cervix (Pierson and Ginther 1984). To achieve this, it is possible to utilize the landmarks provided by the pubis and urinary bladder.

Once the transducer has been positioned correctly, the operator should begin to scan the uterus systematically. This entails a gradual progression of the transducer along the uterine length, commencing at the cervix and progressing toward the uterine horns. It is incumbent upon the operator to adjust the angle of the transducer to visualize the uterus in different planes, thereby ensuring a comprehensive examination. The sonographer will display images in real time, thereby enabling the operator to identify uterine structures and any abnormalities (Pierson and Ginther 1984; Sheldon and Dobson 2004; Suthar et al. 2018).

The ultrasound image of the uterus reveals an echogenic structure, with the different layers of the uterine horn displayed through varying echotextures. The echotexture of the endometrium is characterized by non-specular reflections, which result in the appearance of both dark and bright signals within the ultrasound image (Boyd and Omran 1991). In the context of ultrasound imaging, a normal bovine uterus is observed to present as a symmetrically coiled, tubular structure situated within the pelvic cavity. The uterine horns are typically observed extending laterally from the uterine body, which is situated just cranial to the cervix. The uterine wall, or myometrium, is observed to present as a relatively homogeneous structure with medium to high echogenicity, encircling the central uterine lumen (Fig. 4). The visibility of the lumen may be contingent upon the presence of fluid (Pierson and Ginther 1984, 1987, 1988).

The endometrium, which lines the uterine lumen, typically exhibits a slightly higher echogenicity than the myometrium. In a uterus that is not pregnant, the lumen is typically small and may manifest as a thin, hyperechoic line on an ultrasound image. If the cow is in estrus, the uterus may contain a minimal quantity of anechoic fluid, which is considered to be within the normal range and reflects the increase in uterine glandular secretions that occurs during this phase of the estrous cycle. The presence of this fluid may result in an increased visibility of the uterine lumen on ultrasound imaging (Fricke and Lamb 2002).

The uterus undergoes a series of morphological changes throughout the estrous cycle, driven by hormonal fluctuations that can be observed through ultrasound imaging. It is important to note that the echogenicity of the uterine wall may vary slightly depending on the reproductive status of the cow and the phase of the estrous cycle (Pierson and Ginther 1987; Bonafos et al. 1995; Suthar et al. 2018).

In the proestrus phase, the endometrium undergoes a process of thickening and a concomitant decrease in echogenicity. The uterine lumen dilates, the vascular layer

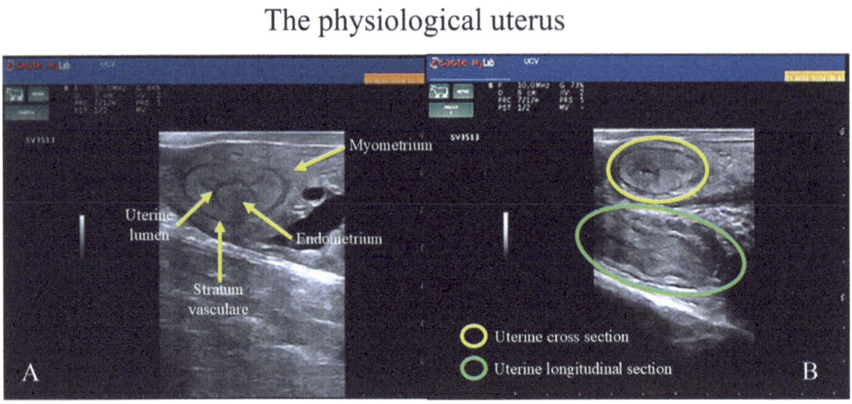

Fig. 4 The ultrasound image of the uterus reveals an echogenic structure, with the different layers of the uterine horn displayed through varying echotextures. (**a**) The uterine myometrium is observed to present as a relatively homogeneous structure with medium to high echogenicity, encircling the central uterine lumen. The visibility of the lumen may be contingent upon the presence of fluid. The stratum vasculare and the myometrium are located surrounding the myometrium. (**b**) The image displays various cuts, including cross sections and longitudinal sections

thickens, and the myometrium thins, with increased echogenicity (Pierson and Ginther 1987; Suthar et al. 2018).

During estrus, the endometrium undergoes edema and appears with low echogenicity, presenting small anechoic spots (blood vessels, edema). The vascular layer is thicker and separates the endometrium from the myometrium. In cross section, it appears as an anechoic ring. The uterine lumen displays a minor anechoic dilation, and specular reflexes may be discernible. The myometrium displays a contracted appearance and exhibits heightened receptivity to external stimuli (DesCôteaux et al. 2006; Suthar et al. 2018).

In the luteal phase, the uterine wall may exhibit slight thickening and increased echogenicity due to the effects of progesterone. These variations are to be expected and should be recognized to avoid misinterpretation of sonographic findings (Pierson and Ginther 1987, 1988; Pieterse et al. 1990a). At the onset of metestrus, the uterus exhibits a resemblance to the uterus in estrus. However, as estrogen levels decline following ovulation, the endometrium, uterine lumen, and vascular layer undergo thinning, while the myometrium undergoes thickening and specular reflexes diminish. In diestrus, the uterus exhibits variations contingent on whether it is in the early, intermediate, or late phase. By day 10, the endometrium and myometrium exhibit comparable thicknesses, separated by a thin vascular layer (anechoic), with no discernible uterine lumen (unless endometritis is present). This reflects a hormonal profile dominated by progesterone (Pierson and Ginther 1987; Bonafos et al. 1995).

Nevertheless, it is not feasible to ascertain the precise phase of the cycle solely based on the ultrasound characteristics of the uterus. A comprehensive assessment should invariably encompass follicular mapping of both ovaries. In cows exhibiting two follicular waves, this assessment is expeditious and precise. However, instances of codominance (multiple dominant follicles within a single wave) or persistent dominant follicles can potentially complicate the assessment (Fricke 2002).

2.1.2 The Pathological Uterus

During ultrasonography, the operator should pay close attention to the echogenicity of the uterine tissue, as alterations in density, tissue thickness, or increased lumen may indicate the presence of pathological conditions. Furthermore, any indications of reproductive disorders, such as endometritis or pyometra, should be documented and subjected to further investigation. For instance, augmented echogenicity may indicate the presence of fibrosis, whereas lumen expansion accompanied by the detection of fluid with some degree of echogenicity may point toward an infectious process (Kähn and Ludlow 1989; Ribadu and Nakao 1999).

Subsequently, after the tenth postpartum day, ultrasound becomes an indispensable instrument for uterine monitoring, given the inherent limitations of manual examination. From the 20th postpartum day onward, ultrasound can be used to diagnose a range of clinical conditions, including purulent and mucopurulent endometritis, pyometra, mucometra, as well as subclinical endometritis and endometrial fibrosis (Sheldon et al. 2009; Quintela et al. 2012).

Purulent and mucopurulent endometritis can be readily identified. The echogenicity of the uterine contents varies depending on the concentration of polymorphonuclear neutrophils (PMN), and a typical ultrasound pattern is observed (Fig. 5). Pyometra shares certain characteristics with clinical endometritis (Fig. 6), but it is typically accompanied by the presence of one or two persistent corpora lutea (CLs). Similarly, ultrasound can detect subclinical endometritis, which presents with an increase in endometrial thickness and uterine lumen, although these criteria are not particularly sensitive (Quintela et al. 2012). In cases of mucometra, the uterine contents are anechoic, and it is crucial to be able to differentiate this pathology from gestation, which requires a meticulous diagnosis (Ashitha et al. 2020).

2.2 Ovaries

Ultrasound examination of bovine ovaries is a fundamental procedure in veterinary practice, particularly for the assessment of reproductive health and the optimization of breeding strategies. This noninvasive method provides comprehensive data on ovarian structures, including follicles, CL, and any pathological alterations. The following guide delineates the methodology for the efficacious performance of a bovine ovarian ultrasound (Torres-Lechuga and Gonzáles-Maldonado 2022).

Before commencing the procedure, it is essential to securely restrain the cow with the appropriate restraint device, such as a clamp or sleeve. Appropriate restraint is essential for the safety of both the operator and the animal, reducing the risk of injury and movement that could interfere with the procedure. It is imperative that the operator thoroughly cleans the rectal area to minimize contamination and enhance the clarity of the ultrasound images (Fricke and Lamb 2002).

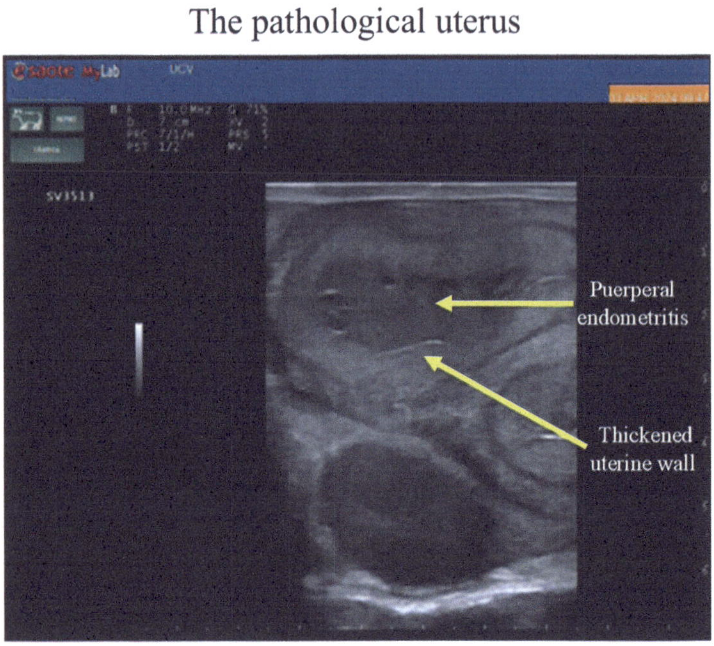

Fig. 5 The uterus exhibits signs of endometritis in a postpartum cow. Lumen expansion is accompanied by the detection of fluid that exhibits a degree of echogenicity, which varies depending on the concentration of polymorphonuclear cells

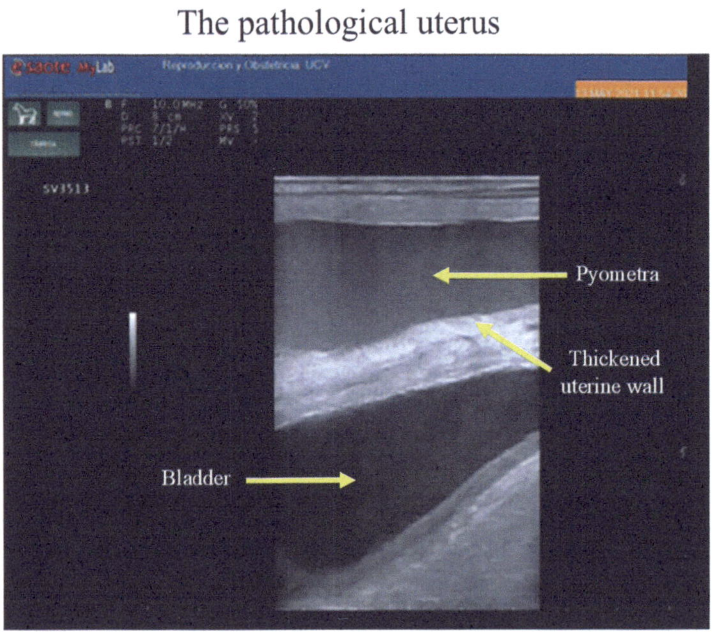

Fig. 6 The uterus of a cow exhibits signs of purulent content, indicative of a possible diagnosis of pyometra

Once the transducer has been positioned correctly, the operator should begin to scan the ovaries systematically. The ovaries are typically situated laterally within the pelvic cavity, near the uterine horns. The transducer must be moved deliberately and methodically across the region to accurately identify each ovary. Once an ovary is visualized, the operator must conduct a thorough examination, identifying and measuring the follicles, which appear as fluid-filled anechoic structures, and the CL, which is denser and appears as a hyperechoic structure. This evaluation is of great importance for determining the phase of the estrous cycle and planning the timing of breeding or other reproductive interventions (Pierson and Ginther 1984, 1988; Pieterse et al. 1990b).

The ultrasound features of normal bovine ovaries allow veterinarians to evaluate ovarian structures, including follicles and the CL. It is imperative to comprehend these typical appearances to distinguish between healthy ovaries and those that are affected by pathological conditions (Dixit and Haloi 2022).

In ultrasound imaging, the ovaries of a healthy bovine appear as small oval structures situated laterally within the pelvic cavity, close to the uterine horns. The size of the ovaries can vary according to breed, age, and phase of the estrous cycle, with a typical diameter range of 15–25 mm. The ovaries exhibit a relatively homogeneous echotexture, except for the presence of follicles and the CL, which represent the primary structures of interest in reproductive evaluations (Pierson and Ginther 1984, Pierson et al., 1988; Pieterse 1990b).

The structures of the ovary are the CL and the follicles. Follicles of the normal bovine ovary are fluid-filled structures that appear as anechoic (black) areas on ultrasound (Fig. 7). They can be distinguished from blood vessels in cross section by the elongated appearance and thicker wall of the latter (Omran et al. 1988; Boyd and Omran 1991). The size of the follicles varies throughout the estrous cycle, with the largest follicles measuring up to 20 mm in diameter during the follicular phase. Additionally, multiple smaller follicles may be observed, particularly during the initial stages of the cycle. The appearance of these follicles is of great importance in determining the phase of the estrous cycle and predicting the optimal time for reproduction (Pierson and Ginther 1988; Fricke 2002).

In a study by Singh et al. (1998) examining the ultrasound echotexture of follicles in relation to functional and endocrine status in heifers, it was demonstrated that the echotexture characteristics of the follicle's antrum and wall (after ovariectomy) were associated with the follicle's functional and endocrine state. The mean pixel values (gray scale: black = 0, white = 255) for the follicle wall and stroma demonstrated a progressive increase as the dominant follicle transitioned from the growing to the regressing phase during the first follicular wave. Furthermore, a negative correlation was observed between pixel heterogeneity in the antrum and wall and estradiol levels and the estradiol-to-progesterone ratio in the follicular fluid.

The CL represents another crucial structure that can be observed in a typical bovine ovary. On ultrasound, the CL is observed to have a higher echogenicity (grayish) compared to the surrounding ovarian tissue and can be identified as early as day 3 after ovulation. A developing CL is visible on ultrasound as an ill-defined, irregular, grayish-black structure with echogenic spots within the ovary. A

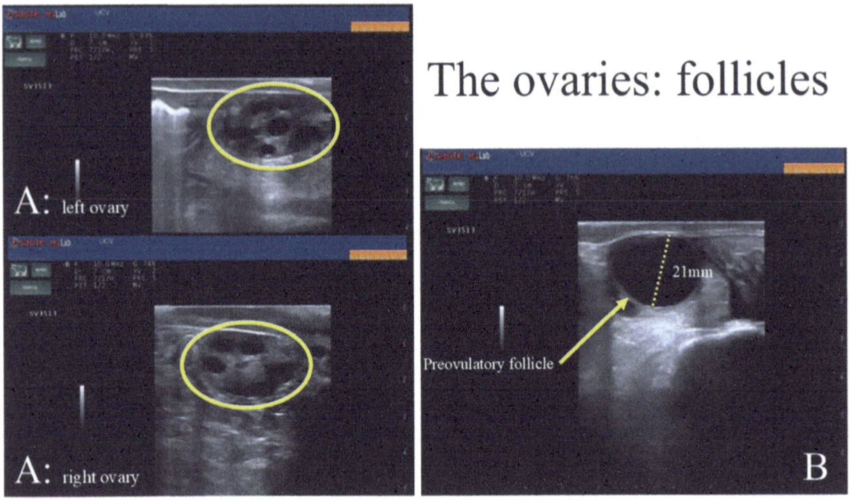

Fig. 7 Follicles of the normal bovine ovary are fluid-filled structures that appear as anechoic (black) areas on ultrasound. (**a**) The image shows heifer ovaries with follicles ranging in diameter from 4 to 6 mm. (**b**) The image shows a cow preovulatory follicle with a diameter of 21 mm

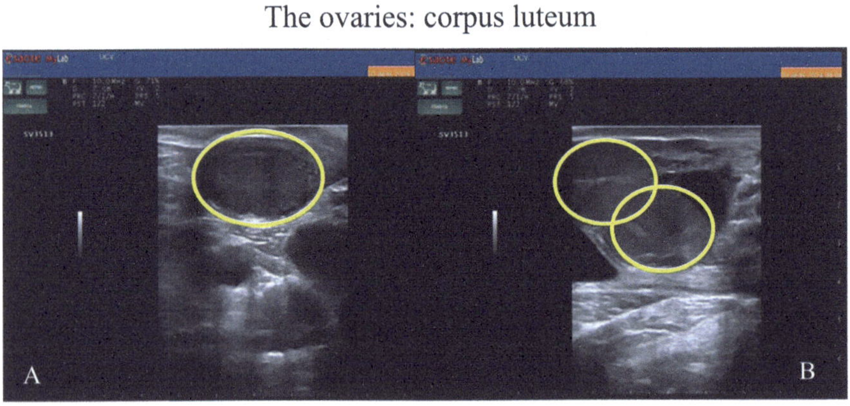

Fig. 8 The corpus luteum (CL) is observed in the middle cycle of a cow. (**a**) In the mid-cycle CL, the structure is well-defined, granular, and grayish echogenic, with a distinct demarcation line separating it from the ovarian stroma. (**b**) A cow ovary containing two CLs is depicted

mid-cycle CL presents as a well-defined, granular, grayish echogenic structure with a clear demarcation line separating it from the ovarian stroma (Fig. 8). In a regressing CL, the demarcation line becomes indistinct due to the minimal difference in echogenicity between the tissues (Pieterse et al. 1990a, b; Fields and Fields 1996; Tom et al. 1998).

The CL is typically larger than follicles, measuring between 20 and 30 mm in diameter. Its presence is indicative of the luteal phase of the estrous cycle, indicating that the cow is not currently in estrus. The detection of a CL is crucial for determining gestational status and the optimal timing for interventions such as hormonal treatments (Pierson and Ginther 1988).

The CL may be observed to be compact or, in some cases, may have a central fluid-filled cavity, which is considered to be normal and should not be confused with a pathologic cyst. This is frequently designated a luteal cavity and is frequently observed in numerous cows, particularly during the initial stages of pregnancy. The size and shape of the cavities are subject to change depending on the cyclic stages of the CL (Kito et al. 1986; Kastelic et al. 1990a). The presence of a cavity does not affect the functionality of the CL, as it continues to produce progesterone in a manner that is consistent with its typical physiological processes. A cavitary CL does not affect the rate of conception or the probability of embryonic or fetal demise, nor does it modify the duration of the estrous cycle (Fig. 9). The size, shape, or number of cavities of the CL does not influence its functionality (Kito et al. 1986; Kastelic et al. 1990a).

The results of ultrasonographic examination are more sensitive and specific than those of rectal palpation. In particular, ultrasonography demonstrated 95% sensitivity and 100% specificity, while rectal palpation exhibited 95% sensitivity and 95.7% specificity. Ribadu et al. (1994a, b) reported an 84% sensitivity and 71% specificity, while Bicalho et al. (2008) documented 79% sensitivity and 40% specificity for this method. These discrepancies are likely attributable to the fact that a considerable number of CL do not protrude to a sufficient extent from the ovarian surface to be palpated. Additionally, hemorrhagic or cavitary CL with a soft consistency can be misidentified as follicles (Hanzen et al. 2000; Quintela et al. 2006).

Ultrasonographic studies of dairy heifers have demonstrated that 79% of otherwise normal CLs develop cavities ranging in diameter from less than 2 mm to over

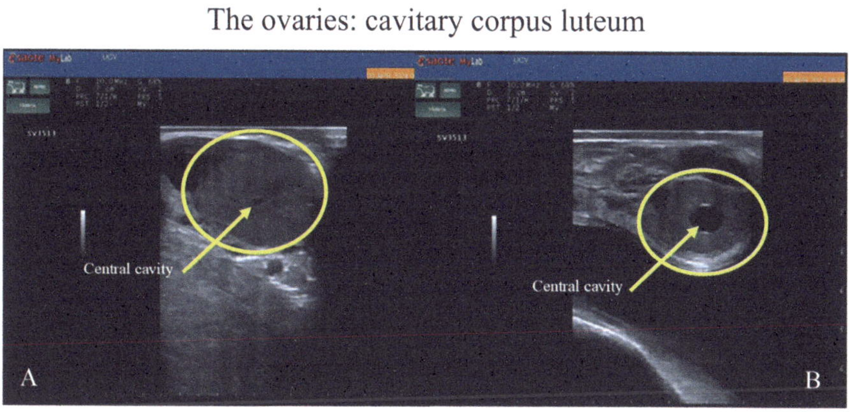

Fig. 9 The presence of a central fluid-filled cavity (CL) has been observed in a considerable number of bovines, particularly during the nascent stages of pregnancy. The dimensions and morphology of these cavities are subject to variation depending on the phase of the CL's cycle. The images below illustrate two examples of CLs with different-sized central cavities: (**a**) CL with a diminutive central cavity, (**b**) CL with a medium-sized central cavity

10 mm at some point during the estrous cycle and early pregnancy (Kastelic et al. 1990a; Singh et al. 1998). Pérez-Marín (2009) observed that 28.6% of CLs exhibited the presence of cavities 4 days following ovulation, with this figure increasing to 50% by day 8. Similarly, Kahn (2010) found that the occurrence of cavitated CLs can range from 25.2% to 78.8% during diestrus, with a decline as the estrous cycle progresses. In the period ranging from the 10th to the 15th day, 70% of the cavitary CLs undergo a process of cavitary collapse. This is subsequently replaced by new luteal tissue exhibiting greater echogenicity compared to the tissue that previously constituted the cavity wall. In some instances, an echogenic scar may persist following the collapse of the cavity. However, there is a lack of consensus among researchers and sonographers regarding the composition of the tissue that fills the CL cavity. Some have proposed that it may consist of fibrin and red blood cells. Approximately 30% of cavitary CLs retain their cavity throughout the estrous cycle, often with a visible cavity bounded by a thin hyperechogenic bright line (Quintela et al. 2012).

It can thus be inferred that a CL with a cavity is more likely to be in the first half of the estrous cycle, while a non-cavitated CL is more likely to be in the second half (Quintela et al. 2012).

In general, the presence of echogenic tissue within the CL, a scar within it, or a cavity surrounded by a thin, shiny hyperechogenic line indicates that the cow is likely in the second decade of the estrous cycle. It should be noted, however, that this pattern is not absolute, as such features can sometimes appear as early as 4–6 days post-ovulation. To ascertain the age and functionality of the CL and the timing of the cow's estrous cycle with greater precision, it is essential to examine the follicular map of both ovaries, evaluate the uterus, or utilize Doppler ultrasonography (Siqueira et al. 2019; Fontes and Oosthuizen 2022).

2.2.1 Follicular Dynamics

A comprehensive reproductive examination in cattle entails the inspection of both ovaries for the identification of major structures or their absence. Although rectal palpation is an effective method for diagnosing pregnancy, it is not a reliable approach for evaluating ovarian follicles (Pieterse et al. 1990b). Conversely, ultrasonic imaging provides an accurate and expedient approach to evaluating ovarian structures (Griffin and Ginther 1992).

Advances in ultrasonography have elucidated the dynamics of follicular development, corroborating the hypothesis proposed by Rajakoski in 1960. The application of ovarian ultrasonography has revealed that the development of follicles in the bovine ovary follows a distinctive and regular pattern. Each wave is characterized by the simultaneous appearance of a cohort of follicles measuring 3 mm or more in diameter (Sirois and Fortune 1988; Adams et al. 2008). In a matter of days, one follicle will exhibit growth beyond that of the others, becoming the dominant follicle (Ireland and Roche 1987; Fortune et al. 1988). It has been demonstrated that heifers undergo either two (Pierson and Ginther 1988; Ginther et al. 1989) or three (Savio et al. 1988; Ginther 2016) follicular waves per estrous cycle.

Each dominant follicle progresses through a growth phase and a static phase, with each phase lasting approximately 5–6 days. The dominant follicle in the initial wave is anovulatory, maintaining its dominant status for approximately 4–5 days. By day 11 or 12 of the estrous cycle, the follicle begins to regress, a process that lasts approximately 5–7 days. Meanwhile, a second wave of follicles emerges, with a new dominant follicle that ovulates. In estrous cycles comprising three follicular waves, the second dominant follicle undergoes regression, and the third dominant follicle is responsible for ovulation (Knopf et al. 1989; Figueiredo et al. 1997; Vassena et al. 2003).

The findings of follicular dynamics studies indicate that non-ovulatory dominant follicles may persist in their largest size for several days following the onset of the subsequent wave. This suggests that a dominant follicle may retain its size dominance for a longer period than its functional dominance, which is responsible for the suppression of other follicles (Fortune et al. 1991; Vassena et al. 2003).

The research on follicular dynamics in early pregnancy demonstrates a wave-like pattern of growth and regression of large follicles during the initial 60–70 days, which is analogous to the pattern observed in non-pregnant cows (Ginther et al. 1989; Taylor and Rajamanhendran 1991). Bergfelt et al. (1991) discovered that the anovulatory follicular waves observed in pregnant heifers were associated with sustained progesterone production, a pattern also evident in non-bred, progesterone-treated heifers.

The use of ultrasound has led to enhanced outcomes in estrus synchronization protocols, with improved detection of CL and dominant follicles and greater accuracy in synchronization programs. Furthermore, Doppler ultrasound has the potential to elucidate the physiological and pathological processes occurring in the ovary and uterus. This technique facilitates the assessment of the degree of CL vascularization, which correlates closely with circulating progesterone levels (Herzog et al. 2010; Rauch et al. 2008). Consequently, non-pregnancy can be diagnosed as early as 20 days after insemination, and the efficacy of synchronization plans can be enhanced by more precise ultrasound diagnostics (Bagley et al. 2023).

In dairy cattle, ultrasonography has markedly enhanced estrus and ovulation synchronization programs by verifying the presence of CL after the volsuntary waiting period, thereby informing the optimal pre-synchronization strategy (Fricke and Lamb 2005).

2.2.2 Ovarian Pathologies: Ovarian Cysts

Additionally, the ultrasound operator should be alert for any anomalies in the ovaries. For instance, follicular cysts are a prevalent condition that can be identified by the presence of large, persistent follicles. These structures are typically larger than normal follicles and have the potential to disrupt the normal estrous cycle. Additionally, an ultrasound examination of the ovary provides greater clarity regarding ovarian function, particularly in differentiating between anestrus and anovulation when persistent luteal structures are absent for a minimum of 10 days (Peter et al. 2009).

Ovarian cysts represent a significant cause of infertility in cattle, characterized by the presence of one or more follicular structures larger than 25 mm in diameter for a minimum of 10 days in the absence of a CL (Seguin 1980). These cysts are classified as either follicular (thin-walled) or luteal (thick-walled) (Fig. 10). It is well-documented that accurately distinguishing between the two through rectal palpation is challenging (Blowey 1992; Dobson and Nanda 1992). Correctly identifying the type of cyst is crucial, as the recommended treatment differs depending on the diagnosis. Ultrasonography represents a reliable methodology for differentiating between follicular and luteal cysts (Farin et al. 1990, 1992; Ribadu et al. 1994a, b).

Follicular cysts, as observed via ultrasonography, present as uniformly non-echogenic ovarian structures exceeding 25 mm in diameter with a wall measuring less than 3 mm in thickness. Luteal cysts, however, present as non-echogenic structures larger than 25 mm in diameter, with gray patches inside the antrum or along the inner cyst wall and a wall thicker than 3 mm. Furthermore, it is imperative to differentiate between luteal cysts and cystic CL, as the latter is not pathological (Ribadu and Nakao 1999; Douthwaite and Dobson 2000; Ambrose et al. 2004).

To distinguish between the two with clarity, it is essential to consider several factors, including size, cavity shape, wall thickness, and internal echogenicity. Cystic CLs are typically no larger than 3 cm in diameter, with a wall thickness of 5–10 mm, and are often oval in shape. The appearance of the cavities is dependent on the direction of the ultrasound beam and may be observed as either circular or oval. Fluid-filled cavities in CL rarely reflect sound waves and are typically homogeneous and near-black, whereas luteal cysts frequently exhibit reflections (Ribadu et al. 1994a, b).

Furthermore, ultrasonography has been demonstrated to be a valuable tool for monitoring the formation and dynamics of experimental ovarian cysts in cows and heifers following steroid administration (Hamilton et al. 1995; Kawate et al. 1996),

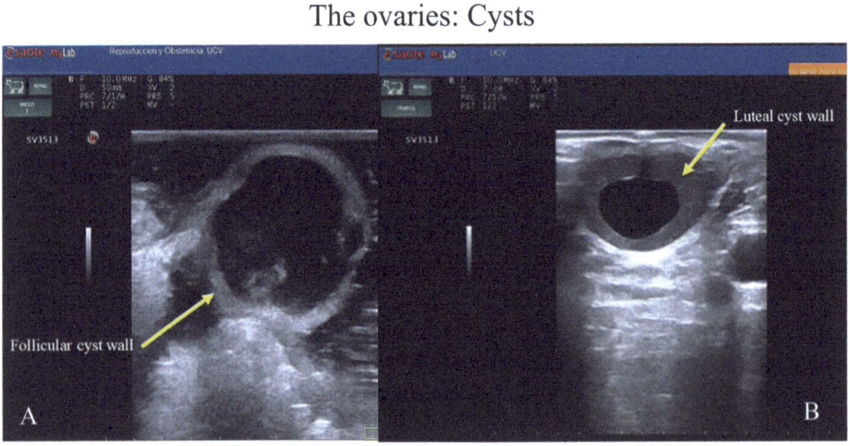

Fig. 10 Ovarian cysts are classified into two primary categories: follicular cysts and luteal cysts. (**a**) Follicular cysts manifest as uniformly non-echogenic ovarian structures exceeding 25 mm in diameter with a wall measuring less than 3 mm in thickness. (**b**) Luteal cysts present as non-echogenic structures larger than 25 mm in diameter, with gray patches inside the antrum or along the inner cyst wall and a wall thicker than 3 mm

as well as for tracking spontaneous ovarian cyst turnover in cows (Yoshioka et al. 1996).

At present, the most common method of diagnosing cysts in dairy cattle is through routine postpartum rectal examinations conducted by bovine practitioners. A large, fluid-filled structure identified through rectal palpation is frequently indicative of a follicular cyst. However, distinguishing between follicular and luteal cysts through rectal palpation can be challenging, even for experienced practitioners (Dawson 1975; Farin et al. 1992). The accuracy of diagnosis is significantly enhanced through the utilization of transrectal ultrasonography, with an identification rate of over 90% for luteal cysts and nearly 75% for follicular cysts (Farin et al. 1990, 1992). Manual diagnosis of ovarian cysts is often inaccurate due to the coexistence of approximately 30% of follicular cysts with a CL, which indicates that they are not functionally problematic (Al-Dahash et al. 1977; Farin et al. 1992).

The significance of luteal cysts is a topic of ongoing debate in the scientific community. Some putative cysts may be cavitary CL cysts or follicular cysts undergoing luteinization. The differentiation between these two entities is based on the thickness of the luteal wall. A wall thickness of greater than 3 mm indicates a CL, whereas a wall thickness of less than 3 mm suggests a luteal cyst. However, some researchers have proposed a wall thickness of 5 mm (3–10 mm) and a cavity size of no less than 30 mm (Hanzen et al. 2000; Torres-Lechuga and Gonzales-Maldonado 2022).

Other ovarian pathologies detectable by ultrasound are ovarian bursa abscesses and hematomas, which allow differential diagnosis with CL. Bursal cysts, which can reach 10–15 cm, do not adversely affect fertility and are distinguishable from ovarian cysts. Ultrasonography can also detect ovarian neoplasms such as granulosa cell tumors, although a biopsy is needed to confirm this. Ovarian dysplasia, sometimes related to advanced age or considered a pretumorous condition, can also be identified (Suthar et al. 2018).

2.3 Pregnancy Diagnosis

A significant number of veterinarians and veterinary technicians who utilize ultrasound in their practice often restrict its application to early pregnancy diagnosis, which represents a narrow approach. In practice, it is of greater importance to diagnose non-pregnancy at an early stage, as these cows represent the most significant contributors to economic losses due to prolonged days open (Suthar et al. 2018).

Chaffaux et al. (1982) were among the first to report the use of real-time ultrasonography for pregnancy diagnosis in cattle. Utilizing a 3.5 MHz transducer, irregularly shaped, non-echogenic structures within the uterine lumen were identified on day 28 post-insemination. Subsequently, numerous other researchers also investigated the potential of ultrasonography for pregnancy diagnosis in cattle (Pierson and Ginther 1984; White et al. 1985; Curran et al. 1986a; Kähn 1990; Baxter and Ward 1997).

Visualization of the bovine fetus can be initiated at day 20 post-breeding (days 19–24) and maintained throughout gestation. In typical on-farm conditions, ultrasound can diagnose pregnancy as early as 26 days post-artificial insemination (AI) with high accuracy (Kastelic et al. 1991; Filteau and DesCôteaux 1998). The sensitivity and specificity of ultrasound for pregnancy diagnosis were 68.8% and 71.7%, respectively, when performed between 29 and 35 days post-AI (Badtram et al. 1991). However, these rates increased to 96.7% and 94.4% or 96.3% and 91.3% for heifers or cows, respectively, when performed between 24 and 29 days post-AI (Romano et al. 2006).

The sensitivity, specificity, positive predictive value, and negative predictive value of ultrasonographic diagnoses made between days 27 and 34 after AI have been found to range from 84.4% to 100% and 92.3% to 100%, respectively (Szenci 2021). The detection of an embryo with a beating heart was found to result in a significantly higher number of false negatives and a lower number of false positives (p <0.001) compared to the use of chorioallantoic fluid as the indicator (Szenci et al. 1998). Conversely, Silva et al. (2007) found that detecting chorioallantoic fluid in the uterine lumen in conjunction with a CL could result in a higher incidence of false positives than when an embryo was visualized using a 5–10 MHz linear array transducer on day 27 after timed AI. The accuracy of the test is contingent upon the transducer frequency and the operator's expertise (Badtram et al. 1991; Fosgate et al. 2017) (Fig. 11).

Different criteria have been used for a positive pregnancy diagnosis (Pieterse et al. 1990a, b; Szenci et al. 1990; Muñoz del Real et al. 2006; Giordano and Fricke 2012; Dufour et al. 2017). However, a confirmed pregnancy diagnosis is based on the presence of one or more CL in one or both ovaries, fluid in the uterus, and, most importantly, the presence of one or more embryos. Therefore, the mere visualization of a CL and fluid in the uterus only indicates the possibility of pregnancy (Ribadu and Nakao 1999; Suthar et al. 2018). While early pregnancy detection is feasible, it

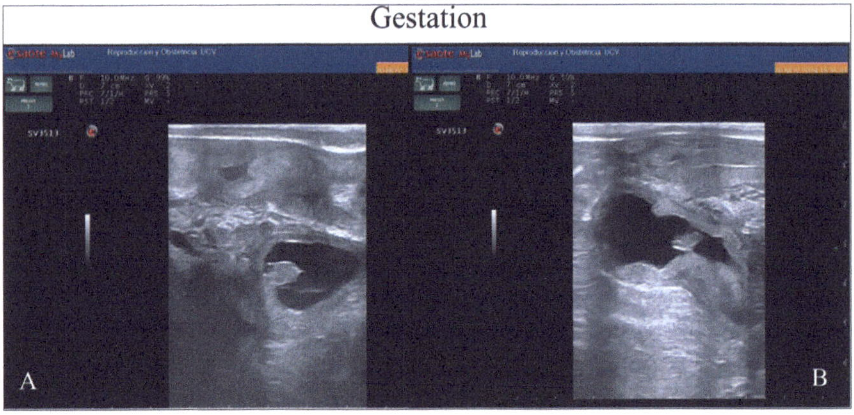

Fig. 11 The visualization of the bovine fetus can be initiated at day 20 post-breeding and maintained throughout gestation. In typical on-farm conditions, ultrasound can diagnose pregnancy as early as 26 days post-breeding or artificial insemination with high accuracy. (**a**) Cross section of a 23-day cow pregnancy. (**b**) Longitudinal section of a 28-day cow pregnancy

is imperative to ensure accuracy. For example, a false-negative diagnosis was more probable when the uterus was situated cranially to the pelvic inlet in beef cattle than when it was within the pelvic cavity (Szenci et al. 1995).

Nevertheless, early diagnosis is more feasible around day 25 of gestation. At this stage, the detection of the embryo by ultrasound may necessitate the manipulation of the uterus (Ginther 1998; Fricke and Lamb 2005). Uterine manipulation may result in the release of prostaglandin F2α (PGF2α), which has been associated with an increased risk of late embryonic death, particularly in young embryos. Accordingly, the diagnosis of pregnancy or non-pregnancy should be initiated at approximately day 25 post-insemination, without the use of uterine manipulation. By day 28 of gestation, the diagnosis of non-pregnancy or pregnancy becomes more straightforward, faster, and more accurate, with fewer cases requiring further investigation and improved detection of twin gestations. In the case of non-pregnant cows that have undergone the estrous cycle, a 6- to 8-day CL is typically observed, which is more readily discernible than a hemorrhagic CL (Velez et al. 1991; Ribadu and Nakao 1999; Scenna et al. 2005; Suthar et al. 2018).

The use of ultrasound for pregnancy detection allows for earlier identification of pregnancy than is possible with palpation. However, this increased early detection of pregnancy may result in a higher incidence of pregnancy loss when a positive diagnosis is made shortly after breeding. In cows diagnosed as pregnant at 28 days post-AI, the incidence of early embryonic loss by 56 days post-AI has been reported to range from 10% to 16% (Mee et al. 1994; Vasconcelos et al. 1997; Fricke et al. 1998). Therefore, cows diagnosed as pregnant at 28 days post-artificial insemination with ultrasound should undergo a follow-up examination around 60 days post-artificial insemination, when the rate of embryonic loss per day significantly decreases (Vasconcelos et al. 1997).

At present, there is no effective method for the prevention of early embryonic loss in lactating dairy cows. Nevertheless, an understanding of the occurrence and extent of early embryonic loss can facilitate the implementation of management strategies that capitalize on new reproductive technologies, which have the potential to enhance artificial insemination (AI) service rates in dairy herds (Barrón Bravo et al. 2023). The routine use of transrectal ultrasonography has the potential to enhance reproductive efficiency by reducing the interval from artificial insemination AI to pregnancy diagnosis to 26–28 days with high diagnostic accuracy (Pieterse et al. 1990a). Nevertheless, the enhancement of reproductive efficiency through the early diagnosis of pregnancy necessitates the implementation of a subsequent management strategy aimed at expediting the return of cows to the AI service. The impact of such strategies on conception rates has yet to be empirically tested. Moreover, cows diagnosed as pregnant in early ultrasound examinations are at an elevated risk of early embryonic loss, necessitating subsequent pregnancy checks to identify and rebreed cows that experience loss. Failure to identify embryonic loss after an early pregnancy diagnosis may result in a reduction in reproductive efficiency, due to an extended calving interval. This issue is pertinent not only to ultrasound but to any method employed to ascertain pregnancy status promptly following breeding (Vaillancourt et al. 1979; Vanroose et al. 2000).

The decision of when to commence gestation diagnosis is contingent upon some factors, including the heat detection rate of the farm, the utilization of resynchronization systems, and the frequency of veterinary visits. In dairy farms where gynecological examinations are conducted every week, initiating gestation diagnosis before day 32 does not result in a reduction in the number of days open, particularly when a Resynch program is employed. In the case of farms with examinations conducted every 15 days, particularly in instances where the heat detection rate is low (≤40%), it may be necessary to commence the diagnosis at an earlier stage, specifically at 25–26 days. Nevertheless, this results in an increased number of cows with uncertain status, a more challenging diagnosis of twin gestation, and greater complexity in the detection and assessment of embryo viability (Poock and Wilson 2011; Barrón Bravo et al. 2023; Bagley et al. 2023).

In addition to confirming pregnancy, it is essential to evaluate the quality of the pregnancy itself during the process of diagnosis (Fig. 12). Embryonic death represents a significant concern in the management of dairy and beef cattle, with a substantial impact on farm profitability (Vasconcelos et al. 1997; Fricke and Lamb 2005). The following criteria may be employed to evaluate gestation viability (Ginther 1998; Buczinski et al. 2007; Suthar et al. 2018; Colloton 2021; Szenci 2021):

- Heart rate.
- Echogenicity of amniotic fluid and/or allantoic fluid.
- Separation of the chorioallantoic membrane.
- Amniotic membrane integrity.
- Embryo size.

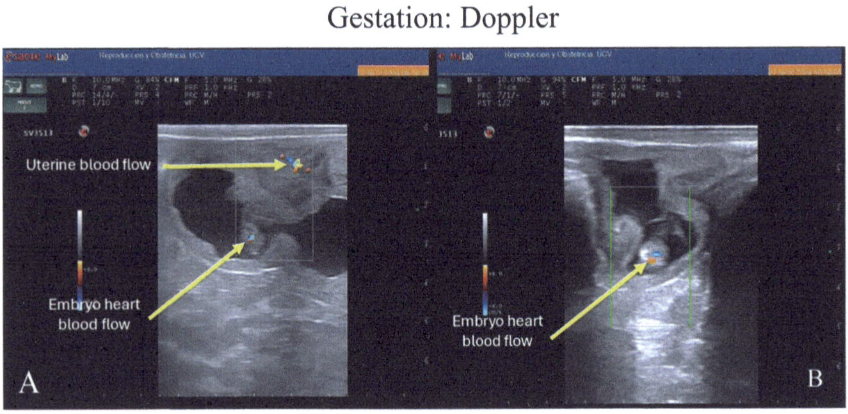

Fig. 12 The confirmation of pregnancy is imperative for evaluating its quality, as embryonic death constitutes a substantial concern in the management of dairy and beef cattle. The utilization of the color Doppler mode facilitates the assessment of cardiac activity during pregnancy. (**a**) Twenty-eight-day cow embryo specimen demonstrating cardiac and uterine blood flow as indicated by color Doppler ultrasound. (**b**) Thirty-three-day cow embryo specimen exhibiting cardiac blood flow as indicated by color Doppler ultrasound

At the earliest stages of development (27–28 days), the most direct method for assessing the viability of the embryo is through observation of the heartbeat, which appears on ultrasound as a series of rapidly moving echogenic dots. At this stage, the normal heart rate is typically between 120 and 130 beats per minute (Ginther 1998; DesCôteaux et al. 2010). Although measuring heart rate in the field can be challenging, a general guideline is to consider a heart rate that is slow enough to count as an indicator of potential complications. Conversely, if the heartbeat is too rapid to be counted, the embryo is likely to be in good health (Lénárt et al. 2019).

2.3.1 Diagnosis of Twin Gestations

Twinning is becoming an increasingly prevalent issue in high-producing dairy cattle. The incidence of twinning has increased in the past years, and with rigorous monitoring, the occurrence of twins can be more accurately identified (López-Gatius et al. 2002; Silva del Rio et al. 2007; Colloton 2021).

To accurately diagnose twins at an early stage, it is essential to conduct a comprehensive scan of the entire length of both uterine horns to avoid any potential for missed embryos. The majority of twin pregnancies in dairy cattle are the result of double ovulations (Wiltbank et al. 2000). Therefore, the presence of two or more CLs in the ovaries during pregnancy diagnosis is indicative of an elevated risk for twins. Fricke et al. (1998) observed that the incidence of double ovulation in lactating dairy cows following ovulation synchronization (Ovsynch) was 14.1%. Furthermore, cows with higher milk production near AI exhibited a rate of double ovulation nearly three times that observed in cows with lower milk production (Fricke and Wiltbank 1999).

The majority of twin gestations in cattle are heterozygous, with two CLs present in the same ovary or different ovaries, resulting in the production of two males, two females, or one male and one female. Monozygotic twins account for only 1% of all twin pregnancies, with one CL and the potential for two males or two females (Ginther 1998; Fricke and Lamb 2005). The incidence of late embryonic death is approximately 30% higher in twin pregnancies, and the rate of miscarriage is also elevated. Ultrasonography is essential for accurate diagnosis, as manual methods are less reliable. Once a twin pregnancy has been diagnosed, the animal must be monitored frequently (at least four times). The assistance of the animal during parturition is essential, and nutritional supplementation during the dry period may be necessary to reduce the risk of postpartum conditions such as retained placenta, metritis, ketosis, or displaced abomasum.

In the case of male and female twins, the female is at high risk of freemartin, a disease in which the female is sterile. This risk persists even in instances where only one embryo was identified during sexing and it was female, as the male embryo may have perished after 30 days of gestation (Andreu-Vázquez et al. 2012).

2.4 Fetal Sexing

The initial approach to fetal sex determination was pioneered by Dr. Sandra Curran in 1989 (Curran et al. 1989; Stroud 2006b). This method relies on the observation that by approximately days 53–56 of gestation, the genital tubercles have moved to their respective positions in male and female fetuses (Curran et al. 1989; Ribadu and Nakao 1999; Jones and Beal 2003; Fricke and Lamb 2005; Stroud 2006b). Fetal sexing is a valuable service for all types of livestock farms, regardless of whether the primary focus is on genetic production. The determination of fetal sex has numerous implications in the animal breeding industry, including the following:

1. The value of cows with female or male gestations is increased.
2. Contracts can be arranged for pregnancies of a specific sex.
3. The management of calving assistance in heifers is more efficient.
4. Calving management can be optimized, especially in beef cattle.
5. Nursery management is improved, including space and labor.
6. The overall effectiveness of breeding plans is improved.

The external genitalia initially develop on the caudoventral surface of the abdominal wall, between the hind limbs (DesCôteaux et al. 2010). These initial structures, which are observed as poorly defined elevations, consist of the genital tubercle (GT), the urogenital fold, and the genital swellings, which undergo evolution in both male and female embryos. In males, the GT elongates to form the penis, the urogenital fold encloses the penis to create the prepuce, and the genital swellings enlarge to develop into the scrotum. In females, the GT becomes the clitoris, the genital swellings disappear, and the urogenital folds enclose the GT to form the labia (Noden and De la Hunta 1985; DesCôteaux et al. 2010).

Before the differentiation of the sexual organs, the structure that will give rise to these is referred to as GT in both males and females. This structure, located in the abdomen, commences a cranial (male) or caudal (female) migration around days 40–45 of gestation, concluding around days 50–55 (Fig. 13). Accordingly, fetal sexing is not advised before day 55 (Ginther 1998; Stroud 2005). Fetal sexing is conducted by discerning the morphology and position of the GT between days 55 and 60 of gestation (Curran and Ginther 1991; Ali 2004). In males, the GT is situated close to the umbilicus, whereas in females, it is located below the tail (Figs. 14 and 15). The TG is visible on ultrasound as a bright, double-structured echogenic feature in both sexes. Following 70 days of gestation, the GT of the male becomes trilobed as a result of the development of the corpus cavernosum of the penis. Additionally, the scrotum and mammae can be observed (Curran 1992; Ginther 1998).

The skill of the technician is of particular importance between days 55 and 60, while between days 60 and 90, the ease of the procedure increases. However, after day 90 and up to day 110, the technician's experience becomes a significant factor. Between days 110 and 130, the position of the fetus may impact the efficacy of sexing. As early as the 1990s, Viana and Marx (1994) employed ultrasonography to ascertain the sex of bovine fetuses between the 60th and 90th days of gestation in cows and heifers. Verification tests conducted on the calves following their birth demonstrated a 96.3% accuracy rate in the diagnoses rendered. Fetal sexing has been demonstrated to predict the sex of offspring with a high degree of accuracy, with a success rate of 92–100% (Lamb and Fricke 2005; Fricke and Lamb 2005;

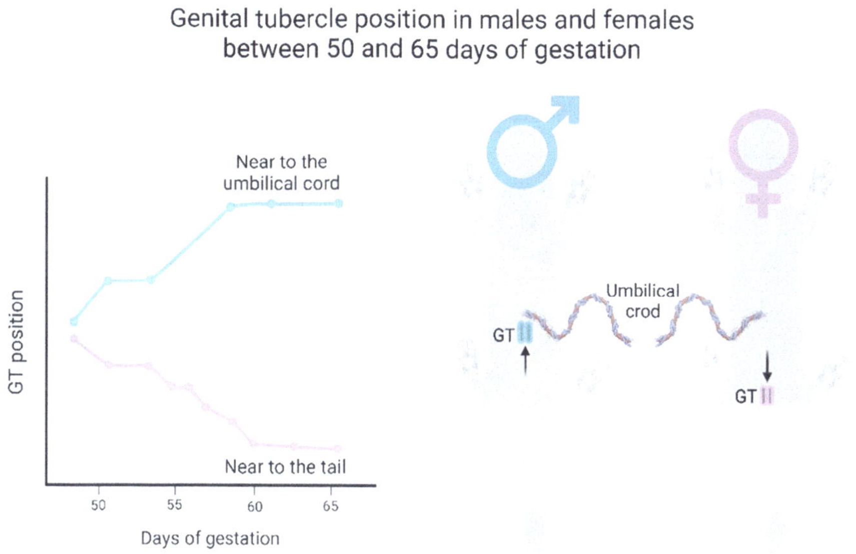

Fig. 13 The genital tubercle (GT) comprises the initial structures, which are observed as poorly defined elevations that undergo evolution in both male and female embryos. In males, the GT elongates to form the penis, the urogenital fold closes the penis to create the prepuce, and the genital swellings enlarge to develop into the scrotum. In the female, the GT develops into the clitoris, the genital swellings regress, and the urogenital folds enclose the GT to form the labia. The GT, located in the abdomen, initiates a cranial (male) or caudal (female) migration around days 40–45 of gestation, concluding around days 50–55

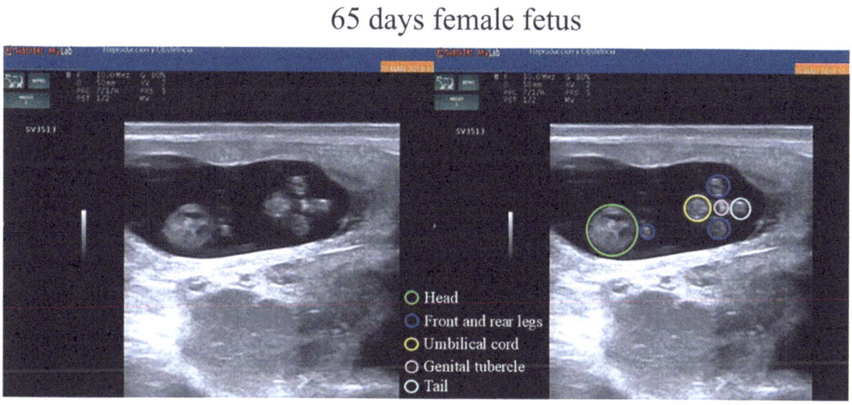

Fig. 14 The longitudinal view of a female fetus at 65 days of gestation is presented herewith. The genital tubercle (GT) is located caudal, in proximity to the tail

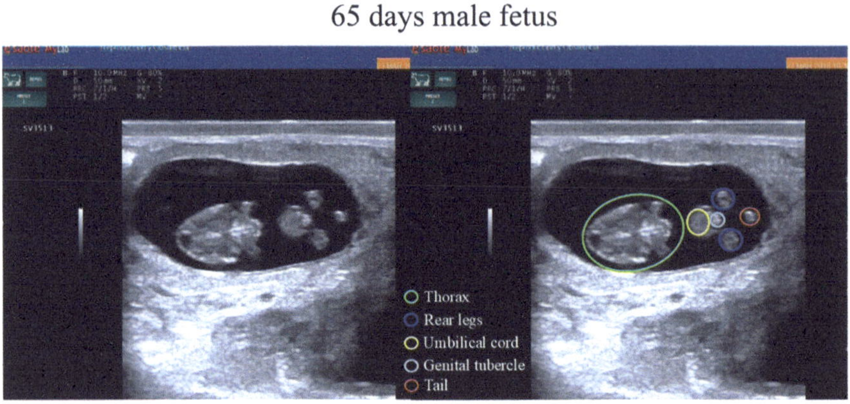

Fig. 15 A longitudinal view of a male fetus at 65 days of gestation is presented. The genital tubercle (GT) is located caudal, close to the umbilical cord

Ribadu and Nakao 1999; Ali 2004; Stroud 2006b). On the other hand, Fetal sexing can be applied in those females with heterosexual twin pregnancies, allowing the gestation of the offspring of the desired sex to continue without affecting its viability (López-Gatius and Garcia-Ispierto 2023).

Nevertheless, attaining this degree of precision necessitates a substantial investment of time and practice (Poock and Wilson 2011). The process of learning fetal sex prediction involves two key steps. The first is identifying the genital tubercles (Stroud 2006b). The second is obtaining high-quality images of the fetus in either transverse or longitudinal sections (Poock and Wilson 2011). As pregnancy advances beyond the first trimester, the accuracy of fetal sexing declines. In later stages, two factors impede gender determination: the growing fetus makes it more challenging to maneuver the transducer for the optimal ultrasound image, and the gravid horn descends ventrally into the abdominal cavity, hindering visualization of the fetal gender without gravid horn retraction (Beal et al. 1992).

2.5 Practical Applications of Doppler Ultrasonography in Cattle Reproduction

The operation of the Doppler ultrasound is based on the principle of the Doppler effect, which describes the change in frequency of a sound wave as it is reflected by a moving object, either toward or away from the probe. In the field of medical ultrasound, the frequency shift is detected in the echo from blood cells and is utilized to visualize blood flow within tissues or organs (Ginther 2007). Doppler ultrasound provides valuable insights into the functional state of tissues or organs, making it a widely utilized technique in both human and veterinary medicine (Herzog and Bollwein 2007; Matsui and Miyamoto 2009; Bollwein et al. 2016).

Despite the evident advantages in diagnostic quality and accuracy, Doppler-equipped devices and the requisite licenses are typically considerably more expensive than standard B-mode-only systems. The initial price differential may have constrained the broader adoption of Doppler technology among field practitioners. However, a significant shift occurred when it was demonstrated that Doppler ultrasound could serve as a practical tool in herd reproductive management, particularly for the early detection of non-pregnancy through corpus luteum blood perfusion (CLBP) (Siqueira et al. 2013; Pugliesi et al. 2014).

2.5.1 Evaluation of Uterine Blood Flow

Doppler ultrasonography has been widely used in research to evaluate changes in uterine blood flow (UBF) throughout the estrous cycle (Bollwein et al. 2000, 2016; Honnens et al. 2008), as well as fetal and placental development (Camacho et al. 2014; Brockus et al. 2016; Lemley et al. 2018), and blood flow alterations related to postpartum uterine involution (Heppelmann et al. 2013). The method has also been applied to assess fetal and placental growth (Camacho et al. 2014; Brockus et al. 2016; Lemley et al. 2018) and changes in UBF linked to postpartum uterine recovery (Heppelmann et al. 2013). Estimating UBF is relatively simple, with detailed explanations available in previous studies (Bollwein et al. 2016). Fluctuations in uterine blood flow have been observed in non-pregnant cows during the estrous cycle, with variations linked to estrogen and progesterone levels. Blood flow velocity increases during proestrus and estrus and decreases during diestrus (Bollwein et al. 2000; Honnens et al. 2008). Uterine blood flow begins to rise shortly after pregnancy is established, detectable within the first 21 days (Honnens et al. 2008; Silva and Ginther 2010). Notably, early pregnancy sees localized increases in endometrial blood supply due to the conceptus's positioning, observable via Doppler ultrasonography (Silva and Ginther 2010). However, due to significant individual variation among cows, UBF is not yet reliable for early pregnancy diagnosis in cattle (Honnens et al. 2008).

Significant increases in blood flow to the reproductive tract are essential to support the exchange of nutrients and waste between maternal and fetal circulations during pregnancy. In fact, placental nutrient transport efficiency is directly tied to placental blood flow (Reynolds and Redmer 1995). Therefore, sufficient maternal blood flow to the reproductive system is vital for proper fetal development. Doppler ultrasonography has been used as a noninvasive method in reproductive studies to measure UBF and assess the potential effects of different management strategies on fetal development and pregnancy outcomes (Bollwein et al. 2002; McCarty et al. 2018; Gard Schnuelle et al. 2021). For instance, color Doppler detected differences in UBF between cows subjected to nutrient restriction during early and mid-gestation and control cows (Camacho et al. 2014; Lemley et al. 2018). These UBF variations were supported by historical findings on placentome capillary differences and macroscopic vascular density (Lemley et al. 2018), demonstrating that color Doppler can provide insights into uterine and placental vascular changes in response to environmental stressors (Lemley 2017). Ongoing research continues to explore the use of color Doppler technology for estimating UBF during pregnancy, furthering our understanding of placental development and prenatal programming of offspring (Bollwein et al. 2002; Panarace et al. 2006; Sharma et al. 2021).

2.5.2 Evaluation of Luteal Blood Flow

The CL is a temporary endocrine gland that is responsible for the production of progesterone (P4), which is crucial for the establishment of pregnancy in ruminants and other farm animals (Mann and Lamming 1999). Given its brief lifespan, the CL must rapidly develop both physically and functionally to support pregnancy or, in the event of unsuccessful conception, to initiate a new estrous cycle. This rapid development is driven by intense angiogenesis (Acosta and Miyamoto 2004; Reynolds et al. 2000), which is necessary for luteal cell growth during the early luteal phase and for providing nutrients and steroid precursors to sustain progesterone secretion during the mid-luteal phase, given the CL's highly vascular nature (Fraser and Wulff 2003).

The most accurate method for assessing CL function is through the measurement of circulating progesterone levels, as the CL's primary role is to secrete progesterone during diestrus and pregnancy. However, the analysis of progesterone concentrations is a time-consuming and costly process, which limits its routine use in reproductive management. B-mode ultrasonography is a rapid and reliable method that is commonly used to assess the presence and size of the CL. The diameter of the CL is a strong indicator of its function (Spell et al. 2017). Moreover, the diameter of the CL (Berger et al. 2017; Pugliesi et al. 2019) and its area (Berger et al. 2017; Velho et al. 2021) have been demonstrated to correlate with blood progesterone levels. However, this correlation is not as strong during luteolysis, as progesterone levels decline more rapidly than CL size (Kastelic et al. 1990b; Assey et al. 1993). Consequently, Doppler ultrasonography has emerged as a valuable technique for indirectly assessing CL function by visualizing blood perfusion (Herzog et al. 2010; Bollwein et al. 2016; Dubuc et al. 2020).

Color Doppler flow imaging has been successfully employed to quantify blood flow to the ovaries of ruminants throughout the estrous cycle (Honnens et al. 2008; Herzog et al. 2010). In cattle and other species, ovarian blood flow to the CL significantly increases from estrus to mid-cycle (diestrus) (Herzog et al. 2010; Niswender et al., 1975; Miyazaki et al. 1998; Lüttgenau and Bollwein 2014) and then declines toward the end of the cycle (Siqueira et al. 2019). In addition to the typical alterations in CL blood flow throughout the estrous cycle, research has revealed a gradual enhancement in CL blood perfusion in pregnant females (Herzog et al. 2010; Siqueira et al. 2019). This augmentation is presumably attributable to the presence of the conceptus and luteotropic factors.

The use of color Doppler has been employed in several studies to characterize the morphological and functional changes occurring in the CL during diestrus (Miyamoto et al. 2005; Herzog et al. 2010; Lüttgenau and Bollwein 2014; Rocha et al. 2019; Siqueira et al. 2019). It is noteworthy that a reduction in luteal blood perfusion occurs before structural regression (Niswender et al. 1994; Herzog et al. 2010; Siqueira et al. 2019). Additionally, the correlation between CL blood perfusion (CLBP) and circulating concentrations of progesterone is more pronounced during luteolysis than the correlation between CL area and progesterone (Rocha et al. 2019). If conception does not occur, luteolysis is triggered by prostaglandin F2α from the uterus (Niswender et al. 2000). Luteolysis is often divided into two phases: functional luteolysis (loss of function) and structural luteolysis (physical regression of luteal tissue). Initially, a reduction in CLBP (Siqueira et al. 2019) quickly suppresses P4 production. This is followed by apoptosis and necrosis of luteal cells, leading to tissue remodeling (Pate 1994). These phases happen at different times, with functional luteolysis (P4 reduction) occurring before physical regression (size reduction) (Siqueira et al. 2009, 2019). CLBP is strongly correlated with plasma P4 levels, while the correlation between luteal tissue area and P4 is weaker during static and regression phases (Herzog et al. 2010). Additionally, blood flow to the CL-bearing ovary is closely linked to circulating P4 levels in cattle (Herzog et al. 2010).

The most straightforward method for evaluating low CLBP is through subjective scoring by the evaluator, who assesses the extent of the colored area in the Doppler image (Siqueira et al. 2013; Guimarães et al. 2015). As with other subjective scoring systems, the reliability of the evaluation is contingent upon the evaluator's experience, with consistency improving over time. Indeed, there is a high degree of agreement among evaluators with similar knowledge, and there is a strong correlation between subjective scores and objective measurements (Siqueira et al. 2013). While technicians are capable of detecting minor fluctuations in blood perfusion, the use of subjective scoring may impede the capacity to discern these subtle distinctions (Fig. 16).

The accuracy of the diagnosis is contingent upon the anticipated differences between categories. For example, CLBP exhibits significant differences between pregnant and non-pregnant cattle from day 16 after AI, and subjective scoring can

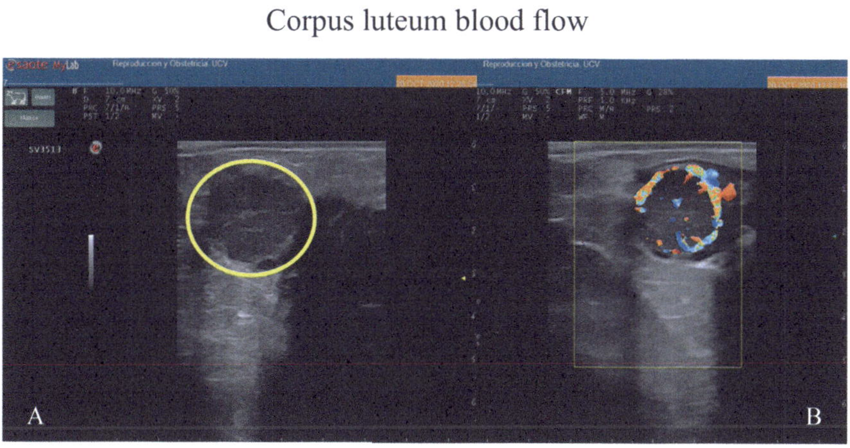

Fig. 16 B-mode ultrasonography is a rapid and reliable method that is commonly used to assess the presence and size of the CL. However, blood flow to the CL-bearing ovary is closely linked to circulating P4 levels in cattle. Therefore, the use of color Doppler to evaluate CL functional status has increased substantially. (**a**) B-mode ultrasound of a mid-cycle CL of a cow. (**b**) Color Doppler ultrasound of the same CL, showing high vascularization

accurately predict non-pregnancy (Siqueira et al. 2013; Pugliesi et al. 2014). Nevertheless, studies examining the relationship between CLBP and P4 production may necessitate the detection of more nuanced alterations in CLBP (Herzog et al. 2010; Pugliesi et al. 2019). For these reasons, the use of color Doppler to evaluate CL functional status during luteolysis has increased substantially in research settings over the past few decades, providing valuable insights into temporal changes to CL function.

2.5.3 Evaluation of Follicular Blood Flow

Vascularization represents a pivotal process in the development of ovarian follicles. The expansion of the capillary network in the theca layer is a necessary step in the formation of the follicular antrum (Rodgers and Irving-Rodgers 2010). This process affects the composition and oxygen levels in follicular fluid, which in turn influences oocyte developmental potential (Sutton et al. 2003).

Research has investigated alterations in follicular blood perfusion throughout their growth and development (Acosta et al. 2005). Adequate blood supply is essential for a follicle to achieve dominance, as evidenced by the observation that dominant and preovulatory follicles exhibit high vascularization, while atretic follicles do not (Acosta et al. 2005; Miyamoto et al. 2006). Furthermore, a positive correlation has been identified between follicular blood perfusion and steroid hormone levels in follicular fluid (Grazul-Bilska et al. 2007; Pancarci et al. 2012; de Tarso et al. 2017). Additionally, studies have demonstrated a correlation between follicular vascularity and the estradiol ratio in dairy cows, as identified by color Doppler ultrasonography (Pancarci et al. 2012).

Moreover, the blood perfusion of preovulatory follicles is associated with the diameter and vascularization of the subsequent CL, as well as its capacity for progesterone secretion (Acosta et al. 2003). This suggests that follicle vascularization may serve as an indicator of its functionality (de Tarso et al. 2017). It has been demonstrated that larger dominant follicles, which exhibit greater vascularization and estradiol secretion, are more functional, as evidenced by higher ovulation rates and larger CL areas (Acosta et al. 2003; Pancarci et al. 2012; Perry et al. 2014). Additionally, larger follicles have been linked to higher pregnancy rates compared to smaller ones (Perry et al. 2007; Mesquita et al. 2014; Pugliesi et al. 2016a, b).

Increased follicular vascularization has been linked to higher pregnancy rates, and follicular blood perfusion has been used to predict pregnancy success after in vitro fertilization and embryo transfer in humans (Chui et al. 1997; Coulam et al. 1999). In cattle, heifers that became pregnant showed greater preovulatory blood perfusion than those that did not, as measured by color Doppler (Siddiqui et al. 2009a). The blood supply to the follicle plays a crucial role in follicle selection, dominance, and ovulation (Matsui and Miyamoto 2009).

In the field of veterinary medicine, a strong relationship has been identified between follicular vascularization and the success of in vitro fertilization and pregnancy rates in cattle (Siddiqui et al. 2009a, b). Ovarian blood supply is closely linked to follicle development and oocyte quality, with color Doppler ultrasonography offering the potential to predict in vitro fertilization outcomes (Arashiro et al.

2013). Moreover, higher vascular perfusion of preovulatory follicles was observed in follicles that produced cleaved oocytes compared to those that did not (Siddiqui et al. 2009b). Pancarci et al. (2012) found that cumulus-oocyte complexes of higher quality were obtained from follicles with detectable blood flow, further supporting the link between greater follicular vascularization, follicle function, and in vitro fertilization and pregnancy outcomes in cattle.

2.5.4 Early Pregnancy Diagnosis

The diagnosis of pregnancy by ultrasound is typically conducted using B-mode and is only recommended after 28–30 days of gestation (Pieterse et al. 1990a, b; Filteau and DesCôteaux 1998). This is because the accuracy in detecting the embryo at this stage is nearly 100% (Nation et al. 2003). In cattle that are not pregnant, estrus typically resumes at regular intervals (18–24 days) following mating or artificial insemination once the CL regresses (Matsui and Miyamoto 2009). Nevertheless, the identification of spontaneous CL regression through ultrasound in non-pregnant cows may facilitate an earlier evaluation of pregnancy status (Pugliesi et al. 2013, 2014; Scully et al. 2014, 2015; Siqueira et al. 2019).

The assessment of CL size has been demonstrated to be a highly accurate method for determining the status of non-pregnant cattle after the 20th day of gestation (Kastelic et al. 1991). Nevertheless, a considerable number of false negatives are observed when relying exclusively on the CL area for the detection of luteolysis in non-pregnant beef cows, as some cows retain a minor yet functional CL that maintains pregnancy (Pugliesi et al. 2018). Furthermore, research has demonstrated that plasma/serum P4 levels are more closely correlated with CLBP than with CL size during the luteolytic phase in ruminants (Herzog et al. 2010; Balaro et al. 2017; Rocha et al. 2019). This indicates that CLBP diminishes before structural luteolysis, as early as day 16 following timed AI in dairy cows and heifers (Siqueira et al. 2019). Consequently, color Doppler ultrasound has been proposed as an earlier method for pregnancy diagnosis compared to conventional ultrasound, whereby cows undergoing CL regression are identified (Siqueira et al. 2013; Pugliesi et al. 2014; Scully et al. 2015). This is because the majority of non-pregnant cows experience luteolysis. It is our understanding that the earliest attempt at diagnosing pregnancy in beef cows through the detection of luteolysis using Doppler imaging was made by Utt et al. (2009). However, the approach employed to assess CLBP was both time-consuming and lacked the desired level of accuracy. Subsequent studies on spontaneous luteolysis examined CLBP differences between pregnant and non-pregnant cows (Matsui and Miyamoto 2009; Pugliesi et al. 2014), which helped define CLBP during luteolysis and established a foundation for further research in dairy (Siqueira et al. 2013) and beef cattle (Pugliesi et al. 2014). The objective of these studies was to establish criteria for the early identification of functional and nonfunctional CL complexes before pregnancy can be detected using traditional ultrasound.

In dairy cows, Siqueira et al. (2013) achieved high accuracy and a strong negative predictive value using only CLBP at 20 days post-timed artificial insemination. In beef cows, Pugliesi et al. (2014) reported an accuracy of 91% and a negative predictive value of 100% for pregnancy status at the same stage. Although the studies employed slightly disparate criteria for defining nonfunctional CLs, their methodologies for detecting non-pregnant females were analogous, and early diagnoses were validated against the conventional method at 30 days of gestation. Both techniques rely on subjective, real-time evaluations, which are simpler and more practical than pixel-based measurements of colored areas to determine CL functionality (Pugliesi et al. 2018). It should be noted that these methods do not allow visualization of the embryo, and therefore the color Doppler ultrasound technique is the only method that allows for the indirect identification and diagnosis of non-pregnant females.

Several studies were conducted on females of various breeds, confirming the high accuracy and null or extremely low false-negative rate of the proposed cutoff points in dairy and beef cattle. In conclusion, these results support the assertion that CLBP determination is an accurate diagnostic method for the detection of pregnancy status. The probability of incorrectly detecting a pregnant female as non-pregnant is low, resulting in a high negative predictive value when applied after 19 days of gestation (Utt et al. 2009; Siqueira et al. 2013; Pugliesi et al. 2014, 2018; Andrade et al. 2019; Dubuc et al. 2020; Holton et al. 2022; Madoz et al. 2022; Ferraz et al. 2022).

The proportion of false positives (cows diagnosed as pregnant via Doppler but not pregnant) varies considerably across published studies from 5.3% to 46.3% (Pugliesi et al. 2018; Andrade et al. 2019; Dubuc et al. 2020; Holton et al. 2022; Madoz et al. 2022; Ferraz et al. 2022). Several factors may contribute to these false-positive results. These may include the presence of an active CL during diagnosis based on CLBP, late ovulation in fixed-time artificial insemination programs, and differences in the timing of physiological luteolysis between breeds and parity numbers. Nevertheless, the primary cause of false positives probably is early embryo loss occurring between the Doppler assessment and the subsequent confirmatory diagnosis (Pugliesi et al. 2023).

Recent studies have investigated the phenomenon of false positives in cattle. These include studies involving Nelore cattle (Dalmaso de Melo et al. 2020), *Bos taurus* beef cattle (Holton et al. 2022), and Holstein cows (Madoz et al. 2022). These last authors (Madoz et al. 2022) have demonstrated that at least one-third of false-positive results are attributable to early embryonic loss occurring between days 19/20 and 30–34 of gestation. In conclusion, the aforementioned findings collectively indicate that the majority of false-positive results are attributable to a prolonged luteal phase rather than embryonic death after maternal recognition of pregnancy. This assertion is further substantiated by the observation that 10–15% of heifers and Nelore cows that were synchronized for fixed-time artificial insemination but not inseminated failed to undergo luteolysis until 22 days post-estrus (Ataide Junior et al. 2021).

Furthermore, no notable discrepancies in CLBP were observed between true positives and false positives on days 20 and 22 (Holton et al. 2022). However, a reduction in CLBP and P4 levels was observed on day 20 in females exhibiting indications of embryonic loss between days 20 and 25 of gestation (Dalmaso de

Melo et al. 2020). Furthermore, a recent study (Pugliesi et al. 2023) indicates that the prevalence of CLBP at day 20 is inversely associated with the occurrence of false-positive outcomes.

The probability of a false-positive result is greater than 25% in females with CLBP at or below 40% but drops to less than 5% in those with CLBP at or above 75% at the time of pregnancy diagnosis (Reese et al. 2020). Moreover, the impact of parity (nulliparous vs. multiparous) on false positives seems to be confined to beef cattle (Pugliesi et al. 2023). In dairy cattle, the incidence of false positives is comparable between lactating cows and heifers (Siqueira et al. 2013; Ferraz et al. 2022). However, in beef cattle, heifers exhibit a higher prevalence of false positives and lower accuracy compared to lactating cows (Dalmaso de Melo et al. 2020; Holton et al. 2022). In *Bos indicus* beef heifers, this phenomenon is likely attributable to delayed spontaneous luteolysis in comparison to lactating cows (Ataide Junior et al. 2021).

Despite the well-documented physiological differences between *Bos indicus* and *Bos taurus* females (Sartori et al. 2016), the accuracy of pregnancy diagnosis using Doppler at days 20 and 22 post-fixed-time artificial insemination is comparable between *Bos taurus* beef cattle (87%; Holton et al. 2022) and *Bos indicus* (92%; Hassan et al. 2019; Dalmaso de Melo et al. 2020). The accuracy of pregnancy diagnosis improves over time following insemination, as a greater number of non-pregnant cows are identified around the time of luteolysis (Pugliesi et al. 2014; Scully et al. 2015; Siqueira et al. 2019). For example, Holton et al. (2022) reported a 40% reduction in false positives when the diagnosis was performed on day 22 versus day 20 post-fixed-time AI in beef cows. In general, the findings of the studies indicate that CLBP-based non-pregnancy diagnosis is both reliable and accurate when conducted after day 20 of pregnancy in cattle (Pugliesi et al. 2023).

The majority of studies that assess the functional status of the CL and predict non-pregnancy rely on quantifying the colored area in CL Doppler images. It has been demonstrated that discrepancies in ultrasound settings, which alter the quantity of signal displayed as colored pixels, can impact diagnostic accuracy, resulting in elevated rates of false positives or false negatives (Pugliesi et al. 2023). In B-mode ultrasound, experienced operators can intuitively adjust settings, including transducer frequency, overall gain, and grayscale curve, to produce images of greater clarity. Furthermore, additional attributes, such as pixel value or heterogeneity, can assist in the assessment of the CL (Siqueira et al. 2009). In routine examinations, the majority of assessments of CL function are based on its size or presence. Consequently, slight variations in B-mode image settings are less likely to result in errors. However, changes in color Doppler settings can have a greater effect on diagnostic accuracy. Of these, pulse repetition frequency (PRF) is of particular importance, as it determines how well the Doppler signal is displayed (Pugliesi et al. 2023).

Pugliesi et al. (2023) reported the results of a study conducted on cows 20 days after timed artificial insemination. Three PRF settings (0.7, 1.0, and 1.5 kHz) were tested, and the results were compared to pregnancy diagnoses at day 30. A reduction in the PRF from 1.0 to 0.7 kHz resulted in a decline in the accuracy of detecting pregnant animals and specificity (from 79.2% and 72.2% to 67.9% and 50.0%, respectively). Conversely, increasing the PRF from 1.0 to 1.5 KHz resulted in a reduction in the accuracy of detecting non-pregnant animals and sensitivity (from 100% to 81.3% and 84.2%, respectively). These findings underscore the importance of maintaining consistent and standardized Doppler settings when using color Doppler to diagnose non-pregnancy based on CL blood flow, as even minor adjustments in these settings can lead to misinterpretations.

2.5.5 Puerperium Evaluation

B-mode ultrasound examination enables direct evaluation of the postpartum uterus (Fig. 17), thereby facilitating improved diagnosis of pathologies such as clinical metritis and clinical endometritis (Bollwein et al. 2016). Doppler technology is a valuable tool for assessing genital blood flow, as evidenced by its utility in evaluating the flow of specific vessels through the measurement of vessel resistance, as demonstrated in previous studies (Abo El-Maaty and Abdelnaby 2017; Cheong et al. 2017; Abdelnaby et al. 2018). The Doppler ultrasound values are calculated from two consecutive blood flow waves captured by the probe and are configured according to the equipment's software.

Krueger et al. (2009) assessed uterine blood flow and the uterine artery pulsatility index from the time of calving to the 12th postpartum week in dairy cows. Similarly, Heppelmann et al. (2013) observed alterations in uterine artery diameter, time-averaged maximum velocity, and pulsatility index during the initial 2 weeks postpartum in Simmental cows. The most significant alterations in uterine blood flow occur during the initial 4 days following calving (Heppelmann et al. 2013), a period that coincides with a pronounced reduction in the size and weight of the uterus. As the postpartum period progresses, there is a decrease in blood flow volume while the pulsatility index (PI) increases. By Day 28 postpartum, these changes

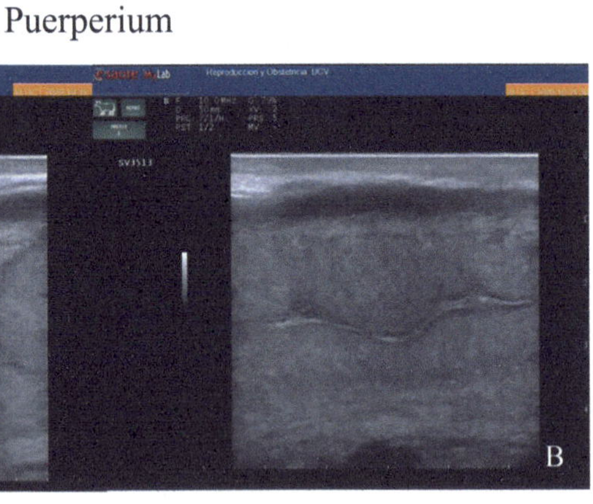

Fig. 17 B-mode ultrasound examination is a direct evaluative modality for the postpartum uterus. As illustrated in cross section (**a**) and longitudinal section (**b**), a 20-day postpartum dairy cow reveals the presence of puerperal content and an increasing thickness of the endometrium

have stabilized. This is due to the completion of changes in the caruncular vascular bed around Day 30. Although clinical and histological uterine involution is typically complete by day 47 postpartum (Gier and Marion 1968; Van Camp 1991), blood flow volume continues to decline until day 65 (Heppelmann et al. 2014) and PI until day 86 (Krueger et al. 2009). This indicates that alterations in the uterine vascular bed are more prolonged than the histological modifications in the endometrium.

The findings of various studies indicate a correlation between uterine vascularity and any defects in the mechanism of uterine involution and the uterine and that the ovarian changes in normal cows occur rapidly during the postpartum period (Jaffa et al. 1996; Nakai et al. 1997; Krueger et al. 2009; Paliulyte et al. 2017). However, the uterine artery blood flow parameters differ between cows that develop postpartum uterine disease and those that remain healthy (Henao-Gonzalez et al. 2023).

In a study conducted to compare the physiological characteristics of cycling and anestrus dairy cows following the conclusion of the puerperal period (days 39 and 70 postpartum), Abdelnaby (2020) investigated the potential correlation between Doppler indices and blood flow volume. The findings of the study indicate that uterine and ovarian morphology exhibited slight alterations in cyclic cows relative to cows in anestrus. However, in the latter, the vascular system of the ovary and uterus underwent more pronounced alterations. The Doppler indices in the ipsilateral and contralateral ovarian and uterine arteries in postpartum anestrus cows demonstrated that the vascular system of the ovary, as well as the uterus, underwent significant alterations, with the most notable being the velocities and volume of blood flow. These vascular changes indicated that the anestrus cows exhibited a deficiency in reproductive functionality, as evidenced by the marked reduction in blood supply to the reproductive organs.

Also, postpartum uterine conditions such as retained fetal membranes (RFM) and metritis have been demonstrated to impede uterine involution (Fonseca et al. 1983; Mateus et al. 2002), and uterine blood flow is similarly affected by these conditions. In cows with RFM, the uterine arteries exhibit higher resistance index (RI) values in the days preceding calving. This is likely due to a failure in the normal detachment of fetal and maternal tissues (Hartmann et al. 2013). By day 8 postpartum, cows with metritis exhibited higher blood flow volume and lower PI values in the uterine arteries compared to healthy cows. In cows with uterine diseases, the decline in blood flow volume ceases after day 45 postpartum, whereas in healthy cows it persists (Heppelmann et al. 2014). These altered blood flow parameters in cows with uterine disease reflect a delay in uterine involution.

In addition to assessing uterine involution, transrectal Doppler sonography can be employed to indirectly evaluate uterine contractility. The contraction of the uterus causes the endometrial arteries to constrict, which results in increased vascular resistance and a reduction in blood flow velocity (Fleischer et al. 1987; Janbu and Nesheim 1987). In healthy cows, administration of oxytocin on day 2 post-calving has been observed to reduce blood flow volume and increase the PI. However, in cows with RFM, there is no discernible response to oxytocin, indicating that the uterus is unable to respond to the hormone during this period (Magata et al. 2013).

2.5.6 Methods for Estimating Gestational Age

2.5.6.1 The Estimation of Gestational Age Is Based on Fetal Parameters

Historically, the study of prenatal development in cattle relied heavily on the examination of postmortem or slaughterhouse samples, as well as conceptuses obtained through induced abortion or delivery (Boyd 1974). However, these approaches fall short in analyzing the dynamic processes of intrauterine development, which are crucial for gynecological practice. Ultrasonography has proven to be a valuable tool for detecting, assessing, and monitoring viable embryos by measuring fetal dimensions, allowing for accurate determination of gestational age (Curran et al. 1986a, b; Omran 1989).

A key reason for estimating or confirming gestational age in dairy cattle is its use in determining the dry-off period and planning for calving (Doize et al. 1997). In more extensive management systems, gestational age also helps in selecting cows that can aid in herd management during specific times of the year (Jephcott and Norman 2004).

In 1985, White et al. were the first to link fetal age in cattle with measurements of fetal lengths and diameters across various body parts and organs. Using a real-time ultrasonic scanner with a 3.5 MHz rectal probe, they found that crown-rump length provided the most accurate estimation of fetal age, while uterine diameter was the least reliable. Other measurements, such as head and trunk length, as well as head and snout diameter, showed moderate accuracy.

Later, Kähn (1989) assessed the effectiveness of 25 different fetal parameters in estimating gestational age in cattle from 30 days to 10 months. The strongest correlations with fetal age were noted for crown-rump length, as well as the femur, metacarpus, tibia, scapula, ischium, eye, brain, trunk, and umbilical cord measurements. The crown-rump length was the easiest to measure during the first 3 months of gestation, showing minimal variation and the highest correlation. Kahn also observed that the growth rate increased from 1.4 mm per day during the first 2 months to 2.2 mm per day by the end of the second month. In the third month, the growth rate further accelerated to 2.5–3.0 mm per day, leading to a linear length of 12 cm by the month's end.

To fully appreciate the role of fetal measurements in determining gestational age, it is crucial to understand fetal growth and the factors that affect it. Fetal development is a complex process shaped by both genetic and environmental influences (Bellows et al. 1993). In bovine, genetic factors include breed-level traits and genotype variations within breeds. Environmental factors encompass maternal influences such as nutritional level, lactation status, and parity, some of which may also have genetic underpinnings. Additionally, placental development and blood flow play significant roles in fetal development. These factors, along with fetal growth itself, are affected by both maternal and fetal components.

Fetal growth is largely driven by the genetic potential of the fetus, as well as the maternal genetic profile (Bellows et al. 1993; Ferrell 1991). During early and mid-gestation, the fetal genotype has the greatest influence on growth, whereas in the later stages, when the majority of growth occurs, the maternal genotype plays a more prominent role (Greenwood et al. 2010).

The physiological condition of the dam also influences fetal growth, which may be tied to placental development. Heifers, for instance, have a smaller cotyledonary surface area compared to mature cows (Van Eetvelde et al. 2016). Another key factor affecting fetal growth is the sex of the fetus, with noticeable differences between male and female fetuses emerging around 100 days of gestation (Eley et al. 1978). Additionally, gestational age plays a role in fetal growth rates, which steadily increase, peaking at around 230 days of gestation (Ferrell et al. 1976; Eley et al. 1978). Eley et al. (1978) suggested that the slowing of fetal growth in the third trimester is due to the placenta's inability to expand fast enough to support the high growth rates seen before 230 days.

Numerous parameters measured via ultrasonography show strong correlations with gestational age. These include crown-rump length ($r = 0.91$), biparietal diameter ($r = 0.95$), head length ($r = 0.94$), trunk diameter ($r = 0.95$), uterine diameter ($r = 0.93$), or femur length ($r = 0.99$) (White et al. 1985; Kähn 1989). Of these, crown-rump length is the most commonly used parameter. It is defined as the distance, either straight or curved, between the top of the head (crown) and the base of the fetal spine (tailbone or perineum). However, pinpointing these exact landmarks early in development through ultrasound can be difficult, leading to variability in results between studies (Bunyaga 2015).

Research has also highlighted the complexities of typical prenatal growth in cattle and the wide range of intrinsic and extrinsic factors related to the fetus, dam, and environment that influence this process (Richardson et al. 1990; Holland and Odde 1992). As a result, the accuracy of gestational age estimation depends on the species and the specific measurement parameters used. Despite these variables, crown-rump length remains the most accurate measure of fetal age in cattle. For instance, Wright et al. (1988) found that when ultrasound was performed at 35 and 125 days of gestation in beef cattle, the predicted calving date was accurate within 0.9 days of the actual date.

In terms of in vivo embryonic development, Pierson and Ginther (1984) observed that the embryonic vesicle in cattle steadily lengthens until day 26, after which it begins extending into the opposite uterine horn. By day 32, the vesicle fully occupies both horns. The fetal heartbeat becomes detectable between days 26 and 29. Additionally, Curran et al. (1986b) reported the ultrasonographic detection of various features of the bovine conceptus between days 20 and 60, including the fetus itself, fetal heartbeat, allantois, spinal cord, limb buds, amnion, optic vesicle, optic lens, split hooves, ribs, and fetal movement. Researchers have highlighted the value of ultrasound in monitoring embryonic growth, suggesting that observing the embryo closely could help study early embryonic loss in cattle (Pierson and Ginther 1984; Curran et al. 1986a, b; Bunyaga 2015; Kouamo et al. 2018).

In *Bos indicus* cattle, the embryonic vesicle becomes visible between days 18 and 20 of gestation, and the fetal heartbeat is detectable by day 22 as well as fetal movements by day 50. The embryo grows steadily until day 39, after which growth accelerates. However, *Bos indicus* embryos grow at a slower rate compared to *Bos taurus* embryos (Bergamaschi et al. 2004; Kouamo et al. 2018).

The biparietal diameter (BPD), which measures the widest distance across the fetal skull at a 90° angle to its long axis, is normally employed to assess fetal size. Studies across many species, including Murrah buffalo (Ferreira et al. 2012), goats (Haibel and Perkins 1989; Abdelghafar et al. 2011), sheep (Sergeev et al. 1990), and cattle (White et al. 1985), have shown a strong correlation between BPD and gestational age.

BPD also correlates strongly with crown-rump length (Riding et al. 2008; Ferreira et al. 2012), a useful parameter in early gestation but difficult to measure in later stages due to fetal size. BPD, however, is easier to measure and has lower measurement error compared to other fetal parameters (Wu et al. 2012). It is also unaffected by fetal sex or the sire's influence in Neroli dams (Bergamaschi et al. 2004).

As Kähn (1989) demonstrated, even though the percentage of successful BPD measurements declines with fetal age (from over 95% in months 3 and 4 to about 50% in months 9 and 10), BPD remains the most consistently measurable fetal parameter throughout gestation.

The head is one of the earliest identifiable embryonic body parts, with a clear distinction from the rest of the body visible by the fifth week of pregnancy. The dark area of the developing eye, a key marker for BPD measurement, becomes visible around day 40 of gestation (Kähn 2004). This makes BPD measurement useful across a broad range of gestational ages.

2.5.6.2 The Estimation of Gestational Age Is Based on the Measurement of the Placentomes

A major challenge in estimating gestational age through fetal measurements is the difficulty in locating the fetus during late gestation, particularly after the 120th day (Kähn 1989). In cattle, an alternative approach is to measure the bovine placentome, as its average size is strongly correlated with gestational age (Adeyinka et al. 2014; Laven and Peters 2001). Both placentome weight and length significantly increase as gestation progresses, with placentome length showing a relatively linear growth throughout pregnancy (Laven and Peters 2001). Some research suggests that placentome growth continues throughout gestation (Abdel-Raouf and Badawi 1966; Reynolds et al. 1990), but Laven and Peters (2001) and Liu et al. (2010) propose that this growth plateaus around day 200.

However, issues other than gestational age can alter placentome size. For example, the placentomes in the non-pregnant uterine horn are shorter than those in the pregnant horn (Laven and Peters 2001). Additionally, placentome size is influenced by its proximity to the site of fetal attachment, with larger placentomes found closer to the attachment site (Youngquist 1997).

Breed differences also impact placentome weight and size. Reynolds et al. (1990) found that Charolais cows had heavier placentomes than Brahman cows, while Van Eetvelde et al. (2016) observed that Belgian Blue cows had a larger caruncular surface area than Holstein-Friesian cows. Fetal sex is another influencing factor, with placentomes associated with male fetuses being larger than those with female fetuses (Arthur et al. 2009).

Similarly, Mendoza Bautista et al. (2024) reported that the placentome size in *Bos taurus* × *Bos indicus* cows was of an intermediate magnitude when compared to the two parent breeds. These findings lend support to the notion that the current reference tables for purebred *Bos taurus* and *Bos indicus* breeds are inaccurate and should not be applied to crossbred animals.

Studies using transabdominal ultrasonography to determine the relationship between placentome size and gestational age (Hunnam et al. 2009; Adeyinka et al. 2014; Lazim et al. 2016) found no significant correlation with placentome height or length. In contrast, Adeyinka et al. (2014), using transrectal ultrasound, and Lazim et al. (2016), using transabdominal ultrasound, reported a strong correlation between placentome size and gestational age. Adeyinka et al. (2014) attributed these differences to variability in the consistency of placentomes assessed in the cervical region (Fig. 18).

Bunyaga (2015) reported that BPD provided the most accurate estimates of gestational age. However, the accuracy of BPD decreases as gestation advances, and by 120 days, the precision of BPD is similar to that of measuring multiple placentomes. Therefore, after this point, placentome measurements become easier and potentially more accurate for estimating gestational age.

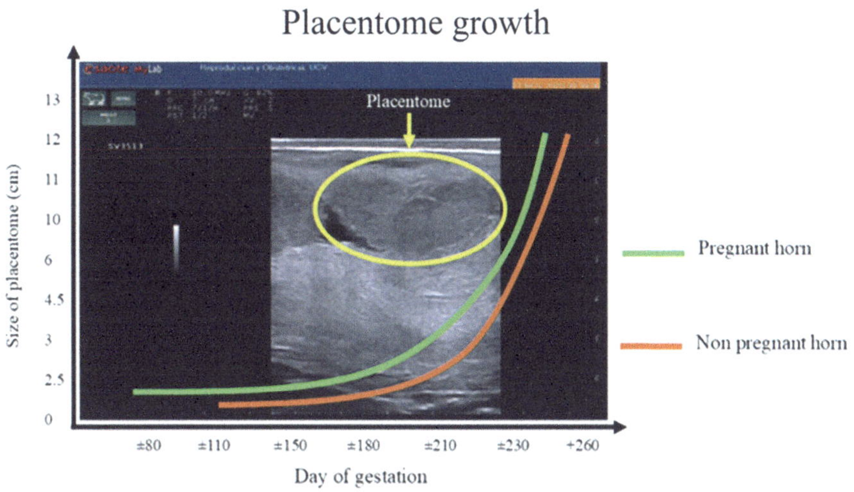

Fig. 18 In the field of bovine reproduction, an alternative approach involves the measurement of placentomes in cattle. The rationale for this method is that the average size of placentomes exhibits a strong correlation with gestational age. It is important to acknowledge that placentome size is influenced by several factors, including its proximity to the site of fetal attachment, fetal sex, and breed, among others. The figure illustrates the average growth rate of placentomes, commencing around days 70–80 (in the case of pregnant horns) or days 100–115 (in the case of non-pregnant horns)

2.5.7 Embryo Transfer and Superovulation Protocols

Ultrasonography represents a valuable instrument for the monitoring of superovulation protocols and treatments. In embryo transfer programs, the primary objective of inducing superovulation is to maximize the number of transferable embryos (Torres-Simental et al. 2021). Initially, superovulation protocols were based on the observation that the optimal response was achieved when treatments commenced between 8 and 12 days after the onset of estrus. However, these initial studies did not take into account the specific follicular status of the animals at the commencement of treatment, as ultrasound technology for cattle was not widely available (Bó and Mapletoft 2014). The literature indicates that the superovulation response is most effective when treatments are initiated precisely at the onset of the follicular wave. Indeed, studies have demonstrated that initiating gonadotropin treatment even 1 day before or after this point significantly decreases the response (Nasser et al. 1993; Sawyer et al. 1995; Bergfelt et al. 1997; Cirit et al. 2019). In light of the multifactorial nature of superovulation in cows, it is of paramount importance to consider these factors when designing protocols. Ultrasonography serves as a crucial tool for monitoring treatment progress (Mikkola et al. 2020).

In 1991, Cattoni conducted a study to assess the efficacy of ultrasound in monitoring follicular development, ovulation, and CL formation in superovulated cows. The findings were then compared with those of embryo recovery rates and postmortem analysis of the ovaries. While ultrasound proved effective in monitoring follicular growth during superovulation, it was challenging to accurately measure ovulation or CL numbers, rendering it unreliable for predicting embryo recovery.

Sawyer et al. (1995) report that the regular monitoring of bovine ovaries by ultrasonography has confirmed that the patterns of ovarian follicle growth, selection, and atresia persist under the exogenous maintenance of progesterone levels by intravaginal devices. Moreover, the timing of progesterone device insertion in relation to the emergence of the follicular wave or phase of the estrous cycle has been demonstrated to have significant implications for the recruitment and maturation of follicular populations in response to superovulation treatments (Fig. 19).

The timing of ovulation can be monitored using ultrasound technology. The results demonstrated that aligning the onset of the follicular wave with estradiol could be effectively incorporated into a timed AI protocol for donors, obviating the necessity to detect estrus (Bó et al. 2006).

The utilization of Doppler ultrasonography to assess the functionality of the CL throughout the estrous cycle has facilitated the development of techniques for integrating CL evaluation into ET programs. One of the pivotal factors in determining whether a recipient should be utilized in ET programs is the assessment of CL quality. This can be accomplished by selecting the recipient through transrectal palpation or grayscale ultrasonography to quantify the CL's size. However, these methodologies may result in embryo transfers to a non-receptive uterus if the luteal phase is not functioning optimally, as these assessments are unable to discern luteal functionality (Pugliesi et al. 2018).

To address this issue, Pinaffi et al. (2015) employed Doppler ultrasonography on recipients on the day of ET, categorizing them based on CL blood flow perfusion as

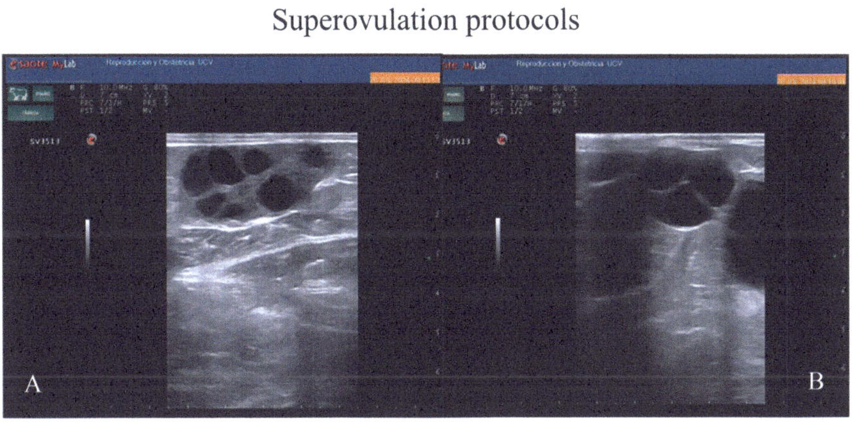

Fig. 19 B-mode ultrasound can be utilized to monitor the recruitment and maturation of follicular populations in response to superovulation treatments. (**a**) Illustrates a cow ovary that demonstrated a robust response to the superovulation protocol. (**b**) Depicts a cow ovary that exhibited a medium response after being treated with the same protocol as the ovary in (**a**)

either low (<40% colored signals) or high (>40% colored signals). Although the study included a limited sample size, none of the recipients with low CL vascularity were found to be pregnant, while the pregnancy rate was 48.4% in those with high luteal vascularity. Interestingly, despite the differences in vascularization, the average CL diameter was similar between cows with low (17.5 mm) and high (17 mm) CL vascularity.

These studies indicate that Doppler ultrasound can be employed to exclude recipients with a nonfunctional CL and to identify those with enhanced receptivity for ET. Pugliesi et al. (2016b) investigated the impact of luteal size and blood flow perfusion, assessed via color Doppler ultrasonography at the time of ET, on pregnancy rates in crossbred recipients with synchronized estrous cycles. The recipients were classified into distinct groups based on their CL size and blood flow perfusion. The CL size was classified into two categories: small (less than 3 cm^2) and large (greater than 3 cm^2). The luteal blood perfusion was divided into three categories: low (≤40% colored signals), medium (45–50% colored signals), and high (>55% colored signals). The findings indicated that only luteal blood flow perfusion influenced pregnancy rates, with higher luteal blood flow perfusion resulting in progressively higher pregnancy rates: 45.1% (low), 55.9% (medium), and 62.3% (high).

This suggests an alternative method for improving pregnancy success with higher-quality embryos. Furthermore, Doppler ultrasound could enhance reproductive efficiency in ET programs by reducing embryo loss, as it would prevent the transfer of embryos to females with nonfunctional CLs. In cases with excess recipients, priority could be given to those with medium to high luteal vascularization (Pugliesi et al. 2018).

2.6 Practical Applications of OPU in Cattle Reproduction (as Described in Ruiz 2021)

The technique of ultrasound-guided transvaginal follicular aspiration for ovum pick-up (OPU) is a noninvasive procedure for recovering *cumulus* oocyte complexes (COCs) from antral follicles in live animals, especially in cows and mares. Unlike conventional multiple ovulation and embryo transfer (MOET), OPU does not interfere with the normal reproduction and production cycles of the donor. It is a technique that can be used in adult cows in various physiological states (cyclical, noncyclic, in the first third of gestation and in which they do not respond to hormonal stimuli), in old animals with reproductive disorders of nongenetic origin (Galli et al. 2001; 2014), and in calves and heifers from 6 months of age (Taneja et al. 2000), and even soon, animals after calving (2–3 weeks) could be a suitable donor.

The association of reproductive efficiency and genetic selection is strategic for the success of the dairy and beef industries. Reproductive technologies, such as OPU and in vitro embryo production (IVEP), can rapidly enhance the genetics of cattle through both the female and male lineage (Watanabe et al. 2017). The combination of OPU and IVEP has been successfully and widely commercially used worldwide. The main advantages of the OPU-IVEP technique are a greater number of embryos and pregnancies per unit of time, accelerated genetic progress due to the expedited turnaround time for donors, and an expanded array of elite sire mating combinations. Additionally, a more extensive spectrum of female age (calves, prepuberal, heifers, cows) and condition (open, pregnant) from which to retrieve oocytes is available. Other advantages include a reduced number of sperm (even sexed) required to fertilize the oocytes (Ferré et al. 2023).

According to the International Embryo Transfer Society (IETS) statistics, the number of embryos produced in vivo and transferred to recipients has been increasing exponentially over the last 7 years to reach more than four times the number of embryos produced in vivo by MOET (1,595,204 in vitro-produced embryos transferred vs. 368,783 in vivo-derived embryos transferred, in 2022) (Viana 2023). This outstanding growth of OPU/IVEP seen in the last years is a consequence of several factors, such as the significant improvement of in vitro culture procedures and the successful use of sex-sorted semen in the IVEP programs enabling the manipulation of the proportion of male and female embryos produced (Galli et al. 2014; Blondin 2015).

2.6.1 OPU Equipment and Procedure

The OPU system consists of three major components: an ultrasonographic scanner with an appropriate transducer (probe), a vacuum pump, and a needle guidance system connected to an oocyte collecting tube (Bols and Stout 2018; Ferré et al. 2023). For visualization of the ovarian follicles, native endovaginal probes (5–7.5 MHz) are available essentially for human use. Several manufacturers of ultrasound equipment have provided custom-made plastic holders or handgrips that can house generic convex array transducers together with the needle guide. At present, there are already several commercial companies (Minitube, IMV, etc.) that manufacture specific holders for OPU in cattle.

Inside the handgrip close to the transducer, the puncture needle can be visualized on the ultrasound screen when it is advanced into the sonographic field to enter a follicle; to facilitate visualization, it is helpful to have a biopsy guide on the ultrasound screen. This biopsy line programmed into the scanner's software is displayed on the screen and indicates where the follicle needs to be positioned for a successful puncture.

The protocol for sedating cows before OPU involves the administration of xylazine to mitigate stress and facilitate the procedure. The cows are also treated with anti-inflammatory drugs to mitigate discomfort and enhance reproductive efficiency. Subsequently, epidural anesthesia is induced using lidocaine 2%. (Ruiz et al. 2013a). Through the rectum, the operator fixes the ovary and positions it against the head of the transducer such that the ovary and follicles can be visualized on the ultrasound screen. The operator then advances the needle slowly forward until the vaginal wall is pierced and the needle is visualized entering the ultrasound field. By monitoring the needle's position and simultaneously manipulating the ovary transrectally, the needle can be directed into a follicle. Once the needle enters the follicle, the aspiration pump is activated using the foot pedal, and the follicular fluid and COCs are collected into the embryo filter which contains an oocyte collection medium. An assistant should wash the system every time three to four follicles are punctured to avoid excessive oocyte loss and improve the recovery rate (Ruiz 2021).

2.6.2 Factors Affecting OPU Results

2.6.2.1 Technical Factors

OPU success rate is quantified firstly in terms of the oocyte recovery rate (RR = number of COCs per 100 follicles punctured), which is influenced by factors including needle diameter, aspiration pressure, ultrasound equipment, operator experience, and other technical factors. The knowledge of the technical variables is essential for maximizing follicle aspiration efficiency (Da Silva et al. 2016, 2017).

Needles with diameters larger than 18G are associated with higher recovery rates as well as higher percentages of denuded oocytes. The reason for this is that a large diameter needle enables quick flow of the oocytes and granulosa cells from the *cumulus oophorus*, which may easily detach from the zona pellucida. Another problem is the large ovarian area that is damaged during the puncture, which also increases the degree of blood aspiration. On the other hand, a needle with too small a diameter (<21G) provides a low-speed flow, which increases the possibility of the oocyte being lost during the procedure and falling into the abdominal cavity. The highest oocyte recovery rates were obtained when using the thickest needle (18G), and the thinner needles resulted in a higher proportion of recovered COCs with a compact *cumulus*, regardless of the aspiration vacuum. Considering all these aspects, it is well-accepted that the best diameters are 18–20G (reviewed by Da Silva et al. 2016).

Another important point related to the needle is the *bevel length*. Bols et al. (1997) studied the effects of needle tip bevel on the morphology and developmental capacity of bovine COCs. The results showed that the length of the needle bevel has a significant effect on oocyte recovery, in favor of the long-beveled needle. The probable explanation for this is the sharp edge of the long bevel allows rapid follicular penetration, whereas short bevels are usually less sharp. Regarding the needle length, 40–75 mm is considered ideal. Needles <40 mm do not cover the ovary during the puncture and excessive manipulation is required to reach all follicles; on the other hand, needles >75 mm are too flexible and bend easily.

The exact *aspiration pressure* exerted through the tip of the needle depends on the aspiration device, the length and diameter of the tubing, the size and type of collection vessel, as well as the needle diameter. To make comparisons possible, the aspiration pressure needs to be expressed in terms of the amount of fluid (in mL) that can be aspirated per minute, rather than in mmHg exerted from the vacuum pump (Bols and Stout 2018). A considerable variation of 4.4–40 mL/min has been reported, although the range of 10–25 mL/min has been most widely used since it demonstrated high efficiency of oocyte recovery and minimal damage to the *cumulus oophorus* during OPU (Bungartz et al. 1995; Galli et al. 2001, 2014).

Transducer frequency and *sensitivity* of the *ultrasound equipment* constitute a significant variable in the oocyte recovery process and represent another parameter of extreme relevance for the OPU efficiency. Reports involve frequencies of 3.5–7.5 MHz. Considering that the operator will manipulate the ovary very close to the transducer, the adoption of high frequencies of 6–8 MHz is recommended since they provide the best image definition of follicles and other ovarian structures (Da Silva et al. 2016). The vast majority of OPU procedures are performed with convex or microconvex transducers, which are commonly used in humans as well. At present, several commercial companies have designed their own scanners, transducers, and holders for performing OPU in cows, e.g., *Easi-Scan* from BCF-Technologies, "All in one" and *Exago* or *Exapad* from Echo Control Medical (ECM); both companies form a new imaging division "IMV Imaging" within the IMV Technology Group (Ruiz 2021).

A final very important factor determining OPU outcome is the *experience* of the *operator* or the *team* that is retrieving the oocytes (reviewed by Bols and Stout 2018). The OPU technique is a process performed by one or two individuals. Within some companies, these individuals have been responsible for developing the technique in a similar way for years, using the techniques under the same conditions. The experience of the operator (in a one-person system) or the operator and the assistant (in a system of two people) has a significant effect on the number and quality of collected oocytes. Proper selection of operators and/or assistants can help improve the results (Merton et al. 2009).

2.6.2.2 Biological Factors

A substantial body of literature is available on biological factors that might influence the likelihood of blastocyst formation when IVEP is based on COCs recovered via OPU. Between these biological factors, we can include the frequency and timing of follicle puncture, the physiological status and body condition of the donor, individual variations, the breed and age of the donor, the hormone stimulation to prepare donors for OPU, and finally the influence of climate and season (reviewed by Ruiz 2021).

The *frequency of sampling* represents a crucial point. OPU is routinely performed twice a week (usually Monday and Thursday), which allows the maximum recovery of oocytes of suitable quality for embryo production in each time interval compared to once-a-week OPU because no dominant follicle develops when all visible follicles are aspirated in the OPU process. In most once-a-week collections, a dominant follicle develops at the successive collection, which causes the regression and degeneration of the subordinate follicles. As a result, oocytes collected by this scheme are relatively less and with low quality in each time interval. In the once-a-week collection schedule, the donors also can come into estrus, while this is never the case in a twice-a-week schedule (Galli et al. 2001). In conclusion, a twice-weekly OPU schedule resulted in an increased follicular wave frequency, and an arrest of the oestrous cycle, follicle maturation, and ovulation. Animals submitted to this sampling regimen entered a para-physiological status in which follicular waves were uncoupled from the estrous cycle. However, as soon as OPU sampling ceased, ovulation took place within 6 days (Kruip et al. 1994).

Regarding *physiological status*, in cattle breeding programs, OPU is generally performed on selected healthy heifers with excellent genetic potential for production traits that could in themselves be predictive of oocyte yield and the number of blastocysts produced (Merton et al. 2009). The reproductive potential of females can be maximized if OPU/IVEP embryos are applied during nonproductive periods such as pregnancy, anestrus, and postpartum. OPU can be performed at various stages of a cow's reproductive life; even pregnancy does not exclude OPU, since oocytes can successfully be retrieved during the first 3 months of gestation (Bungartz et al. 1995). The only exceptions are pregnant animals after the third or fourth month of pregnancy and animals with severe ovarian hypoplasia or in the immediate postpartum before ovarian activity is restored (Galli et al. 2001).

In relation to *body condition*, there is consensus about malnourished donors having reduced capacity for oocytes to develop embryos. Effectively, undernutrition has a negative effect on the developmental competence of recovered oocytes in vitro, as illustrated by the decreasing percentage of blastocysts associated with decreasing body condition score of the donor and an increasing proportion of good-quality oocytes with increasing body condition score (reviewed by Bols and Stout 2018).

The *donor animal* is a major source of variation from both the management and genetic point of view. The efficiency of OPU/IVEP techniques has been impaired by the large variability among donors in their response to these procedures (Merton et al. 2009; Pontes et al. 2009). Therefore, the success of IVEP technologies has been associated primarily with physiological characteristics, such as ovarian antral follicle population (AFP) and oocyte competence to reach the blastocyst stage (Pontes et al. 2011). The AFP has been related to several substances, including the concentrations of circulating insulin, insulin-like growth factor I, and anti-mullerian hormone (AMH) (Batista et al. 2014). In human medicine, plasma AMH levels have become an important predictive parameter for the success of assistant reproductive technologies, which include OPU/IVEP (Dewailly et al. 2014). However, in veterinary reproductive medicine, this parameter is not used very often. During the last few years, strong linear regressions between plasma AMH levels and the number of antral follicles have also been described in different breeds of cattle (Guerreiro et al. 2014; Batista et al. 2016). In opposition, Vernunft et al. (2015) analyzed the plasma AMH levels in Holstein-Friesian heifers, and their results suggest that correlations between AMH and outcomes of an OPU/IVEP program are too low to use AMH as a precise predictive parameter for the success of a particular OPU procedure in cattle. However, AMH can help to identify groups of very good or very poor oocyte donors. In conclusion, more studies must be conducted to clarify these different results and to determine the role of AMH levels in OPU/IVEP procedures in cattle.

The OPU efficiency did not show a significant variation among *cow breeds* if *Bos taurus* breeds are concerned (Kruip et al. 1994). If this evaluation is extended to *B. indicus* breeds and *indicus-taurus* donors, significant differences were observed. *B. indicus* breeds tend to have, on average, more follicular waves (Figueiredo et al. 1997, b) as well as a greater number of follicles >5 mm per wave compared to *B. taurus* breeds. As there is a greater efficiency of oocyte retrieval from follicles <4 mm in diameter (Seneda et al. 2001), this is another reason that why more oocytes are obtained from *B. indicus* than from *B. taurus* donors. However, the physiological basis for the number of follicles in *B. indicus* Nelore cattle has not been established, and despite the physiological importance of these differences, the high number of oocytes obtained via OPU seems to be a unique characteristic of Nelore cows (Pontes et al. 2009). Despite the great differences in the total of oocytes retrieved per OPU procedure across breeds, outstanding donors can be found in all breeds, and the number of oocytes retrieved per individual donor seems consistent across time. In fact, it was previously shown the occurrence of high within-cow repeatability in oocyte production over time (Monteiro et al. 2017). In conclusion, breed donor influences the OPU process because different breeds have various numbers of follicles in the ovary during the follicular wave cycle. *B. indicus* cows tend to have more numerous follicular waves and a larger number of smaller follicles than *B. taurus*. This feature has resulted in very fast OPU/IVEP popularity growth in Brazil since the Brazilian herd is around 200 million animals about 85% of which are Nelore breeds. Although in vivo embryo production continues to be widely used in *indicus* donors, the use of OPU/IVEP enables the production of more embryos in the same period (reviewed by Da Silva et al. 2016).

Age is a commonly discussed factor. The very young and very old donor ages are generally reported as problematic situations. Prepubertal and aging donors commonly have poor-quality oocytes. Elderly cows also have fewer available gametes (Da Silva et al. 2016). Virtually all female cattle starting from 2 months of age can be oocyte donors by follicular aspiration. The only exceptions are pregnant animals after the third or fourth month of pregnancy and animals with severe ovarian hypoplasia or in the immediate postpartum before ovarian activity is restored. The major problem with prepubertal donors is the impaired in vitro developmental capacity of the recovered oocytes (Taneja et al. 2000), resulting in a low overall efficiency of the procedure.

Many attempts of *hormonal stimulation* have been evaluated to improve the follicle population in animals submitted to OPU. Several investigators have used gonadotropin stimulation in OPU protocols ranging from a full superovulatory dose to shorter treatments for 2–3 days with one or two injections per day in the presence of a progesterone-releasing device or a CL. The positive effect of the gonadotropins can be attributed to the increase in size of small follicles and to the acquisition of a higher developmental competence of the oocytes as it occurs with in vivo matured oocytes (Merton et al. 2009).

In prepubertal calves, gonadotropin stimulation is required to obtain an acceptable level of developmental competence (Taneja et al. 2000; Galli et al. 2001). However, the use of gonadotropin stimulation does not seem to be effective in *B. indicus* donors; it is not used in large-scale programs (Pontes et al. 2009, 2010).

De Roover et al. (2008) published a retrospective study about the effects of OPU frequency and follicle-stimulating hormone (FSH) stimulation of beef cattle (Belgium Blue) IVEP. Data from 7 years of work were analyzed comparing the number of follicles, COCs, and IVP embryos obtained from 1396 non-stimulated OPU sessions on 81 donor animals (twice-weekly OPU scheme) with 640 sessions following FSH-LH superstimulation, on 112 donors (OPU once every 2 weeks). The results of the study demonstrated that, expressed per session, follicle-stimulating hormone (FSH) stimulation prior to OPU increased production efficiency, with a significantly higher number of follicles punctured and oocytes retrieved. However, when overall results during comparable 2-week periods are considered (4 non-stimulated sessions vs. 1 stimulated), more follicles are punctured, and more oocytes are retrieved using the non-stimulated protocol. The absence of significant differences between these two regimens in the number of in vitro-produced embryos within the 2-week period suggests a positive effect on in vitro oocyte developmental competence in the treated animals.

The influence of *temperature* and *climate* during the bovine follicle formation and development may also interfere with oocyte quality and embryonic development. Heat stress suppresses follicular dominance, causing a series of follicle growth-related changes (Qi et al. 2013). *B. indicus* cattle have shown better reproductive performance than *B. taurus* in tropical and subtropical regions. This adaptation may involve different mechanisms. A severe decrease in the quality of Holstein cow oocytes was observed in the warmer season compared to the cold season, whereas no differences were detected in the oocytes obtained from Brahman cows (*B. indicus*) in different seasons. On the other hand, lymphocytes of Brahman and Senepol cows were less susceptible to apoptosis induced by the heat, suggesting a mechanism of protection of cells in *B. indicus* animals. One example of such mechanisms is heat shock proteins (HSP), similar to chaperones, which promote cell protection against heat damage by preventing protein denaturation and blocking apoptosos (reviewed by De Ondiz and Ruiz 2011).

2.6.3 Other Uses of OPU

OPU can be used to eliminate the deleterious effect of the presence of a dominant follicle during superovulation in cows, by puncturing the dominant follicle 38–46 h before superovulation treatment (Merton et al. 2009). The OPU application efficiently resets the >2 mm diameter follicular population to zero which in effect guarantees that the new population will be uniformly renewed with less atresia (Boni 2012).

OPU can have its own therapeutic effect on infertile donors, especially those affected by ovarian cysts. Virtually all empty donors come on cycle within 2 weeks from the last oocyte collection and can be inseminated successfully (Galli et al. 2001). Single transvaginal-guided needle aspiration of ovarian follicular cysts is an easy and good method for the treatment of follicular cysts, and it is a safe and good alternative method for the manual, active rupturing of cysts during rectal palpation (Lievaart et al. 2006).

OPU has been used as a source of oocytes for studies of in vitro fertilization (IVF) (Ruiz et al. 2009, 2013b) or IVEP (Pontes et al. 2010) with sex-sorted semen and in vitro embryo production after IVF with frozen-thawed, sex-sorted, re-frozen-thawed bull sperm (Underwood et al. 2010) or using sexed sperm from previously frozen doses (reverse-sorted semen) (Morotti et al. 2014). Some researchers have employed the OPU technique as a tool together with another reproductive biotechnology for the recovery and conservation of endangered bovine breeds (Ruiz et al. 2013a). Finally, special modifications of OPU have been used to collect ovarian tissue for primary follicle isolation (Aerts et al. 2005).

2.6.4 Donor Health: OPU Sequelae

Follicular aspiration is a technique with pros and cons. The risk of sequelae must be considered. The procedure must be performed carefully to minimize the possible damage allowing the results to outweigh the expected results (Da Silva et al. 2016, 2017). During ultrasound-guided oocyte collection, cattle undergo repeated restraint, transrectal palpation, the introduction of an intravaginal device, epidural anesthesia, and repeated ovarian punctures. Among these manipulations, transrectal palpation and the introduction of an intravaginal device, commonly practiced by veterinarians and inseminators, have been shown to induce moderate stress. Epidural anesthesia is a less common practice. However, even when repeated associated lesions seem minor and occur in only a few animals. The most questionable aspect of OPU is the multiple transvaginal and transovarian punctures, which constitute the more invasive and the most specific part of the OPU.

3 Conclusion

Ultrasonography is a fundamental component of bovine female reproductive management, providing unparalleled benefits in the domains of diagnosis, monitoring, and intervention. It facilitates early pregnancy detection, precise evaluation of ovarian structures, and the diagnosis of uterine pathologies, all of which are crucial for optimizing fertility and herd productivity. The integration of advanced techniques, including Doppler ultrasonography and bovine ovum pick-up (OPU), further enhances reproductive outcomes by facilitating detailed assessments and efficient oocyte collection for in vitro fertilization. The noninvasive, precise nature of ultrasonography, in conjunction with ongoing technological advancements, ensures its pivotal role in improving reproductive performance and advancing bovine herd management practices.

References

Abdelghafar RM, Ahmed BH, Ibrahim MT, Mantis P (2011) Prediction of gestational age by transabdominal real-time ultrasonographic measurements in Saanen goats (Capra hircus). Glob Vet 6(4):346–351

Abdelnaby EA (2020) Hemodynamic changes evaluated by Doppler ultrasonographic technology in the ovaries and uterus of dairy cattle after the puerperium. Reprod Biol 20(2):202–209. https://doi.org/10.1016/j.repbio.2020.03.001

Abdelnaby EA, El-Maaty AMA, Ragab RSA, Seida AA (2018) Dynamics of uterine and ovarian arteries flow velocity waveforms and their relation to follicular and luteal growth and blood flow vascularization during the estrous cycle in Friesian cows. Theriogenology 121:112–121. https://doi.org/10.1016/j.theriogenology.2018.08.003

Abdel-Raouf M, Badawi HM (1966) Morphological study of uterine caruncles in Egyptian buffalo cows. Zentralbl Veterinarmed Reihe A 13(3):252–263

Abo El-Maaty AM, Abdelnaby EA (2017) Follicular blood flow, antrum growth and angiogenic mediators in mares from ovulation to deviation. Anim Reprod 2017(14):1043–1056. https://doi.org/10.21451/1984-3143-AR848

Acosta TJ, Miyamoto A (2004) Vascular control of ovarian function, ovulation, corpus luteum formation and regression. Anim Reprod Sci 82–83:127–140. https://doi.org/10.1016/j.anireprosci.2004.04.022

Acosta TJ, Hayashi KG, Ohtani M, Miyamoto A (2003) Local changes in blood flow within the preovulatory follicle wall and early corpus luteum in cows. Reproduction 125:759–767. https://doi.org/10.1530/rep.0.1250759

Acosta TJ, Hayashi KG, Matsui M, Miyamoto M (2005) Changes in follicular vascularity during the first follicular wave in lactating cows. J Reprod Dev 51:273–280. https://doi.org/10.1262/jrd.16092

Adams GP, Jaiswal R, Singh J, Malhi P (2008) Progress in understanding ovarian follicular dynamics in cattle. Theriogenology 69:72–80. ISSN 0093-691X. https://doi.org/10.1016/j.theriogenology.2007.09.026

Adeyinka FD, Laven RA, Lawrence KE, van Den Bosch M, Blankenvoorde G, Parkinson TJ (2014) Association between placentome size, measured using transrectal ultrasonography, and gestational age in cattle. N Z Vet J 62(2):51–56. https://doi.org/10.1080/00480169.2013.832620

Aerts JM, Oste M, Bols PEJ (2005) Development and practical applications of a method for repeated transvaginal, ultrasound-guided biopsy collection of the bovine ovary. Theriogenology 64:947–957. https://doi.org/10.1016/j.theriogenology.2005.01.011

Ali A (2004) Effect of gestational age and fetal position on the possibility and accuracy of ultrasonographic fetal gender determination in dairy cattle. Reprod Domest Anim 39:190–194. https://doi.org/10.1111/j.1439-0531.2004.00502.x

Ambrose DJ, Schmitt EJP, Lopes FL, Mattos RC, Thatcher WW (2004) Ovarian and endocrine responses associated with the treatment of cystic ovarian follicles in dairy cows with gonadotropin releasing hormone and prostaglandin F2α, with or without exogenous progesterone. Can Vet J 45(11):931–937. http,//www.ncbi.nlm.nih.gov/pmc/articles/pmc545983/

Andrade JPN, Andrade FS, Guerson YB, Domingues RR, Gomez-León VE, Cunha TO, Jacob JCF, Sales JN, Martins JPN, Mello MRB (2019) Early pregnancy diagnosis at 21 days post artificial insemination using corpus luteum vascular perfusion compared to corpus luteum diameter and/or echogenicity in Nelore heifers. Anim Reprod Sci 209:106144. https://doi.org/10.1016/j.anireprosci.2019.106144

Andreu-Vázquez C, García-Ispierto I, Ganau S, Fricke PM, López-Gatius F (2012) Effects of twinning on the subsequent reproductive performance and productive lifespan of high-producing dairy cows. Theriogenology 78:2061–2070. https://doi.org/10.1016/j.theriogenology.2012.07.027

Arashiro EKN, Palhao MP, Santos JRL, Fontes RC, Siqueira LGB, Henry M, Viana JHM (2013) Three-dimensional modeling of color Doppler images, a new approach to study follicular vascularization in cattle. Anim Reprod 10:662–669. https://doi.org/10.1016/j.theriogenology.2011.10.020

Arthur GH, Noakes DE, Pearson H, Parkinson TJ (2009) Veterinary reproduction and obstetrics (theriogenology). Saunders Elsevier, London, pp 80–100

Ashitha S, Saran S, Umamageswari J, Kavitha K, Thangapandiyan M, Arunmozhi N, Sarath T (2020) Mucometra due to follicular cyst with cystic dilation of endometrial glands in a Holstein Friesian crossbred cow. Int J Livestock Res 10(4):111–114. https://doi.org/10.5455/ijlr.20191105055853

Assey RJ, Purwantara B, Greve T, Hyttel P, Schmidt MH (1993) Corpus luteum size and plasma progesterone levels in cattle after cloprostenol-induced luteolysis. Theriogenology 39:1321–1330. https://doi.org/10.1016/0093-691X(93)90234-V

Ataide Junior GA, Kloster A, de Moraes ÉG, Motta IG, Claro JI, Vasconcelos JLM, Ferraz PA, de Paula Nogueira G, Pugliesi G (2021) Early resynchronization of follicular wave emergence among Nelore cattle using injectable and intravaginal progesterone for three timed artificial inseminations. Anim Reprod Sci 229:106759. https://doi.org/10.1016/j.anireprosci.2021.106759

Badtram GA, Gaines JD, Thomas CB, Bosu WTK (1991) Factors influencing the accuracy of early pregnancy detection in cattle by real-time ultrasound scanning of the uterus. Theriogenology 35:1153–1167. https://doi.org/10.1016/0093-691X(91)90362-H

Bagley JE, Richter MP, Lane TJ (2023) The role of transrectal sonography in pregnancy diagnosis in cattle. J Diagn Med Sonogr 39(1):50–60. https://doi.org/10.1177/87564793221120260

Balaro MFA, Santos AS, Moura LFGM, Fonseca JF, Brandão FZ (2017) Luteal dynamic and functionality assessment in dairy goats by luteal blood flow, luteal biometry, and hormonal assay. Theriogenology 95:118–126. https://doi.org/10.1016/j.theriogenology.2017.02.021

Barrón Bravo O, Avilés Ruiz R, Fraga Escamilla E, Bautista Martínez Y (2023) Los procesos reproductivos en vacas y el uso de la ultrasonografía. Abanico Veterinario 13:1–34. https://doi.org/10.21929/abavet2023.18

Batista EOS, Macedo CG, Sala RV, Ortolan MDDV, Sá Filho MF, Del Valle TA, Jesus EF, Lopes RNVR, Renno FP, Baruselli PS (2014) Plasma antimullerian hormone as a predictor of ovarian antral follicular population in B. indicus (Nelore) and B. taurus (Holstein) heifers. Reprod Domest Anim 49:448–452. https://doi.org/10.1111/rda.12304

Batista EOS, Guerreiro BM, Freitas BG, Silva JCB, Vieira LM, Ferreira RM, Rezende RG, Basso AC, Lopes RNVR, Rennó FP, Souza AH, Baruselli PS (2016) Plasma anti-Müllerian hormone as a predictive endocrine marker to select Bos taurus (Holstein) and Bos indicus (Nelore) calves for in vitro embryo production. Domest Anim Endocrinol 54:1–9. https://doi.org/10.1016/j.domaniend.2015.08.001

Baxter SJ, Ward WR (1997) Incidence of fetal loss in dairy cattle after pregnancy diagnosis using an ultrasound scanner. Vet Rec 140:287–288. https://doi.org/10.1136/vr.140.11.287

Beal WE, Perry RC, Corah LR (1992) The use of ultrasound in monitoring reproductive physiology of beef cattle. J Anim Sci 70:924–929. https://doi.org/10.2527/1992.703924x

Bellows RA, Staigmiller RB, Orme LE, Short RE, Knapp BW (1993) Effects of sire and dam on late pregnancy conceptus and hormone traits in beef-cattle. J Anim Sci 71(3):714–723. https://doi.org/10.2527/1993.713714x

Bergamaschi MACM, Vicente WRR, Barbosa RT, Machado R, Marques JA, Freitas AR (2004) Ultrasound assessment of fetal development in Nelore cows. Archivos de Zootecnia 53(204):371–374

Berger H, Lietzau M, Tichy A, Herzog K (2017) Pregnancy outcome is influenced by luteal area during diestrus before successful insemination but not by milk production level. Theriogenology 104:115–119. https://doi.org/10.1016/j.theriogenology.2017.04.020

Bergfelt DR, Kastelic JP, Ginther OJ (1991) Continued periodic emergence of follicular waves in non-bred progesterone-treated heifers. Anim Reprod Sci 24:193–204. https://doi.org/10.1016/S0378-4320(05)80003-X

Bergfelt DR, Bo GA, Mapletoft RJ, Adams GP (1997) Superovulatory response following ablation-induced follicular wave emergence at random stages of the oestrous cycle in cattle. Anim Reprod Sci 49(1):1–12. https://doi.org/10.1016/S0378-4320(97)00064-X

Bicalho RC, Galvão KN, Guard CL, Santos JEP (2008) Optimizing the accuracy of detecting a functional corpus luteum in dairy cows. Theriogenology 70:199–207. https://doi.org/10.1016/j.theriogenology.2008.03.015

Binelli M, Martins T, Arthington J, Waters K, Mercadante V, Moriel P, Gonella-Diaza A (2023) Practical uses for ultrasound in managing beef cattle reproduction. The University of Florida George A. Smathers Libraries. https,//journals.flvc.org/edis/article/view/128129

Blondin P (2015) Status of embryo production in the world. Anim Reprod 12:356–358

Blowey RW (1992) Milk progesterone profiles in untreated cystic ovarian disease. Vet Rec 130(19):429. https://doi.org/10.1136/vr.130.19.429

Bó GA, Mapletoft RJ (2014) Historical perspectives and recent research on superovulation in cattle. Theriogenology 81(1):38–48. https://doi.org/10.1016/j.theriogenology.2013.09.020

Bó GA, Baruselli PS, Chesta P, Martins CM (2006) The timing of ovulation and insemination schedules in superstimulated cattle. Theriogenology 65:89–101. https://doi.org/10.1016/j.theriogenology.2005.10.008

Bollwein H, Meyer HHD, Maierl J, Weber F, Baumgartner U, Stolla R (2000) Transrectal Doppler sonography of uterine blood flow in cows during the estrous cycle. Theriogenology 53(8):1541–1552. https://doi.org/10.1016/S0093-691X(00)00296-X

Bollwein H, Baumgartner U, Stolla R (2002) Transrectal Doppler sonography of uterine blood flow in cows during pregnancy. Theriogenology 57(8):2053–2061. https://doi.org/10.1016/S0093-691X(02)00706-9

Bollwein H, Heppelmann M, Lüttgenau J (2016) Ultrasonographic Doppler use for female reproduction management. Vet Clin N Am Food Anim Pract 32(1):149–164. https://doi.org/10.1016/j.cvfa.2015.09.005

Bols PEJ, Stout TAE (2018) Transvaginal ultrasound-guided oocyte retrieval (OPU: ovum pick-up) in cows and mares. In: Niemann H, Wrenzycki C (eds) Animal biotechnology 1: reproductive biotechnologies. Springer, Cham, pp 209–234. https://doi.org/10.1007/978-3-319-92327-7_10

Bols PEJ, Ysebaeti MT, Van Soom A, de Kruif A (1997) Effects of needle tip bevel and aspiration procedure on the morphology and developmental capacity of bovine compact cumulus oocyte complexes. Theriogenology 47:1221–1236. https://doi.org/10.1016/s0093-691x(97)00102-7

Bonafos LD, Kot K, Ginther OJ (1995) Physical characteristics of the uterus during the bovine estrous cycle and early pregnancy. Theriogenology 43(4):713–721. https://doi.org/10.1016/0093-691X(95)00014-Y

Boni R (2012) Ovum pick-up in cattle: a 25 yr retrospective analysis. Anim Reprod 9:362–369

Boyd JS (1974) Patterns of ossification in feline fetus, a study of the fetal development of the skeleton of the feline using comparative methods. PhD thesis, University of Glasgow

Boyd JS, Omran SN (1991) Diagnostic ultrasonography of bovine female reproductive tract. In Pract 13:109–118. https://doi.org/10.1136/inpract.13.3.109

Brad K, Stroud BK (2005) Bovine fetal sexing using ultrasound. In: Proceedings, applied reproductive strategies in beef cattle, November 12 and 13, 2005. Texas A&M University, College Station

Brockus KE, Hart CG, Gilfeather CL, Fleming BO, Lemley CO (2016) Dietary melatonin alters uterine artery hemodynamics in pregnant Holstein heifers. Domest Anim Endocrinol 55:1–10. https://doi.org/10.1016/j.domaniend.2015.10.006

Buczinski SM, Fecteau G, Lefebvre RC, Smith LC (2007) Fetal well-being assessment in bovine near-term gestations, current knowledge and future perspectives arising from comparative medicine. Can Vet J 48(2):178–183. PMID: 17334032
Bungartz L, Lucas-Hahn A, Rath D, Niemann H (1995) Collection of oocytes from cattle via follicular aspiration aided by ultrasound with or without gonadotropin pretreatment and in different reproductive stages. Theriogenology 43:667–675. https://doi.org/10.1016/0093-691x(94)00072-3
Bunyaga AS (2015) Estimation of gestational age by the use of fetal parameters, placentome, femur length, and biparietal diameter. Master thesis, Institute of Veterinary, Animal and Biomedical Sciences, Massey University, Palmerston North
Camacho LE, Lemley CO, Prezotto LD, Bauer ML, Freetly HC, Swanson KC, Vonnahme KA (2014) Effects of maternal nutrient restriction followed by realimentation during midgestation on uterine blood flow in beef cows. Theriogenology 81:1248–1256. https://doi.org/10.1016/j.theriogenology.2014.02.006
Cattoni J (1991) Ultrasonic evaluation of superovulation in cattle. Thesis Masters Degree in Veterinary Medicine, Faculty of Veterinary Medicine, University of Glasgow
Chaffaux S, Valon F, Martinez J (1982) Evoluation de l'image echographique du produit de conception chez la vache. Bull Acad Vet Fr 55:213–221. https://www.persee.fr/doc/bavf_0001-4192_1982_num_135_2_7914
Cheong SH, Sá Filho OG, Absalon-Medina VA, Schneider A, Butler WR, Gilbert RO (2017) Uterine and systemic inflammation influences ovarian follicular function in postpartum dairy cows. PLoS One 12(5):e0177356. https://doi.org/10.1371/journal.pone.0177356
Chui DKC, Pugh ND, Walker SM, Gregory L, Shaw RW (1997) Follicular vascularity—the predictive value of transvaginal power Doppler ultrasonography in an in vitro fertilization programme, a preliminary study. Hum Reprod 12:191–196. https://doi.org/10.1093/humrep/12.1.191
Cirit Ü, Özmen MF, Küçükaslan İ, Köse M, Kutsal HG, Çinar EM (2019) Effect of the interval from follicle aspiration to initiation of lengthened FSH treatment on follicular superstimulatory and superovulatory responses and embryo production in lactating Simmental cows. Theriogenology 128:218–224. https://doi.org/10.1016/j.theriogenology.2019.02.008
Colloton J (2021) Ultrasound evaluation of the female reproductive tract. In: Hopper RM (ed) Bovine reproduction, 2nd edn. Wiley, New York
Coulam CB, Goodman C, Rinehart JS (1999) Colour Doppler indices of follicular blood flow as predictors of pregnancy after in-vitro fertilization and embryo transfer. Hum Reprod 14(8):1979–1982. https://doi.org/10.1093/humrep/14.8.1979
Curran S (1992) Fetal sex determination in cattle and horses by ultrasonography. Theriogenology 37(1):17–21. https://doi.org/10.1016/0093-691X(92)90244-L
Curran S, Ginther OJ (1991) Ultrasonic determination of fetal gender in horses and cattle under farm conditions. Theriogenology 36:809–814. https://doi.org/10.1016/0093-691X(91)90346-F
Curran S, Pierson RA, Ginther OJ (1986a) Ultrasonographic appearance of the bovine conceptus from days 10 through 20. J Am Vet Med Assoc 189:1289–1294
Curran S, Pierson RA, Ginther OJ (1986b) Ultrasonographic appearance of the bovine conceptus from days 20 through 60. J Am Vet Med Assoc 189:1295–1302. PMID: 3793569
Curran S, Kastelic JP, Ginther OJ (1989) Determining sex of the bovine fetus by ultrasonic assessment of the relative location of the genital tubercle. Anim Reprod Sci 19:217–227. https://doi.org/10.1016/0378-4320(89)90095-X
Da Silva CB, Machado FZ, González SM, Seneda MM (2016) Ovum pick-up. In: Seneda MM, Silva-Santos KC, Rafagnin LS (eds) Recent trends in biotechnology. Biotechnology of animal reproduction. Nova Science, New York, pp 157–169
Da Silva JCB, Ferreira RM, Filho MM, Naves JR, Santin T, Pugliesi G, Madureira EH (2017) Use of FSH in two different regimens for ovarian superstimulation prior to ovum pick up and in vitro embryo production in Holstein cows. Theriogenology 90:65–73. https://doi.org/10.1016/j.theriogenology.2016.11.016
Dalmaso de Melo G, Mello BP, Ferreira CA, Souto Godoy Filho CA, Rocha CC, Silva AG, Reese ST, Madureira EH, Pohler KG, Pugliesi G (2020) Applied use of interferon-tau stimulated genes

expression in polymorphonuclear cells to detect pregnancy compared to other early predictors in beef cattle. Theriogenology 152:94–105. https://doi.org/10.1016/j.theriogenology.2020.04.001
Dawson FLM (1975) Accuracy of rectal palpation in the diagnosis of ovarian function in the cow. Vet Rec 96:218–220. https://doi.org/10.1136/vr.96.10.218
De Ondiz A, Ruiz S (2011) Aspiración folicular transvaginal guiada por ultrasonografía (OPU), producción in vitro de embriones (PIV) y semen sexado en la reproducción bovina. In: González-Stagnaro C, Madrid N, Soto E (eds) Innovación y Tecnologías en la ganadería doble propósito. Astrodata, Maracaibo, pp 797–805
De Roover R, Feugang JMN, Bols PEJ, Genicot G, Hanzen C (2008) Effects of ovum pick-up frequency and FSH stimulation: a retrospective study on seven years of beef cattle in vitro embryo production. Reprod Domest Anim 43:239–245. https://doi.org/10.1111/j.1439-0531.2007.00873.x
de Tarso SGS, Gastal GDA, Bashir ST, Gastal MO, Apgar GA, Gastal EL (2017) Follicle vascularity coordinates corpus luteum blood flow and progesterone production. Reprod Fertil Dev 29:448–457. https://doi.org/10.1071/RD15223
DesCôteaux L, Carriere PD, Durocher J (2006) Ultrasonography of the reproductive system of the cow—basic principles, practical uses and economic aspects of this diagnostic tool in dairy production. World Buiatrics Congress, Nice
DesCôteaux L, Picard-Hagen N, Durocher J, Buczinski S, Colloton J, Chastant-Maillard S, Curran S (2010) Bovine fetal development after 55 days, fetal sexing, anomalies, and well-being. In: DesCôteaux L, Gnemmi G, Colloton J (eds) Practical atlas of ruminant and camelid reproductive ultrasonography. Blackwell Publishing, Iowa. https://doi.org/10.1002/9781119265818.ch7
Dewailly D, Andersen CY, Balen A, Broekmans F, Dilaver N, Fanchin R, Griesinger G, Kelsey TW, La Marca A, Lambalk C, Mason H, Nelson SM, Visser JA, Wallace WH, Anderson RA (2014) The physiology and clinical utility of anti-Mullerian hormone in women. Hum Reprod Update 20:370–385. https://doi.org/10.1093/humupd/dmt062
Dixit CP, Haloi S (2022) Clinical application of ultrasound in bovine reproduction, a review. Pharma Innov J SP11(8S):1311–1314. https://www.thepharmajournal.com/archives/2022/vol11issue8S/PartQ/S-11-8-171-672.pdf
Dobson H, Nanda AS (1992) Reliability of cyst diagnosis and effect of energy status on LH released by estradiol or GnRH in cows with ovarian cysts. Theriogenology 37:466–472. https://doi.org/10.1016/0093-691X(92)90203-4
Doize F, Vaillancourt D, Carabin H, Belanger D (1997) Determination of gestational age in sheep and goats using transrectal ultrasonographic measurement of placentomes. Theriogenology 48(3):449–460. https://doi.org/10.1016/s0093-691x(97)00254-9
Douthwaite R, Dobson H (2000) Comparison of different methods of diagnosis cyst ovarian disease in cattle and an assessment of its treatment with a progesterone-releasing intravaginal device. Vet Rec 147(13):355–359. https://doi.org/10.1136/vr.147.13.355
Dubuc J, DuFeld TF, Leslie KE, Walton JS, Leblanc SJ (2011) Effects of postpartum uterine diseases on milk production and culling in dairy cows. J Dairy Sci 94(3):1339–1346. https://doi.org/10.3168/jds.2010-3758
Dubuc J, Houle J, Rousseau M, Roy JP, Buczinski S (2020) Short communication, accuracy of corpus luteum color flow Doppler ultrasonography to diagnose nonpregnancy in dairy cows on day 21 after insemination. J Dairy Sci 103:2019–2023. https://doi.org/10.3168/jds.2019-17234
Dufour S, Durocher J, Dubuc J, Dendukuri N, Hassan S, Buczinski S (2017) Bayesian estimation of sensitivity and specificity of a milk pregnancy-associated glycoprotein-based ELISA and of transrectal ultrasonographic exam for diagnosis of pregnancy at 28-45 days following breeding in dairy cows. Prev Vet Med 140:122–133. https://doi.org/10.1016/j.prevetmed.2017.03.008
Eley RM, Thatcher WW, Bazer FW, Wilcox CJ, Becker RB, Head HH, Adkinson RW (1978) Development of conceptus in bovine. J Dairy Sci 61(4):467–473. https://doi.org/10.3168/jds.S0022-0302(78)83622-4
Elmetwally MA (2018) Uterine involution and ovarian activity in postpartum Holstein dairy cows. A review. J Vet Healthcare 1(4):29–40. https://doi.org/10.14302/issn.2575-1212.jvhc-18-2447

Farin PW, Youngquist RS, Parfet JR, Gaverick HA (1990) Diagnosis of luteal and follicular cysts in dairy cows by sector scan ultrasonography. Theriogenology 34:633–642. https://doi.org/10.1016/0093-691X(90)90019-P

Farin PW, Youngquist RS, Parfet JR, Garverick H (1992) Diagnosis of luteal and follicular ovarian cysts by palpation per rectum and linear array ultrasonography in dairy cows. J Am Vet Med Assoc 200:1065–1089. https://doi.org/10.2460/javma.1992.200.08.1085

Ferraz PA, Poit DAS, Neto AL, Luiz F, Azrak AJ, Carneiro AL, Pardo FJD, Pohler KG, Baruselli PS, Pugliesi G (2022) Determination of pregnancy loss at different moments of early gestation in dairy cattle subjected to TAI. Anim Reprod 19:22136. (Abstract)

Ferré LB, Alvarez-Gallardo H, Romo S, Fresno C, Stroud T, Stroud B, Lindsey B, Kjelland ME (2023) Transvaginal ultrasound-guided oocyte retrieval in cattle: state-of-the-art and its impact on the in vitro fertilization embryo production outcome. Reprod Domest Anim 58:363–378. https://doi.org/10.1111/rda.14303

Ferreira JCP, Martin I, Irikura CR, Gimenes LU, Fujiraha CJ, Jorge AM, Oba E (2012) Ultrasonic monitoring of early pregnancy development in Murrah buffalo heifers (Bubalus bubalis). J Anim Plant Sci 22:377–383

Ferrell CL (1991) Maternal and fetal influences on uterine and conceptus development in the cow: I. Growth of tissues of the gravid uterus. J Anim Sci 69(5):1945–1953. https://doi.org/10.2527/1991.6951945x

Ferrell CL, Garrett WN, Hinman N (1976) Growth, development and composition of udder and gravid uterus of beef heifers during pregnancy. J Anim Sci 42(6):1477–1489. https://doi.org/10.2527/jas1976.4261477x

Fields MJ, Fields PA (1996) Morphological characteristics of the bovine corpus luteum during estrous cycle and pregnancy. Theriogenology 45:1295–1325. https://doi.org/10.1016/0093-691X(96)00099-4

Figueiredo RA, Barros CM, Pinheiro OL, Sole JMP (1997) Ovarian follicular dynamics in Nelore breed (Bos indicus) cattle. Theriogenology 47:1489–1505. https://doi.org/10.1016/S0093-691X(97)00156-8

Filteau V, DesCôteaux L (1998) Predictive values of early pregnancy diagnosis by ultrasonography in dairy cattle. In: 31st annual convention of the American Association of Bovine Practitioners, Spokane, WA, USA, pp 170–171

Fleischer A, Anyaegbunam AA, Schulman H, Farmakides G, Randolph G (1987) Uterine and umbilical artery velocimetry during normal labor. Am J Obstet Gynecol 157:40–43. https://doi.org/10.1016/s0002-9378(87)80342-3

Fonseca FA, Britt JH, McDaniel BT, Wilk JC, Rakes AH (1983) Reproductive traits of Holsteins and Jerseys. Effects of age, milk yield, and clinical abnormalities on involution of cervix and uterus, ovulation, estrous cycles, detection of estrus, conception rate, and days open. J Dairy Sci 66:1128–1147. https://doi.org/10.3168/jds.S0022-0302(83)81910-9

Fontes P, Oosthuizen N (2022) Applied use of Doppler ultrasonography in bovine reproduction. Front Anim Sci 3:912854. https://doi.org/10.3389/fanim.2022.912854

Fortune JE, Sirios J, Quirk SM (1988) The growth and differentiation of ovarian follicles during the bovine estrous cycle. Theriogenology 29:95–109. https://doi.org/10.1016/0093-691X(88)90034-9

Fortune JE, Sirios J, Turzillo AM, Lavoir M (1991) Follicular selection in domestic ruminants. J Reprod Fertil Suppl 43:187–198. PMID: 1843340

Fosgate GT, Motimele B, Ganswindt A, Irons PC (2017) Bayesian latent class model to estimate the accuracy of pregnancy diagnosis by transrectal ultrasonography and laboratory detection of pregnancy-associated glycoproteins in dairy cows. Prev Vet Med 145:100–109. https://doi.org/10.1016/j.prevetmed.2017.07.004

Fraser HM, Wulff C (2003) Angiogenesis in the corpus luteum. Reprod Biol Endocrinol 1:1–8. https://doi.org/10.1186/1477-7827-1-88

Fricke PM (2002) Scanning the future--ultrasonography as a reproductive management tool for dairy cattle. J Dairy Sci 85(8):1918–1926. https://doi.org/10.3168/jds.S0022-0302(02)74268-9

Fricke PM, Lamb GC (2002) Practical applications of ultrasound for reproductive management of beef and dairy cattle. In: The applied reproductive strategies in beef cattle workshop, September 5–6, 2002, Manhattan, Kansas
Fricke PM, Lamb GC (2005) Potential applications and pitfalls of reproductive ultrasonography in bovine practice. Vet Clin Food Anim Pract 21:419–436. https://doi.org/10.1016/j.cvfa.2005.02.005
Fricke PM, Wiltbank MC (1999) Effect of milk production on the incidence of double ovulation in dairy cows. Theriogenology 52:1133–1143. https://doi.org/10.1016/S0093-691X(99)00205-8
Fricke PM, Guenther JN, Wiltbank MC (1998) Efficacy of decreasing the dose of GnRH used in a protocol for synchronization of ovulation and timed AI in lactating dairy cows. Theriogenology 50:1275–1284. https://doi.org/10.1016/S0093-691X(98)00226-X
Galli C, Crotti G, Notari C, Turini P, Duchi R, Lazzari G (2001) Embryo production by ovum pick up from live donors. Theriogenology 55:1341–1357. https://doi.org/10.1016/s0093-691x(01)00486-1
Galli C, Duchi R, Colleoni S, Lagutina I, Lazzari G (2014) Ovum pick up, intracytoplasmic sperm injection and somatic cell nuclear transfer in cattle, buffalo and horses: from the research laboratory to clinical practice. Theriogenology 81:138–151. https://doi.org/10.1016/j.theriogenology.2013.09.008
Gard Schnuelle J, Blythe E, Cole R, Taylor S, Alfaro G, Muntifering R, Pacheco WJ, Rodning SP, Moisa SJ (2021) Evaluation of alterations in uterine blood flow using Doppler ultrasonography in pregnant, genotyped beef cows and heifers consuming endophyte-infected tall fescue seeds and supplemented with rumen-protected niacin. J Vet Sci Res 6:1–8. https://doi.org/10.23880/oajvsr-16000207
Gier HT, Marion GB (1968) Uterus of cow after parturition, involutional changes. Am J Vet Res 29:83–96. PMID: 5688807
Ginther OJ (1998) Ultrasonic imaging and animal reproduction, cattle. Equiservices Publishing, Cross Plains, pp 29–58
Ginther OJ (2007) Ultrasonic imaging and animal reproduction, color-Doppler ultrasonography. Equiservices Publishing, Cross Plains
Ginther OJ (2016) The theory of follicle selection in cattle. Domest Anim Endocrinol 57:85–99. https://doi.org/10.1016/j.domaniend.2016.06.002
Ginther OJ, Kastelic P, Knopf L (1989) Composition and characteristics of follicular waves during the bovine estrous cycle. Anim Reprod Sci 20:187–200. https://doi.org/10.1016/0378-4320(89)90084-5
Giordano JO, Fricke PM (2012) Accuracy of pregnancy diagnosis outcomes using transrectal ultrasonography 29 days after artificial insemination in lactating dairy cows. J Dairy Sci 95(2):75
Grazul-Bilska AT, Navanukraw C, Johnson ML, Vonnahme KA, Ford SP, Reynolds LP, Redmer DA (2007) Vascularity and expression of angiogenic factors in bovine dominant follicles of the first follicular wave. J Anim Sci 85:1914–1922. https://doi.org/10.2527/jas.2007-0044
Greenwood PL, Bell AW, Vercoe PE, Viljoen GJ (2010) Managing the prenatal environment to enhance livestock productivity. Springer, Berlin, pp 3–36
Griffin PG, Ginther OJ (1992) Research applications of ultrasonic imaging in reproductive biology. J Anim Sci 70:953–972. https://doi.org/10.2527/1992.703953x
Guerreiro BM, Batista EOS, Vieira LM, Sá Filho MF, Rodrigues CA, Castro Netto A, Silveira CRA, Bayeux BM, Dias EAR, Monteiro FM, Accorsi M, Lopes RNVR, Baruselli PS (2014) Plasma anti-mullerian hormone: an endocrine marker for in vitro embryo production from B. taurus and B. indicus donors. Domest Anim Endocrinol 49:96–104. https://doi.org/10.1016/j.domaniend.2014.07.002
Guimarães CR, Oliveira ME, Rossi JR, Fernandes CA, Viana JH, Palhao MP (2015) Corpus luteum blood flow evaluation on Day 21 to improve the management of embryo recipient herds. Theriogenology 84:237–241. https://doi.org/10.1016/j.theriogenology.2015.03.005
Haibel GK, Perkins NR (1989) Real-time ultrasonic biparietal diameter of 2nd trimester Suffolk and Finn sheep fetuses and prediction of gestational-age. Theriogenology 32(5):863–869. https://doi.org/10.1016/0093-691x(89)90474-3

Hamilton SA, Garverick HA, Kesler DJ, Xu ZZ, Loss K, Youngquist RS, Salfen BE (1995) Characterization of ovarian follicular cysts and associated endocrine profile in dairy cows. Biol Reprod 53:890–895. https://doi.org/10.1095/biolreprod53.4.890
Hanzen CH, Pieterse M, Scenzi O, Drost M (2000) Relative accuracy of the identification of ovarian structures in the cow by ultrasonography and palpation per rectum. Vet J 159:161–170. https://doi.org/10.1053/tvjl.1999.0398
Hartmann D, Honnens A, Piechotta M, Lüttgenau J, Niemann H, Rath D, Bollwein H (2013) Effects of a protracted induction of parturition on the incidence of retained placenta and assessment of uterine artery blood flow as a measure of placental maturation in cattle. Theriogenology 80:176–184. https://doi.org/10.1016/j.theriogenology.2013.02.001
Hassan M, Arshad U, Bilal M, Sattar A, Avais M, Bollwein H, Ahmad N (2019) Luteal blood flow measured by Doppler ultrasonography during the first three weeks after artificial insemination in pregnant and non-pregnant Bos indicus dairy cows. J Reprod Dev 65(1):29–36. https://doi.org/10.1262/jrd.2018-084
Henao-Gonzalez M, Ferrer MS, Jiménez-Escobar C, Palacio-Baena LG, Maldonado-Estrada JG (2023) Ultrasonographic screening of dairy cows with normal uterine involution or developing postpartum uterine disease using B-mode, color, and spectral Doppler. Vet Med Int 9:1–14. https://doi.org/10.1155/2023/2597332
Heppelmann M, Krüger L, Leidl S, Bollwein H (2013) Transrectal Doppler sonography of uterine blood flow during the first two weeks after parturition in Simmental heifers. J Vet Sci 14:323–327. https://doi.org/10.4142/jvs.2013.14.3.323
Heppelmann M, Weinert M, Brömmling A, Piechotta M, Hoedemaker M, Bollwein H (2014) The effect of puerperal uterine disease on uterine involution in cows assessed by Doppler sonography of the uterine arteries. Anim Reprod Sci 143:1–7. https://doi.org/10.1016/j.anireprosci.2013.11.003
Herzog K, Bollwein H (2007) Application of Doppler ultrasonography in cattle reproduction. Reprod Domest Anim 42(2):51–58. https://doi.org/10.1111/j.1439-0531.2007.00903.x
Herzog K, Brockhan-Lüdemann M, Kaske M, Beindorff N, Paul V, Niemann H, Bollwein H (2010) Luteal blood flow is a more appropriate indicator for luteal function during the bovine estrous cycle than luteal size. Theriogenology 73:691–697. https://doi.org/10.1016/j.theriogenology.2009.11.016
Holland MD, Odde KG (1992) Improved gestational age estimation of the fetal calf. ISVP Newslett 5(1):10–11
Holton MP, Oosthuizen N, de Melo GD, Davis DB, Stewart RL Jr, Pohler KG, Lamb GC, Fontes PLP (2022) Luteal color Doppler ultrasonography and pregnancy-associated glycoproteins as early pregnancy diagnostic tools and predictors of pregnancy loss in Bos taurus postpartum beef cows. J Anim Sci 100:1–9. https://doi.org/10.1093/jas/skac018
Honnens A, Niemann H, Paul V, Meyer HHD, Bollwein H (2008) Doppler sonography of the uterine arteries during a superovulatory regime in cattle, uterine blood flow in superovulated cattle. Theriogenology 70(5):859–867. https://doi.org/10.1016/j.theriogenology.2008.05.054
Hunnam JC, Parkinson TJ, Lopez-Villalobos N, McDougall S (2009) Association between gestational age and bovine fetal characteristics measured by transcutaneous ultrasound over the right flank of the dairy cow. Aust Vet J 87(9):379–383. https://doi.org/10.1111/j.1751-0813.2009.00468.x
Ireland JJ, Roche JF (1987) Hypothesis regarding development of dominant follicles during the bovine estrous cycle. In: Roche JF, O'Callaghan D (eds) Follicular growth and ovulation rate in farm animals. Martin Nijhoff Publishers, Boston, pp 1–18
Jaffa AJ, Wolman I, HarToov J, Amster R, Peyser MR (1996) Changes in uterine artery resistance to blood flow during puerperium—a longitudinal study. J Matern Fetal Invest 6:27–30
Janbu T, Nesheim BI (1987) Uterine artery blood velocities during contractions in pregnancy and labour related to intrauterine pressure. Br J Obstet Gynaecol 94:1150–1155. https://doi.org/10.1111/j.1471-0528.1987.tb02314.x
Jephcott S, Norman S (2004) Pregnancy diagnosis in cattle. Australian Association of Cattle Veterinarians (AACV), Eight Mile Plains, p 94

Jones AL, Beal WE (2003) Reproductive applications of ultrasounds in the cow. Bovine Pract 37(1):1–9. https://doi.org/10.21423/bovine-vol37no1p1-9

Kähn W (1989) Sonographic fetometry in the bovine. Theriogenology 31:1105–1121. https://doi.org/10.1016/0093-691X(89)90494-9

Kähn W (1990) Sonographic imaging of the bovine fetus. Theriogenology 33:385–396. https://doi.org/10.1016/0093-691X(90)90497-H

Kähn W (2004) Veterinary reproductive ultrasonography, horse, cattle, sheep, goat, pig, dog, cat. Schlutersche GmbH & Co. KG, Hannover, pp 83–129, 175

Kähn W, Ludlow W (1989) Characteristics of pathological conditions of the bovine uterus and ovaries. In: Taverne MM, Willemse AH (eds) Diagnostic ultrasound and animal reproduction. Kluwer Academic Publishers, Dordrecht, pp 53–65

Kamimura S, Ohgi T, Takahashi M, Tsukamoto T (1993) Postpartum resumption of ovarian activity and uterine involution monitored by ultrasonography in Holstein cows. J Vet Med Sci 55:643–647

Kamimura S, Kondoh S, Enomoto S, Sameshima H, Goto K, Hamana K (1994) Ultrasonic diagnosis of bovine fetal sex by the relative location of the genital tubercle in early pregnancy. J Reprod Dev 40:343–347. https://doi.org/10.1016/0093-691X(92)90244-L

Kastelic JP, Pierson RA, Ginther OJ (1990a) Ultrasonic morphology of corpora lutea and central luteal cavities during the estrous cycle and early pregnancy in heifers. Theriogenology 34:487–498. https://doi.org/10.1016/0093-691X(90)90006-F

Kastelic JP, Bergfelt DR, Ginther OJ (1990b) Relationship between ultrasonic assessment of the corpus luteum and plasma progesterone concentrations in heifers. Theriogenology 33:1269–1278. https://doi.org/10.1016/0093-691X(90)90045-U

Kastelic JP, Bergfelt DR, Ginther OJ (1991) Ultrasonic detection of the conceptus and characterization of intrauterine fluid on Days 10 to 22 in heifers. Theriogenology 35:569–581. https://doi.org/10.1016/0093-691X(91)90453-K

Kawate N, Inaba T, Mori J (1996) Changes in plasma concentrations of gonadotropins and steroid hormones during the formation of bovine follicular cysts induced by the administration of ACTH. J Vet Med Sci 58:141–144. https://doi.org/10.1292/jvms.58.141

Kirberger RM (1995) Imaging artifacts in diagnostic ultrasound—a review. Vet Radiol Ultrasound 36:297–306. https://doi.org/10.1111/j.1740-8261.1995.tb00266.x

Kito S, Okuda K, Miyazawa K, Sato K (1986) Study on the appearance of the cavity in the corpus luteum of cows by using ultrasonic scanning. Theriogenology 25(2):325–333. https://doi.org/10.1016/0093-691X(86)90068-3

Knopf L, Kastelic JP, Schallenberger E, Ginther OJ (1989) Ovarian follicular dynamics in heifers, test of two wave hypothesis by ultrasonically monitoring individual follicles. Domest Anim Endocrinol 6:111–119. https://doi.org/10.1016/0739-7240(89)90040-4

Kouamo J, Saague AMN, Zoli AP (2018) Determination of age and weight of bovine fetus (Bos indicus) by biometry. J Livestock Sci 9:9–15

Krueger L, Koerte J, Tsousis G, Herzog K, Flachowsky G, Bollwein H (2009) Transrectal Doppler sonography of uterine blood flow during the first 12 weeks after parturition in healthy dairy cows. Anim Reprod Sci 114:23–31. https://doi.org/10.1016/j.anireprosci.2008.09.006

Kruip TAM, Boni R, Wurth YA, Roelofsen MW, Pieterse MC (1994) Potential use of ovum pick-up for embryo production and breeding in cattle. Theriogenology 42:675–684. https://doi.org/10.1016/0093-691x(94)90384-u

Lamb GC, Fricke PM (2005) Ultrasound-early pregnancy diagnosis and fetal sexing. In: Proceedings, applied reproductive strategies in beef cattle, pp 253–262

Laven RA, Peters AR (2001) Gross morphometry of the bovine placentome during gestation. Reprod Domest Anim 36(6):289–296. https://doi.org/10.1046/j.1439-0531.2001.00297.x

Lazim E, Alrawi H, Aziz D (2016) Relationship between gestational age and transabdominal ultrasonographic measurements of fetus and uterus during the 2nd and 3rd trimester of gestation in cows. Asian Pac J Reprod 5(4):326–330. https://doi.org/10.1016/j.apjr.2016.06.010

LeBlanc SJ, Osawa T, Dubuc J (2011) Reproductive tract defense and disease in postpartum dairy cows. Theriogenology 76(9):1610–1618. https://doi.org/10.1016/j.theriogenology.2011.07.017

Lemley CO (2017) Investigating reproductive organ blood flow and blood perfusion to ensure healthy offspring. Anim Front 7:18–24. https://doi.org/10.2527/af.2017-0124

Lemley CO, Hart CG, Lemire RL, King EH, Hopper RM, Park SB, Rude BJ, Burnett DD (2018) Maternal nutrient restriction alters uterine artery hemodynamics and placentome vascular density in Bos Indicus and Bos Taurus. J Anim Sci 96(11):4823–4834. https://doi.org/10.1093/jas/sky329

Lénárt L, Taverne M, Wolleswinkel P, Gubik Z, Molnár L, Szenci O (2019) Reference values for fetal heart rate in cattle in the first trimester of pregnancy. Acta Vet Hung 67(2):274–281. https://doi.org/10.1556/004.2019.028

Lievaart JJ, Parlevliet JM, Dieleman SJ, Rientjes S, Bosman E, Vos PL (2006) Transvaginal aspiration as first treatment of ovarian follicular cysts in dairy cattle under field circumstances. Tijdschr Diergeneeskd 131:438–442. https://pubmed.ncbi.nlm.nih.gov/16830510/

Ligtvoet CM, Bom N, Gussenhoven WI (1989) Technical principles of ultrasound. In: Taverne MAM, Willemse AH (eds) Diagnostic ultrasound and animal reproduction. Kluwer Academic Publishers, Dordrecht, pp 1–9

Liu B, Cui Y, Yang B, Fan JF, Zhao ZW, Yu SJ (2010) Morphometric analysis of yak placentomes during gestation. Anat Rec 293(11):1873–1879. https://doi.org/10.1002/ar.21231

López-Gatius F, Garcia-Ispierto I (2023) Sexing of embryos at the time of twin reduction, a clinical approach. Animals (Basel) 13(8):1326. https://doi.org/10.3390/ani13081326

López-Gatius F, Santolaria P, Yániz J, Rutllant J, López-Béjar M (2002) Factors affecting pregnancy loss from gestation day 38 to 90 in lactating dairy cows from a single herd. Theriogenology 57:1251–1261. https://doi.org/10.1016/S0093-691X(01)00715-4

Lüttgenau J, Bollwein H (2014) Evaluation of bovine luteal blood flow by using color Doppler ultrasonography. Reprod Biol 14(2):103–109. https://doi.org/10.1016/j.repbio.2014.03.003

Madoz LV, Lorenti SN, Rearte R, Quintero-Rodriguez L, Migliorisi AL, Jaureguiberry M, Gabler C, Drillich M, de la Sota RL (2022) Detection of nonpregnant cows and potential embryo losses by color Doppler ultrasound and interferon-stimulated gene expression in grazing dairy cows. J Dairy Sci 105:6973–6984. https://doi.org/10.3168/jds.2021-21171

Magata F, Hartmann D, Ishii M, Miura R, Takahashi H, Matsui M, Kida K, Miyamoto A, Bollwein H (2013) Effects of exogenous oxytocin on uterine blood flow in puerperal dairy cows, the impact of days after parturition and retained fetal membranes. Vet J 196:76–80. https://doi.org/10.1016/j.tvjl.2012.08.010

Mann GE, Lamming GE (1999) Influence of progesterone during early pregnancy in cattle. Reprod Domest Anim 34:269–274. https://doi.org/10.1111/j.1439-0531.1999.tb01250.x

Mateus L, Lopes da Costa L, Bernardo F, Robalo Silva J (2002) Influence of puerperal uterine infection on uterine involution and postpartum ovarian activity in dairy cows. Reprod Domest Anim 37:31–35. https://doi.org/10.1046/j.1439-0531.2002.00317.x

Matsui M, Miyamoto A (2009) Evaluation of ovarian blood flow by colour Doppler ultrasound, practical use for reproductive management in the cow. Vet J 181:232–240. https://doi.org/10.1016/j.tvjl.2008.02.027

Mattoon JS, Berry CR (2021) Fundamentals of diagnostic ultrasound. In: Mattoon JS, Sellon RK, Berry CR (eds) Small animal diagnostic ultrasound, 4th edn. Elsevier, Amsterdam, pp 1–48

McCarty KJ, Owen MPT, Hart CG, Thompson RC, Burnett DD, King EH, Hopper RM, Lemley CO (2018) Effect of chronic melatonin supplementation during mid to late gestation on maternal uterine artery blood flow and subsequent development of male offspring in beef cattle. J Anim Sci 96:5100–5111. https://doi.org/10.1093/jas/sky363

Mee JF, Ryan DP, Condon T (1994) Ultrasound diagnosis of pregnancy in cattle. Vet Rec 134:532

Mendoza Bautista N, Saavedra-Jiménez L, Sánchez-Santillán P, Bottini-Luzardo M (2024) Placentome size in Bos taurus x Bos indicus crosses in the Mexican tropics differs from the source breeds. Stud Environ Anim Sci 5:e5070. https://doi.org/10.54020/seasv5n2-002

Merton JS, Ask B, Onkundi DC, Mullaart E, Colenbrander B, Nielen M (2009) Genetic parameters for oocyte number and embryo production within a bovine ovum pick-up-in vitro production embryo-production program. Theriogenology 72:885–893. https://doi.org/10.1016/j.theriogenology.2009.06.003

Mesquita FS, Pugliesi G, Scolari SC, França MR, Ramos RS, Oliveira M, Papa PC, Bressan FF, Meirelles FV, Silva LA, Nogueira GP, Membrive CM, Binelli M (2014) Manipulation of the periovulatory sex steroidal milieu affects endometrial but not luteal gene expression in early diestrus Nelore cows. Theriogenology 81:861–869. https://doi.org/10.1016/j.theriogenology.2013.12.022

Mikkola M, Hasler JF, Taponen J (2020) Factors affecting embryo production in superovulated Bos taurus cattle. Reprod Fertil Dev 32(2):104–124. https://doi.org/10.1071/RD19279

Miyamoto A, Shirasuna K, Hayashi KG, Kamada D, Awashima C, Kaneko E, Acosta TJ, Matsui M (2006) A potential use of color ultrasound as a tool for reproductive management, new observations using color ultrasound scanning that were not possible with imaging only in black and white. J Reprod Dev 52:153–160. https://doi.org/10.1262/jrd.17087

Miyazaki T, Tanaka M, Miyakoshi K, Minegishi K, Kasai K, Yoshimura Y (1998) Power and colour Doppler ultrasonography for the evaluation of the vasculature of the human corpus luteum. Hum Reprod 13:2836–2841. https://doi.org/10.1093/humrep/13.10.2836

Monteiro FM, Batista EOS, Vieira LM, Bayeux BM, Accorsi M, Campanholi SP, Dias EAR, Souza AH, Baruselli PS (2017) Beef donor cows with high number of retrieved COC produce more in vitro embryos compared with cows with low number of COC after repeated ovum pick-up sessions. Theriogenology 90:54–58. https://doi.org/10.1016/j.theriogenology.2016.11.002

Morotti F, Sanches BV, Pontes JHF, Basso AC, Siqueira ER, Lisboa LA, Seneda MM (2014) Pregnancy rate and birth rate of calves from a large-scale IVF program using reverse-sorted semen in B. indicus, B. indicus-taurus, and B. taurus cattle. Theriogenology 81:696–701. https://doi.org/10.1016/j.theriogenology.2013.12.002

Muñoz del Real LM, Valencia GL, Corral A, Renteria T (2006) Evaluation of ultrasound for pregnancy diagnosis between 20- and 40-days post insemination in dairy cows. J Anim Vet Adv 5:949–957

Nakai Y, Imanaka M, Nishio J, Maeda T, Ozaki A, Sun TT, Ogita S (1997) Uterine blood flow velocity waveforms during early postpartum course following caesarean section. Eur J Obstet Gynecol Reprod Biol 74:121–124. https://doi.org/10.1016/S0301-2115(97)02760-7

Nasser L, Adams GP, Bó GA, Mapletoft RJ (1993) Ovarian superstimulatory response relative to follicular wave emergence in heifers. Theriogenology 40:713–724. https://doi.org/10.1016/0093-691X(93)90207-L

Nation DP, Malmo J, Davis GM, Macmillan KL (2003) Ultrasonography at 28 to 35 days after insemination. Aust Vet J 81:6365. https://doi.org/10.1111/j.1751-0813.2003.tb11435.x

Niswender GD, Juengel JL, McGuire WJ, Belfiore CJ, Wiltbank MC (1994) Luteal function, the estrous cycle and early pregnancy. Biol Reprod 50:239–247. https://doi.org/10.1095/biolreprod50.2.239

Niswender GD, Juengel JL, Silva PJ, Rollyson MK, McIntush EW (2000) Mechanisms controlling the function and life span of the corpus luteum. Physiol Rev 80:1–29. https://doi.org/10.1152/physrev.2000.80.1.1

Noden DM, De la Hunta A (1985) The embryology of domestic animals. Development mechanisms and malformation. Williams and Wilkins, Baltimore, pp 330–335

Okano A, Tomizuka T (1987) Ultrasonic observation of postpartum uterine involution in the cow. Theriogenology 27:369–376. https://doi.org/10.1016/0093-691X(87)90225-1

Omran SN (1989) Ultrasonographic imaging of the reproductive events in the cow. PhD thesis, University of Glasgow

Omran SN, Ayliffe TR, Boyd JS (1988) Preliminary observations of bovine ovarian structures using B-mode real-time ultrasound. Vet Rec 112:465–466. https://doi.org/10.1136/vr.122.19.465

Paliulyte V, Drasutiene GS, Ramasauskaite D, Bartkeviciene D, Zakareviciene J, Kurmanavicius J (2017) Physiological uterine involution in primiparous and multiparous women, ultrasound study. Obstet Gynecol Int 2017:6739345. 10 pages. https://doi.org/10.1155/2017/6739345

Panarace M, Garnil C, Marfil M, Jauregui G, Lagioia J, Luther E, Medina M (2006) Transrectal Doppler sonography for evaluation of uterine blood flow throughout pregnancy in 13 cows. Theriogenology 66:2113–2119. https://doi.org/10.1016/j.theriogenology.2006.03.040

Pancarci SM, Ari UÇ, Atakisi O, Güngör Ö, Cigremis Y, Bollwein H (2012) Nitric oxide concentrations, estradiol-17β progesterone ratio in follicular fluid, and COC quality with respect to perifollicular blood flow in cows. Anim Reprod Sci 130:9–15. https://doi.org/10.1016/j.anireprosci.2011.12.013

Perry GA, Smith MF, Roberts AJ, MacNeil MD, Geary TW (2007) Relationship between size of the ovulatory follicle and pregnancy success in beef heifers. J Anim Sci 85:684–689. https://doi.org/10.2527/jas.2006-519

Perry GA, Swanson OL, Larimore EL, Perry BL, Djira GD, Cushman RA (2014) Relationship of follicle size and concentrations of estradiol among cows exhibiting or not exhibiting estrus during a fixed-time AI protocol. Domest Anim Endocrinol 48:15–20. https://doi.org/10.1016/j.domaniend.2014.02.001

Peter AT, Jakovljevic S, Pierson RA (1992) Use of real-time ultrasonography in bovine and equine reproduction. Compend Contin Educ Pract Vet 14(8):1116–1124

Pierson RA, Ginther OJ (1984) Ultrasonography for the detection of pregnancy and study embryonic development in heifers. Theriogenology 22:225–233. https://doi.org/10.1016/0093-691X(84)90435-7

Pierson RA, Ginther OJ (1987) Ultrasonographic appearance of the bovine uterus during the estrous cycle. J Am Vet Med Assoc 190(8):995–1001. PMID: 3570959

Pierson RA, Ginther OJ (1988) Ultrasonic imaging of the ovaries and uterus in cattle. Theriogenology 29:21–37. https://doi.org/10.1016/0093-691X(88)90029-5

Pierson RA, Kastelic JP, Ginther OJ (1988) Basic principles and techniques of transrectal ultrasonography in cattle and horses. Theriogenology 29:3–20. https://doi.org/10.1016/0093-691X(88)90028-3

Pieterse MC, Taverne MAM, Kruip TAM, Willemse AH (1990a) Detection of corpora lutea and follicles in cows, a comparison of transvaginal ultrasonography and rectal palpation. Vet Rec 126:552–554

Pieterse MC, Szenci O, Willemse AH, Bajcsy CSA, Dieleman SJ, Taverne MAM (1990b) Early pregnancy diagnosis in cattle by means of linear-array real-time ultrasound scanning of the uterus and a qualitative and quantitative milk progesterone test. Theriogenology 33:697–707. https://doi.org/10.1016/0093-691X(90)90547-7

Pinaffi FLV, Santos ES, Silva MG, Maturana Filho M, Madureira EH, Silva LA (2015) Follicle and corpus luteum size and vascularity as predictors of fertility at the time of artificial insemination and embryo transfer in beef cattle. Pesqui Vet Bras 35:470–476. https://doi.org/10.1590/S0100-736X2015000500015

Pontes JHF, Nonato-Junior I, Sanches BV, Ereno-Junior JC, Uvoa S, Barreiros TRR, Oliveira JA, Hasler JF, Seneda MM (2009) Comparison of embryo yield and pregnancy rate between in vivo and in vitro methods in the same Nelore (B. indicus) donor cows. Theriogenology 71:690–697. https://doi.org/10.1016/j.theriogenology.2008.09.031

Pontes JHF, Silva KC, Basso AC, Rigo AG, Ferreira CR, Santos GM, Sanches BV, Porcionato JP, Vieira PH, Faifer FS, Sterza FA, Schenk JL, Seneda MM (2010) Large-scale in vitro embryo production and pregnancy rates from Bos taurus, Bos indicus, and indicus-taurus dairy cows using sexed sperm. Theriogenology 74:1349–1355. https://doi.org/10.1016/j.theriogenology.2010.06.004

Pontes JHF, Melo Sterza FA, Basso AC, Ferreira CR, Sanches BV, Rubin KCP, Seneda MM (2011) Ovum pick up, in vitro embryo production, and pregnancy rates from a large-scale commercial program using Nelore cattle (B. indicus) donors. Theriogenology 75:1640–1646. https://doi.org/10.1016/j.theriogenology.2010.12.026

Poock SE, Wilson DJ (2011) A review of the use of ultrasound for reproductive purposes in beef cattle. In: Proceedings, applied reproductive strategies in beef cattle, August 31—September 1, 2011, Joplin, MO

Pugliesi G, Pinaffi FLV, Beg MA, Ginther OJ (2013) Use of corpus luteum area as a predictor of ongoing functional luteolysis in dairy heifers. Reprod Fertil Dev 25:235. https://doi.org/10.1071/RDv25n1Ab173

Pugliesi G, Miagawa BT, Paiva YN, França MR, Silva LA, Binelli M (2014) Conceptus-induced changes in the gene expression of blood immune cells and the ultrasound-accessed luteal function in beef cattle, how early can we detect pregnancy? Biol Reprod 91:1–12. https://doi.org/10.1095/biolreprod.114.121525

Pugliesi G, Santos FB, Lopes E, Nogueira É, Maio JR, Binelli M (2016a) Improved fertility in suckled beef cows ovulating large follicles or supplemented with long-acting progesterone after timed-AI. Theriogenology 85:1239–1248. https://doi.org/10.1016/j.theriogenology.2015.12.006

Pugliesi G, Silva JCB, Nishimura T, Miyai D, Silva LA, Binelli M (2016b) Use of color-Doppler ultrasonography to improve selection of higher fertility beef recipient cows for embryo transfer. Anim Reprod 13:454. (Abstract)

Pugliesi G, de Melo GD, Ataíde GA, Pellegrino CAG, Silva JB, Rocha CC, Motta IG, Vasconcelos JLM, Binelli M (2018) Use of Doppler ultrasonography in embryo transfer programs, feasibility and field results. Anim Reprod 15:239–246. https://doi.org/10.21451/1984-3143-AR2018-0059

Pugliesi G, de Melo GD, Silva JB, Carvalhêdo AS, Lopes E, de Siqueira Filho E, Silva LA, Binelli M (2019) Use of color-Doppler ultrasonography for selection of recipients in timed-embryo transfer programs in beef cattle. Theriogenology 135:73–79. https://doi.org/10.1016/j.theriogenology.2019.06.006

Pugliesi G, Guimarães da Silva A, Viana JHM, Siqueira LGB (2023) Review, current status of corpus luteum assessment by Doppler ultrasonography to diagnose non-pregnancy and select embryo recipients in cattle. Animal 17(1):100752. https://doi.org/10.1016/j.animal.2023.100752

Qi M, Yao Y, Ma H, Wang J, Zhao X, Liu L, Tang X, Zhang L, Zhang S, Sun F (2013) Transvaginal ultrasound-guided ovum pick-up (OPU) in cattle. J Biomimetics Biomater Tissue Eng 18:118. https://doi.org/10.4172/1662-100X.1000118

Quintela LA, Díaz C, Herradón PG, Peña AI, Becerra JJ (2006) Ecografía y reproducción en la vaca. Ed. Servicio de publicaciones e intercambio científico of the University of Santiago de Compostela, Santiago de Compostela, p 92

Quintela LA, Barrio M, Peña AI, Becerra JJ, Cainzos J, Herradón PG, Díaz C (2012) Use of ultrasound in the reproductive management of dairy cows. Reprod Domest Anim 47:34–44. https://doi.org/10.1111/j.1439-0531.2012.02032.x

Rajakoski E (1960) The ovarian follicular system in sexually mature heifers with special reference to seasonal, cyclical, and left-right variations. Acta Endocrinol 34(3):S7–S68. https://doi.org/10.1530/acta.0.XXXIVS0007

Rantanen NW (1986) General considerations for ultrasound examinations. Vet Clin North Am 2:29–32. https://doi.org/10.1016/S0749-0739(17)30730-7

Rantanen NW, Ewing RL (1981) Principles of ultrasound application in animals. Vet Radiol 22:196–203. https://doi.org/10.1111/j.1740-8261.1981.tb01373.x

Rauch A, Krüger L, Miyamoto A, Bollwein H (2008) Colour Doppler sonography of cystic ovarian follicles in cow. J Reprod Dev 54:447–453. https://doi.org/10.1262/jrd.20025

Reese ST, Franco GA, Poole RK, Hood R, Fernadez Montero L, Oliveira Filho RV, Cooke RF, Pohler KG (2020) Pregnancy loss in beef cattle, a meta-analysis. Anim Reprod Sci 212:106251. https://doi.org/10.1016/j.anireprosci.2019.106251

Reynolds LP, Redmer DA (1995) Utero-placental vascular development and placental function. J Anim Sci 73:1839–1851. https://doi.org/10.2527/1995.7361839x

Reynolds LP, Millaway DS, Kirsch JD, Infeld JE, Redmer DA (1990) Growth and in-vitro metabolism of placental tissues of cows from day 100 to day 250 of gestation. J Reprod Fertil 89(1):213–222. https://doi.org/10.1530/jrf.0.0890213

Reynolds LP, Grazul-Bilska AT, Redmer DA (2000) Angiogenesis in the corpus luteum. Endocrine 12:1–9. https://doi.org/10.1385/ENDO:12:1:1

Ribadu AI, Nakao T (1999) Bovine reproductive ultrasonography, a review. J Reprod Dev 45:13–28. https://doi.org/10.1262/jrd.45.13

Ribadu AY, Dobson H, Ward WR (1994a) Ultrasound and progesterone monitoring of ovarian follicular cysts in cows treated with GnRH. Br Vet J 150:489–497. https://doi.org/10.1016/S0007-1935(05)80202-7

Ribadu AY, Ward WR, Dobson H (1994b) Comparative evaluation of ovarian structures in cattle by palpation per rectum, ultrasonography and plasma progesterone concentration. Vet Rec 135:452–457. https://doi.org/10.1136/vr.135.19.452

Richardson C, Jones P, Barnard V, Hebert C, Terlecki S, Wijeratne W (1990) Estimation of the developmental age of the bovine fetus and newborn calf. Vet Rec 126:279–284

Rocha CC, Martins T, Cardoso BO, Silva LA, Binelli M, Pugliesi G (2019) Ultrasonography-accessed luteal size endpoint that most closely associates with circulating progesterone during the estrous cycle and early pregnancy in beef cows. Anim Reprod Sci 201:12–21. https://doi.org/10.1016/j.anireprosci.2018.12.003

Rodgers RJ, Irving-Rodgers HF (2010) Formation of the ovarian follicular antrum and follicular fluid. Biol Reprod 82:1021–1029. https://doi.org/10.1095/biolreprod.109.082941

Romano JE, Thompson JA, Forrest DW, Westhusin ME, Tomaszweski MA, Kraemer DC (2006) Early pregnancy diagnosis by transrectal ultrasonography in dairy cattle. Theriogenology 66:1034–1041. https://doi.org/10.1016/j.theriogenology.2006.02.044

Ruiz S (2021) Ovum pick-up (OPU) in cattle: an update. In: Gardón JC, Satué K (eds) Biotechnologies applied to animal reproduction. Current trends and practical applications for reproductive management. Apple Academic Press & CRC Press, Palm Bay, pp 139–183. https://doi.org/10.1201/9780367817527

Ruiz S, Zaraza J, De Ondiz A, Rath D (2009) Early fertilization events in bovine IVF employing OPU-oocytes and sex-sorted frozen/thawed sperm. Reprod Domest Anim 44:123

Ruiz S, Romero J, Astiz S, Peinado B, Almela L, Poto A (2013a) Application of reproductive biotechnology for the recovery of endangered breeds: birth of the first calf of Murciana-Levantina bovine breed derived by OPU, in vitro production and embryo vitrification. Reprod Domest Anim 48:81–84. https://doi.org/10.1111/rda.12179

Ruiz S, Camisao J, Zaraza J, De Ondiz A, Romero J, Rodrigues R, Rath D (2013b) Use of sex-sorted and unsorted frozen/thawed sperm and in vitro fertilization events in bovine oocytes derived from ultrasound guided aspiration. Rev Bras Zootec 42:721–727. https://doi.org/10.1590/S1516-35982013001000005

Sartori R, Gimenes LU, Monteiro PLJ, Melo LF, Baruselli PS, Bastos MR (2016) Metabolic and endocrine differences between Bos taurus and Bos indicus females that impact the interaction of nutrition with reproduction. Theriogenology 86:32–40. https://doi.org/10.1016/j.theriogenology.2016.04.016

Savio JD, Keenan L, Boland MP, Roche JF (1988) Pattern of growth of dominant follicles during the oestrous cycle of heifers. J Reprod Fertil 83:663–671. https://doi.org/10.1530/jrf.0.0830663

Sawyer GJ, Broadbent PJ, Dolman DF (1995) Ultrasound-monitored ovarian responses in normal and superovulated cattle given exogenous progesterone at different stages of the oestrous cycle. Anim Reprod Sci 38(3):187–201. https://doi.org/10.1016/0378-4320(94)01363-Q

Scenna FN, Hockett ME, Towns TM, Saxton AM, Rohrbach NR, Wehrman ME, Schrick FN (2005) Influence of a prostaglandin synthesis inhibitor administered at embryo transfer on pregnancy rates of recipient cows. Prostaglandins Other Lipid Mediat 78(1–4):38–45. https://doi.org/10.1016/j.prostaglandins.2005.02.003

Scully S, Evans ACO, Duffy P, Crowe MA (2014) Characterization of follicle and CL development in beef heifers using high resolution three-dimensional ultrasonography. Theriogenology 81:407–418. https://doi.org/10.1016/j.theriogenology.2013.10.012

Scully S, Evans ACO, Carter F, Duffy P, Lonergan P, Crowe MA (2015) Ultrasound monitoring of blood flow and echotexture of the corpus luteum and uterus during early pregnancy of beef heifers. Theriogenology 83:449–458. https://doi.org/10.1016/j.theriogenology.2014.10.009

Seguin BE (1980) Ovarian cysts in dairy cattle. In: Morrow DA (ed) Current therapy in theriogenology, 1st edn. W.B. Saunders, Philadelphia, pp 199–204

Seneda MM, Esper CR, Garcia JM, de Oliveira JA, Vantini R (2001) Relationship between follicle size and ultrasound-guided transvaginal oocyte recovery. Anim Reprod Sci 67:37–43. https://doi.org/10.1016/s0378-4320(01)00113-0

Sergeev L, Kleemann DO, Walker SK, Smith DH, Grosser TI, Mann T, Seamark RF (1990) Real-time ultrasound imaging for predicting ovine fetal age. Theriogenology 34(3):593–601. https://doi.org/10.1016/0093-691X(90)90014-K

Sharma I, Singh M, Sharma A, Kumar P, Dogra PK (2021) Doppler sonography for evaluation of haemodynamic changes of uterine arteries and umbilicus during different months of gestation in dairy cows. Veterinarija ir Zootechnika 79:52–58. https,//vetzoo.lsmuni.lt/data/vols/2021/7901/pdf/sharma2.pdf
Sheldon IM, Dobson H (2004) Postpartum uterine health in cattle. Anim Reprod Sci 82:295–306
Sheldon IM, Cronin J, Goetze L, Donofrio G, Schuberth HJ (2009) Defining postpartum uterine disease and the mechanisms of infection and immunity in the female reproductive tract in cattle. Biol Reprod 81:1025–1032. https://doi.org/10.1095/biolreprod.109.077370
Siddiqui MAR, Almamun M, Ginther OJ (2009a) Blood flow in the wall of the preovulatory follicle and its relationship to pregnancy establishment in heifers. Anim Reprod Sci 113:287–292. https://doi.org/10.1016/j.anireprosci.2008.07.008
Siddiqui MAR, Gastal EL, Gastal MO, Almamun M, Beg MA, Ginther OJ (2009b) Relationship of vascular perfusion of the wall of the preovulatory follicle to in vitro fertilization and embryo development in heifers. Reproduction 137:689–697. https://doi.org/10.1530/rep-08-0403
Silva del Rio N, Stewart S, Rapnicki P, Chang YM, Fricke PM (2007) An observational analysis of twin births, calf sex ratio, and calf mortality in Holstein dairy cattle. J Dairy Sci 90:1255–1264. https://doi.org/10.3168/jds.S0022-0302(07)71614-4
Silva LA, Ginther OJ (2010) Local effects of the conceptus on uterine perfusion during early pregnancy in heifers. Reproduction 139(2):453–463. https://doi.org/10.1530/REP-09-0363
Silva E, Sterry RA, Kolb D, Mathialagan N, McGrath MF, Ballam JM, Fricke PM (2007) Accuracy of a pregnancy associated glycoprotein ELISA to determine pregnancy status of lactating dairy cows twenty-seven days after timed artificial insemination. J Dairy Sci 90:4612–4622. https://doi.org/10.3168/jds.2007-0276
Singh J, Pearson RA, Adams GP (1998) Ultrasound image attributes of bovine ovarian follicles and endocrine and functional correlates. J Reprod Fertil 112(1):19–29. https://doi.org/10.1530/jrf.0.1120019
Siqueira LGB, Torres CAA, Amorim LS, Souza ED, Camargo LSA, Fernandes CAC, Viana JHM (2009) Interrelationships among morphology, echotexture, and function of the bovine corpus luteum during the estrous cycle. Anim Reprod Sci 115:18–28. https://doi.org/10.1016/j.anireprosci.2008.11.009
Siqueira LGB, Areas VS, Ghetti AM, Fonseca JF, Palhao MP, Fernandes CAC, Viana JHM (2013) Color Doppler flow imaging for the early detection of nonpregnant cattle at 20 days after timed artificial insemination. J Dairy Sci 96:6461–6472. https://doi.org/10.3168/jds.2013-6814
Siqueira LGB, Arashiro EK, Ghetti AM, Souza ED, Feres LF, Pfeifer LF, Fonseca JF, Viana JH (2019) Vascular and morphological features of the corpus luteum 12 to 20 days after timed artificial insemination in dairy cattle. J Dairy Sci 102:5612–5622. https://doi.org/10.3168/jds.2018-15853
Sirois J, Fortune JE (1988) Ovarian follicular dynamics during the estrous cycle in heifers monitoring by real-time ultrasonography. Biol Reprod 39(2):308–317. https://doi.org/10.1095/biolreprod39.2.308
Spell AR, Beal WE, Corah LR, Lamb GC (2001) Evaluating recipient and embryo factors that affect pregnancy rates of embryo transfer in beef cattle. Theriogenology 56:287–297. https://doi.org/10.1016/S0093-691X(01)00563-5
Stroud B (1996) Using ultrasonography to determine bovine fetal sex. Vet Med 91:663–672
Stroud B (2006a) The use of ultrasound in cow/calf applications. Am Assoc Bovine Pract Proc 39:3–7. https://doi.org/10.21423/aabppro20064651
Stroud B (2006b) Guidelines for ultrasound fetal sexing in a cow/calf operation. Am Assoc Bovine Pract Proc 39:66–69. https://doi.org/10.21423/aabppro20064674
Suthar VS, Momin T, Mahesh P, Kelawala DN, Patil DB (2018) Reproductive ultrasonography in bovine practice—a review. Intas Polivet 19(1):22–28. https,//link.gale.com/apps/doc/A563571881/AONE?u=anon~7c1bdde4&sid=googleScholar&xid=e20c1fd9
Sutton ML, Gilchrist RB, Thompson JG (2003) Effects of in vivo and in vitro environments on the metabolism of the cumulus-oocyte complex and its influence on oocyte development capacity. Hum Reprod Update 9:35–48. https://doi.org/10.1093/humupd/dmg009

Szenci O (2021) Recent possibilities for the diagnosis of early pregnancy and embryonic mortality in dairy cows. Animals (Basel) 11(6):1666. https://doi.org/10.3390/ani11061666

Szenci O, Piros A, Kovács L (1990) Early bovine pregnancy diagnosis by a battery operated portable ultrasonic scanner the ultra-scan. In: Proceedings of the 16th World Buiatrics Congress, Salvador, Brazil, 13-17 August 1990, pp 219—223

Szenci O, Gyulai G, Nagy P, Kovács L, Varga J, Taverne MAM (1995) Effect of uterus position relative the pelvic inlet on the accuracy of early bovine pregnancy diagnosis by means of ultrasonography. Vet Q 17:37–39. https://doi.org/10.1080/01652176.1995.9694528

Szenci O, Beckers JF, Humblot P, Sulon J, Sasser G, Taverne MAM, Varga J, Baltusen R, Schekk G (1998) Comparison of ultrasonography bovine pregnancy-specific protein B, and bovine pregnancy-associated glycoprotein 1 tests for pregnancy detection in dairy cows. Theriogenology 50:77–88. https://doi.org/10.1016/S0093-691X(98)00115-0

Taneja M, Bols PEJ, Van de Velde A, Ju JC, Schreiber D, Tripp MW (2000) Developmental competence of juvenile calf oocytes in vitro and in vivo: influence of donor animal variation and repeated gonadotropin stimulation. Biol Reprod 62:206–213. https://doi.org/10.1095/biolreprod62.1.206

Taylor C, Rajamanhendran R (1991) Follicular dynamics and corpus luteum growth and function in pregnant versus nonpregnant dairy cows. J Dairy Sci 74:115–123. https://doi.org/10.3168/jds.S0022-0302(91)78151-4

Tom JW, Pierson RA, Adams GP (1998) Quantitative echotexture analysis of bovine corpora lutea. Theriogenology 49:1345–1352. https://doi.org/10.1016/S0093-691X(98)00081-8

Torres-Simental JF, Peña-Calderón C, Avendaño-Reyes L, Correa-Calderón A, Macías-Cruz U, Rodríguez-Borbón A, Leyva-Corona JC, Rivera-Acuña F, Thomas MG, Luna-Nevárez P (2021) Predictive markers for superovulation response and embryo production in beef cattle managed in northwest Mexico are influenced by climate. Livest Sci 250:e104590. https://doi.org/10.1016/j.livsci.2021.104590

Torres-Lechuga ME, González-Maldonado J (2022) Ultrasonography and physiological description of essential events for reproductive management in dairy cattle. Review. Revista Mexicana De Ciencias Pecuarias, 13(2), 452–472. https://doi.org/10.22319/rmcp.v13i2.5789

Underwood SL, Bathgate R, Pereira DC, Castro A, Thomson PC, Maxwell WMC, Evans G (2010) Embryo production after in vitro fertilization with frozen-thawed, sex-sorted, re-frozen-thawed bull sperm. Theriogenology 73:97–102. https://doi.org/10.1016/j.theriogenology.2009.08.005

Utt MD, Johnson GL III, Beal WE (2009) The evaluation of corpus luteum blood flow using color-flow Doppler ultrasound for early pregnancy diagnosis in bovine embryo recipients. Theriogenology 71:707–715. https://doi.org/10.1016/j.theriogenology.2008.09.032

Vaillancourt D, Vierschwal CJ, Ogwu D, Elmore RG, Martin CE, Sharp AJ, Youngquist RS (1979) Correlation between pregnancy diagnosis by membrane slip and embryonic mortality. J Am Vet Med Assoc 175(5):466–468. PMID: 500480

Vallejo-Timaran DA, Arango-Sabogal JC, Reyes-Vélez J, Maldonado-Estrada JG (2020) Postpartum uterine diseases negatively impact the time to pregnancy in grazing dairy cows from high-altitude tropical herds. Prev Vet Med 185:105202. https://doi.org/10.1016/j.prevetmed.2020.105202

Van Camp SD (1991) Understanding the process of placental separation and uterine involution. Vet Med 86:642–646

Van Eetvelde M, Kamal M, Hostens M, Vandaele L, Fiems LO, Opsomer G (2016) Evidence for placental compensation in cattle. Animal 10(8):1342–1350. https://doi.org/10.1017/S1751731116000318

Vanroose G, de Kruif A, Van Soom A (2000) Embryonic mortality and embryo–pathogen interactions. Anim Reprod Sci 60–61:131–143. https://doi.org/10.1016/S0378-4320(00)00098-1

Varshney JP, Chaudhary PS (2022a) Basics of ultrasound. In: Varshney JP (ed) Ultrasound in veterinary medicine fundamentals and applications. NIPA Genx Electronic Resources & Solutions P. Ltd, New Delhi, pp 1–24

Varshney JP, Chaudhary PS (2022b) Ultrasonography and ultrasound machine. In: Varshney JP (ed) Ultrasound in veterinary medicine fundamentals and applications. NIPA Genx Electronic Resources & Solutions P. Ltd, New Delhi, pp 25–46

Vasconcelos JLM, Silcox RW, Lacerda JA, Pursley JR, Wiltbank MC (1997) Pregnancy rate, pregnancy loss, and response to heat stress after AI at two different times from ovulation in dairy cows. Biol Reprod 56(1):140. (Abstr.)

Vassena R, Adams GP, Mapletoft RJ, Pierson RA, Singh J (2003) Ultrasound image characteristics of ovarian follicles in relation to oocyte competence and follicular status in cattle. Anim Reprod Sci 76:25–41. https://doi.org/10.1016/S0378-4320(02)00234-8

Velez JS, Randel RD, Neuendorff DA (1991) Effect of uterine manipulation on postpartum fertility and plasma 13,14-dihydro-15-keto-prostaglandin F2α in Brahman cows and first-calf heifers. Theriogenology 36(6):987–998. https://doi.org/10.1016/0093-691X(91)90323-6

Velho G, Rovani MT, Ferreira R, Gasperin BG, Dalto AGC (2021) Blood perfusion and diameter of bovine corpus luteum as predictors of luteal function in early pregnancy. Reprod Domest Anim 57(3):246–252. https://doi.org/10.1111/rda.14046

Vernunft A, Schwerhoff M, Viergutz T, Diederich M, Kuwer A (2015) Anti-Muellerian hormone levels in plasma of Holstein-Friesian heifers as a predictive parameter for ovum pick-up and embryo production outcomes. J Reprod Dev 61:74–79. https://doi.org/10.1262/jrd.2014-091

Viana JHM (2023) 2022 statistics of embryo production and transfer in domestic farm animals. International Embryo Transfer Society (IETS). Embryo Technol Newslett 41:4

Viana GI, Marx D (1994) Sex determination of bovine fetuses by ultrasonography. Tierarztliche-Umschan 49:484–486

Watanabe YF, de Souza AH, Mingoti RD, Ferreira RM, Batista EOS, Dayan A, Watanabe O, Meirelles FM, Nogueira MFG, Ferraz JBS, Baruselli PS (2017) Number of oocytes retrieved per donor during OPU and its relationship with in vitro embryo production and field fertility following embryo transfer. Anim Reprod 14:635–644

White IR, Russel AJF, Wright IA, Whyte TK (1985) Real-time ultrasonic scanning in the diagnosis of pregnancy and the estimation of gestational age in cattle. Vet Rec 117:5–8

Wiltbank MC, Fricke PM, Sangsritavong S, Sartori R, Ginther OJ (2000) Mechanisms that prevent and produce double ovulations in dairy cattle. J Dairy Sci 83(12):2998–3007. https://doi.org/10.3168/jds.S0022-0302(00)75201-5

Wright IA, White IR, Russel AJF, Whyte TK, McBean AJ (1988) Prediction of calving date in beef-cows by real-time ultrasonic-scanning. Vet Rec 123(9):228–229

Wu FSY, Hwu YM, Lee RKK, Li SH, Sun FJ, Lin MH, Lin SY (2012) First trimester ultrasound estimation of gestational age in pregnancies conceived after in vitro fertilization. Eur J Obstet Gynecol Reprod Biol 160(2):151–155. https://doi.org/10.1016/j.ejogrb.2011.11.019

Yoshioka K, Iwamura S, Kamomae H (1996) Ultrasonic observations on the turnover of ovarian follicular cysts and associated changes of plasma LH, FSH, progesterone and oestradiol-17 in cows. Res Vet Sci 61:240–244. https://doi.org/10.1016/S0034-5288(96)90071-5

Youngquist RS (1997) Pregnancy diagnosis. In: Youngquist RS, Threlfall WR (eds) Current therapy in large animal theriogenology. WB Saunders Co., Philadelphia, pp 295–303

Equine Cloning and Its Impact on Production

Micaela Navarro and Adrián Mutto

Abstract

Horses hold significant economic value due to their diverse roles in sports, recreation, and agriculture. Cloning has emerged as a prominent reproductive biotechnology in equines, proving more reliable than other techniques such as superovulation or in vitro fertilization. Cloning offers advantages like genetic replication of elite individuals, preservation of valuable genetics, and accelerated breeding programs. Various techniques exist for equine cloning, including traditional somatic cell nuclear transfer (SCNT) and zona pellucida-free (ZP-free) embryo production. In this work, we will summarize the methodology of each technique, discuss their advantages and disadvantages, and provide a brief overview of the epigenetic deficiencies that still persist in cloned embryos, showing that there is still much work to be done to understand the mechanisms underlying cloning in equids.

Keywords

Equine · Artificial reproductive techniques · Somatic cell nuclear transfer · Zona pellucida-free cloning · Epigenetic barriers

M. Navarro · A. Mutto (✉)
Laboratory of Biotechnologies Applied to Animal Reproduction, Institute of Biotechnology Research 'Dr Rodolfo Ugalde', UNSAM-CONICET, Buenos Aires, Argentina
e-mail: minavarro@iibintech.com.ar; amutto@iib.unsa.edu.ar

J. C. Gardón, K. Satué Ambrojo (eds.), *Assisted Reproductive Technologies in Animals Volume 2*, https://doi.org/10.1007/978-3-031-87198-6_2

1 Introduction

Equids encompass a group of herbivorous mammals of significant scientific and economic interest. This group includes species such as horses, donkeys, and zebras, which have garnered global attention due to their profound economic impact across various industries. Since ancient times, these animals have been pivotal in agriculture, transportation, and warfare. Today, their influence extends to modern sectors such as the food industry, livestock management, and high-performance sports like polo. Their adaptability and ability to perform diverse functions make them essential protagonists in the contemporary global economy. Due to their economic importance, these species have been subject to various assisted reproductive technologies (ARTs) aimed at increasing the number of individuals with desirable genetics or preserving elite animal genetics. Numerous reproductive biotechnologies have been tested in equids, although their efficacy varies among species. Specifically in horses, techniques like superovulation or in vitro fertilization (IVF) have not been achieved efficiently (Cabeza and Gambini 2023). However, more complex techniques such as cloning have emerged as an alternative.

The production of clone individuals refers to the creation of genetically identical individuals to an extant one. The origins of this technique date back to 1935 when Hans Spemann (1928) demonstrated that the nucleus directs cell division by separating the cells of a salamander embryo with a hair and obtaining individuals from each cell. Based on this, he proposed an experiment involving replacing the nucleus of an oocyte with the nucleus of a differentiated cell. This experiment was carried out in 1962 when John Gurdon successfully created frog clones by transferring the nucleus of a frog intestinal epithelial cell into the enucleated oocyte cytoplasm (Gurdon 1962). Later, in 1996, with the birth of the sheep "Dolly," it was demonstrated that this technique could be applied to other species, including mammals (Campbell et al. 1996). The birth of "Dolly" the sheep marked a groundbreaking milestone in biotechnology as the first cloned mammal through somatic cell nuclear transfer (SCNT). Dolly's creation underscored the remarkable ability of an oocyte to reprogram a somatic cell into an embryonic state, highlighting the plasticity inherent in cells. Following Dolly's success, cloning technologies advanced rapidly, encompassing more than 20 species such as cows, sheep, and goats, among others (Matoba and Zhang 2018).

In equids, cloning initially found success in mules, resulting in the birth of three healthy individuals (Woods et al. 2003). Subsequently, this technology was adapted for use in horses, culminating in the historic birth of the first cloned horse in 2003 (Galli et al. 2003). In both species, foals are born healthy and fertile and reach adulthood. Overall, the availability of this technology not only expanded our understanding of reproductive biology but also opened new avenues for the conservation of endangered species, the creation of genetically modified or transgenic organisms, preserving valuable genetic traits, and enhancing the productivity of equine populations worldwide. In this study, we summarize the different approaches for producing equine embryos through cloning, highlighting the advantages and disadvantages of each method. Finally, we briefly described some of the epigenetic barriers that are still observed in horses, showing the need to continue studying the process underlying cellular reprogramming.

2 Artificial Reproductive Biotechnologies

Artificial reproductive biotechnologies comprise the technique or set of techniques that allow for increasing the reproductive efficiency of animals. Thanks to a greater understanding of developmental biology and scientific-technological advancement, these techniques have evolved over the years in their degree of complexity and have been classified into five generations, ranging from artificial insemination to somatic cell nuclear transfer and stem cells (Palma 2001) (Fig. 1).

The first generation began in the twelfth century with the application of artificial insemination (AI), which involves introducing semen into the female reproductive tract to achieve fertilization and subsequently embryo development (Palma 2001). In horses, AI is the most widely used reproductive technique worldwide due to its high effectiveness and numerous advantages. The efficiency of AI typically yields conception rates between 70% and 80% per cycle, which is influenced by factors such as semen quality and the timing of the mare's estrus cycle (Morrell et al. 2014; Kowalczyk et al. 2019). AI offers significant benefits including the ability to use high-value stallions without the need for physical transportation, thereby enhancing genetic dissemination. It also reduces the risk of transmitting reproductive diseases and optimizes semen use by allowing for lower doses and multiple inseminations

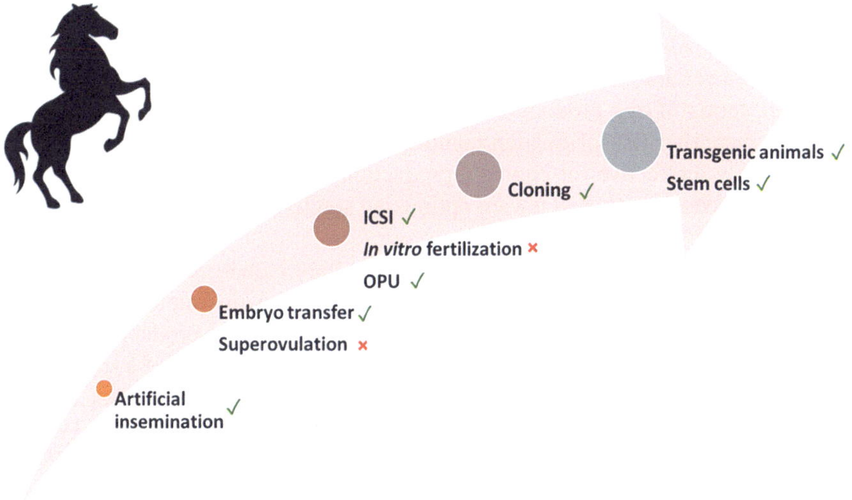

Fig. 1 Development of artificial reproductive techniques in horses. The development of different artificial reproductive techniques has evolved through different years. In this scheme, the main ARTs were included, highlighting which of those are successfully applied in equines and which have not yet been developed

from a single semen collection. Additionally, AI provides flexibility in managing reproduction, making it possible to inseminate mares at different locations and times, thus improving synchronization and overall reproductive efficiency (Allen et al. 1976; Yates and Whitacre 1988). This technique was even more efficient when combined with semen cryopreservation (a technique used to store semen long term through freezing), where the last one has become an essential tool in equine reproduction. It allows for the preservation of valuable stallion genetics, with survival rates for cryopreserved semen generally ranging from 30% to 60% (Hernández-Avilés et al. 2024). This method minimizes the risks associated with handling live animals and ensures that high-quality genetics can be accessed even after the stallion's death, making AI an indispensable technique in modern equine breeding (Clulow and Gibb 2022).

The second generation was characterized by the application of technologies that allowed for controlling the reproductive activity of females through ovulation programs, superovulation, and embryo transfer (ET) (Palma 2001). These technologies enable synchronization of estrus and more efficient use of AI, facilitate the reintegration of lactating or postpartum anestrous females into the reproductive cycle, and allow for the scheduling of parturition (Cooper 1974; Leung and Armstrong 1980). In horses, embryo transfer is a highly effective reproductive technology, involving the transfer of a fertilized embryo from a donor mare to a recipient mare, which then carries the pregnancy to term. This technique enables the efficient use of high-value genetics, allowing mares to produce multiple offspring in a single season and maximizing the reproductive potential of both donor and recipient mares (Squires et al. 1999). In most livestock species, the combination of superovulation with AI and ET has revolutionized breeding efficiency (Hinrichs 2018). However, in horses, superovulation —a method aimed at inducing mares to ovulate multiple eggs in one cycle—has not proven effective. Mares often show a variable response to superovulation protocols, leading to inconsistent results in terms of egg production and embryo quality. This variability, coupled with the increased risk of ovarian complications, has resulted in superovulation being largely ineffective for equine reproduction (Maclellan et al. 2002). Consequently, the focus has shifted toward optimizing single-embryo transfer techniques rather than relying on superovulation.

The third generation of reproductive biotechnologies emerged around the 1980s and was characterized by sperm and embryo sexing and in vitro embryo production (IVP) (Palma 2001). Sex selection proved to be highly useful in production systems where profitability relies on having a higher number of individuals of a specific sex (e.g., mares for polo) (Aurich and Schneider 2014). On the other hand, in vitro embryo production encompasses techniques such as in vitro fertilization (IVF) or intracytoplasmic sperm injection (ICSI) (Palma 2001; Carnevale et al. 2019). The first report of in vitro embryo production was in 1970 when in vitro matured oocytes were fertilized with pre-incubated sperm (Sreenan 1970). However, the first fully in vitro process was not performed until 1987 (Xu et al. 1987). The objective of IVF is to produce embryos in vitro by mimicking the physiological events that occur during fertilization and preimplantation development in vivo. These events are summarized in three stages: in vitro oocyte maturation (IVM), in vitro fertilization (IVF), and in vitro embryo culture, where the success or failure of each stage depends on the previous one.

In most livestock species, IVF revolutionized the agricultural sector by enabling the mass production of embryos, increasing the number of offspring per animal per year, and allowing the generation of embryos from oocytes obtained through transvaginal ultrasound-guided follicle aspiration (OPU, "ovum pickup") or from slaughtered animals (Palma 2001). However, this technique has been a long-term challenge in horses mostly because sperm capacitation cannot be accurately imitated in vitro (Hinrichs et al. 2002; Hinrichs 2010). Recently, a novel work has achieved in vitro fertilization with good developmental rates using fresh semen in horses (Felix et al. 2022). Although this work set the basis for developing IVF in horses, it is still evident that further research is needed to apply this technique routinely.

Due to the lack of a feasible IVF procedure in horses, techniques like ICSI have been applied in equine reproduction. ICSI involves injecting a single sperm directly into the cytoplasm of an egg to achieve fertilization. ICSI is particularly useful in cases where there are sperm quality issues or when other assisted reproductive methods have not been successful. This technique was initially developed for its use in humans but it has also been adapted for use in animals, including horses (Briski and Salamone 2022). The technique began to be applied to equines in the early 2000s, where viable blastocysts were produced (Hinrichs et al. 2005; Colleoni et al. 2022). Although relatively new in equine reproduction, ICSI has shown significant potential for addressing fertility challenges. The success rates for ICSI in horses can vary based on several factors, including the quality of the eggs and sperm, the expertise of the veterinary team, and the conditions in the laboratory (Hinrichs 2018; Gonzalez-Castro and Carnevale 2018; Ramírez-Agámez et al. 2023). Generally, success rates for ICSI in horses are lower compared to assisted reproductive technologies in other species, such as humans (Benammar et al. 2021). However, ICSI has proven effective in cases where conventional fertilization methods have failed. Similar to IVF, one of the main advantages of this technique is that oocytes can be obtained in slaughtered animals or by OPU.

OPU is an assisted reproductive technology used in horses to collect eggs from the ovaries of live mares without the need for open surgery. The technique involves the use of transvaginal ultrasound to visualize the ovaries and a needle guided by ultrasound to aspirate eggs from ovarian follicles (Hinrichs 2018). The collected eggs are then processed in the laboratory for evaluation and can be fertilized or cryopreserved. OPU is less invasive than open surgery, which reduces the risk of complications and recovery time and allows for the collection of eggs from live mares, which is particularly useful for preserving the fertility of valuable mares or managing reproductive issues. Additionally, it provides flexibility in the use of eggs, enabling embryo creation at more convenient times and facilitating genetic preservation. However, the procedure can be costly due to the specialized equipment and expertise required, and success rates can vary based on the quality and quantity of the collected eggs. The technique also demands a high level of skill and experience from the veterinary team and can cause some degree of stress or discomfort to the mare (Hinrichs 2018). Despite these challenges, OPU remains a valuable tool in equine reproduction, offering significant benefits for genetic preservation and reproductive management.

Two different samples can be collected when doing OPU: oocytes from dominant follicles and already matured or immature oocytes from secondary follicles. Depending on the sample, the oocyte competence and future developmental rate can be affected.

Oocytes obtained from dominant follicles (24–35 h after gonadotropin stimulation) are usually easier to collect due to the large size of the follicle and the oocytes being already detached from the follicle parenchyma (Carnevale and Ginther 1995; Riera et al. 2016). Once collected, these oocytes can be directly fertilized without the need for an IVM procedure. In equine and several species, using oocytes collected from dominant follicles, and thus skipping the maturation step in vitro, has ended up in more competent oocytes and better development rates (70% of ICSI-derived blastocysts vs 30% when oocytes were in vitro matured) (Foss et al. 2013). However, some risks are associated with working with this sample. On the one hand, hormonal synchronization and routine follicle growth control have to be precise to avoid ovulation and the loss of the oocyte. On the other hand, the quantity of the sample is limited because superovulation is not working in horses, making it possible to aspirate one oocyte per mare (or a maximum of two if two dominant follicles are available).

Aspirating immature oocytes is also an alternative when doing OPU. In this case, mares are usually followed by an ultrasound to visualize follicle growth and ensure an adequate number of follicles, and only follicles between 5 and 8 mm in diameter without the presence of a preovulatory follicle are aspirated (Galli et al. 2014). Interestingly, mares in nonbreeding season can be aspirated, without notorious differences in the number of follicles or blastocysts' developmental rates (Hinrichs and Schmidt 2000; Choi et al. 2016a). The major disadvantage of working with immature oocytes is that they require in vitro maturation to achieve meiotic resumption and become ready for fertilization. This process often results in a ~ 60% rate of mature oocytes that are not as competent as those obtained mature directly from the mare (Hinrichs et al. 2005). Additionally, the aspiration itself is also more difficult because follicles are smaller and oocytes are tightly attached to the follicle parenchyma, being necessary to flush the follicle multiple times and scrape the follicle wall to ensure the recovery of the oocyte (Hawley et al. 1995).

Somatic cell nuclear transfer (SCNT), or cloning, falls within the fourth generation of reproductive biotechnologies (Palma 2001). This technique involves removing the genetic material from an oocyte and then introducing a cell from the individual to be cloned into the perivitelline space of the oocyte. This process, also known as the "genetic photocopier," allows the production of an embryo that is genetically identical to the donor cell from an individual, without the need to mate an egg with a sperm (Ross and Cibelli 2010). Cloning is a widely used technique in horses, particularly in polo, where it has been observed that clones of highly successful mares continue to exhibit the same level of athletic performance (Maserati and Mutto 2016). This approach is also valuable for expanding desirable genetics, especially in this species where superovulation does not produce the desired results. It allows for the preservation of exceptional genetic material from horses with outstanding traits, ensuring that these qualities can be maintained and propagated. It also helps overcome reproductive challenges, enabling the production of offspring from animals with known fertility issues or those no longer able to reproduce naturally. Moreover, cloning facilitates the replication of unique or rare traits that may be difficult to achieve through conventional breeding methods (Maserati and Mutto 2016).

The creation of transgenic organisms, coupled with advancements in stem cell technology, has led to the emergence of the fifth generation of reproductive biotechnologies (Palma 2001). On the one hand, transgenic animals offer numerous advantages across various fields, including medicine, agriculture, and research. In medical research, they enable the creation of models for studying human diseases, which helps in understanding disease mechanisms and developing new treatments. These animals can also be engineered to produce valuable pharmaceuticals, such as hormones and antibodies, in their milk or tissues, making drug production more cost-effective and scalable. In agriculture, transgenic animals can exhibit traits like increased growth rates or disease resistance, which can enhance food security and agricultural efficiency (Lee et al. 2020; Fischer and Schnieke 2023).

On the other hand, stem cells—defined as cells capable of dividing to produce a copy of themselves while also generating a differentiated cell—offer a range of significant advantages across various scientific and medical fields (Sell 2004). They are crucial for regenerative medicine, as they have the potential to repair or replace damaged tissues and organs, which can be transformative for treating injuries and degenerative diseases. Stem cells also play a vital role in developmental biology by providing insights into cell differentiation and organ development, helping researchers understand and potentially correct developmental disorders. In drug testing, stem cells are used to model diseases and test the effects of new medications in a controlled environment, improving the safety and efficacy of treatments before clinical trials. In agriculture, stem cells are being explored for enhancing livestock health and productivity, potentially leading to improved breeds with desirable traits (Navarro et al. 2019). In horses, stem cells were derived from preimplantation blastocysts which have shown their capability to differentiate into primordial germ cells, cells that are the precursor of gametes (Yu et al. 2021). The possibility of producing gametes in vitro will revolutionize traditional methods of breeding (Goszczynski et al. 2019, 2023).

3 Somatic Cell Nuclear Transfer in Horses

The process of cloning consists of three critical stages, followed by a cautious gestation period until the foal reaches full term (Galli et al. 2008; Salamone and Maserati 2023). The first stage includes oocyte maturation, oocytes that are typically obtained from mares without significant genetic value and used to carry out the reprogramming of the cell of the animal to be cloned. During oocyte maturation, a series of nuclear and cytoplasmic rearrangements occur, involving the reduction of the chromosome number, the reorganization of organelles, and the storage of proteins and nucleic acids (Hatırnaz et al. 2018). These preparations are crucial to ensure embryonic development once the oocyte is fertilized. This process can be performed either in vitro or in vivo, with reported variations in maturation efficiency depending on the methodology used. Similar to other species (Lonergan and Fair 2016), in equines, in vivo maturation has resulted in oocytes of better quality and higher developmental competence compared to in vitro matured oocytes (Scott et al. 2001). Many factors can affect oocyte maturation, such as the configuration of the meiotic spindle, the interaction with surrounding cumulus cells, and hormonal environment, among others (Carnevale 2008; Hinrichs 2010; Lewis et al. 2020). Studies have explored methods such as prolonged maturation times; supplementation with antioxidants, follicular fluid, and hormones; and variations in the levels of pH and osmolarity to enhance oocyte competence (Dell'Aquila et al. 1997; Hinrichs et al. 2002; Choi et al. 2004a, 2015a; Hinrichs 2010; Fernández-Hernández et al. 2020). Research indicates that extended maturation periods can improve oocyte quality and subsequent embryo development (Choi et al. 2004a; Choi et al. 2015). Additionally, the inclusion of antioxidants like ascorbic acid has shown promise in reducing oxidative stress during maturation (Gomes et al. 2015). Furthermore, the use of melatonin and insulin growth factor 1 hormone has shown beneficial effects on oocyte maturation (Choi et al. 2003; Clérico et al. 2021). Finally, studies on the use of equine follicular fluid have highlighted its role in supporting oocyte maturation and fertilization outcomes (Dell'Aquila et al. 1997; Hinrichs et al. 2002). Altogether, these approaches underscore ongoing efforts to refine assisted reproduction techniques in horses, aiming to enhance reproductive success through better oocyte maturation protocols.

Once oocyte maturation is completed, next consists of the enucleation of the matured oocyte. The enucleation process consists of removing the genetic material from the mature oocyte, meaning that chromosomes configured in the metaphase II of meiosis II (MII) and the first polar body (PB1) are eliminated. With that purpose, oocytes were incubated in the presence of cytochalasin B and stained with a DNA dye to relax the cytoskeleton and identify the genetic material, respectively. Thereafter, oocytes are briefly illuminated with UV light, and with the help of a micromanipulator and a sharp pipette, genetic material is removed (Galli et al. 2008; Salamone and Maserati 2023). During the extraction of genetic material, it is important to preserve the higher proportion of cytoplasm, since it contains essential organelles and structures necessary for future embryonic development. Precise manipulation and technical skill are required to perform enucleation without damaging the oocyte. Particularly in equines, they possess a highly flexible zona pellucida (ZP), which complicates penetration with a standard pipette (Hinrichs et al. 2000). Additionally, the metaphase plate and the first polar body are typically distant from each other, sometimes necessitating multiple punctures of the ZP to completely remove all DNA (Hinrichs et al. 2000). Furthermore, the polar body is usually firmly attached to the ZP, increasing the risk of oocyte damage during its removal (Hinrichs et al. 2000). While oocytes from other species seem to be more resistant to manipulation and allow multiple punctures, equine oocytes seem to be more fragile, resulting in lower survival rates after enucleation (Hinrichs et al. 2000). Alternatives to this method have been developed, including the use of ZP-free oocytes as discussed later.

After enucleation, the next step consists of the somatic cell nuclear transfer to the perivitelline space of the enucleated oocyte. Importantly, the donor cells have to be synchronized in the G0/G1 phase of the cell cycle (ensuring diploid genomic content is synchronized with the oocyte cell cycle). For cell synchronization, different approaches have been tested although there is not a faithful procedure that ensures a proper future development. Some of them include serum starvation, growth to confluence, or the use of roscovitine to induce the exit to the cell cycle and, thus, the entry into a quiescent G0 stage (Hinrichs et al. 2006). Additionally, different cell types have been tested for nuclear transfer in horses, including adult or fetal fibroblasts, mesenchymal stem cells, induced pluripotent stem cells (iPSCs), and cumulus cells, although most reports use fibroblasts due to their easy obtention and simple culture conditions (Lagutina et al. 2005; Olivera et al. 2016). While there is no robust information on which cell type results in better development efficiency, it seems mandatory to use cells in the lowest passage possible to ensure better nuclear reprogramming (Olivera et al. 2016).

After the cell transfer, it is necessary to achieve membrane fusion between the donor cell and the oolemma. Fusion results in the nuclear membrane of the donor cell rupturing, and with the aid of oocyte cytoplasmic factors, chromosomes quickly adopt a distribution similar to metaphase (Campbell et al. 1996). In equines, several methods are employed to facilitate this process. Electrofusion, commonly used in SCNT procedures, involves applying brief and controlled electric pulses to induce membrane fusion. These pulses transiently alter membrane permeability, allowing the membranes to merge and integrate. This method is favored for its effectiveness in fusing large cells like oocytes with donor cells, ensuring precise integration of genetic material (Galli et al. 2008; Salamone and Maserati 2023). Another alternative is the use of chemical reagents such as polyethylene glycol or fusogenic proteins like HVJ (hemagglutinating virus of Japan) to promote membrane fusion and form a single cellular entity. These are less invasive methods compared to electrofusion but require careful optimization to enhance fusion efficiency and minimize cellular stress. While these fusion techniques have been adapted and optimized in species like cattle and mice (Yang and Shen 2006; Kim et al. 2010; Greggains et al. 2014), their application in horses remains largely unexplored in the scientific literature. This gap highlights the need for further research to assess the feasibility, efficacy, and potential advantages of these methods in cell fusion in equines, particularly in enhancing the success rates of SCNT.

Following successful membrane fusion, the reconstructed oocyte undergoes activation. This technique involves stimulating an oocyte to undergo activation and start embryonic development without fertilization by sperm. This process aims to simulate the natural fertilization process and typically involves chemical or electric treatments for calcium release and oocyte cytoplasmic factor degradations (Nixon et al. 2002). The activation process, also known as parthenogenetic activation, serves as a robust positive control during SCNT procedures (Hisey et al. 2021). Firstly, parthenogenetic activation confirms the developmental competency of the manipulated oocyte and its ability to initiate embryogenesis in the absence of sperm DNA. This ensures that the oocyte itself is capable of supporting embryo development following activation, which is a critical prerequisite in SCNT experiments, mostly in horses where IVF cannot be conducted. Secondly, parthenogenetic activation provides a benchmark for evaluating the effectiveness of the nuclear transfer procedure. If the manipulated oocyte fails to activate parthenogenetically, it suggests potential issues with the manipulation technique, such as damage during enucleation or cell transfer or improper fusion of donor cell and oocyte membranes. Additionally, parthenogenetic activation allows researchers to assess the quality and viability of the reconstructed embryo before proceeding with SCNT. Successful activation indicates that the oocyte cytoplasm and cellular machinery are intact and functional, setting a baseline for comparing developmental outcomes following SCNT. In summary, employing parthenogenetic activation as a positive control in SCNT experiments ensures the reliability of the procedure, verifies oocyte competency, and aids in troubleshooting potential technical challenges. In equines, parthenogenetic activation methods have been explored to assess oocyte competency and facilitate SCNT procedures (Choi et al. 2013). Several techniques are utilized to induce parthenogenetic activation in equine oocytes or reconstructed embryos. Electrical activation involves applying electrical pulses to the oocyte to mimic natural fertilization-induced calcium oscillations, triggering intracellular calcium release and initiating embryonic development. Chemical activation employs agents like ionomycin and 6-dimethylaminopurine (6-DMAP) to induce calcium influx and inhibit the maturation-promoting factor (MPF) activity, respectively, promoting meiosis resumption and nuclear envelope breakdown (Choi et al. 2001; Choi et al. 2004b; Hinrichs et al. 2007). Other drugs such as cytochalasin B disrupt actin filaments, and cycloheximide inhibits protein synthesis, enhancing activation efficiency (Choi et al. 2001, 2004b; Galli et al. 2007). Ethanol is also used to stimulate oocyte activation by altering membrane permeability and calcium release pathways (Li 2000). Additionally, stallion sperm extract has also been shown to be effective in the activation of equine oocytes (Choi et al. 2002; Bedford et al. 2003; Hinrichs et al. 2007). These activation techniques are crucial in equine reproductive research, providing essential controls for evaluating SCNT success and enhancing cloning in horses.

As a result of activation, the activated oocyte proceeds with the integration of donor genetic material. The donor cell's nucleus enters the G1 phase of the cell cycle, followed by the formation of a nuclear membrane around the genetic material. A pseudo-pronucleus expands due to the incorporation of maternal proteins, altering chromatin structure. Subsequently, genetic material replication begins, initiating successive mitotic divisions. These mitotic divisions lead to the formation of blastomeres and embryo development (Prather et al. 2000) (Fig. 2).

The culture media employed to support embryonic development must be appropriate to ensure successful embryonic development. Equine embryos are characterized by high glucose consumption and metabolic activity at late embryonic stages,

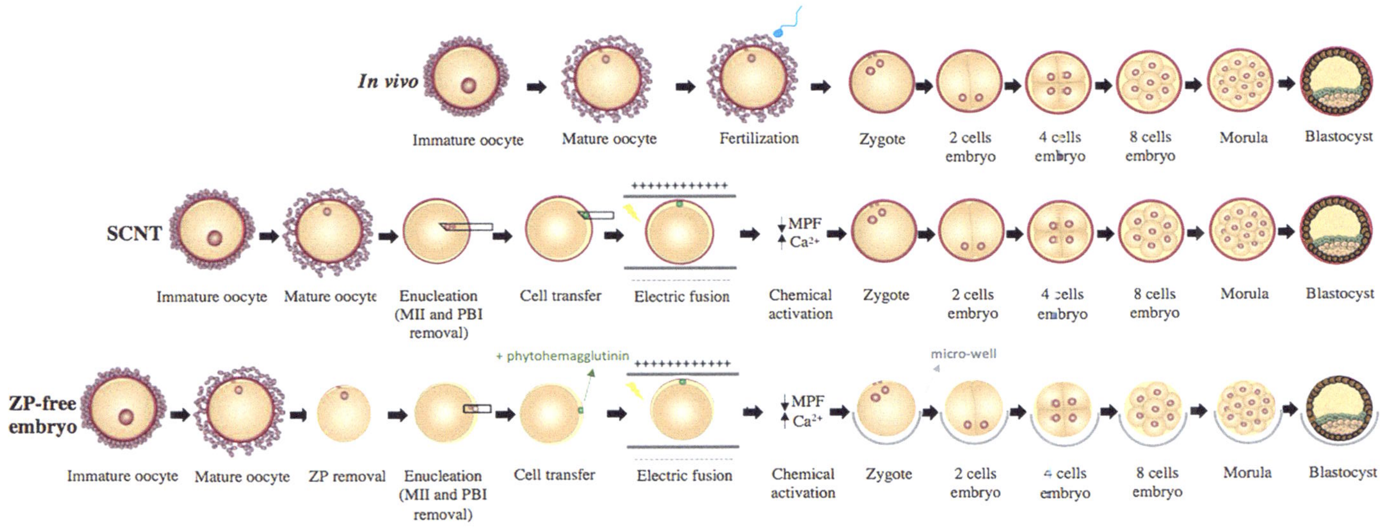

Fig. 2 Comparison between in vivo development and cloning by SCNT or ZP-free embryo reconstruction. A scheme that shows fertilization and embryonic development in vivo, compared to what occurs in cloning through SCNT or ZP-free embryo reconstruction

necessitating adjustments in culture media throughout their development. During early development, typically on days 1–3 post-activation, media are often supplemented with low glucose to meet the energetic demands of rapid cell division and growth. As development progresses, usually from days 4 to 7, media composition may be modified to provide additional nutrients and support as the embryo undergoes differentiation and morphogenesis (Azuma et al. 1995; Tu et al. 2001; Choi et al. 2016b; Hu and Yu 2017). These adjustments in media formulation ensure that equine embryos receive adequate metabolic support tailored to their changing developmental needs, thereby optimizing their viability and potential for successful development. Commonly used media include Tissue Culture Medium 199 (TCM-199) and DMEM-F12, which are a balanced salt solution supplemented with amino acids, vitamins, and glucose, providing a suitable environment for embryo culture (Choi et al. 2015b). Modified synthetic oviduct fluid (mSOF) is also employed to mimic the oviduct environment, containing essential ions, amino acids, and energy substrates necessary for equine embryo development (Lagutina et al. 2005). Additionally, G1 medium and global medium have been adapted from humans to equine embryo culture, providing essential nutrients and growth factors required for cell growth and division (Choi et al. 2015a, b; Leisinger et al. 2016). These media are meticulously formulated to maintain pH balance and osmolarity to provide appropriate levels of nutrients and growth factors critical for embryonic development. Regular monitoring and optimization of culture conditions are essential to maximize embryo viability and developmental potential in equine reproductive research and assisted reproductive technologies.

4 Zona Pellucida-Free Embryo Reconstruction

As mentioned before, the zona pellucida of equine oocytes is highly flexible. Therefore, thicker gauge needles or stronger technologies such as piezo are often used. Because of this, there is a higher likelihood of equine oocyte lysis during micromanipulation (Hinrichs et al. 2000). To overcome this obstacle, techniques have been developed to prevent such occurrences. Based on that, the method of cloning in zona pellucida-free oocytes was developed (Gambini et al. 2012).

ZP-free cloning involves removing the zona pellucida from oocytes using enzymes like pronase. Without the zona pellucida, blunt-ended needles are used for enucleation. A single micromanipulation needle suffices because genetic material is aspirated and then cut by moving the microscope stage. Next, donor cell transfer is performed, facilitated by phytohemagglutinin, a natural substance found in certain plants like beans (*Phaseolus vulgaris*), known for its ability to agglutinate or bind cells together. Phytohemagglutinin is used to aid membrane adhesion during this process (Du et al. 2006). The enucleated oocyte is soaked in a phytohemagglutinin solution before placing it over a cell. After that, electrical fusion and activation follow, as previously described. For embryo culture, embryos produced by this technique require special culture surfaces because they lack the zona pellucida, which normally provides structural support to the embryo. Therefore, they need to be cultured individually in special plates with small wells (well of wells) where the embryo can be supported and structured by the plate's own walls (Gambini et al. 2012; Salamone and Maserati 2023) (Fig. 2).

Thanks to the cell-cytoplasm tight contact, this method improves fusion rates (Lagutina et al. 2005, 2007; Gambini et al. 2012, 2014). Additionally, this method offers the advantage of embryo aggregation, thereby increasing the number of blastomeres and potentially compensating for epigenetic deficiencies. In horses, embryo aggregation improved developmental rates compared to single embryo transfers (Gambini et al. 2012). Studies suggest that aggregating up to four embryos can be effective (Gambini et al. 2014). However, the success of aggregation can depend on factors such as embryo quality and specific procedural conditions (Gambini et al. 2014). While notorious advantages are carried by this technique, ZP-free embryo pregnancy rates are still lower than embryos produced by conventional SCNT (Lagutina et al. 2005; Gambini et al. 2012, 2014; Choi et al. 2015a). Continued research is essential to fully understand the advantages and disadvantages of each technique in terms of reproductive success in the equine species.

5 Reprogramming Deficiencies

Over 80 years ago, Conrad Waddington published his famous description of embryonic development, in which he likened the process of cellular differentiation to a landscape of mountains and valleys. Undifferentiated cells were depicted as occupying the peaks of mountains, descending like a ball during differentiation into valleys where they settled as differentiated cells (Waddington 1957). For a long time, it has been postulated that an embryo in early developmental stages is composed of cells with the capacity to differentiate into all cell types that make up an organism. As development progresses, these cells commit to lineages, gradually losing their potentiality. However, with the first report of SCNT and later with the derivation of iPSCs (Takahashi and Yamanaka 2006), it was demonstrated that a fully differentiated cell could be "re-differentiated" or "reprogrammed" to generate a cell population capable of differentiating into multiple tissues. Thus, it became evident that it was necessary to reformulate Waddington's model, dismissing the notion that cellular differentiation was a unidirectional and irreversible process and instead illustrating it as a bidirectional movement, with the cells ascending and descending through the hill (Fig. 3).

During the reprogramming process, epigenetic marks present in the original cell must be removed, and a new epigenetic configuration compatible with the pluripotent state is established. However, this reprogramming, whether carried out by the oocyte (during SCNT) or exogenous factors (during iPSC derivation), often occurs incompletely, leaving behind a residual "epigenetic memory." This memory includes genes essential for development or pluripotency that fail to be correctly activated, or specific genes of the reprogrammed cell that fail to be silenced, thereby affecting reprogramming efficiency, gene expression, and cellular phenotype (Hochedlinger and Jaenisch 2006; Reik 2007; Sridharan et al. 2013). As a consequence, in most

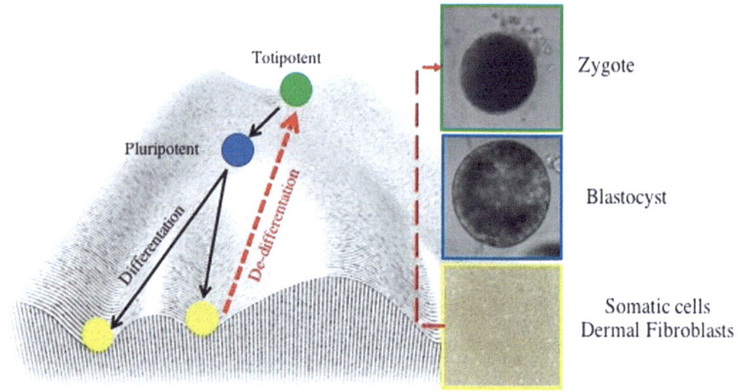

Fig. 3 Reversion of Waddington's model. Cell differentiation is a dynamic and reversible process. According to Waddington's model, cells with the highest level of pluripotency are at the top of the hill (highlighted in green) and exemplified with an equine zygote, followed by pluripotent cells (highlighted in blue) and exemplified with an equine blastocyst were, pluripotent cells form the inner cell mass; and finally, at the base of the hill, fully-differentiated cells are found (highlighted in yellow) and exemplified with equine dermal fibroblasts

mammal species, the efficiency of SCNT ranges between 0 and 10% of transferred embryos resulting in viable offspring (Kues and Niemann 2004; Rodríguez-Osorio et al. 2012), since many defects are observed in cloned embryos which can be manifested at pre- or postimplantation development including during embryonic genome activation, during implantation due to problems in placental formation, and even after birth causing respiratory, cardiac, or metabolic disorders in neonates (Ogura et al. 2013; Loi et al. 2016; Matoba and Zhang 2018). While much remains to be understood about the epigenetic barriers encountered during cellular reprogramming, some of these barriers have been identified and were associated with post-transduction modification of histone marks such as trimethylation on histone 3 lysine 4 (H3K4me3) (Liu et al. 2016; Zhang et al. 2018a; Zhou et al. 2020), trimethylation on histone 3 lysine 9 (H3K9me3) (Matoba et al. 2014; Chung et al. 2015; Liu et al. 2018a; Ruan et al. 2018; Liu et al. 2018b; Zhang et al. 2018b), trimethylation on histone 3 lysine 27 (H3K27me3) (Inoue et al. 2010; Okae et al. 2014; Xie et al. 2016; Matoba et al. 2018; Zhou et al. 2019), as well as alterations in histone variants, DNA methylation, and noncoding RNAs (Long et al. 2014; Matoba and Zhang 2018).

In equines, research into epigenetic barriers has been hindered by several fundamental challenges. Firstly, there is a scarcity of embryos available for study due to their high value and the substantial costs involved in their production. Additionally, the limited availability of quality oocytes further restricts the number of embryos that can be obtained and studied. Consequently, there is a lack of comprehensive studies that have thoroughly examined the epigenetic barriers potentially present in horses. However, some studies have investigated the use of epigenetic modifiers that have worked in other species in equine embryonic development. For instance, the histone deacetylase inhibitor Scriptaid was tested but did not improve the development of SCNT embryos (Choi et al. 2013). Chetomin, an inhibitor of H3K9me3, was also tested in the culture of somatic cells intended for SCNT. Additionally, caffeine was applied during enucleation and until embryo activation, and the histone deacetylase inhibitor trichostatin A (TSA) was used during activation and the initial 6 h of embryo culture. While chetomin treatment resulted in a live birth, none of these treatments (chetomin, caffeine, nor TSA) significantly increased the developmental rate of cloned embryos produced by ZP-free embryo aggregation (Damasceno Teixeira et al. 2019). These findings underscore the need for further research into epigenetic barriers in equine embryogenesis and whether modifying these barriers could enhance the efficiency of cloning techniques in this species.

6 Conclusions

In conclusion, the economic importance of equines underscores a pressing need for continued advancements in reproductive technologies to enhance breeding efficiency and genetic management. The introduction of sophisticated techniques such as cloning by SCNT or ZP-free embryo reconstruction represents a significant leap forward in equine reproductive science. These technologies have shown considerable promise in terms of their potential applications; however, their efficiency can vary based on several factors, including the specific methodologies employed and the inherent biological characteristics of the animals involved.

To fully leverage these advanced reproductive techniques, it is crucial to conduct extensive research aimed at understanding and overcoming the epigenetic barriers that currently limit their effectiveness. Epigenetic modifications, which involve changes in gene expression that do not alter the DNA sequence itself, play a critical role in the development and viability of embryos. By elucidating how these epigenetic factors influence reproductive outcomes, researchers can develop more refined approaches to optimize oocyte usage and improve the overall cost-effectiveness of reproductive technologies.

Improving the efficiency of these techniques involves not only enhancing the success rates of embryo development and implantation but also addressing the economic aspects, such as reducing the costs associated with cloning and other advanced reproductive procedures. Achieving these goals will significantly increase the feasibility and practicality of equine cloning, making it a more accessible option for industries that rely on high-quality equine genetics. This advancement will, in turn, support sustainable breeding practices by ensuring that valuable genetic lineages are preserved and efficiently utilized.

Furthermore, the ongoing investigation into epigenetic modulation during equine embryogenesis holds great promise for advancing reproductive outcomes. By gaining a deeper understanding of how epigenetic changes affect embryo development, researchers can develop strategies to enhance cloning technologies and improve overall reproductive success. This progress is crucial not only for the economic viability of equine cloning but also for contributing to broader goals in sustainable agriculture and the preservation of genetic diversity.

Overall, the future of equine reproductive technologies depends on the continued exploration and refinement of these innovative methods. As research advances and our understanding of epigenetic mechanisms deepens, the potential benefits of these technologies will expand, offering new opportunities for improving reproductive efficiency and achieving greater economic gains in the equine industry. The commitment to advancing these technologies will ensure that they meet the evolving needs of breeders and contribute to the sustainable management of equine genetics.

References

Allen WR, Bowen JM, Frank CJ, Jeffcott LB, Rossdale PD (1976) The current position of A.I. in horse breeding. Equine Vet J 8:72–74. https://doi.org/10.1111/J.2042-3306.1976.TB03295.X

Aurich C, Schneider J (2014) Sex determination in horses—current status and future perspectives. Anim Reprod Sci 146:34–41. https://doi.org/10.1016/J.ANIREPROSCI.2014.01.014

Azuma T, Choi YH, Hochi S, Oguri N (1995) Effect of glucose in the culture medium on development of horse oocytes matured and microfertilized in vitro. Reproduct Fertil Dev 7:1067–1071. https://doi.org/10.1071/RD9951067

Bedford SJ, Kurokawa M, Hinrichs K, Fissore RA (2003) Intracellular calcium oscillations and activation in horse oocytes injected with stallion sperm extracts or spermatozoa. Reproduction 126:489–499. https://doi.org/10.1530/REP.0.1260489

Benammar A, Derisoud E, Vialard F, Palmer E, Ayoubi JM, Poulain M, Chavatte-Palmer P (2021) The mare: a pertinent model for human assisted reproductive technologies? Animals 11(8):2304. https://doi.org/10.3390/ANI11082304

Briski O, Salamone DF (2022) Past, present and future of ICSI in livestock species. Anim Reprod Sci 246:106925. https://doi.org/10.1016/J.ANIREPROSCI.2022.106925

Cabeza JP, Gambini A (2023) Advancements and challenges in in vitro reproductive technologies for the conservation of equine species. Theriogenol Wild 2:100036. https://doi.org/10.1016/J.THERWI.2023.100036

Campbell KHS, Loi P, Otaegui PJ, Wilmut I (1996) Cell cycle co-ordination in embryo cloning by nuclear transfer. Rev Reprod 1:40–46. https://doi.org/10.1530/ror.0.0010040

Carnevale EM (2008) The mare model for follicular maturation and reproductive aging in the woman. Theriogenology 69:23–30. https://doi.org/10.1016/J.THERIOGENOLOGY.2007.09.011

Carnevale EM, Ginther OJ (1995) Defective oocytes as a cause of subfertility in old mares. Biol Reproduct Mono 1:209–214. https://doi.org/10.1093/biolreprod/52.monograph_series1.209

Carnevale EM, Maclellan LJ, Stokes JAE (2019) In vitro culture of embryos from horses. Methods Mol Biol 2006:219–227. https://doi.org/10.1007/978-1-4939-9566-0_16

Choi YH, Love CC, Varner DD, Thompson JA, Hinrichs K (2001) Activation of cumulus-free equine oocytes: effect of maturation medium, calcium ionophore concentration and duration of cycloheximide exposure. Reproduction 122:177–183. https://doi.org/10.1530/REP.0.1220177

Choi YH, Love CC, Chung YG, Varner DD, Westhusin ME, Burghardt RC, Hinrichs K (2002) Production of nuclear transfer horse embryos by Piezo-driven injection of somatic cell nuclei and activation with stallion sperm cytosolic extract. Biol Reprod 67:561–567. https://doi.org/10.1095/BIOLREPROD67.2.561

Choi YH, Chung YG, Walker SC, Westhusin ME, Hinrichs K (2003) In vitro development of equine nuclear transfer embryos: effects of oocyte maturation media and amino acid composition during embryo culture. Zygote 11:77–86. https://doi.org/10.1017/S0967199403001102

Choi YH, Love LB, Varner DD, Hinrichs K (2004a) Factors affecting developmental competence of equine oocytes after intracytoplasmic sperm injection. Reproduction 127:187–194. https://doi.org/10.1530/REP.1.00087

Choi YH, Love LB, Westhusin ME, Hinrichs K (2004b) Activation of equine nuclear transfer oocytes: methods and timing of treatment in relation to nuclear remodeling. Biol Reprod 70:46–53. https://doi.org/10.1095/BIOLREPROD.103.018200

Choi YH, Norris JD, Velez IC, Jacobson CC, Hartman DL, Hinrichs K (2013) A viable foal obtained by equine somatic cell nuclear transfer using oocytes recovered from immature follicles of live mares. Theriogenology 79:791–796.e1. https://doi.org/10.1016/J.THERIOGENOLOGY.2012.12.005

Choi YH, Vélez IC, Macías-García B, Hinrichs K (2015a) Timing factors affecting blastocyst development in equine somatic cell nuclear transfer. Cell Reprogram 17:124–130. https://doi.org/10.1089/CELL.2014.0093

Choi YH, Ross P, Vélez IC, Macías-García B, Riera FL, Hinrichs K (2015b) Cell lineage allocation in equine blastocysts produced in vitro under varying glucose concentrations. Reproduction 150:31–41. https://doi.org/10.1530/REP-14-0662

Choi YH, Vélez IC, Macías-García B, Riera FL, Ballard CS, Hinrichs K (2016a) Effect of clinically-related factors on in vitro blastocyst development after equine ICSI. Theriogenology 85:1289–1296. https://doi.org/10.1016/J.THERIOGENOLOGY.2015.12.015

Choi YH, Gibbons JR, Canesin HS, Hinrichs K (2016b) Effect of medium variations (zinc supplementation during oocyte maturation, perifertilization pH, and embryo culture protein source) on equine embryo development after intracytoplasmic sperm injection. Theriogenology 86:1782–1788. https://doi.org/10.1016/J.THERIOGENOLOGY.2016.05.037

Chung YG, Matoba S, Liu Y, Eum JH, Lu F, Jiang W, Lee JE, Sepilian V, Cha KY, Lee DR, Zhang Y (2015) Histone demethylase expression enhances human somatic cell nuclear transfer efficiency and promotes derivation of pluripotent stem cells. Cell Stem Cell 17:758–766. https://doi.org/10.1016/j.stem.2015.10.001

Clérico G, Taminelli G, Veronesi JC, Polola J, Pagura N, Pinto C, Sansinena M (2021) Mitochondrial function, blastocyst development and live foals born after ICSI of immature vitrified/warmed equine oocytes matured with or without melatonin. Theriogenology 160:40–49. https://doi.org/10.1016/J.THERIOGENOLOGY.2020.10.036

Clulow J, Gibb Z (2022) Liquid storage of stallion spermatozoa—past, present and future. Anim Reprod Sci 247:107088. https://doi.org/10.1016/J.ANIREPROSCI.2022.107088

Colleoni S, Barbacini S, Necchi D, Duchi R, Lazzari G, Galli C (2022) Application of ovum pick-up, intracytoplasmic sperm injection and embryo culture in equine practice. Proc Am Assoc Equine Pract 53:554–559

Cooper MJ (1974) Control of oestrous cycles of heifers with a synthetic prostaglandin analogue. Vet Rec 95:200–203. https://doi.org/10.1136/VR.95.10.200

Damasceno Teixeira TV, Fry RC, McKinnon A, Fry KL, Kelly JM, Verma PJ, Burden C, Salamone DF, Gambini A (2019) Targeting epigenetic nuclear reprogramming in aggregated cloned equine embryos. Reproduct Fertil Dev 31:1885–1893. https://doi.org/10.1071/RD19239

Dell'Aquila ME, Cho YS, Minoia P, Traina V, Lacalandra GM, Maritato F (1997) Effects of follicular fluid supplementation of in-vitro maturation medium on the fertilization and development of equine oocytes after in-vitro fertilization or intracytoplasmic sperm injection. Hum Reprod 12:2766–2772. https://doi.org/10.1093/HUMREP/12.12.2766

Du F, Shen PC, Xu J, Sung LY, Jeong BS, Lucky Nedambale T, Riesen J, Cindy Tian X, Cheng WTK, Lee SN, Yang X (2006) The cell agglutination agent, phytohemagglutinin-L, improves the efficiency of somatic nuclear transfer cloning in cattle (Bos taurus). Theriogenology 65:642–657. https://doi.org/10.1016/J.THERIOGENOLOGY.2005.05.052

Felix MR, Turner RM, Dobbie T, Hinrichs K (2022) Successful in vitro fertilization in the horse: production of blastocysts and birth of foals after prolonged sperm incubation for capacitation. Biol Reprod 107:1551–1564. https://doi.org/10.1093/BIOLRE/IOAC172

Fernández-Hernández P, Sánchez-Calabuig MJ, García-Marín LJ, Bragado MJ, Gutiérrez-Adán A, Millet Ó, Bruzzone C, González-Fernández L, Macías-García B (2020) Study of the metabo-

lomics of equine preovulatory follicular fluid: a way to improve current in vitro maturation media. Animals 10:883. https://doi.org/10.3390/ANI10050883
Fischer K, Schnieke A (2023) How genome editing changed the world of large animal research. Front Genome Ed 5:1272687. https://doi.org/10.3389/FGEED.2023.1272687
Foss R, Ortis H, Hinrichs K (2013) Effect of potential oocyte transport protocols on blastocyst rates after intracytoplasmic sperm injection in the horse. Equine Vet J Suppl. 45:39–43. https://doi.org/10.1111/EVJ.12159
Galli C, Lagutina I, Crotti G, Colleoni S, Turini P, Ponderato N, Duchi R, Lazzari G (2003) A cloned horse born to its dam twin. Nature 424:635. https://doi.org/10.1038/424635a
Galli C, Colleoni S, Duchi R, Lagutina I, Lazzari G (2007) Developmental competence of equine oocytes and embryos obtained by in vitro procedures ranging from in vitro maturation and ICSI to embryo culture, cryopreservation and somatic cell nuclear transfer. Anim Reprod Sci 98:39–55. https://doi.org/10.1016/J.ANIREPROSCI.2006.10.011
Galli C, Lagutina I, Duchi R, Colleoni S, Lazzari G (2008) Somatic cell nuclear transfer in horses. Reprod Domest Anim 43:331–337. https://doi.org/10.1111/J.1439-0531.2008.01181.X
Galli C, Duchi R, Colleoni S, Lagutina I, Lazzari G (2014) Ovum pick up, intracytoplasmic sperm injection and somatic cell nuclear transfer in cattle, buffalo and horses: from the research laboratory to clinical practice. Theriogenology 81:138–151. https://doi.org/10.1016/J.THERIOGENOLOGY.2013.09.008
Gambini A, Jarazo J, Olivera R, Salamone DF (2012) Equine cloning: in vitro and in vivo development of aggregated embryos. Biol Reprod 87:1–9. https://doi.org/10.1095/biolreprod.112.098855
Gambini A, De Stefano A, Bevacqua RJ, Karlanian F, Salamone DF (2014) The aggregation of four reconstructed zygotes is the limit to improve the developmental competence of cloned equine embryos. PLoS One 9(11):e110998. https://doi.org/10.1371/JOURNAL.PONE.0110998
Gomes RG, Lisboa LA, Silva CB, Max MC, Marino PC, Oliveira RL, González SM, Barreiros TRR, Marinho LSR, Seneda MM (2015) Improvement of development of equine preantral follicles after 6 days of in vitro culture with ascorbic acid supplementation. Theriogenology 84:750–755. https://doi.org/10.1016/J.THERIOGENOLOGY.2015.05.006
Gonzalez-Castro RA, Carnevale EM (2018) Association of equine sperm population parameters with outcome of intracytoplasmic sperm injections. Theriogenology 119:114–120. https://doi.org/10.1016/J.THERIOGENOLOGY.2018.06.027
Goszczynski DE, Cheng H, Demyda-Peyrás S, Medrano JF, Wu J, Ross PJ (2019) In vitro breeding: application of embryonic stem cells to animal production†. Biol Reprod 100:885–895. https://doi.org/10.1093/biolre/ioy256
Goszczynski DE, Navarro M, Mutto AA, Ross PJ (2023) Review: embryonic stem cells as tools for in vitro gamete production in livestock. Animal 17(Suppl 1):100828. https://doi.org/10.1016/J.ANIMAL.2023.100828
Greggains GD, Lister LM, Tuppen HAL, Zhang Q, Needham LH, Prathalingam N, Hyslop LA, Craven L, Polanski Z, Murdoch AP, Turnbull DM, Herbert M (2014) Therapeutic potential of somatic cell nuclear transfer for degenerative disease caused by mitochondrial DNA mutations. Sci Rep 41 4:1–10. https://doi.org/10.1038/srep03844
Gurdon JB (1962) The developmental capacity of nuclei taken from intestinal epithelium cells of feeding tadpoles. Development 10:622–640. http://dev.biologists.org/content/develop/10/4/622.full.pd
Hatırnaz Ş, Ata B, Hatırnaz ES, Dahan MH, Tannus S, Tan J, Tan SL (2018) Oocyte in vitro maturation: a systematic review. Turk J Obstetr Gynecol 15:112. https://doi.org/10.4274/TJOD.23911
Hawley LR, Enders AC, Hinrichs K (1995) Comparison of equine and bovine oocyte-cumulus morphology within the ovarian follicle. Biol Reprod 52:243–252. https://doi.org/10.1093/BIOLREPROD/52.MONOGRAPH_SERIES1.24
Hernández-Avilés C, Ramírez-Agámez L, Varner DD, Love CC (2024) Factors affecting the analysis and interpretation of sperm quality in frozen/thawed stallion semen. Theriogenology 218:35–44. https://doi.org/10.1016/J.THERIOGENOLOGY.2024.01.039

Hinrichs K (2010) The equine oocyte: factors affecting meiotic and developmental competence. Mol Reprod Dev 77:651–661. https://doi.org/10.1002/MRD.21186
Hinrichs K (2018) Assisted reproductive techniques in mares. Reprod Domest Anim 53:4–13. https://doi.org/10.1111/RDA.13259
Hinrichs, k. (2010) In vitro production of equine embryos: state of the art. Reprod Domest Anim 45:3–8. https://doi.org/10.1111/J.1439-0531.2010.01624.X
Hinrichs K, Schmidt AL (2000) Meiotic competence in horse oocytes: interactions among chromatin configuration, follicle size, cumulus morphology, and season. Biol Reprod 62:1402–1408. https://doi.org/10.1095/BIOLREPROD62.5.1402
Hinrichs K, Shin T, Love CC, Varner DD, Westhusin ME (2000) Comparison of bovine and equine oocytes as host cytoplasts for equine nuclear transfer. In: Katila T, Wade J (eds) Proceedings of the 5th international symposium on equine embryo transfer, Saari, Finland (abstract), pp 43–44
Hinrichs K, Love CC, Brinsko SP, Choi YH, Varner DD (2002) In vitro fertilization of in vitro-matured equine oocytes: effect of maturation medium, duration of maturation, and sperm calcium ionophore treatment, and comparison with rates of fertilization in vivo after oviductal transfer. Biol Reprod 67:256–262. https://doi.org/10.1095/BIOLREPROD67.1.256
Hinrichs K, Choi YH, Love LB, Varner DD, Love CC, Walckenaer BE (2005) Chromatin configuration within the germinal vesicle of horse oocytes: changes post mortem and relationship to meiotic and developmental competence. Biol Reprod 72:1142–1150. https://doi.org/10.1095/BIOLREPROD.104.036012
Hinrichs K, Choi YH, Love CC, Chung YG, Varner DD (2006) Production of horse foals via direct injection of roscovitine-treated donor cells and activation by injection of sperm extract. Reproduction 131:1063–1072. https://doi.org/10.1530/REP.1.01095
Hinrichs K, Choi YH, Varner DD, Hartman DL (2007) Production of cloned horse foals using roscovitine-treated donor cells and activation with sperm extract and/or ionomycin. Reproduction 134:319–325. https://doi.org/10.1530/REP-07-0069
Hisey EA, Ross PJ, Meyers S (2021) Genetic manipulation of the equine oocyte and embryo. J Equine Vet Sci 99:103394. https://doi.org/10.1016/J.JEVS.2021.103394
Hochedlinger K, Jaenisch R (2006) Nuclear reprogramming and pluripotency. Nature 441:1061–1067. https://doi.org/10.1038/NATURE04955
Hu K, Yu Y (2017) Metabolite availability as a window to view the early embryo microenvironment in vivo. Mol Reprod Dev 84:1027–1038. https://doi.org/10.1002/MRD.22868
Inoue K, Kohda T, Sugimoto M, Sado T, Ogonuki N, Matoba S, Shiura H, Ikeda R, Mochida K, Fujii T, Sawai K, Otte AP, Tian XC, Yang X, Ishino F, Abe K, Ogura A (2010) Impeding Xist expression from the active X chromosome improves mouse somatic cell nuclear transfer. Science 330(6003):496–499. https://doi.org/10.1126/SCIENCE.1194174
Kim S-U, Song B-S, Kim J-S, Kim Y-H, Yoon S-B, Baek S-J, Koo D-B, Chang K-T (2010) Sendai virus-mediated fusion improves the early development of the cloned bovine embryos through precise nuclear remodeling. Biol Reprod 83:244–244. https://doi.org/10.1093/BIOLREPROD/83.S1.244
Kowalczyk A, Czerniawska-Piatkowska E, Kuczaj M (2019) Factors influencing the popularity of artificial insemination of mares in Europe. Animals 9:460. https://doi.org/10.3390/ANI9070460
Kues WA, Niemann H (2004) The contribution of farm animals to human health. Trends Biotechnol 22:286–294. https://doi.org/10.1016/J.TIBTECH.2004.04.003
Lagutina I, Lazzari G, Duchi R, Colleoni S, Ponderato N, Turini P, Crotti G, Galli C (2005) Somatic cell nuclear transfer in horses: effect of oocyte morphology, embryo reconstruction method and donor cell type. Reproduction 130:559–567. https://doi.org/10.1530/REP.1.00772
Lagutina I, Lazzari G, Duchi R, Turini P, Tessaro I, Brunetti D, Colleoni S, Crotti G, Galli C (2007) Comparative aspects of somatic cell nuclear transfer with conventional and zona-free method in cattle, horse, pig and sheep. Theriogenology 67:90–98. https://doi.org/10.1016/J.THERIOGENOLOGY.2006.09.011
Lee K, Uh K, Farrell K (2020) Current progress of genome editing in livestock. Theriogenology 150:229–235. https://doi.org/10.1016/j.theriogenology.2020.01.036

Leisinger CA, Markle ML, Paccamonti DL, Cramer E, Pinto CRF (2016) Production of equine embryos in vitro using conventional intracytoplasmic sperm injection and a complete human embryo culture system. J Equine Vet Sci 41:78. https://doi.org/10.1016/J.JEVS.2016.04.076

Leung PC, Armstrong DT (1980) Interactions of steroids and gonadotropins in the control of steroidogenesis in the ovarian follicle. Annu Rev Physiol 42:71–82. https://doi.org/10.1146/ANNUREV.PH.42.030180.000443

Lewis N, Hinrichs K, Leese HJ, McG Argo C, Brison DR, Sturmey R (2020) Energy metabolism of the equine cumulus oocyte complex during in vitro maturation. Sci Rep 10:3493. https://doi.org/10.1038/S41598-020-60624-Z

Li X (2000) Effects of different activation treatments on fertilization of horse oocytes by intracytoplasmic sperm injection. J Reprod Fertil 119:253–260. https://doi.org/10.1530/REPROD/119.2.253

Liu W, Liu X, Wang C, Gao Y, Gao R, Kou X, Zhao Y, Li J, Wu Y, Xiu W, Wang S, Yin J, Liu W, Cai T, Wang H, Zhang Y, Gao S (2016) Identification of key factors conquering developmental arrest of somatic cell cloned embryos by combining embryo biopsy and single-cell sequencing. Cell Discov 2:1–15. https://doi.org/10.1038/celldisc.2016.10

Liu Z, Cai Y, Wang Y, Nie Y, Zhang C, Xu Y, Zhang X, Lu Y, Wang Z, Poo M, Sun Q (2018a) Cloning of macaque monkeys by somatic cell nuclear transfer. Cell 172:881–887.e7. https://doi.org/10.1016/J.CELL.2018.01.020

Liu X, Wang Y, Gao Y, Su J, Zhang J, Xing X, Zhou C, Yao K, An Q, Zhang Y (2018b) H3K9 demethylase KDM4E is an epigenetic regulator for bovine embryonic development and a defective factor for nuclear reprogramming. Development 145:dev158261. https://doi.org/10.1242/dev.158261

Loi P, Iuso D, Czernik M, Ogura A (2016) A new, dynamic era for somatic cell nuclear transfer? Trends Biotechnol 34:791–797. https://doi.org/10.1016/J.TIBTECH.2016.03.008

Lonergan P, Fair T (2016) Maturation of oocytes in vitro. Ann Rev Anim Biosci 4:255–268. https://doi.org/10.1146/ANNUREV-ANIMAL-022114-110822/CITE/REFWORKS

Long CR, Westhusin ME, Golding MC (2014) Reshaping the transcriptional frontier: epigenetics and somatic cell nuclear transfer. Mol Reprod Dev 81(2):183–193. https://doi.org/10.1002/mrd.22271

Maclellan LJ, Carnevale EM, Coutinho da Silva MA, Scoggin CF, Bruemmer JE, Squires EL (2002) Pregnancies from vitrified equine oocytes collected from super-stimulated and non-stimulated mares. Theriogenology 58:911–919. https://doi.org/10.1016/S0093-691X(02)00920-2

Maserati M, Mutto A (2016) In vitro production of equine embryos and cloning: today's status. J Equine Vet Sci 41:42–50. https://doi.org/10.1016/J.JEVS.2016.04.004

Matoba S, Zhang Y (2018) Somatic cell nuclear transfer reprogramming: mechanisms and applications. Cell Stem Cell 4:471–485. https://doi.org/10.1016/j.stem.2018.06.018

Matoba S, Liu Y, Lu F, Iwabuchi KA, Shen L, Inoue A, Zhang Y (2014) Embryonic development following somatic cell nuclear transfer impeded by persisting histone methylation. Cell 4:884–895. https://doi.org/10.1016/j.cell.2014.09.055

Matoba S, Wang H, Jiang L, Lu F, Iwabuchi KA, Wu X, Inoue K, Yang L, Press W, Lee JT, Ogura A, Shen L, Zhang Y (2018) Loss of H3K27me3 imprinting in somatic cell nuclear transfer embryos disrupts post-implantation development. Cell Stem Cell 23:343–354.e5. https://doi.org/10.1016/j.stem.2018.06.008

Morrell JM, Richter J, Martinsson G, Stuhtmann G, Hoogewijs M, Roels K, Dalin AM (2014) Pregnancy rates after artificial insemination with cooled stallion spermatozoa either with or without single layer centrifugation. Theriogenology 82:1102–1105. https://doi.org/10.1016/J.THERIOGENOLOGY.2014.07.028

Navarro M, Soto DA, Pinzon CA, Wu J, Ross PJ (2019) Livestock pluripotency is finally captured in vitro. Reproduct Fertil Dev 32:11–39. https://doi.org/10.1071/RD19272

Nixon VL, Levasseur M, McDougall A, Jones KT (2002) Ca(2+) oscillations promote APC/C-dependent cyclin B1 degradation during metaphase arrest and completion of meiosis in fertilizing mouse eggs. Curr Biol 12:746–750. https://doi.org/10.1016/S0960-9822(02)00811-4

Ogura A, Inoue K, Wakayama T (2013) Recent advancements in cloning by somatic cell nuclear transfer. Philos Trans R Soc B 36820110329:20110329. https://doi.org/10.1098/RSTB.2011.0329

Okae H, Matoba S, Nagashima T, Mizutani E, Inoue K, Ogonuki N, Chiba H, Funayama R, Tanaka S, Yaegashi N, Nakayama K, Sasaki H, Ogura A, Arima T (2014) RNA sequencing-based identification of aberrant imprinting in cloned mice. Hum Mol Genet 23(4):992–1001. https://doi.org/10.1093/hmg/ddt495

Olivera R, Moro LN, Jordan R, Luzzani C, Miriuka S, Radrizzani M, Xavier Donadeu F, Vichera G (2016) In vitro and in vivo development of horse cloned embryos generated with iPSCs, mesenchymal stromal cells and fetal or adult fibroblasts as nuclear donors. PLoS One 11:e0164049. https://doi.org/10.1371/JOURNAL.PONE.0164049

Palma Gustavo A (2001). Biotecnología de la Reproducción. Editor: Gustavo A Palma ISBN, 9874337796, 9789874337795

Prather RS, Kühholzer B, Lai L, Park KW (2000) Changes in the structure of nuclei after transfer to oocytes. Cloning 2:117–122. https://doi.org/10.1089/152045500750039815

Ramírez-Agámez L, Hernández-Avilés C, Varner DD, Love CC (2023) Sperm factors associated with the production of equine blastocysts by intracytoplasmic sperm injection (ICSI) using frozen/thawed semen. Theriogenology 195:85–92. https://doi.org/10.1016/J.THERIOGENOLOGY.2022.10.014

Reik W (2007) Stability and flexibility of epigenetic gene regulation in mammalian development. Nature 4477143(447):425–432. https://doi.org/10.1038/nature05918

Riera FL, Roldán JE, Gomez J, Hinrichs K (2016) Factors affecting the efficiency of foal production in a commercial oocyte transfer program. Theriogenology 85:1053–1062. https://doi.org/10.1016/J.THERIOGENOLOGY.2015.11.016

Rodríguez-Osorio N, Urrego R, Cibelli JB, Eilertsen K, Memili E (2012) Reprogramming mammalian somatic cells. Theriogenology 9:1869–1886. https://doi.org/10.1016/j.theriogenology.2012.05.030

Ross PJ, Cibelli JB (2010) Bovine somatic cell nuclear transfer. Methods Mol Biol 636:155–177. https://doi.org/10.1007/978-1-60761-691-7_10

Ruan D, Peng J, Wang X, Ouyang Z, Zou Q, Yang Y, Chen F, Ge W, Wu H, Liu Z, Zhao Y, Zhao B, Zhang Q, Lai C, Fan N, Zhou Z, Liu Q, Li N, Jin Q, Shi H, Xie J, Song H, Yang X, Chen J, Wang K, Li X, Lai L (2018) XIST derepression in active X chromosome hinders pig somatic cell nuclear transfer. Stem Cell Rep 10:494. https://doi.org/10.1016/J.STEMCR.2017.12.015

Salamone D, Maserati M (2023) Horse somatic cell nuclear transfer using zona pellucida-enclosed and zona-free oocytes. Methods Mol Biol 2647:269–281. https://doi.org/10.1007/978-1-0716-3064-8_15

Scott TJ, Carnevale EM, Maclellan LJ, Scoggin CF, Squires EL (2001) Embryo development rates after transfer of oocytes matured in vivo, in vitro, or within oviducts of mares. Theriogenology 55:705–715. https://doi.org/10.1016/S0093-691X(01)00438-1

Sell S (2004) Stem cell origin of cancer and differentiation therapy. Crit Rev Oncol Hematol 51(1):1–28. https://doi.org/10.1016/j.critrevonc.2004.04.007

Spemann H (1928) Die entwicklung seitlicher und dorso-ventraler Keimhälften bein verzögerter keimversorgung. Zeitschr Zoo, pp 105–134

Squires EL, McCue PM, Vanderwall D (1999) The current status of equine embryo transfer. Theriogenology 51:91–104. https://doi.org/10.1016/S0093-691X(98)00234-9

Sreenan J (1970) In vitro maturation and attempted fertilization of cattle follicular oocytes. J Agric Sci 75:393–396. https://doi.org/10.1017/S0021859600025016

Sridharan R, Gonzales-Cope M, Chronis C, Bonora G, McKee R, Huang C, Patel S, Lopez D, Mishra N, Pellegrini M, Carey M, Garcia BA, Plath K (2013) Proteomic and genomic approaches reveal critical functions of H3K9 methylation and heterochromatin protein-1γ in reprogramming to pluripotency. Nat Cell Biol 157(15):872–882. https://doi.org/10.1038/ncb2768

Takahashi K, Yamanaka S (2006) Induction of pluripotent stem cells from mouse embryonic and adult fibroblast cultures by defined factors. Cell 126:663–676. https://doi.org/10.1016/j.cell.2006.07.024

Tu ZJ, Pan W, Gong Z, Kiang DT (2001) Assessment of metabolism of equine morulae and blastocysts. Mol Reprod Dev 59:33–37. https://doi.org/10.1002/MRD.1004

Waddington CH (1957) The strategy of the genes. A discussion of some aspects of theoretical biology. George Allen & Unwin, London

Woods GL, White KL, Vanderwall DK, Li GP, Aston KI, Bunch TD, Meerdo LN, Pate BJ (2003) A mule cloned from fetal cells by nuclear transfer. Science 80:301–1063. https://doi.org/10.1126/SCIENCE.1086743/SUPPL_FILE/WOODS.SOM.REV.PDF

Xie B, Zhang H, Wei R, Li Q, Weng X, Kong Q, Liu Z (2016) Histone H3 lysine 27 trimethylation acts as an epigenetic barrier in porcine nuclear reprogramming. Reproduction 151:9–16. https://doi.org/10.1530/REP-15-0338

Xu KP, Greve T, Callesen H, Hyttel P (1987) Pregnancy resulting from cattle oocytes matured and fertilized in vitro. J Reprod Fertil 81:501–504. https://doi.org/10.1530/JRF.0.0810501

Yang J, Shen MH (2006) Polyethylene glycol-mediated cell fusion. Methods Mol Biol 325:59–66. https://doi.org/10.1385/1-59745-005-7:59

Yates DJ, Whitacre MD (1988) Equine artificial insemination. Vet Clin North Am Equine Pract 4:291–304. https://doi.org/10.1016/S0749-0739(17)30642-9

Yu L, Wei Y, Sun HX, Mahdi AK, Pinzon Arteaga CA, Sakurai M, Schmitz DA, Zheng C, Ballard ED, Li J, Tanaka N, Kohara A, Okamura D, Mutto AA, Gu Y, Ross PJ, Wu J (2021) Derivation of intermediate pluripotent stem cells amenable to primordial germ cell specification. Cell Stem Cell 28:550–567.e12. https://doi.org/10.1016/j.stem.2020.11.003

Zhang Z, Zhai Y, Ma X, Zhang S, An X, Yu H, Li Z (2018a) Down-regulation of H3K4me3 by MM-102 facilitates epigenetic reprogramming of porcine somatic cell nuclear transfer embryos. Cell Physiol Biochem 45:1529–1540. https://doi.org/10.1159/000487579

Zhang Y, Wang Q, Liu K, Gao E, Guan H, Hou J (2018b) Treatment of donor cells with recombinant KDM4D protein improves preimplantation development of cloned ovine embryos. Cytotechnology 70(5):1469–1477. https://doi.org/10.1007/S10616-018-0224-6

Zhou C, Wang Y, Zhang J, Su J, An Q, Liu X, Zhang M, Wang Y, Liu J, Zhang Y (2019) H3K27me3 is an epigenetic barrier while KDM6A overexpression improves nuclear reprogramming efficiency. FASEB J 33:4638–4652. https://doi.org/10.1096/fj.201801887R

Zhou C, Zhang J, Zhang M, Wang D, Ma Y, Wang Y, Wang Y, Huang Y, Zhang Y (2020) Transcriptional memory inherited from donor cells is a developmental defect of bovine cloned embryos. FASEB J 34:1637–1651. https://doi.org/10.1096/FJ.201900578RR

Stored and Frozen Semen in Swine

Felipe Pezo, Raúl Sanchez, Ana Carolina Pedrosa, and Andre Furugen Cesar de Andrade

Abstract

Artificial insemination (AI) in swine was initially performed during the 1920s by Ilya Ivanov. However, its widespread use only became popular in the 1940s and 1960s, with advancements in semen refrigeration techniques. Nowadays, AI has established itself as one of the most essential techniques for swine reproduction, enabling the broader dissemination of superior genetic material and contributing to improved productivity in pig farming. In this context, the refrigeration of boar semen has become a standard practice, given that spermatozoa cryopreservation, widely used in other species, is not yet commercially viable in the swine industry due to the high sensitivity of boar spermatozoa to freezing. Both techniques include a series of steps that subject the spermatozoa to thermal, osmotic, and oxidative shock that will cause a reduction in the quality of the preserved seminal

F. Pezo (✉)
Faculty of Agricultural and Environmental Sciences, University of La Frontera, Temuco, Chile

Faculty of Medicine, Center of Excellence in Reproductive Biotechnology (CEBIOR-BIOREN), University of La Frontera, Temuco, Chile
e-mail: felipe.pezo@ufrontera.cl

R. Sanchez
Department of Preclinical Sciences, Faculty of Medicine, University of La Frontera, Temuco, Chile

Faculty of Medicine, Center of Excellence in Translational Medicine (CEMT—BIOREN), University of La Frontera, Temuco, Chile
e-mail: raul.sanchez@ufrontera.cl

A. C. Pedrosa · A. F. C. de Andrade
Department of Animal Reproduction, School of Veterinary Medicine and Animal Science, University of São Paulo, São Paulo, Brazil
e-mail: pedrosa_carolina@usp.br; andrefc@usp.br

J. C. Gardón, K. Satué Ambrojo (eds.), *Assisted Reproductive Technologies in Animals Volume 2*, https://doi.org/10.1007/978-3-031-87198-6_3

sample that will be reflected in reproductive indices after AI with semen preserved in liquid and that will be accentuated with the use of frozen/thawed spermatozoa.

The objective of this book chapter is to describe the protocols for refrigeration and freezing/thawing boar spermatozoa, the effects of in vitro manipulation and the preservation techniques of a seminal sample, the strategies used to optimize the protocol considering those variables that could have a greater impact on the conservation of cellular functionality, and finally the future perspectives of these assisted reproduction biotechnologies.

Keywords

Spermatozoa preservation · Refrigeration · Cryopreservation · Boar spermatozoa · Oxidative stress

1 Stored Boar Semen in Liquid State

The semen refrigeration involves collecting, evaluating, diluting, and storing at controlled temperatures between 15 °C and 18 °C. This temperature range was chosen based on studies demonstrating the ability to maintain the structural and functional integrity of spermatozoa for up to a week, depending on the extender used and storage conditions. Refrigeration allows semen to be safely transported and stored until the time of insemination, which is crucial for logistics in large-scale pig farms and semen exportation (Namuncura et al. 2020; Wiebke et al. 2022).

The effectiveness of refrigerated semen storage is directly related to the initial quality of the semen, the type of extender used, and the temperature conditions during storage. In this context, various commercial extenders have proven effective in preserving spermatozoa motility and protecting against oxidative and microbial damage. These extenders contain components that stabilize the spermatozoa membrane and inhibit premature spermatozoa capacitation, ensuring that they remain viable and capable of fertilization throughout the storage period (Namuncura et al. 2020; Paschoal et al. 2021; Waberski et al. 2011; Wiebke et al. 2022).

Despite its advantages, the refrigeration of boar semen is not without challenges; spermatozoa motility tends to decrease over the storage period, with studies showing a significant reduction after 5 to 7 days. Additionally, refrigeration can induce changes in spermatozoa physiology, such as altered membrane permeability, which may compromise fertilization capacity. Therefore, strict control of storage conditions is essential to minimize these adverse effects and maximize fertility rates (Mellagi et al. 2023; Waberski et al. 2011; Wiebke et al. 2022).

The longevity and effectiveness of refrigerated semen depend on the cooling rates applied and temperature stability during storage. Inadequate cooling rates can damage the spermatozoa plasma membrane, negatively affecting their viability. Therefore, standardizing refrigeration protocols, including using appropriate extenders and controlling storage conditions, is crucial for ensuring the success of the process (Hensel et al. 2020; Wiebke et al. 2022).

Overall, advancements in boar semen refrigeration, driven by new extender formulations and technological improvements, have enabled greater efficiency in utilizing high-quality genetic material. However, significant challenges remain, such as the variability in individual spermatozoa response to refrigerated storage and the need for customized strategies to maximize the reproductive potential of each ejaculate. These aspects remain active research areas, primarily aiming to improve AI practices and increase productivity in global pig farming (Hensel et al. 2020; Mellagi et al. 2023; Wiebke et al. 2022).

2 Semen Collection and Initial Processing

2.1 Methods of Boar Semen Collection

The collection of boar semen is an essential step to ensure an excellent seminal sample. Boar semen collection involves a series of carefully executed procedures to ensure the quality of the ejaculate and to avoid bacterial contamination, which could compromise semen viability and fertility. The first essential step is to ensure that all personnel involved in the process follow strict hygiene practices. Personnel involved must wear protective gloves throughout semen handling. Personal hygiene measures such as proper hand washing and using caps and masks, which help prevent contamination from respiratory secretions or hair, are also necessary. Before beginning the collection process, the boar should be prepared appropriately. The ventral abdomen must be clean and dry, and, if necessary, the hair around the preputial orifice should be trimmed to avoid the accumulation of organic material, which could contaminate the ejaculate. If the area around the prepuce is moist or contains visible organic material, it should be cleaned with a disposable wipe to ensure optimal hygiene (Althouse 2008).

The most widely used method for semen collection is the "gloved-hand" method. This method involves using a glove to hold the boar's glans penis, applying light pressure during ejaculation, allowing the semen to be collected directly into a clean and sterilized container. The boar ejaculation generally occurs in three fractions: (1) The pre-sperm fraction, the first to be expelled, mainly contains urine and epithelial cells and is discarded to avoid contamination. (2) The rich fraction, which is the second and most valuable, contains the highest concentration of spermatozoa and is the one that truly matters for producing AI doses. (3) The post-sperm fraction contains fewer sperm cells and a large amount of fluid derived from seminal vesicles, prostate, and bulbourethral glands (Rodríguez-Martínez et al. 2005). Bulbourethral gland secretion produces tapioca-like flocculant; this secretion is separated using a filter and discarded before semen processing (Paschoal et al. 2021; Pedrosa et al. 2024).

Recently, studies (Paschoal et al. 2021) have explored methods for penis fixation during semiautomatic semen collection. These methods aim to reduce bacterial contamination and improve the quality of the collected semen, decreasing the need for subsequent antibiotic treatment and preserving spermatozoa integrity. The technique suggested by these authors has shown effectiveness in reducing bacterial load, which can be essential for maintaining semen quality during storage and increasing fertilization rates.

The collection environment is critical to ensure semen quality. Animal stress can negatively affect the ejaculate quality, decreasing spermatozoa motility and viability. Therefore, it is essential that the environment is calm and that the animal is accustomed to the collection process. Dummies or estrus females can simulate natural mating behavior, facilitating the collection and increasing process efficiency (Pedrosa et al. 2024).

The preparation of the animal before collection is also a determining factor in semen quality. Depleting the preputial diverticulum before collection helps reduce the risk of bacterial contamination, which could compromise semen viability during storage. Additionally, sterilized or preferably disposable materials throughout the collection process are essential to ensure that the ejaculate remains free from contamination (López Rodríguez et al. 2017; Paschoal et al. 2021; Pedrosa et al. 2024).

Semen collection requires training and experience on the part of the technicians. The collector's skill can directly influence the quantity and quality of the collected semen. Therefore, professionals involved in the process must be well-trained and familiar with the animal's behavior and proper handling techniques to minimize stress and maximize the production of high-quality semen (López Rodríguez et al. 2017; Pedrosa et al. 2024).

Additionally, the frequency of collection can affect semen quality. Frequent collections can decrease spermatozoa concentration and motility, while excessively long intervals between collections may increase dead or abnormal spermatozoa. Studies suggest that 3 to 4 days between collections is ideal for maintaining optimal spermatozoa quality (Namuncura et al. 2020; Pezo et al. 2019; Roca et al. 2011). After a successful semen collection, spermatozoa evaluation can be performed using conventional or advanced analyses. One of the parameters usually analyzed is spermatozoa motility, which can help ensure that only high-quality samples are processed for AI spermatozoa; motility analysis is especially important in commercial breeding programs, where semen quality is critical for reproductive success (Bustamante-Filho and Torres 2024; Namuncura et al. 2020; Pedrosa et al. 2024).

2.2 Handling and Initial Processing of Semen After Collection

After collection, boar semen is processed, aiming to preserve spermatozoa viability and quality until it is used in artificial insemination. Proper handling of the semen starts during collection, with the removal of the gel fraction from the ejaculate to avoid issues during insemination, such as catheter blockage and uneven distribution of spermatozoa in the female reproductive tract (Pedrosa et al. 2024).

Once the gel is removed, the semen is diluted in an appropriate extender. The temperature of the extender and the semen should be close at the time the mixing occurs. In practice, it is suggested to remove the extender from the water bath (37–38 °C) as soon as the semen is collected and leave it at room temperature (approximately 24 °C), as well as the raw semen (personal communication). The correct choice of extender is essential, as it must provide the necessary nutrients, maintain the proper pH, and protect the spermatozoa from oxidative stress and microbial contamination. The most used extenders include BTS (Beltsville Thawing Solution), formulated to prolong spermatozoa viability during storage at controlled temperatures (Namuncura et al. 2020). The initial dilution must be performed carefully to avoid osmotic shock, which can damage the spermatozoa plasma membrane and compromise its viability. Therefore, it is expected that both (semen and extender) are at approximately 30–32 °C at the time they are mixed. It is recommended that the semen be added to the extender and not vice versa, as this facilitates sample homogenization (Schulze et al. 2013, 2017).

Antioxidants such as glutathione, vitamin E, and ascorbic acid, discussed by Jofré et al. (2019), can help protect spermatozoa from oxidative damage during storage. However, it is important to carefully control the balance between antioxidant protection and the maintenance of spermatozoa viability. When used excessively, antioxidants such as resveratrol can adversely affect fertility, as Torres et al. (2021) demonstrated. Therefore, antioxidant additives must be carefully managed (Paschoal et al. 2021).

The dilution process is generally performed at a ratio of 1:1 to 1:3 (semen), depending on the initial spermatozoa concentration and the processing objectives. The semen is diluted to a concentration of 20 to 30 × 10^6 spermatozoa per milliliter. Gentle handling is sufficient to ensure the homogeneity of the sample and the proper distribution of spermatozoa in the extender (Pavaneli et al. 2019). However, as mentioned previously, the reverse method (adding semen to the extender) results in a more uniform distribution of spermatozoa in the semen mixing cylinder (when doses are produced in an automated manner), with more minor differences in spermatozoa concentration between the top and bottom of the cylinder. This indicates a more homogeneous distribution, which can improve the quality of semen doses (Schulze et al. 2017). After dilution, the semen is packaged in containers that can be bottles or blisters, with volumes varying between 100 and 40 mL depending on whether they are doses for intracervical or intrauterine artificial insemination, respectively. As soon as semen is packaged, it is gradually cooled to the storage

Fig. 1 Semen conservation unit with automatic trays for homogenizing doses (BT 1500/140 Biotecno®, RS, Brazil). The temperature varies between 16 and 17 °C

temperature, usually between 15 and 18 °C. This cooling process should occur slowly over several hours to prevent thermal shock, which can reduce motility and cause spermatozoa death. Cooling should be done without excessive agitation to ensure uniform temperature distribution in the sample (Pavaneli et al. 2019). First, the semen is diluted and kept protected from light in an environment with temperatures close to 24–25 °C for up to 90 min. After this period, extended semen is then transferred to a semen conservation unit that will have temperatures between 15 and 18 °C (Fig. 1).

Another important aspect concerning boar semen storage is centrifugation, which can concentrate spermatozoa or remove seminal plasma (SP), depending on the processing protocol. Studies such as those by Pavaneli et al. (2019, 2020) have shown that removing seminal plasma can improve the fertilizing capacity of spermatozoa when semen is stored at 17 °C for up to 72 h. However, centrifugation must be performed carefully, using appropriate speeds to minimize mechanical stress on the spermatozoa (Paschoal et al. 2021). Centrifugation with subsequent removal of seminal plasma, despite its beneficial effects, is not widely practiced, as it adds another step to the preparation of doses of stored liquid semen.

Quality control during the initial handling of semen is crucial for the success of the process. Samples are frequently tested for motility, concentration, and membrane integrity to ensure that only high-quality semen is used for artificial insemination. Computer-assisted analysis tools, such as CASA (Computer-Assisted Sperm Analysis) systems, provide fast and accurate assessments of semen quality, enabling informed decisions about processing (Bustamante-Filho and Torres 2024).

After spermatozoa analysis, the processed semen is divided into insemination doses, stored in specific containers or tubes, and maintained under controlled conditions until insemination. Proper storage includes maintaining the temperature and protecting against light and oxygen to preserve semen quality until use (Pedrosa et al. 2024; Thema et al. 2023).

2.3 Evaluation of Fresh Semen Quality

The evaluation of semen quality is a critical step in the artificial insemination process, as it determines the viability of semen for reproductive use. Semen quality is assessed through various parameters, including motility, morphology, and spermatozoa concentration, which together provide a comprehensive view of the fertilizing capacity of the ejaculate (Bustamante-Filho and Torres 2024; Namuncura et al. 2020).

Sperm motility is one of the primary indicators of semen quality. Spermatozoa with good motility can swim through the female reproductive tract and fertilize the oocyte. Motility is typically evaluated using computerized systems, such as CASA (Computer-Assisted Sperm Analysis), which allow a detailed analysis of spermatozoa movement, including total motility, progressive motility, speed, and other kinetic parameters (Bustamante-Filho and Torres 2024; Paschoal et al. 2021).

In addition to motility, spermatozoa morphology is another essential parameter in semen quality assessment. Sperm with morphological defects, such as head, midpiece, or tail abnormalities, may have compromised fertilization ability. Morphological analysis is performed using or without staining and observation under a microscope. The sample should be stained when phase-contrast or differential interference contrast (DIC) microscopes are unavailable. The phase-contrast or differential interference contrast (DIC) microscopes allow the evaluation of morphology using the wet chamber technique, in which spermatozoa are only fixed in a solution with formol-saline (2 to 5%). Typically, an anomaly percentage below 20% is considered acceptable for most AI programs (Jofré et al. 2019; Passarelli et al. 2024).

Sperm concentration, measured in millions of sperm per milliliter, is another critical factor determining semen efficiency. Adequate concentration is necessary to ensure that enough viable sperm are present in each insemination dose. Techniques such as spectrophotometry or direct counting under a microscope are used to determine sperm concentration, and adjustments are made in the extender volume to achieve the desired concentration for AI (Passarelli et al. 2024; Pavaneli et al. 2019; Quirino et al. 2023).

In addition to these basic parameters, advanced evaluations can be performed for a more in-depth semen quality analysis. Plasma membrane integrity tests, using dyes such as eosin-nigrosin, are frequently performed to differentiate live from dead sperm. Osmotic resistance tests and mitochondrial activity assessments can also evaluate the sperm's ability to withstand osmotic stress and energy efficiency, respectively (Bustamante-Filho and Torres 2024; Namuncura et al. 2020).

Sperm DNA integrity is another parameter that can be assessed, particularly in high-performance breeding programs. Damage to sperm DNA can lead to fertilization failures or issues in embryonic development. Tests such as the comet assay or the TUNEL assay evaluate sperm DNA fragmentation and ensure that only semen with intact DNA is used for AI (Bustamante-Filho and Torres 2024; Paschoal et al. 2021).

Semen quality assessment is not limited to the laboratory. The reproductive performance of the semen, measured by fertilization rates and embryonic development after insemination, is the ultimate test of its quality. Therefore, correlating laboratory parameters with reproductive outcomes is essential to validate semen evaluation techniques and ensure AI success (Pavaneli et al. 2020).

Moreover, the storage and transport conditions of semen can significantly impact its quality. Strict temperature control during transport and protection from light and oxidation are critical factors in preserving sperm viability from processing to insemination (Namuncura et al. 2020).

The success of artificial insemination largely depends on the quality of the semen used. Therefore, continuous and accurate evaluation of all semen quality parameters is essential to maximize fertility rates and reproductive success in pig farming.

3 Composition and Function of Semen Extenders

3.1 Types of Extenders Used for Boar Semen

Semen extenders are fundamental to preserving sperm viability and functionality during storage and transport. They are classified according to their capacity to maintain semen quality into short-, medium-, and long-term extenders (Table 1). Short-term extenders maintain semen quality for up to 3 days and medium-term for 4 to

Table 1 Effect of the addition of antioxidants to the freezing extender on in vitro functionality parameters

Company	Diluent name	Usage time
Magapor	Duragen	Short term (3 to 5 days)
	Duragen+	Long term (up to 10 days)
	Vital+	Long term (up to 10 days)
Minitube	Androhep	Short term (3 to 5 days)
	Androstar Plus	Long term (up to 7 days)
	Androstar Premium	Long term (up to 10 days)
IMV	BTS	Short term (3 to 5 days)
	Modena	Long term (up to 7 days)
	GTL	Long term (up to 10 days)
Kubus	Kubus Semen Extender KMT	Short term (3 to 5 days)
	Kubus Semen Extender KK	Long term (up to 10 days)

T-AOC total antioxidant capacity, *SOD* superoxide dismutase, *GSH* glutathione peroxidase, *CAT* catalase, *LPO* lipoperoxidation

5 days, and long-term extenders preserve semen for up to 12 days. These extenders contain a combination of components such as glucose, fructose, osmotic stabilizers (e.g., NaCl, KCl), and pH regulators such as bicarbonate and Tris, which ensure medium stability during storage (Pezo et al. 2019; Roca et al. 2011).

In recent years, new extenders have been developed to allow semen refrigeration at lower temperatures, such as 5 °C, which reduces bacterial proliferation and minimizes the need for antibiotics (Waberski and Luther 2024). These extenders include amino acids and antioxidants that help stabilize the plasma membrane of sperm cells, preventing damage caused by oxidative stress and cold shock during storage. This approach has gained attention due to growing concerns about antimicrobial resistance and pursuing more sustainable practices (Paschoal et al. 2021; Waberski and Luther 2024).

Medium- and long-term extenders often have compositions that are not disclosed for commercial reasons. However, it is known that these extenders contain energy sources to support the cellular metabolism of sperm cells, agents that protect against cold shock, such as bovine serum albumin, as well as pH and osmotic pressure stabilizers like Hepes and Tris. The inclusion of antibiotics in many of these extenders is standard, but there is a growing trend to replace antibiotics with safer alternatives, such as using lower temperatures during storage (Pezo et al. 2019; Waberski and Luther 2024).

3.2 The Role of Extenders in Maintaining Semen Viability During Refrigeration

Refrigerating semen at temperatures between 15 °C and 18 °C is widely used to slow down sperm metabolism and extend the spermatozoa's lifespan. However, this practice is only effective when extenders provide conditions for preserving cellular functions. In addition to providing nutrients, such as glucose and fructose, which ensure sperm motility during storage, extenders contain membrane stabilizers and pH regulators that protect sperm cells against osmotic and thermal variations occurring during the refrigeration process (Johnson et al. 2000; Roca et al. 2011).

Another important factor is controlling the medium's osmolarity. Boar spermatozoa tolerate a wide range of osmolality, varying between 240 and 380 mOsM (Johnson et al. 2000). Fluctuations in osmolarity can cause damage to the plasma membrane, resulting in the loss of sperm viability. Extenders containing bicarbonate and buffers like Tris and Hepes help stabilize pH and osmolarity, preventing membrane rupture and ensuring spermatozoa remain viable throughout storage. Recent studies indicate that using extenders with these characteristics, combined with controlled cooling protocols, can improve the success rate of boar semen preservation over long periods (Paschoal et al. 2021).

3.3 Additives and Their Effects on Semen Quality

The additives present in extenders have a significant impact on semen quality preservation during storage. Antioxidants such as glutathione, vitamin E, and ascorbic acid are frequently included in extenders to protect sperm cells from oxidative stress. Oxidative stress, caused by free radicals, can compromise the integrity of sperm membrane lipids, reducing motility and, consequently, fertilization capacity. The inclusion of antioxidants neutralizes these free radicals, maintaining cellular integrity during storage (Jofré et al. 2019; Pezo et al. 2019; Torres et al. 2021).

The use of antibiotics in extenders is another common practice to prevent bacterial growth during semen storage. Bacteria can compromise semen quality by altering the medium's pH and producing toxins that negatively affect sperm cells. However, with increasing concerns about antimicrobial resistance, alternatives to the use of antibiotics are being explored. One of these alternatives is storage at 5 °C, which inhibits bacterial growth without antibiotics. Studies show that combining extenders formulated with natural antimicrobial additives and refrigeration at 5 °C has proven effective in maintaining sperm quality for extended periods (Paschoal et al. 2021; Waberski and Luther 2024).

3.4 Seminal Plasma: Retain or Remove?

The use of seminal plasma (SP) in stored boar semen is widely debated due to its ambiguous effects on sperm functionality during storage (Leal et al. 2018; Pavaneli et al. 2019). SP contains beneficial factors, such as proteins and antioxidants, which can protect spermatozoa from oxidative stress and detrimental changes to the cell membrane induced by the cooling process. However, prolonged exposure to SP during storage can decrease sperm motility and viability, especially after 72 h of storage (Andrade et al. 2022; Höfner et al. 2020a, b; Luther and Waberski 2019).

Studies show that, although SP contains essential components for maintaining sperm viability, such as antioxidants that minimize oxidative stress and lipid peroxidation, its prolonged presence can compromise sperm membrane integrity and induce premature capacitation. This early capacitation process negatively affects the fertilization ability of spermatozoa (Andrade et al. 2022; Höfner et al. 2020a, b; Luther and Waberski 2019). Additionally, extracellular vesicles present in SP have been identified as contributors to these effects, particularly during prolonged storage (De Lazari et al. 2020; Höfner et al. 2020a, b; Xu et al. 2024).

Variations in SP composition between different ejaculates also affect sperm's resistance to stored semen. Proteomic studies have identified specific proteins in SP, such as PSP-I, that are associated with greater sperm resistance to storage at 17 °C. These proteins help stabilize the sperm membrane and inhibit premature capacitation, contributing to the preservation of sperm quality (De Lazari et al. 2020; Höfner et al. 2020a, b; Luther and Waberski 2019).

However, partial or total removal of SP may be beneficial in some cases, mainly to avoid the harmful effects of prolonged exposure. Conversely, the controlled addition of SP or its purified components, such as exosomes, can improve sperm motility and membrane integrity, delaying premature capacitation and enhancing viability (De Lazari et al. 2020; Du et al. 2016; Höfner et al. 2020a, b; Luther and Waberski 2019).

Therefore, the decision to maintain or remove SP in stored boar semen should consider the specific composition of SP and the resistance of sperm to storage (Andrade et al. 2022). While some ejaculates may benefit from the controlled presence of SP, others may perform better with its removal, requiring careful management to optimize preservation and fertility (Höfner et al. 2020a, b; Song et al. 2023).

3.5 Storage Time and Longevity of Refrigerated Spermatozoa

The storage time is one of the main factors affecting the longevity of stored sperm, directly influencing semen quality and, consequently, fertilization rates. The use of stored boar semen is a common practice in artificial insemination, but sperm viability can decline significantly after 72 h of storage. Several studies indicate that prolonged storage, even under controlled temperature conditions (typically between 15 °C and 17 °C), can lead to a progressive decrease in motility, membrane integrity, and the fertilization capacity of sperm (Roca et al. 2015; Waberski et al. 2019).

The decline in sperm quality during storage is related to several factors, including oxidative stress and changes in membrane permeability. During storage, the sperm membrane becomes more susceptible to calcium ion influx and the production of reactive oxygen species (ROS), which leads to lipid peroxidation and, eventually, cellular apoptosis. Studies have shown that using extenders containing antioxidants and substances that stabilize the sperm membrane can mitigate these effects, thereby prolonging sperm longevity during storage (Li et al. 2018a).

Sperm longevity also varies significantly between ejaculates from different males, and this variation can be attributed to the individual composition of seminal plasma and the sperm's resistance to the stress induced by storage. Recent research suggests that specific seminal plasma proteins, such as spermadhesins and immunoglobulin-binding proteins, are associated with greater resistance to storage and the preservation of motility after prolonged periods (De Lazari et al. 2020; Fraser et al. 2021).

Moreover, the storage time directly affects sperm viability after artificial insemination. Studies demonstrate that fertilization rates and embryo quality decrease considerably after 5 days of storage, emphasizing the need to use semen doses with shorter storage times to maximize reproductive success (Roca et al. 2015). Therefore, optimizing storage conditions, including the type of extender and the addition of protective components, remains a priority in reproductive biotechnology.

3.6 Impact of Cooling on Sperm Physiology

The cooling process directly impacts the physiology of boar sperm, especially during long-term storage. The reduction in temperature to 17 °C, widely used in semen preservation, slows down sperm metabolism, preserving cell viability for longer. However, this temperature drop also induces physiological stress that can alter sperm functionality. One of the main adverse effects of cooling is the alteration of the plasma membrane permeability, which becomes more susceptible to damage caused by reactive oxygen species (ROS) and uncontrolled ion influx, particularly calcium, which can trigger premature capacitation (Roca et al. 2015).

Furthermore, cooling can induce lipid peroxidation, a process that deteriorates the phospholipids of the sperm membrane. The increased ROS concentration during stored semen contributes to oxidative stress, leading to DNA fragmentation and loss of motility. Studies indicate that the use of extenders containing antioxidants, such as superoxide dismutase (SOD) and catalase, can reduce oxidative damage, protecting sperm membrane integrity and increasing the longevity of stored cells (Li et al. 2018a).

Another meaningful impact of cooling is altering mitochondrial membrane potential, which compromises the mitochondria's ability to generate energy for sperm motility. Slowing mitochondrial metabolism during storage affects ATP production, which is essential for sperm movement. Consequently, sperm motility decreases progressively with storage time, even when cell viability is preserved (De Lazari et al. 2020).

Finally, the thermal stress generated by cooling can compromise the integrity of the acrosome, an essential structure for sperm penetration of the oocyte. Damage to the acrosome during storage can reduce the fertilizing capacity of semen, even when motility and viability are maintained. Thus, preserving the acrosome and protecting the membrane and mitochondria is crucial for maintaining fertility in stored semen (Roca et al. 2015).

3.7 Advances in Boar Semen Cooling

In recent years, advances in boar semen cooling techniques have been driven by detailed ultrastructural analyses, such as lipidomics, proteomics, metabolomics, and miRNA studies. These approaches provide a deeper understanding of the molecular mechanisms involved in preserving sperm quality during low-temperature storage, allowing for the identification of potential biomarkers that may help optimize cooling protocols.

Pedrosa et al. (2021) investigated the role of miRNAs in sperm cells and extracellular vesicles from seminal plasma. These miRNAs may be associated with cryotolerance and the ability of sperm to withstand cooling. Although the research primarily focused on cryopreservation, the results suggest that specific miRNAs, such as ssc-miR-503 and ssc-miR-130a, may also be linked to sperm viability during refrigerated storage. These miRNAs are involved in energy metabolism regulation and preservation of cell integrity, making them potential biomarkers to predict sperm quality in refrigerated semen.

The study by Torres et al. (2022) analyzed the metabolomic profile of spermatozoa and seminal plasma during cooling (holding time (HT)). The research showed that energy-related metabolites, such as creatine, hypoxanthine, and niacinamide, are involved in maintaining sperm quality during storage. Although these findings were related to cryopreservation, the same metabolic mechanisms could have implications for refrigerated semen, as energy metabolism is critical to maintaining motility and cell integrity during cooling. The authors suggest regulating these metabolic pathways may be a promising strategy for improving semen quality during low-temperature storage.

Proteomic analyses have also demonstrated the importance of seminal plasma proteins in sperm resistance to cooling. De Lazari et al. (2020) identified proteins such as spermadhesins, which play a crucial role in stabilizing the plasma membrane and preventing damage during cold storage. While this study focused on cryopreserved semen, these proteins may also be relevant in preserving sperm during refrigerated storage by improving resistance to oxidative stress and maintaining viability and motility.

Additionally, lipidomic has proven to be a valuable tool for analyzing the lipid composition of sperm membranes and its relationship with cryotolerance. Yeste (2016) showed that lipids such as cholesterol are essential for maintaining membrane fluidity and integrity during cooling. These findings suggest that lipid composition may directly influence sperm's ability to resist cold-induced damage, which could apply to both cryopreserved and refrigerated semen.

These advances in molecular and ultrastructural analyses provide a solid foundation for developing new boar semen cooling protocols. By identifying molecular biomarkers and understanding the physiological changes that occur during cooling, these studies pave the way for more efficient strategies to preserve fertility in refrigerated semen. While many of these findings were initially applied to cryopreservation, they may also have relevant implications for semen stored at low temperatures.

3.8 Challenges and Limitations of Using Refrigerated Semen in Swine Production

The sperm functionality declines considerably after 72 h of storage, even under controlled refrigeration conditions, which reduces the efficient usage window of refrigerated semen. Oxidative stress is also a key factor compromising the quality of refrigerated semen. The production of reactive oxygen species (ROS) during storage leads to lipid peroxidation of membranes and sperm DNA damage, impacting motility and membrane integrity. Koppers et al. (2008) observed that increased ROS levels significantly contribute to the deterioration of semen quality. While the use of extenders with antioxidants shows promise, their effectiveness still depends on factors such as the initial semen composition and storage time.

Another significant limitation is the variation between ejaculates from different boars. Differences in seminal plasma composition and cooling resistance mean that some ejaculates are more susceptible to quality loss during storage (Li et al. 2018a).

Maintaining temperature during semen storage and transport is crucial to preserving its quality. Even slight temperature variations can accelerate the decline in sperm quality. Roca et al. (2015) emphasize that maintaining a constant temperature between 15 °C and 17 °C is essential to ensure the viability of semen during transport and prolonged storage. This requires proper refrigeration and transport equipment, which can increase operational costs.

Finally, the cost of refrigerated semen is another challenge to consider. Although the method is more economical than cryopreservation, the need for high-quality extenders and reliable refrigeration systems increases operating costs, especially in large-scale production systems. Roca et al. (2015) note that increased reproductive efficiency can offset these additional costs if protocols are optimized to minimize quality losses during storage.

4 Cryopreservation of Boar Spermatozoa

To this day, freezing boar spermatozoa remains a challenge. Studies in the area have determined a negative effect during in vitro analysis, and despite being more efficient in other species of domestic animals on in vivo trials (pregnancy rates) comparing the use of fresh and preserved ones at 17 °C, freezing/thawing causes a reduction in reproductive parameters, evident in the number of piglets born; this issue would be one of the main limitations for its massive implementation in pig herds. Cryopreservation is the only way to preserve the genetic material of domestic or endangered animals (e.g., autochthonous pig breed Gochu Asturcelta; Caamaño et al. 2021), valuable individuals, or genetically relevant breeds for indefinite periods of time; the storage in frozen state is necessary to maintain germplasm, preserving genetic diversity and providing efficiency to the breeding management. It is important to consider that a loss of sperm suspension quality has been reported after long-term freezing storage; a study evaluated post-thaw sperm quality after 2, 4, or 8 years of cryostorage, compared to thawed spermatozoa 15 d after freezing, reporting that sperm viability was not affected by length of cryostorage, but total and progressive sperm motility were lower in sperm samples cryostored for 4 or 8 years compared to the control group (Li et al. 2018b). Similarly, evidence of aging processes in frozen/thawed boar spermatozoa was reported after comparing short (1 week) and long (42–48 months) freezing storage in liquid nitrogen (LN_2): long storage induced a marked reduction in post-thaw sperm motility, mitochondrial function, and plasma membrane and acrosomal integrity, and they reported cryosusceptibility to lipoperoxidation (LPO) and DNA fragmentation (Fraser et al. 2014b). This could be relevant when planning reproductive management using long-storage straws; it could be necessary to increase the number of straws to use during artificial insemination.

For the porcine species, freezing spermatozoa is carried out by exposing them to LN_2; attempts have been made using vitrification (Serrano Albal et al. 2023); however, hypertonic solutions cause irreversible damage to the cellular structure, significantly reducing the quality of the vitrified spermatozoa sample; therefore conventional or automated slow freezing is the most used techniques to preserve boar spermatozoa. Conventional slow freezing is carried out by exposure to liquid nitrogen vapors; therefore, it is not possible to control the speed at which the straws freeze, and automated slow freezing requires freezing equipment in which it is possible to control the temperature drop. Not many studies compared conventional and automated methods for slow freezing. Automated freezing using a cooling rate of 3 °C/min^{-1} from 5 to −6 °C, 1 min holding at −6 °C, and then freezing rates of either 20 °C, 40 °C, or 60 °C/min^{-1} from −6 °C to −140 °C improves motility, membrane and acrosomal integrity, and mitochondrial membrane potential compared to conventional freezing method (Baishya et al. 2014). The heterogeneous composition of the plasma membrane imply different ice nucleation points; therefore, the critical period occurs between −5 and −80 °C, a period in which the different cellular components freeze; for this reason using a temperature drop curve of 6 °C/min^{-1} for 100 s (from 5 to −5 °C), 40 °C/min^{-1} for 113 s (from −5 to −80 °C), along with holding time at −80 °C for 30 s, and − 60 °C·min^{-1} for 70 s (from −80 to −150 °C) reduces free radical levels (superoxide anion and peroxynitrite) and with it the lipoperoxidation of the membrane, reducing phosphatidylserine translocation and preserving motility over time compared to the use of conventional freezing method (Pezo et al. 2023). Similarly, using 40 °C/min^{-1} from −5 to −80 °C demonstrates that the automated cooling curve of the programmed freezing machine achieved higher values of sperm motility, vigor, and viability (Shiomi et al. 2015). Throughout this manuscript, the boar spermatozoa freezing protocol is described; the variables that could affect cell survival, those effects that alter spermatic function in vitro and in vivo, and strategies to optimize the freezing extender are also included.

5 Effect of Slow Freezing on Boar Spermatozoa Functionality

Recent review publications have delved into preservation techniques such as cryopreservation (Waberski et al. 2019), proteomics of frozen/thawed boar spermatozoa that have a key role for cryotolerance and when they undergo alterations because of cryopreservation lose their functions (Delgado-Bermúdez 2024), and strategies to minimize the harmful effects of in vitro manipulation such as the use of antioxidants at different stages of freezing (Ribas-Maynou et al. 2021) and thawing (Pezo et al. 2024). Freezing boar sperm involves a series of steps that include thermal holding and equilibration times (ET), spermatozoa resuspension in extenders, exposure to freezing temperatures, and restoration of cellular metabolism through thawing. Each of these steps will cause irreversible damage to the spermatozoa functionality, which is caused by osmotic, thermal, and oxidative shock, and induces changes associated with cryo-capacitation along the cooling curve on spermatozoa exposed to 5 °C (Torres et al. 2024).

Deleterious effects are associated with the cryopreservation process, some of them can be related to the extraction of the seminal plasma (SP), a necessary step in the freezing protocol, because it has been demonstrated that certain SP antioxidants are positively involved in boar sperm cryotolerance, minimizing the oxidative stress imposed by cryogenic handling the sperm-rich fraction (SRF; the first 10 mL) is recommended to be cryopreserved, other sperm fractions negatively impacts sperm cryotolerance (Li et al. 2018a), and exhibited significantly higher percentages of spermatozoa with total motility, mitochondrial function, and acrosome integrity (Zasiadczyk et al. 2024). An increase in reactive oxygen species (ROS) levels and, with it, a decrease in the plasma membrane and DNA integrity and the appearance of apoptosis markers (Bax, caspase-3, caspase-8, and caspase-9), all these, inhibit mitochondrial function and, with it, impair motility (Zhang et al. 2021). Proteomic analyses reported that 76 proteins were identified to be differentially expressed, among which 9 and 67 proteins showed higher and lower expression comparing fresh and frozen/thawed boar spermatozoa; the classified functions of these proteins included oxidative phosphorylation, mitochondrial inner membrane and matrix, pyruvate metabolic processes, and sperm flagellum and motile cilium (Kim et al. 2023), which are in accordance with the structural and functional effects that are caused by the arrest and restoration of cellular metabolism after the freezing/thawing process, which will be discussed later in the text.

A recent study determined that the different lipid metabolism levels and long-chain polyunsaturated fatty acids (PUFAs) in the plasma membrane may be key factors contributing to differences in spermatozoa resistance to cryopreservation among boars (Zhang et al. 2023). Phospholipids are the most abundant component in spermatozoa membrane, making it the initial target of damage; at body temperature, the membrane is organized in a fluid state, and cholesterol would play a relevant role by increasing the stability of the membrane; for boar species cholesterol to phospholipid ratio is 0.26, while protein to phospholipid ratio is 1.26 (Parks and Lynch 1992). During cryopreservation, the spermatozoa membrane undergoes a phase transition from fluid to a crystalline state, because it reaches the ice nucleation point, which alters membrane function by lipid aggregations, and also there is lipid peroxidation, produced by an increase in the levels of free radicals, which results in a loss of organization and membrane fluidity (Gautier and Aurich 2022). Cryopreservation will have a negative impact on the performance of the cryopreserved spermatozoa sample during in vivo trials, this could be increased, if the immune response of the female is considered, leukocyte migration towards the uterine lumen causes phagocytosis (Matthijs et al. 2003) and NETosis (Zambrano et al. 2021) of spermatozoa, either through natural mating (Katila 2012), or during AI, an increase in levels of IL-23, that promotes cytokine and chemokine production by various cells, with subsequent recruitment of neutrophils (Svensson et al. 2021) and is reflected in a reduction in reproductive parameters after IA with cryopreserved boar spermatozoa (Sánchez-Sánchez et al. 2022).

6 The Freezing Protocol

In this section, the freezing/thawing protocol will be described considering the holding time, the equilibration time, the composition of the freezing extender, and the thawing process.

6.1 The Holding Time

The holding time (HT) is the period during which an ejaculate, in a diluted state, is held at 15–17 °C before cryopreservation. Different studies described the HT as responsible for increasing spermatozoa cryotolerance; this could be due to the extended interaction between seminal plasma (SP) components and spermatozoa plasma membrane. In fact, an HT of 24 h at 17 °C moderate fluidity changes the boar spermatozoa membrane and allows better maintenance of sperm motility without an effect on plasma membrane integrity; this period will improve the ejaculate freezability (Casas and Althouse 2013). The results are controversial in terms of how long the HT is; the first approaches to the application of the HT on sperm characteristics and fertility showed that an increase of the HT from 3 to 24 h at 15 °C had no significant effect on pregnancy rate (Guthrie and Welch 2005). Also, an HT of 2 h resulted in higher sperm motility and beat cross frequency (BCF) compared to 24 h of HT, while no effect was detected on mitochondrial activity, plasma and acrosomal membrane integrity, and DNA fragmentation index; however, freezing results are still satisfactory if semen is processed until 44 h after collection (Schäfer et al. 2017). An HT of 2 to 8 h at 15 °C has no effect in terms of total post-thaw sperm motility and plasma and acrosome membrane integrity, and the semen sample can be preserved for up to 24 h without a detrimental effect on post-thaw sperm quality (Tomás et al. 2014a). Similar results were observed with an HT of 24 h at 15 °C without an effect on spermatozoa motility, viability, acrosome integrity, and capacitation status (Gale et al. 2014). A 24 h HT increases total and progressive motility and plasma and acrosome membrane integrities, without an effect on membrane fluidity, while for mitochondrial membrane potential, a 32 h HT is necessary to improve this parameter; however, the HT was not able to control the superoxide anion levels and, with it, the membrane lipid peroxidation on frozen/thawed boar spermatozoa (Torres et al. 2018). A metabolomic study compared good (GEF) and poor (PEF) ejaculate freezability (SP and spermatozoa metabolomics), extended at a 1:1 (vol/vol) ratio with BTS and held at 17 °C for 24 h before cryopreservation; they reported that the SP did not influence ejaculate freezability, but spermatozoa metabolites were markedly different between GEF and PEF. Several metabolic markers in the sperm cells were associated with energy production (including inosine, hypoxanthine, creatine, ADP, niacinamide, spermine, and 2-methylbutyrylcarnitine) and are directly related to the improvement of ejaculate freezability during HT (Torres et al. 2022). Another study states that a 24 h HT after freeze-thawing boar spermatozoa increases their cryotolerance, by preserving total and progressive motility 240 min post-thawing and reducing the levels of spermatozoa DNA fragmentation and lipid disorder; also mini-array analyses demonstrate that factors present in extended semen confer higher cryotolerance during HT and act through serine phosphorylation of some proteins that are involved in stress, such as HSP70 (Yeste et al. 2014). As we can see, the results are controversial; however, many research groups used an HT of 24 h (Dziekońska et al. 2015).

6.2 The Equilibration Time

Equilibration time (ET) is the period during which sperm cells are in contact with cooling/freezing media components at a temperature from 17 °C to 5 °C, providing a proper osmotic balance between the intra- and extracellular medium. Although a recent study indicated that ET in the straws before freezing has no impact on post-thaw sperm quality parameters except for the percent of total motility (Shepherd and Herickhoff 2022), studies continue to be developed with the aim of defining the duration and effect of equilibrium time on sperm function. The lactose and egg yolk (LEY) extender without glycerol at 5 °C for 2, 4, 24, or 48 h, an ET at 5 °C for 4 h yielded the highest motility and BCF, whereas the equilibration for 48 h impaired sperm motility, while other parameters like membrane integrity, mitochondrial activity, and DNA fragmentation index were not affected by any of the times evaluated (Schäfer et al. 2017). Similarly, an ET of 2 h results in greater sperm cryotolerance in both GEF and PEF samples, which suggests that modifying the freezing protocol led to an increase in post-thaw sperm function and survival (Passarelli et al. 2020).

6.3 Freezing Extenders

The composition of the freezing extender plays a crucial role during the cryopreservation process. All cryopreservation extenders contain distilled water, sugar, cryoprotectant agents (CPAs), egg yolk (EY), detergent, and antibiotics. The most used freezing extender is the lactose and egg yolk (LEY) as a non-penetrating CPA (NpCPAs), followed by glycerol as a penetrating CPA (PCPAs), Equex STM or Orvus Paste as a detergent, and antibiotics. The pre-adjustment of the freezing extender at pH 8 can improve the post-thaw boar, plasma membrane, and acrosome integrity and motility (Tomas et al. 2015). Slow freezing can be done in one or two steps: in the first one, the addition of the freezing extender with its CPAs is carried out at 15–17 °C, which is easier, and cheaper, and no special equipment is required, but increases the toxicity as a product of prolonged spermatozoa exposure to CPAs, while for the two-step protocol, the permeating cryoprotectant agents (PCPAs) are added at 5 °C; therefore, CPA exposure and toxicity are reduced, improving total and progressive motility and the integrity of plasma and acrosomal membrane of spermatozoa with a high mitochondrial membrane potential (MMP) (Monteiro et al. 2022). For the cryopreservation protocol it is necessary, a freezing extender A (LEY) and B (LEYGO), the difference between them is the presence or absence of glycerol and the detergent; for this reason, extender B is a hypertonic solution. The extender A is added at 17 °C after the HT, usually 24 h as described above, but first, the extended semen sample must be centrifugated, a necessary step before freezing, because it facilitates the removal of SP and concentrates the spermatozoa, so it can be re-diluted with LEY extender (Zhang et al. 2012).

The use of short-term centrifugation with a relatively high *g*-force (2400 *g* for 3 min) showed significantly higher post-thaw sperm motility, viability, % of uncapacitated spermatozoa, and oocyte penetrating ability compared to the use of 800 *g* for 3 min (Carvajal et al. 2004). Similar results were reported comparing 800 *g* for 10 min and 2400 *g* for 3 min, where centrifugation for shorter times achieves better results regarding the percentage of morphologically intact sperm (Shiomi et al. 2015). Control of the temperature during centrifugation is an important issue, considering that the diluted semen sample is kept at 15–17 °C; a study compared centrifugation at 800 *g* for 10 min at 15 °C versus 24 °C, and they reported a better preservation of the sperm motility and membrane integrity at 15 °C (Bianchi et al. 2008). We recommend minimizing temperature variations during the manipulation of the seminal sample; therefore, it is necessary to use a thermoregulated centrifuge.

Finally, the spermatozoa pellet is resuspended on the LEY extender and exposed to the ET from 17 to 5 °C, as described in the section above, 2 h of ET is usually used, and then it must be added to the LEYGO extender at 5 °C, straw packaged, and sealed and subjected to cryogenic temperatures either in a controlled manner or through exposure to liquid nitrogen vapors. The components of the freezing media will be described below.

6.3.1 Non-Permeating Cryoprotectant Agents (NpCPAs)

NpCPAs do not pass through the plasma membrane of sperm, and these include sugars, amino acids, various proteins, lipoproteins, and other macromolecules such as methylcellulose, polyvinyl alcohol, and polyvinylpyrrolidone with spermatozoa had a negative. Sugars are the most used, and lactose, a non-reducible disaccharide, is the most common sugar added in an egg yolk-based extender, and many research groups used it as a control group (Dong et al. 2022; Lee and Lee 2023). Lactose has a better cryoprotective effect than monosaccharides like glucose, galactose, or fructose and when the lactose is compared with sucrose, lactulose, trehalose, or melibiose, it was determined that those with glucose in their composition exerted greater cryoprotective effect and of those disaccharide, lactose, and melibiose extenders provided higher percentages of live sperm with intact plasma membrane and motile sperm (Gómez-fernández et al. 2012). Similarly, it was found that the lactose-based extender resulted in higher post-thawed sperm motility and viability and intact acrosome and plasma membrane compared to sorbitol, a polyol (Chanapiwat et al. 2011).

Another disaccharide with a good cryoprotectant effect is trehalose; the use of a trehalose-based extender (100 mm) significantly preserves sperm motility, mitochondrial activity, membrane integrity, and acrosome integrity during the freeze/thaw process (Hu et al. 2009). Other study demonstrated that total and progressive motility, viability, and acrosome integrity were significantly improved using a trehalose-based extender composed of 80% (vol/vol) sugar solution and 20% (vol/vol) egg yolk; it also reported an increase in the penetration and monospermy rates and a better efficiency and blastocyst formation rates compared to the use of lactose- or glucose-based extenders (Malo et al. 2010). The trehalose demonstrated good results alone or in combination with glycerol; a glycerol-free trehalose-based extender allows to maintain good levels of motility, viability, acrosome integrity, mitochondrial membrane potential, and in vitro penetrability of thawed boar spermatozoa (Athurupana et al. 2015). While the use of 250 mM of trehalose and 1% of glycerol (vol/vol) helps to diminish the acrosomal damage observed during cryopreservation; maintaining viability and motility (Gutiérrez-Pérez et al. 2009). The use of trehalose (250 mM) preserves motility and reduces lipid membrane disorders with a significant increase in MMP, and interestingly the levels of $O_2^{\bullet -}$ and peroxynitrite were lower compared to the use of a lactose-based extender (Pezo et al. 2020); this reduction in free radical levels could be because trehalose reduced the oxidative stress by enhancing the antioxidant activity of thawed bovine spermatozoa (Hu et al. 2010). Nontoxic polysaccharides have been used in the cryopreservation of boar spermatozoa; dextran and pentaisomaltose represent promising compounds for the elimination or reduction of glycerol in extenders; however, only pentaisomaltose resulted in a better preservation of spermatozoa mitochondrial activity and DNA integrity (Simonik et al. 2022). Other molecules with cryoprotective properties will be discussed later.

6.3.2 Egg Yolk

The low-density lipoproteins (LDLs) contained in the yolk fraction were defined as the active ingredients that give it its cryoprotective capacity. The use of complete egg yolk (EY) and its insoluble, soluble, and LDL fractions on boar spermatozoa quality during the equilibration time (from 17 °C to 5 °C) demonstrated that cooling without any lipid component or in the presence of the EY insoluble fraction significantly reduced sperm motility (Orrego et al. 2019). The composition of the yolk, particularly the amount of cholesterol, fatty acids, and phospholipids, varies depending on the species; when the chicken EY was compared by either duck or quail EY, chicken EY displayed higher motility immediately after thawing, while chicken or duck EY better preserved intact acrosomes after thawing, and only chicken EY preserved this parameter 6 h after thawing (Bathgate et al. 2006). In a similar study, compared different concentrations of LDLs extracted from the yolk of hen, duck, quail, pigeon, or ostrich egg were added to the extender; showing that the extender supplemented with 0.09 g/mL pigeon EY LDLs has beneficial effects preserving the spermatozoa motility, acrosome, and plasma membrane integrity (Wang et al. 2014). LDLs isolated from ostrich EY (LPFo) provide excellent protection for boar spermatozoa against cold shock in terms of total motility, MMP, plasma membrane integrity, DNA fragmentation and after IA the pregnancy and farrowing rates were 84.2% and 81.6%, respectively, with 11.4 piglets born alive per litter (Fraser et al. 2018). The addition of a 20% LDL can be used to substitute the whole egg yolk as a cryoprotective additive for freezing spermatozoa (Souza et al. 2015). Also, pasteurized egg yolk (PEY) has been used because is easier to standardize and reduces hygienic risks. PEY has no effect on post-thawing motility, membrane fluidity, or morphologically intact sperm and acrosome defects, while mitochondrial activity, membrane integrity, and motility (after 300 min of incubation) were slightly affected on PEY group (Selige et al. 2021). A study was conducted to compare ultracentrifugation (three times at ×10,000 *g*, for 45 min) and sonication of an egg yolk-based extender, methods by which liposomes of the EY are obtained; the use of ultracentrifugation improves the quality of kinematic parameters, membrane integrity, and mitochondria activity (de Brito et al. 2021). Unfortunately, information is scarce for the swine species, and it is necessary to develop new lines of research that allow us to define how the LDL of the EY interacts with the components of the spermatic membrane and what finally confers the cryoprotective capacity to the EY.

6.3.3 Detergents

To achieve better extraction of lipids or LDL from egg yolk to the freezing extender, it has been shown that the use of a detergent such as Equex STM paste or Orvus ES Paste (OEP), the most used ones, significantly improves post-thaw sperm quality and fertilizing capacity in most mammalian spermatozoa. The addition of 0.5% (vol/vol) of Equex to the lactose and egg yolk-based extenders in combination with glycerol or dimethylacetamide (DMA) improves motility, viability, and acrosomal integrity compared to the control group without Equex (Wu et al. 2013). Freezing boar spermatozoa using 1.5% of Equex on a LEY-glycerol-based extender and packed on 0.25 or 0.5 mL straws, improves motility, viability, and acrosomal normal apical ridge (NAR) when Equex was added to the freezing extender, and it has a better performance when a 0.5 mL straw is used in terms of NAR (Buranaamnuay et al. 2009). Similarly, the effect of adding 0.5% of Equex or different concentrations of sodium dodecyl sulfate (SDS; at 0.1%, 0.3%, or 0.5% (*v/v*)), an anionic detergent that also seems to potentialize the cryoprotective effect of the EY, on a Tris-egg yolk-glycerol, shows no differences between the use of 0.1% of SDS or 0.5% of Equex on total and progressive motility, membrane functionality and integrity, and mitochondrial activity (Moreira et al. 2023). Concentrations of 0.25 and 0.5% of OEP added to a LEY plus glycerol extender markedly enhanced post-thaw sperm total motility; however, a decline of post-thaw motile spermatozoa was observed with 0.5% OEP following a 120 min incubation period (Fraser et al. 2014a).

6.3.4 Permeating Cryoprotectant Agents (PCPAs)

PCPAs, like glycerol, confer protection against damage avoiding intracellular ice formation; also PCPAs improve the rearrangement of membrane lipids and proteins, membrane fluidity, and dehydration at lower temperatures that contribute to improved cryo-survival (Sharafi et al. 2022). In fact, an increase in post-thaw sperm DNA damage in spermatozoa frozen in the absence of cryoprotective substances has been reported (Fraser et al. 2009). Glycerol is the most common PCPA used to cryopreserve boar spermatozoa, and the concentration added to the extender was identified as a highly influential factor that affects post-thaw sperm quality (Shepherd and Herickhoff 2022); this is due to the negative impact it can have on different functional sperm parameters such as an increase in alterations of the ultrastructure of the cytoskeleton of the perinuclear theca of boar sperm head (Arenas Núñez et al. 2013), in the presence of high concentrations of glycerol (10% and 14%), a negative impact on motility and viability (Yang et al. 2016), and also a decrease on MMP of boar sperm cooling at 5 °C (Malcervelli et al. 2020).

Different studies have compared the effect of glycerol alone or in combination with other PCPAs. A 3% (vol/vol) is the most used glycerol concentration by different researcher groups. The use of 2–3% glycerol has the best antiapoptotic effects (Zeng et al. 2014), and also, it has been reported that the motility of frozen/thawed spermatozoa in 1–3% glycerol is significantly higher than those treated with 0.5% of glycerol, and a better maintenance of membrane integrity is obtained when glycerol is used at 2% in a lactose-based extender with an osmolarity of 400 mOsm/kg; the conception and implantation rate were significantly higher than those spermatozoa frozen by the conventional freezing extender (lactose and egg yolk-based extender; 300 mOsm/kg + 3% glycerol) (Okazaki et al. 2009). The use of 3% glycerol achieves better results on plasma and acrosome membrane integrity compared with the use of different concentrations (3, 5, and 7%) of DMA or dimethyl sulfoxide (DMSO) (Kim et al. 2011). A concentration of 3% glycerol in combination with lactose or trehalose achieves a better effect on total motility during 3 h after thawing, while the addition of 3% of dimethylformamide (DMF) to lactose- or trehalose-based extender diminished motility immediately and at 1 h after thawing and a DMF (1.5%) combined with glycerol (1.5%) in lactose-based extender decreased motility at 2 h after thawing (Malo et al. 2012). Glycerol alone (3%) or a combination of glycerol (3%) and DMA (1%) improved sperm motility, viability, and acrosome integrity compared to the use of other combinations of concentrations of both PCPAs (Yang et al. 2016). The use of 2% of glycerol in combination with 3% of dimethylacetamide or DMF or 5% of DMF significantly preserves motility and viability values compared to a control group using 2% glycerol (Pinho et al. 2014). The few updated references demonstrate that the strategies focused on the optimization of the freezing diluent do not consider PCPAs as an objective, and therefore, to date, the use of glycerol at a concentration of 3% seems to be the best option.

6.3.5 Antibiotics

Healthy boars' semen usually does not contain bacteria. But the preputial diverticulum, skin, and hair of the boar contain several microorganisms; therefore, the collection must be carried out with the corresponding aseptic measures, because bacterial growth has been associated with deleterious effects on semen quality and shelf life, such as sperm agglutination and decreased sperm motility and viability (Contreras et al. 2022). The freezing medium is an enriched medium that provides the necessary conditions for bacterial growth; however, this has greater relevance in boar semen preservation techniques through refrigeration at 17 and/or 5 °C, above what occurs during freezing/thawing (Waberski and Luther 2024). A study demonstrates that there are *Lactobacillus* species present in the porcine SP and during long-term co-incubation of lactic acid with spermatozoa had a negative influence on functionality (Schulze et al. 2018). There is no unified criterion regarding which antibiotic to use in freezing media; however, the most used by different research groups is kanamycin sulfate (Bang et al. 2022; Tomás-Almenar and de Mercado 2022), but others have also been used, like gentamicin sulfate (Osman et al. 2023) and also combinations of different antibiotics like lincomycin, streptomycin, and penicillin (Simonik et al. 2022).

6.4 Thawing

The thawing extender is a mixture of molecules of ions, energy sources, antibiotics, and other compounds that can be added with the aim of improving spermatozoa functionality. Different commercial thawing extenders are available. Androstar® Plus was the most effective in preserving the integrity of the plasma and acrosomal membranes (De Andrade et al. 2022) compared to Beltsville Thawing Solution (BTS), the most common thawing extender used because its composition is published (Pursel and Johnson 1975), is isosmotic (290–300 mOsmol/kg) and a pH adjustment is recommended to 8, because better achievements were observed on motility until 90 min of incubation at 37 °C (de Mercado et al. 2020).

Lately, there has been interest in optimizing the thawing medium by adding antioxidants. The addition of GSH plus ascorbic acid reduces intracellular peroxide levels preserving motility, viability, acrosome integrity, and nucleoprotein structure during 240 min of incubation at 37 °C (Crespo-Félez et al. 2017). Quercetin, a polyphenol, reduces hydroperoxides allowing for preservation motility and viability (Winn and Whitaker 2020). Resveratrol, a phytoalexin, and epigallocatechin-3-gallate, a polyphenol, added to BTS do not improve spermatozoa quality but optimize IVF achievement having a better total fertilization efficiency (Gadani et al. 2016); similar results were observed by adding tannin-rich vegetal extract to an IVF extender used as a thawing solution (Galeati et al. 2020). Other molecules have been added to the thawing extender demonstrating positive effects like membrane stabilizers such as methyl-b-cyclodextrin preloaded with cholesterol (Tomás et al. 2014b) or activating molecules, like alsterpaullone (Martín-Hidalgo et al. 2018) or caffeine (de Mercado et al. 2020), that improve motility and sperm velocities.

Thawing protocols can be done as a one- or two-step process, necessary to restore spermatozoa metabolism: the first one is performed at 37–38 °C, the normal temperature of the sow relevant to IA performance, while the two-step requires higher and faster thawing rates and a resetting at 37–38 °C as incubation temperature (Pezo et al. 2024). The two-step thawing protocol demonstrates better preservation of plasma and acrosomal membrane integrity and motility (Tomás-Almenar and de Mercado 2022). 70 °C for 8–10 s is the most recommended one (Knox et al. 2015; Tomás et al. 2014a), compared to the use of 37 °C for 20 s or lower temperature rates (Thema et al. 2023). Interestingly, under field conditions, thawing at 37 °C and storage at 17 °C reduced apoptotic and oxidative stress markers for up to 6 h (Li et al. 2023b). Thawing is a crucial step and the temperature, time, and composition of the medium are variables that could determine cell survival.

7 Vitrification

An alternative to slow freezing is vitrification, which involves direct exposure to liquid nitrogen, in which viable cells will undergo a glass-like solidification state; this assisted reproductive technologies (ART) requires large concentrations of cryoprotectants which increase the viscosity of the medium to prevent intracellular ice formation during cooling and devitrification (Barbosa et al. 2023). Few studies have been developed on sperm vitrification using raw semen diluted in media without and with permeable cryoprotectants and vitrified using the cryoloops (Arraztoa et al. 2017b) or spheres (Arraztoa et al. 2016) methods, and two different devitrification protocols were compared (30 s at 37 °C vs. 7 s at 75 °C, followed by 30 s at 37 °C); in both studies, sperm motility, viability, and acrosome integrity were not maintained in any of the vitrified/devitrified samples evaluated, and devitrified using a short time and a higher temperature is better; this method is only applicable in intracytoplasmic sperm injection; in fact, the use of boar sperm vitrified using spheres, in Tyrodes´s albumin lactate pyruvate (TALP) medium with 1% bovine serum albumin (BSA), conserves condensed and intact chromatin capable of producing early embryo development up to the pronuclear stage (Arraztoa et al. 2017a). Also, the spheres method using 0.3 M sucrose as a non-permeable cryoprotectant had a positive effect on the survivability of the samples, reducing sperm DNA damage and preserving motility after thawing (Serrano Albal et al. 2023).

8 Strategies Used for the Optimization of Freezing Media

Nowadays, strategies aim to optimize the freezing medium with the aim of improving the quality of the frozen/thawed spermatozoa suspension, with antioxidant supplementation being the most used. A lot of information regarding the addition of antioxidants to freezing/thawing media are available; therefore, we urge the readers to consider those literature reviews that have been compiled in the last two decades (Bathgate 2011; Pezo et al. 2021a; Ribas-Maynou et al. 2021). Only those articles published since 2020 will be included in this section.

8.1 Antioxidants and Their Effect on In Vitro Spermatozoa Functionality and Fertility Efficiency

8.1.1 In Vitro Free Radical Generation Systems and Their Effect on Sperm Function

Free radicals (FR) are extremely reactive because, in their outermost orbital, they have unpaired electrons; these oxidizing agents can be, for example, reactive species of oxygen or nitrogen, which are all a product of regular cellular metabolism, so oxidative/nitrosative stress happens when the free radical's production exceeds the capacity of the antioxidant system in the cell (Hassan et al. 2024). Not many studies have been conducted to define the effect of free radicals on spermatic functionality; however, some in vitro free radical-generated systems have demonstrated that boar spermatozoa are exposed to an H_2O_2-induced oxidative stress system, resulting in decreases in various sperm parameters including total motility and viability and a reduction of MMP (Wang et al. 2024); another study has shown, under similar conditions, that spermatozoa are highly susceptible to H_2O_2 reducing motility, kinematic parameters, and viability (Ledwaba et al. 2021). This is contrary to what was reported when comparing an in vitro H_2O_2 generation system and freezing, concluding that H_2O_2 is the major free radical mediating direct ROS effects, but not cryopreservation changes (Awda et al. 2009). Using 3-morpholinosydnonimine (SIN-1), a peroxynitrite ($ONOO^-$) generator added to a capacitated and non-capacitated media, significantly increases LPO in both media and potentiates the phosphorylation of pathways leading to sperm motility, which can be associated to the motility impairment (Serrano et al. 2020).

8.1.2 Antioxidants

The addition of astaxanthin (AST) at a concentration of 2 μM achieved positive effects on post-thaw sperm motility preserving membrane and acrosome integrity by reducing LPO, also regulated fatty acid composition of the sperm membrane, and with it good results on IVF capacity, improving sperm penetration ratio and total fertilization by increasing cleavage and blastocyst rates (Guo et al. 2021). Spermidine, a polyamine compound implicated in cellular survival, growth, and proliferation, with an antioxidative effect (Naumova et al. 2023); the addition of 6 mmol/L of spermidine to the boar freezing extender improved motility, viability, and kinetic parameters, membrane and DNA integrity, mitochondrial activity, and total antioxidant capacity (T-AOC) by reducing H_2O_2 levels and LPO; also in vivo results reported a significant increase in the litter number (Li et al. 2023a). Rutin has been shown to possess antiapoptotic, antioxidant, and anti-inflammatory properties which could improve male fertility (Rotimi et al. 2023); in fact, the addition of 0.4 mM and 0.6 mM of rutin to the spermatozoa freezing extenders allows to preserve motility, mitochondrial activity, and plasma membrane and acrosomal integrity which could be associated with a reduction on ROS levels and a reduction of LPO because there is an improving antioxidant enzymatic activity of SOD, CAT, and GSH-Px; the IVF assay shows a higher penetration rate and the increased total efficiency of fertilization with an increase in cleavage and blastocyst rates (Xu et al. 2020). The addition of butylhydroxytoluene (BHT; 1 mM) or reduced glutathione (GSH; 2 mM) or a combination of both antioxidants to a commercial freezing extender (Minitube Cryoguard), demosatrated that GSH improved total and progressive motility, while a reduction in the proportion of capacitated spermatozoa was observed as a combined effect of BHT and GSH, and only BHT improved blastocyst rate (de Andrade et al. 2023). The positive effects of silymarin (SMN) have been reported for the motility, plasma, and acrosomal integrity; the addition of 20 μM into boar spermatozoa freezing extender reduces the levels of total ROS and LPO and also significantly increases the T-AOC, which could have an impact in the upregulated ATP content and enhanced the mitochondrial activity, while for IVF parameters, there is an increase in the penetration rate and total efficiency of fertilization (Yue et al. 2023). The use of myoinositol at 0.5 mg/mL reduced the expression of Bcl-2, a pro-apoptotic gene, and showed higher expression of *SMCP*, a motility-related gene, in addition to better preservation of total and progressive motility, acrosome integrity, and fertilization ability with a significant increase in cleavage and blastocyst rates (Osman et al. 2023). Nobiletin, a bioflavonoid compound, added in concentration of 100 μM or 150 μM to the freezing extender significantly improved sperm progressive motility, and acrosome integrity, while LPO was significantly reduced when 50 μM and 150 μM were used; only 150 μM increased the cleavage and blastocyst formation rates (Almubarak et al. 2024b). The use of 5 μM sulforaphane (SFN), a dietary phytocompound from the isothiocyanate family, whose antioxidant effects would be via activation of the gene *NFE2L2* that modulates the expression of several antioxidants, including GSH-Px, SOD, CAT, NAD(P)H-quinone oxidoreductase (NQO1), and heme oxygenase-1 (HO-1), achieved a better preservation of motility, motion performance, and membrane and acrosome integrity, with a reduction of total ROS and LPO, and an improvement of T-AOC and antioxidant enzyme activity of SOD, CAT, and GPX, as well as a reduction of the presence of apoptotic markers and increase of cleavage rate after IVF (Zhang et al. 2024a). The addition of L-carnitine, a quaternary amine, a natural antioxidant found in mammals, synthesized in the liver and kidneys from the amino acids lysine and methionine to the freezing extender with different concentrations (1, 2.5, or 5 mM) improved sperm motility, mitochondrial and membrane integrities, and the acrosomal proteolytic activity, while boar spermatozoa cryopreserved with 2.5 mM exhibited significantly higher penetrability into matured oocytes after IVF (Tatemoto et al. 2022). The in vitro effects on functional sperm parameters of different antioxidants can be found in Table 2.

8.1.3 Other Molecules

Membrane stabilizers have been added to the boar spermatozoa freezing extender, and the use of cholesterol-loaded cyclodextrin (CLC; 1.0 mg CLC/1.2 × 10^8) enhanced plasma and acrosome integrity and total motility after thawing in the CLC-treated group with a reduction on ROS levels and LPO; these positive effects of CLC treatment were observed in reducing structural and oxidative damage and improving the metabolic state (Zhang et al. 2024b), and when combined with 0.2 mM BHT, post-thaw motility and acrosome integrity improved (Baishya et al. 2018). The addition of 1 μg/mL of antifreeze protein III (AFP III), which primarily prevents ice crystal formation, preserved motility, MMP and cell membrane integrity, and the acrosomal proteolytic activity and maintained DNA quality and in vitro sperm penetrability (Masuda et al. 2023). The 25% (vol/vol) of 3,3-dimethylglutaric anhydride poly-L-lysine (DMGA-PLL) as a cryoprotectant for boar spermatozoa preserves in a better way the motility index and the blastocyst formation rate after IVF, and on in vivo trials, the mean number of total piglets (11.7) was not significantly different from using refrigerated spermatozoa at 17 °C after IA (Jin et al. 2023). The addition of 0.25% (vol/vol) of carboxylated poly-L-lysine (CPLL) to the freezing extender as a NpCPA preserved motility indices after 12 h of incubation, the blastocyst formation rate of embryos was

Table 2 Main diluters for boar semen refrigeration

Antioxidant	Best concentration	Main findings on spermatozoa function	Reference
Riboflavin	10 μM	Increase in total motility and MMP, improving the enzymatic activity of SOD, GSH, and CAT; a better preservation of plasmatic, acrosomal, and DNA integrity; and reduction of the appearance of apoptosis markers	Dong et al. (2022)
L-Proline	10 mM	Achieved a better viability and progressive motility even 1 h after thawing; acrosome and DNA integrity and mitochondrial activity were significantly higher than those in the control or other L-proline concentrations. It doesn't influence LPO or T-AOC	Liu et al. (2023)
Mitoquinone (MitoQ)	10 μM	A better post-thaw sperm motility and kinematics parameters with an increase on T-AOC and glucose transporter abundance (GLUT 3 and 8); also, 189 proteins were differentially expressed, mainly involved in motility, energy metabolism, and oxidative phosphorylation	Shi et al. (2022)
Palm kernel meal protein hydrolysates (PKMPH)	2.5 μg/mL	PKMPH contains bioactive peptides that have antioxidant and antimicrobial activities improving post-thawed spermatozoa in terms of total motility, viability, acrosome integrity, and mitochondrial activity which could be due to a reduction in LPO	Khophloiklang et al. (2023)
Resveratrol	50 μM	Is a natural polyphenolic antioxidant, reduces the LPO, improves sperm motility, viability, acrosome intactness, and MMP in frozen/thawed boar sperm	Kaeoket and Chanapiwat (2023)
Melatonin (MEL)	1 μM	Both antioxidants demonstrated a reduction on LPO associated to lower levels of peroxynitrite and $\cdot O_2^-$ and, with it, a higher viability and acrosome integrity without an effect on motility along 120 min of incubation at 37 °C	Pezo et al. (2021b)
Butylhydroxytoluene (BHT)	2.0 mM		
Camellia oil (CO)	1.5% (vol/vol)	CO has a high content of unsaturated fatty acids, polyphenols, vitamin E, and carotene and shows a positive effect on spermatozoa morphology, motility, acrosome integrity, and mitochondrial activity by reducing the LPO	Khophloiklang et al. (2024)

significantly higher, and there was no effect on the conception rate of sows inseminated, but the mean number of total piglets born using CPLL (10.0) was significantly lower compared to the use of refrigerated semen at 17 °C (13.4) (Jin et al. 2022). Also, other molecules can be added to the freezing extender as you can see in Table 3.

9 Conclusion

The use of refrigerated semen in swine production has been widely adopted in artificial insemination programs due to its practicality and potential to increase reproductive efficiency. However, this method presents challenges and limitations that directly affect sperm quality and fertilization outcomes. One main limiting factor is the reduction in motility and sperm membrane integrity during storage, the period that a semen sample can preserved functionally, and the fertility potential. Exposure of boar sperm to cryogenic temperatures causes a negative effect on sperm function in vitro and in vivo. We recommended a freezing protocol with a holding time of

Table 3 Effect of the addition of different types of molecules to the freezing extender on in vitro functionality parameters

Molecule	Classification	Best concentration	Main findings on spermatozoa function	Reference
κ-Carrageenan	Sulfated polysaccharide	0.2 mg/mL	Increased total and progressive motility for 120 min. Significantly maintains viability and acrosome integrity with high MMP and lower LPO	Almubarak et al. (2024a)
Phosphocreatine	Phosphoamino	10.0 mmol/L	Enhanced motility, kinematic parameters, and viability and preserved plasma, acrosome and DNA integrity, and MMP activity	Wang et al. (2023)
Honey	Natural	0.25% (vol/vol)	Supplementing Androhep plus with 0.25% of honey maintains a better morphology and acrosomes; replacing 50% Equex STM (1%) with honey improves total and progressive motility	Balogun et al. (2023)

(continued)

Table 3 (continued)

Molecule	Classification	Best concentration	Main findings on spermatozoa function	Reference
Poria cocos mushroom	Polysaccharides	600–900 μg/mL	Improved the motility and plasma membrane and acrosomal integrity and enhanced MMP activity. ROS, LPO, and GSH-Px activity significantly decreased. SOD and CAT enzymatic activity were improved	Zhou et al. (2023)
Carboxylated ε-poly-L-lysine (CPLL)	NpCPA	0.25% (vol/vol)	Improved total and progressive motility, kinematic parameters, and membrane and acrosome integrity. Also increased MMP, NADH-CoQ activity, and ATP level. An increase in CAT, GPx, and SOD enzymatic activity and a reduction of LPO	Zhang et al. (2022)
Zardaverine	Phosphodiesterase inhibitor	50 μM	Motility and kinematic parameters were improved without an effect on spermatozoa viability, DNA integrity, and mitochondrial activity	Jeong et al. (2018)

MMP mitochondrial membrane potential, *SOD* superoxide dismutase, *GSH* glutathione peroxidase, *CAT* catalase, *LPO* lipoperoxidation

24 h at 17 °C in BTS or a commercial extender, centrifugation at 2400 *g* for 3 min and resuspension in a lactose or trehalose egg yolk-based extender (LEY), an equilibration time of 1–2 h at 5 °C and resuspension in a LEY medium with 3% (vol/vol) of glycerol and 0.5% (vol/vol) of a detergent such as Orvus Paste, and then exposure of the straws to automated freezing using 40 °C/min as a rate of temperature drop between −5 and −80 °C and thawing in two steps at 70 °C for 8–10 s. The strategies to optimize the refrigeration and freezing protocol described can focus both on the effects caused by thermal and/or osmotic shock by testing the use of new cryoprotectants or membrane stabilizers and on those caused by oxidative stress by supplementing the refrigerated and freezing/thawing extenders with antioxidants. It is necessary to develop new studies that allow the applicability of this cryopreservation at the farm level.

Acknowledgments This work was supported by the Fundación de Apoyo a la Investigación del Estado de São Paulo (FAPESP), Brazil y la Universidad de La Frontera (UFRO), Chile (grant number FPP22–0022 to R.S. and A.F.C.; grant number FAPESP 2022/09573–2 to A.F.C. and R.S.), and Project DI24–0114 (F.P.) of the Dirección de Investigación de la Universidad de La Frontera, Temuco, Chile.

Declaration of Competing Interest The authors declare that they have no known competing financial interests or personal relationships that could have appeared to influence the work reported in this paper.

References

Almubarak A, Kim E, Yu I-J, Park H, Jeon Y (2024a) The effect of κ-carrageenan on porcine sperm cryo-survival. Animals 14:1387. https://doi.org/10.3390/ani14091387

Almubarak A, Lee S, Yu I-J, Jeon Y (2024b) Effects of Nobiletin supplementation on the freezing diluent on porcine sperm cryo-survival and subsequent *in vitro* embryo development. Theriogenology 214:314–322. https://doi.org/10.1016/j.theriogenology.2023.11.006

Althouse G (2008) Sanitary procedures for the production of extended semen. Reprod Domest Anim 43:374–378. https://doi.org/10.1111/j.1439-0531.2008.01187.x

Andrade AFC, Knox RV, Torres MA, Pavaneli APP (2022) What is the relevance of seminal plasma from a functional and preservation perspective? Anim Reprod Sci 246:106946. https://doi.org/10.1016/j.anireprosci.2022.106946

Arenas Núñez MA, Juárez-Mosqueda ML, Gutiérrez-Pérez O, Anzaldúa Arce SR, Izquierdo AC, Rodríguez RM, Trujillo Ortega ME (2013) Glycerol decreases the integrity of the perinuclear theca in boar sperm. Zygote 21:172–177. https://doi.org/10.1017/S0967199411000785

Arraztoa CC, Miragaya MH, Chaves MG, Trasorras VL, Neild MCGDM (2016) Porcine sperm vitrification II: spheres method. Andrologia 49:1–8. https://doi.org/10.1111/and.12738

Arraztoa CC, Baca Castex C, Alvarez GM, Cetica PD, Neild DM (2017a) *In vitro* production of porcine zygotes using intracytoplasmic injection of vitrified sperm. Reprod Domest Anim 52:775–780. https://doi.org/10.1111/rda.12978

Arraztoa CC, Miragaya MH, Chaves MG, Trasorras VL, Gambarotta MC, Péndola CH, Neild DM (2017b) Porcine sperm vitrification I: cryoloops method. Andrologia 49:1–8. https://doi.org/10.1111/and.12706

Athurupana R, Takahashi D, Ioki S, Funahashi H (2015) Trehalose in glycerol-free freezing extender enhances post-thaw survival of boar spermatozoa. J Reprod Dev 61:205–210. https://doi.org/10.1262/jrd.2014-152

Awda BJ, Mackenzie-Bell M, Buhr MM (2009) Reactive oxygen species and boar sperm function. Biol Reprod 81:553–561. https://doi.org/10.1095/biolreprod.109.076471

Baishya SK, Biswas RK, Kadirvel G, Deka BC, Kumar S, Sinha S, Dutta DJ, Saikia GK (2014) Effect of conventional and controlled freezing method on the post thaw characteristics of boar spermatozoa. Anim Reprod Sci 149:231–237. https://doi.org/10.1016/j.anireprosci.2014.06.020

Baishya SK, Biswas RK, Govindasamy K, Deka BC, Sinha S, Singh M (2018) Pre-freezing treatment with butylated Hydroxytoluene and cholesterol-loaded methyl-β-cyclodextrin improves quality of cryopreserved boar semen. Cryo Letters 39:336–344

Balogun KB, Nicholls G, Sokunbi OA, Stewart KR (2023) Cryoprotectant effects of natural honey on spermatozoa quality of pre-freezing and frozen-thawed boar semen. J Anim Sci 101:skac384. https://doi.org/10.1093/jas/skac384

Bang S, Tanga BM, Fang X, Seong G, Saadeldin IM, Qamar AY, Lee S, Kim K-J, Park Y-J, Nabeel AHT, Yu I, Cooray A, Lee KP, Cho J (2022) Cryopreservation of pig semen using a quercetin-supplemented freezing extender. Life 12:1155. https://doi.org/10.3390/life12081155

Barbosa BB, Evangelista ITA, Soares ARB, Leão DL, Pereira RJG, Domingues SFS (2023) Kinetic vitrification: concepts and perspectives in animal sperm cryopreservation. Anim Reprod 20:e20220096. https://doi.org/10.1590/1984-3143-ar2022-0096
Bathgate R (2011) Antioxidant mechanisms and their benefit on post-thaw boar sperm quality. Reprod Domest Anim 46:23–25. https://doi.org/10.1111/j.1439-0531.2011.01826.x
Bathgate R, Maxwell W, Evans G (2006) Studies on the effect of supplementing boar semen cryopreservation media with different avian egg yolk types on *in vitro* post-thaw sperm quality. Reprod Domest Anim 41:68–73. https://doi.org/10.1111/j.1439-0531.2006.00623.x
Bianchi I, Calderam K, Maschio ÉF, Madeira EM, da Rosa Ulguim R, Corcini CD, Bongalhardo DC, Corrêa ÉK, Lucia T, Deschamps JC, Corrêa MN (2008) Evaluation of amides and centrifugation temperature in boar semen cryopreservation. Theriogenology 69:632–638. https://doi.org/10.1016/j.theriogenology.2007.11.012
Buranaamnuay K, Tummaruk P, Singlor J, Rodriguez-Martinez H, Techakumphu M (2009) Effects of straw volume and Equex-STM® on boar sperm quality after cryopreservation. Reprod Domest Anim 44:69–73. https://doi.org/10.1111/j.1439-0531.2007.00996.x
Bustamante-Filho IC, Torres MA (2024) Análises avançadas do sêmen suíno. In: Reprodução animal: suínos e aves, pp 149–167
Caamaño JN, Tamargo C, Parrilla I, Martínez-Pastor F, Padilla L, Salman A, Fueyo C, Fernández Á, Merino MJ, Iglesias T, Hidalgo CO (2021) Post-thaw sperm quality and functionality in the autochthonous pig breed Gochu Asturcelta. Animals 11:1885. https://doi.org/10.3390/ani11071885
Carvajal G, Cuello C, Ruiz M, Vázquez JM, Martínez EA, Roca J (2004) Effects of centrifugation before freezing on boar sperm Cryosurvival. J Androl 25:389–396. https://doi.org/10.1002/j.1939-4640.2004.tb02805.x
Casas I, Althouse GC (2013) The protective effect of a 17 ° C holding time on boar sperm plasma membrane fluidity after exposure to 5 °C. Cryobiology 66:69–75. https://doi.org/10.1016/j.cryobiol.2012.11.006
Chanapiwat P, Kaeoket K, Tummaruk P (2011) Cryopreservation of boar semen by egg yolk-based extenders containing lactose or fructose is better than sorbitol. Theriogenology 74:351–354. https://doi.org/10.1292/jvms.11-0273
Contreras MJ, Núñez-Montero K, Bruna P, García M, Leal K, Barrientos L, Weber H (2022) Bacteria and boar semen storage: progress and challenges. Antibiotics 11:1796. https://doi.org/10.3390/antibiotics11121796
Crespo-Félez I, Castañeda-Sampedro A, Sánchez DI, Fernández-Alegre E, Álvarez-Rodríguez M, Domínguez JC, Morrell JM, Martínez-Pastor F (2017) Effect of single layer centrifugation Porcicoll (70%, 80% and 90%) or supplementation with reduced glutathione, seminal plasma and bovine serum albumin on frozen-thawed boar sperm. Anim Reprod Sci 187:167–173. https://doi.org/10.1016/j.anireprosci.2017.11.002
De Andrade AFC, Grossfeld R, Knox RV (2022) *In vitro* effects of two different commercial freezing and thawing extenders on boar sperm quality. Anim Reprod Sci 236:106906. https://doi.org/10.1016/j.anireprosci.2021.106906
de Andrade A, Balogun K, Machaty Z, Knox R (2023) Effects of supplemental antioxidants on *in vitro* fertility measures for cryopreserved boar spermatozoa. Theriogenology 2:33–42
de Brito CRC, Varela Junior AS, Gheller SMM, Acosta IB, Anciuti AN, Gatti NLS, Silva EA, Knabah NW, Corcini CD (2021) High-speed centrifugation of extender of freeze-thaw boar semen. Reprod Domest Anim 56:821–825. https://doi.org/10.1111/rda.13915
De Lazari FL, Sontag ER, Schneider A, Araripe Moura AA, Vasconcelos FR, Nagano CS, Dalberto PF, Bizarro CV, Mattos RC, Mascarenhas Jobim MI, Bustamante-Filho IC (2020) Proteomic identification of boar seminal plasma proteins related to sperm resistance to cooling at 17 °C. Theriogenology 147:135–145. https://doi.org/10.1016/j.theriogenology.2019.11.023
de Mercado E, Tomás-Almenar C, Gómez-Izquierdo E (2020) Improvement of the motility of boar sperm after cryopreservation. Anim Reprod Sci 222:106610. https://doi.org/10.1016/j.anireprosci.2020.106610

Delgado-Bermúdez A (2024) Insights into crucial molecules and protein channels involved in pig sperm cryopreservation. Anim Reprod Sci 269:107547. https://doi.org/10.1016/j.anireprosci.2024.107547

Dong R, Luo L, Liu X, Yu G (2022) Effects of riboflavin on boar sperm motility, sperm quality, enzyme activity and antioxidant status during cryopreservation. Vet Med Sci 8:1509–1518. https://doi.org/10.1002/vms3.833

Du J, Shen J, Wang Y, Pan C, Pang W, Diao H, Dong W (2016) Boar seminal plasma exosomes maintain sperm function by infiltrating into the sperm membrane. Oncotarget 7:58832–58847. https://doi.org/10.18632/oncotarget.11315

Dziekońska A, Zasiadczyk Ł, Lecewicz M, Strzeżek R, Koziorowska-Gilun M, Fraser L, Mogielnicka-Brzozowska M, Kordan W (2015) Effects of storage in different semen extenders on the pre-freezing and post-thawing quality of boar spermatozoa. Pol J Vet Sci 18:733–740. https://doi.org/10.1515/pjvs-2015-0095

Fraser L, Parda A, Filipowicz K, Strzeżek J (2009) Comparison of post-thaw DNA integrity of boar spermatozoa assessed with the neutral comet assay and sperm-Sus Halomax test kit. Reprod Domest Anim 45:e155. https://doi.org/10.1111/j.1439-0531.2009.01537.x

Fraser L, Jasiewicz E, Kordan W (2014a) Supplementation of different concentrations of Orvus Es paste (OEP) to ostrich egg yolk lipoprotein extender improves post-thaw boar semen quality. Pol J Vet Sci 17:225–230. https://doi.org/10.2478/pjvs-2014-0032

Fraser L, Strzezek J, Kordan W (2014b) Post-thaw sperm characteristics following long-term storage of boar semen in liquid nitrogen. Anim Reprod Sci 147:119–127. https://doi.org/10.1016/j.anireprosci.2014.04.010

Fraser L, Zasiadczyk Ł, Strzeżek J, Strzeżek R, Karpiesiuk K (2018) Freezability and fertility of frozen-thawed boar semen supplemented with ostrich egg yolk lipoproteins. Pol J Vet Sci 21:255–263. https://doi.org/10.24425/119046

Fraser L, Wasilewska-Sakowska K, Zasiadczyk Ł, Piątkowska E, Karpiesiuk K (2021) Fractionated seminal plasma of boar ejaculates analyzed by LC–MS/MS: its effects on post-thaw semen quality. Genes (Basel) 12:1574. https://doi.org/10.3390/genes12101574

Gadani B, Bucci D, Spinaci M, Tamanini C, Galeati G (2016) Resveratrol and Epigallocatechin-3-gallate addition to thawed boar sperm improves *in vitro* fertilization. Theriogenology 90:88–93. https://doi.org/10.1016/j.theriogenology.2016.11.020

Gale I, Gil L, Malo C, González N, Martínez F (2014) Effect of Camellia sinensis supplementation and increasing holding time on quality of cryopreserved boar semen. Andrologia 47:505–513. https://doi.org/10.1111/and.12293

Galeati G, Bucci D, Nerozzi C, Gadani B, Tamanini C, Mislei B, Spinaci M (2020) Improvement of *in vitro* fertilization by a tannin rich vegetal extract addition to frozen thawed boar sperm. Anim Reprod 17:1–11. https://doi.org/10.1590/1984-3143-AR2019-0130

Gautier C, Aurich C (2022) "Fine feathers make fine birds"—the mammalian sperm plasma membrane lipid composition and effects on assisted reproduction. Anim Reprod Sci 246:106884. https://doi.org/10.1016/j.anireprosci.2021.106884

Gómez-fernández J, Gómez-izquierdo E, Tomás C, Mocé E, de Mercado E (2012) Effect of different monosaccharides and disaccharides on boar sperm quality after cryopreservation. Anim Reprod Sci 133:109–116. https://doi.org/10.1016/j.anireprosci.2012.06.010

Guo H, Wang J, Sun L, Jin X, Shi X, Lin J, Yue S, Zhou J (2021) Effects of astaxanthin on plasma membrane function and fertility of boar sperm during cryopreservation. Theriogenology 164:58–64. https://doi.org/10.1016/j.theriogenology.2021.01.007

Guthrie HD, Welch GR (2005) Impact of storage prior to cryopreservation on plasma membrane function and fertility of boar sperm. Theriogenology 63:396–410. https://doi.org/10.1016/j.theriogenology.2004.09.020

Gutiérrez-Pérez O, Juárez-Mosqueda Mde L, Carvajal SU, Ortega ME (2009) Boar spermatozoa cryopreservation in low glycerol/trehalose enriched freezing media improves cellular integrity. Cryobiology 58:287–292. https://doi.org/10.1016/j.cryobiol.2009.02.003

Hassan HA, Ahmed HS, Hassan DF (2024) Free radicals and oxidative stress: mechanisms and therapeutic targets: review article. Hum Antibodies 32:151. https://doi.org/10.3233/HAB-240011

Hensel B, Jakop U, Scheinpflug K, Mühldorfer K, Schröter F, Schäfer J, Greber K, Jung M, Schulze M (2020) Low temperature preservation of porcine semen: influence of short antimicrobial lipopeptides on sperm quality and bacterial load. Sci Rep 10:13225. https://doi.org/10.1038/s41598-020-70180-1

Höfner L, Luther A-M, Palladini A, Fröhlich T, Waberski D (2020a) Tolerance of stored boar spermatozoa to autologous seminal plasma: a proteomic and lipidomic approach. Int J Mol Sci 21:6474. https://doi.org/10.3390/ijms21186474

Höfner L, Luther AM, Waberski D (2020b) The role of seminal plasma in the liquid storage of spermatozoa. Anim Reprod Sci 220:106290. https://doi.org/10.1016/j.anireprosci.2020.106290

Hu J, Li Q, Li G, Jiang Z, Bu S, Yang H, Wang L (2009) The cryoprotective effect of trehalose supplementation on boar spermatozoa quality. Anim Reprod Sci 112:107–118. https://doi.org/10.1016/j.anireprosci.2008.04.009

Hu J-H, Zan L-S, Zhao X-L, Li Q-W, Jiang Z-L, Li Y-K, Li X (2010) Effects of trehalose supplementation on semen quality and oxidative stress variables in frozen-thawed bovine semen. J Anim Sci 88:1657–1662. https://doi.org/10.2527/jas.2009-2335

Jeong J, Sa SJ, Chung H, Baek S, Choi I (2018) A dual phosphodiesterase inhibitor, zardaverine (type 3/4), enhances motility of frozen-thawed boar sperm. Cryo Letters 39:196–200

Jin H, Choi W, Matsumara K, Hyon S, Gen Y, Hayashi M, Kawabata T, Ijiri M, Miyoshi K (2022) Cryopreservation of pig spermatozoa using carboxylated poly-L-lysine as cryoprotectant. J Reprod Dev 68:2022–2058. https://doi.org/10.1262/jrd.2022-058

Jin H, Choi W, Matsumura K, Hyon S, Gen Y, Hayashi M, Kawabata T, Ijiri M, Miyoshi K (2023) Improved fertility of frozen-thawed porcine spermatozoa with 3,3-dimethylglutaric anhydride poly-L-lysine as a novel cryoprotectant. Anim Sci J 94:e13821. https://doi.org/10.1111/asj.13821

Jofré I, Cuevas M, de Castro LS, de Agostini Losano JD, Torres MA, Alvear M, Scheuermann E, Andrade AFC, Nichi M, Assumpção MEO, Romero F (2019) Antioxidant effect of a polyphenol-rich Murtilla (Ugni molinae Turcz) extract and its effect on the regulation of metabolism in refrigerated boar sperm. Oxid Med Cell Longev 2019:2917513. https://doi.org/10.1155/2019/2917513

Johnson LA, Weitze KF, Fiser P, Maxwell WMC (2000) Storage of boar semen. Anim Reprod Sci 62:143–172. https://doi.org/10.1016/S0378-4320(00)00157-3

Kaeoket K, Chanapiwat P (2023) The beneficial effect of resveratrol on the quality of frozen-thawed boar sperm. Animals 13:2829. https://doi.org/10.3390/ani13182829

Katila T (2012) Post-mating inflammatory responses of the uterus. Reprod Domest Anim 47:31–41. https://doi.org/10.1111/j.1439-0531.2012.02120.x

Khophloiklang V, Chanapiwat P, Aunpad R, Kaeoket K (2023) Palm kernel meal protein hydrolysates enhance post-thawed boar sperm quality. Animals 13:3040. https://doi.org/10.3390/ani13193040

Khophloiklang V, Chanapiwat P, Kaeoket K (2024) Camellia oil with its rich in fatty acids enhances post-thawed boar sperm quality. Acta Vet Scand 66:6. https://doi.org/10.1186/s13028-024-00728-y

Kim S, Lee Y-J, Ji D-B, Kim Y-J (2011) Evaluation of different Cryoprotectants (CPAs) in boar semen cryopreservation. J Vet Med Sci 73:961–963. https://doi.org/10.1292/jvms.10-0345

Kim Y-M, Park S-W, Lee M-J, Jeon D-Y, Sa S-J, Jeong Y-D, Seong H-S, Choi J-W, Hochi S, Cho E-S, Chung H-J (2023) Profiling of differentially expressed proteins between fresh and frozen-thawed Duroc boar semen using protein Chip CM10. J Anim Sci Technol 65:401–411. https://doi.org/10.5187/jast.2023.e27

Knox RV, Ringwelski JM, McNamara KA, Aardsma M, Bojko M (2015) The effect of extender, method of thawing, and duration of storage on *in vitro* fertility measures of frozen-thawed boar sperm. Theriogenology 84:407–412. https://doi.org/10.1016/j.theriogenology.2015.03.029

Koppers AJ, De Iuliis GN, Finnie JM, McLaughlin EA, Aitken RJ (2008) Significado das espécies reativas de oxigênio mitocondrial na geração de estresse oxidativo em espermatozoides. J Clin Endocrinol Metab 93(8):3199–207. https://doi.org/10.1210/jc.2007-2616

Leal DF, Torres MA, Ravagnani GM, Martins SMMK, Meirelles FV, de Andrade AFC (2018) Absence of seminal plasma from sperm-rich fraction decreases boar sperm quality characteristics during the course of liquid storage. Anim Reprod Sci 198:20–26. https://doi.org/10.1016/j.anireprosci.2018.08.029

Ledwaba MR, Mphaphathi ML, Thema MA, Pilane CM, Nedambale TL (2021) Kinematic and morphological properties of large white boar sperm under induced oxidative stress. Reprod Fertil Dev 34:280–281. https://doi.org/10.1071/RDv34n2Ab87

Lee S-H, Lee S (2023) Effects of melatonin and Silymarin on reactive oxygen species, nitric oxide production, and sperm viability and motility during sperm freezing in pigs. Animals 13:1705. https://doi.org/10.3390/ani13101705

Li J, Barranco I, Tvarijonaviciute A, Molina MF, Martinez EA, Rodriguez-Martinez H, Parrilla I, Roca J (2018a) Seminal plasma antioxidants are directly involved in boar sperm cryotolerance. Theriogenology 107:27–35. https://doi.org/10.1016/j.theriogenology.2017.10.035

Li J, Parrilla I, Ortega MD, Martínez EA, Rodríguez-Martínez H, Roca J (2018b) Post-thaw boar sperm motility is affected by prolonged storage of sperm in liquid nitrogen. Cryobiology 80:119–125. https://doi.org/10.1016/j.cryobiol.2017.11.004

Li J, Wang H, Guo M, Li T, Zhang H, Zhang Q, Wang Q, Song Y, Feng H, Wei G (2023a) Exogenous spermidine effectively improves the quality of cryopreserved boar sperm. Anim Sci J 94:e13859. https://doi.org/10.1111/asj.13859

Li J, Li J, Wang S, Ju H, Chen S, Basioura A, Ferreira-Dias G, Liu Z, Zhu J (2023b) Post-thaw storage temperature influenced boar sperm quality and lifespan through apoptosis and lipid peroxidation. Animals 14:87. https://doi.org/10.3390/ani14010087

Liu C, Feng H, Han J, Zhou H, Yuan L, Pan H, Wang X, Han X, Qiao R, Yang F, Li X (2023) Effect of L-proline on sperm quality during cryopreservation of boar semen. Anim Reprod Sci 258:107359. https://doi.org/10.1016/j.anireprosci.2023.107359

López Rodríguez A, Van Soom A, Arsenakis I, Maes D (2017) Boar management and semen handling factors affect the quality of boar extended semen. Porcine Health Manag 3:15. https://doi.org/10.1186/s40813-017-0062-5

Luther AM, Waberski D (2019) *In vitro* aging of boar spermatozoa: role of sperm proximity and seminal plasma. Andrology 7:382–390. https://doi.org/10.1111/andr.12600

Malcervelli DM, Torres P, Suhevic JF, Cisale H, Fischman ML (2020) Effect of different glycerol concentrations on phosphatidylserine translocation and mitochondrial membrane potential in chilled boar spermatozoa. Cryobiology 95:97–102. https://doi.org/10.1016/j.cryobiol.2020.05.012

Malo C, Gil L, González N, Cano R, De Blas I, Espinosa E (2010) Comparing sugar type supplementation for cryopreservation of boar semen in egg yolk based extender. Cryobiology 61:17–21. https://doi.org/10.1016/j.cryobiol.2010.03.008

Malo C, Gil L, Cano R, Martínez F, García A, Jerez RA (2012) Dimethylformamide is not better than glycerol for cryopreservation of boar semen. Andrologia 44:605–610. https://doi.org/10.1111/j.1439-0272.2011.01237.x

Martín-Hidalgo D, Bragado MJ, García-Marín LJ, Gil Anaya MC (2018) Supplementation of freezing/thawing media with GSK3 inhibitor alsterpaullone does not bypass the harmful effect of cryopreservation on boar spermatozoa. Anim Reprod Sci 196:176–183. https://doi.org/10.1016/j.anireprosci.2018.08.002

Masuda Y, Kheawkanha T, Nagahama A, Kawasaki K, Konno T, Yamanaka K, Tatemoto H (2023) Antifreeze protein type III addition to freezing extender comprehensively improves post-thaw sperm properties in Okinawan native Agu pig. Anim Reprod Sci 252:107232. https://doi.org/10.1016/j.anireprosci.2023.107232

Matthijs A, Engel B, Woelders H (2003) Neutrophil recruitment and phagocytosis of boar spermatozoa after artificial insemination of sows, and the effects of inseminate volume, sperm dose and specific additives in the extender. Reproduction 125:357–367. https://doi.org/10.1530/REP.0.1250357

Mellagi APG, Will KJ, Quirino M, Bustamante-Filho IC, Ulguim RDR, Bortolozzo FP (2023) Update on artificial insemination: semen, techniques, and sow fertility. Mol Reprod Dev 90:601–611. https://doi.org/10.1002/mrd.23643

Monteiro MS, Torres MA, Passarelli MDS, Martins MP, Ravagnani GM, Papa FO, Alvarenga MA, Dell'Aqua Júnior JA, Yasui GS, Martins SMMK, de Andrade AFC (2022) Impact of cryopreservation protocols (one- and two-step) on boar semen quality at 5 °C and post-thawing. Anim Reprod Sci 247:107093. https://doi.org/10.1016/j.anireprosci.2022.107093

Moreira SSJ, Silva AM, Pereira AG, Santos RP, Dantas MRT, Souza-Júnior JBF, Snoeck PPN, Silva AR (2023) Effect of detergents based on sodium dodecyl sulfate on functional metrics of frozen–thawed collared peccary (Pecari tajacu) semen. Animals 13:451. https://doi.org/10.3390/ani13030451

Namuncura C, Sánchez R, Pezo F, Uribe P, Navarro P, Zambrano F (2020) Rest days and storage of boar semen at 17°C: effect on motility and sperm concentration. Andrologia 52:1–6. https://doi.org/10.1111/and.13578

Naumova N, Koliada A, Kuzub N, Vaiserman AM (2023) Mitochondrial modulators. In: Anti-aging pharmacology. Elsevier, Amsterdam, pp 193–226. https://doi.org/10.1016/B978-0-12-823679-6.00012-6

Okazaki T, Abe S, Shimada M (2009) Improved conception rates in sows inseminated with cryopreserved boar spermatozoa prepared with a more optimal combination of osmolality and glycerol in the freezing extender. Anim Sci J 80:121–129. https://doi.org/10.1111/j.1740-0929.2008.00612.x

Orrego MT, Melian SI, Montenegro J, Cimato AN, Cisale H, Piehl LL (2019) Boar sperm protein tyrosine phosphorylation in the presence of egg yolk soluble and low density lipoprotein fractions during cooling. Theriogenology 123:151–158. https://doi.org/10.1016/j.theriogenology.2018.09.031

Osman R, Lee S, Almubarak A, Han J-I, Yu I-J, Jeon Y (2023) Antioxidant effects of Myo-inositol improve the function and fertility of cryopreserved boar semen. Antioxidants 12:1673. https://doi.org/10.3390/antiox12091673

Parks J, Lynch D (1992) Lipid composition and thermotropic phase behavior stallion, and rooster sperm membranes of boar. Bull Cryobiol 29:255–266. https://doi.org/10.1016/0011-2240(92)90024-v

Paschoal AFL, Mellagi APG, Ferrari CV, Takeuti KL, Oliveira GDS, Bernardi ML, Ulguim RDR, Bortolozzo FP (2021) Adjusted method of penis fixation during boar semi-automatic semen collection aiming to reduce bacterial contamination. Reprod Domest Anim 56:897–904. https://doi.org/10.1111/rda.13932

Passarelli MDS, Pinoti Pavaneli AP, Mouro Ravagnani G, Pasini Martins M, Pedrosa AC, Martins SMMK, Rocha NRA, Bittar Rigo VH, Yasui G, Yeste M, de Andrade AFC (2020) Effects of different equilibration times at 5 °C on boar sperm cryotolerance. Anim Reprod Sci 219:106547. https://doi.org/10.1016/j.anireprosci.2020.106547

Passarelli MS, Pavaneli APP, de Andrade AFC (2024) Análises convencionais do sêmen suíne
Reprodução animal: suínos e aves, pp 139–148 ma-

Pavaneli APP, Passarelli MDS, de Freitas FV, Ravagnani GM, Torres MA, Martins https://
M, de Andrade AFC (2019) Removal of seminal plasma prior to liquid storag net S, De
tozoa: a practice that can improve their fertilizing ability. Theriogenolo orage of pig
doi.org/10.1016/j.theriogenology.2018.10.020 ger Acrosomal

Pavaneli APP, Recuero S, Chaves BR, Garcia-Bonavila E, Llavane s SMM, Coelho da
Andrade AFC, Yeste M (2020) The presence of seminal pl all extracellular vesi-
spermatozoa at 17 °C modulates their ability to elicit *in vi* ogy 174:60–72. https://
exocytosis. Int J Mol Sci 21:4520. https://doi.org/10

Pedrosa AC, Andrade Torres M, Vilela Alkmin D, Pi
Silveira J, de Andrade AFC (2021) Spermat
cles miRNAs as biomarkers of boar sem
doi.org/10.1016/j.theriogenology.20

Pedrosa AC, Passarelli MS, De Andrade AFC (2024) Métodos de colheita seminal de suínos. In: Reprodução animal: suínos e aves. Manole, Santana de Parnaíba, pp 133–138

Pezo F, Zambrano F, Romero F, Sanchez R (2019) Preservation of boar semen: an update. Reprod Domest Anim 54:423–434. https://doi.org/10.1111/rda.13389

Pezo F, Zambrano F, Uribe P, Risopatrón J, Moya C, de Andrade AFC, Burgos RA, Yeste M, Sánchez R (2020) Oxidative and nitrosative stress in frozen-thawed pig spermatozoa. II: effect of the addition of saccharides to freezing medium on sperm function. Cryobiology 97:5–11. https://doi.org/10.1016/j.cryobiol.2020.10.015

Pezo F, Yeste M, Zambrano F, Uribe P, Risopatrón J, Sánchez R (2021a) Antioxidants and their effect on the oxidative/nitrosative stress of frozen-thawed boar sperm. Cryobiology 98:5. https://doi.org/10.1016/j.cryobiol.2020.11.007

Pezo F, Zambrano F, Uribe P, Moya C, de Andrade AFC, Risopatron J, Yeste M, Burgos RA, Sánchez R (2021b) Oxidative and nitrosative stress in frozen-thawed pig spermatozoa. I: protective effect of melatonin and butylhydroxytoluene on sperm function. Res Vet Sci 136:143–150. https://doi.org/10.1016/j.rvsc.2021.02.006

Pezo F, Zambrano F, Uribe P, de Andrade AFC, Sánchez R (2023) Slow freezing of preserved boar sperm: comparison of conventional and automated techniques on post-thaw functional quality by a new combination of sperm function tests. Animals 13:2826. https://doi.org/10.3390/ani13182826

Pezo F, Contreras MJ, Zambrano F, Uribe P, Risopatron J, de Andrade AFC, Yeste M, Sánchez R (2024) Thawing of cryopreserved sperm from domestic animals: impact of temperature, time, and addition of molecules to thawing/insemination medium. Anim Reprod Sci 268:107572. https://doi.org/10.1016/j.anireprosci.2024.107572

Pinho RO, Lima DMA, Shiomi HH, Siqueira JB, Silva HT, Lopes PS, Guimarães SEF, Guimarães JD (2014) Effect of different cryo-protectants on the viability of frozen/thawed semen from boars of the Piau breed. Anim Reprod Sci 146:187–192. https://doi.org/10.1016/j.anireprosci.2014.02.018

Pursel VG, Johnson LA (1975) Freezing of boar spermatozoa: fertilizing capacity with concentrated semen and a new thawing procedure. J Anim Sci 40:99–102. https://doi.org/10.2527/jas1975.40199x

Quirino M, Pereira VN, Tamanini MSC, Ulguim RDR, Schulze M, Mellagi APG, Bortolozzo FP (2023) Sperm concentration of boar semen doses and sperm quality: novel perspectives based on the extender type and sperm resilience. Anim Reprod Sci 255:107293. https://doi.org/10.1016/j.anireprosci.2023.107293

Ribas-Maynou J, Mateo-Otero Y, Delgado-Bermúdez A, Bucci D, Tamanini C, Yeste M, Barranco I (2021) Role of exogenous antioxidants on the performance and function of pig sperm after preservation in liquid and frozen states: A systematic review. Theriogenology 173:279–294. https://doi.org/10.1016/j.theriogenology.2021.07.023

Roca J, Parrilla I, Rodríguez-Martínez H, Gil M, Cuello C, Vázquez J, Martínez E (2011) Approaches towards efficient use of boar semen in the pig industry. Reprod Domest Anim 46:79–83. https://doi.org/10.1111/j.1439-0531.2011.01828.x

Broekhuijse M, Parrilla I, Rodríguez-Martínez H, Martínez E, Bolarin A (2015) Boar dif-
Ro[illegible]s in artificial insemination outcomes: can they be minimized? Reprod Domest Anim
E[illegible] https://doi.org/10.1111/rda.12530
Els[illegible]nez H, Saravia F, Wallgren M, Tienthai P, Johannisson A, Vázquez JM, Martínez
Rotimi [illegible] L, Calvete JJ (2005) Boar spermatozoa in the oviduct. In: Theriogenology.
review[illegible]
Sánchez-S[illegible]m, pp 514–535. https://doi.org/10.1016/j.theriogenology.2004.09.028
Izquierd[illegible] Ojo OA (2023) Therapeutic potential of rutin in male infertility: a mini
E (2022) [illegible] 130–135. https://doi.org/10.1016/j.joim.2023.01.004
ated, cryop[illegible] [illegible]nelo SS, Martín Lluch M, de la Cruz P, Falceto MV, Córdova-
doi.org/10.1[illegible] [illegible]a O, Suárez AE, Montull T, Ansó JG, Ansó T, Gómez-Fidalgo
[illegible] of sows inseminated at single dose fixed time with refriger-
[illegible]d spermatozoa. Reprod Domest Anim 57:90–93. https://

Schäfer J, Waberski D, Jung M, Schulze M (2017) Impact of holding and equilibration time on post-thaw quality of shipped boar semen. Anim Reprod Sci 187:109–115. https://doi.org/10.1016/j.anireprosci.2017.10.014
Schulze M, Henning H, Rüdiger K, Wallner U, Waberski D (2013) Theriogenology temperature management during semen processing: impact on boar sperm quality under laboratory and field conditions. Theriogenology 80:990–998. https://doi.org/10.1016/j.theriogenology.2013.07.026
Schulze M, Ammon C, Schaefer J, Luther A-M, Jung M, Waberski D (2017) Impact of different dilution techniques on boar sperm quality and sperm distribution of the extended ejaculate. Anim Reprod Sci 182:138–145. https://doi.org/10.1016/j.anireprosci.2017.05.013
Schulze M, Schäfer J, Simmet C, Jung M, Gabler C (2018) Detection and characterization of lactobacillus spp. in the porcine seminal plasma and their influence on boar semen quality. PLoS One 13:e0202699. https://doi.org/10.1371/journal.pone.0202699
Selige C, Quirino M, Schröter F, Aepli M, Schulze M (2021) Effect of pasteurized egg yolk on the quality of cryopreserved boar semen. Reprod Domest Anim 56:1152–1155. https://doi.org/10.1111/rda.13977
Serrano Albal M, Aquilina MC, Zak LJ, Ellis PJ, Griffin DK (2023) Successful recovery of motile and viable boar sperm after vitrification with different methods (pearls and mini straws) using sucrose as a cryoprotectant. Cryobiology 113:104583. https://doi.org/10.1016/j.cryobiol.2023.104583
Serrano R, Garrido N, Céspedes JA, Fernández LG, Marín LJG, Bragado MJ (2020) Molecular mechanisms involved in the impairment of boar sperm motility by peroxynitrite—induced nitrosative stress. Int J Mol Sci 21:12–14. https://doi.org/10.3390/ijms21041208
Sharafi M, Borghei-Rad SM, Hezavehei M, Shahverdi A, Benson JD (2022) Cryopreservation of semen in domestic animals: a review of current challenges, applications, and prospective strategies. Animals 12:3271. https://doi.org/10.3390/ani12233271
Shepherd MJ, Herickhoff LA (2022) A novel experimental design for boar sperm cryopreservation. J Anim Sci 100:skac169. https://doi.org/10.1093/jas/skac169
Shi L, Zhang Y, Huang X, Shi M, Sun D, Zhang Y, Li W, Jin T, Feng J, Xing J, Li B, Cao G (2022) Effects of mitoquinone (MitoQ) supplementation during boar semen cryopreservation on sperm quality, antioxidant status and mitochondrial proteomics. Anim Reprod Sci 247:107099. https://doi.org/10.1016/j.anireprosci.2022.107099
Shiomi H, Pinho R, Lima D, Siqueira J, Santos M, Costa E, Lopes P, Guimarães S, Guimarães J (2015) Cryopreservation of Piau-breed wild boar sperm: assessment of cooling curves and centrifugation regimes. Reprod Domest Anim 50:545–553. https://doi.org/10.1111/rda.12520
Simonik O, Bubenickova F, Tumova L, Frolikova M, Sur VP, Beran J, Havlikova K, Hackerova L, Spevakova D, Komrskova K, Postlerova P (2022) Boar sperm cryopreservation improvement using semen extender modification by dextran and pentaisomaltose. Animals 12:868. https://doi.org/10.3390/ani12070868
Song C, Chang L, Wang B, Zhang Z, Wei Y, Dou Y, Qi K, Yang F, Li X, Li X, Wang K, Qiao R, Han X (2023) Seminal plasma metabolomics analysis of differences in liquid preservation ability of boar sperm. J Anim Sci 101:skad392. https://doi.org/10.1093/jas/skad392
Souza ALP, Lima GL, Peixoto GCX, de Souza Castelo T, Oliveira MGC, de Paula VV, Silva AR (2015) Sperm characteristics following freezing in extenders supplemented with whole egg yolk and different concentrations of low-density lipoproteins in the collared peccary (Pecari tajacu). Reprod Biol 15:223–228. https://doi.org/10.1016/j.repbio.2015.10.006
Svensson A, Jiwakanon J, Fossum C, Dalin AM (2021) Expression of IL-23 in gilt endometrium and oviduct after insemination with seminal plasma, spermatozoa or semen extender. BMC Res Notes 14:221. https://doi.org/10.1186/s13104-021-05630-8
Tatemoto H, Osokoshi N, Hirai M, Masuda Y, Konno T, Yamanaka K (2022) Addition of l-carnitine to the freezing extender improves post-thaw sperm quality of Okinawan native Agu pig. Theriogenology 188:170–176. https://doi.org/10.1016/j.theriogenology.2021.12.030
Thema MA, Mphaphathi ML, Ledwaba MR, Nedambale TL (2023) Sperm cryopreservation in Windsnyer boars; principles, technique, and updated outcomes. Anim Reprod 20:e20220100. https://doi.org/10.1590/1984-3143-ar2022-0100

Tomás C, Gómez-Fernández J, Gómez-Izquierdo E, De Mercado E (2014a) Effect of the holding time at 15°C prior to cryopreservation, the thawing rate and the post-thaw incubation temperature on the boar sperm quality after cryopreservation. Anim Reprod Sci 144:115–121. https://doi.org/10.1016/j.anireprosci.2013.12.011

Tomás C, Gómez-Fernández J, Gómez-Izquierdo E, Mocé E, de Mercado E (2014b) Addition of cholesterol-loaded Cyclodextrins to the thawing extender: effects on boar sperm quality. Reprod Domest Anim 49:427–432. https://doi.org/10.1111/rda.12291

Tomas C, Gómez-Fernandez J, Gómez-Izquierdo E, Gómez-Fidalgo E, Sánchez-Sánchez R, González-Bulnes A, de Mecado E (2015) Effect of the pH pre-adjustment in the freezing extender on post-thaw boar sperm quality. Cryo Letters 36:97–103

Tomás-Almenar C, de Mercado E (2022) Optimization of the thawing protocol for Iberian boar sperm. Animals 12:2600. https://doi.org/10.3390/ani12192600

Torres MA, Monteiro MS, Passarelli MS, Papa FO, Antônio J, Aqua D, Alvarenga MA, Martins SMMK, De Andrade FC (2018) The ideal holding time for boar semen is 24 h at 17 °C prior to short-cryopreservation protocols. Cryobiology 86:58–64. https://doi.org/10.1016/j.cryobiol.2018.12.004

Torres MA, Rigo VHB, Leal DF, Pavaneli APP, Muro BBD, de Agostini Losano JD, Kawai GKV, Del Collado M, Perecin F, Nichi M, Martins SMMK, de Andrade AFC (2021) The use of resveratrol decreases liquid-extend boar semen fertility, even in concentrations that do not alter semen quality. Res Vet Sci 136:360–368. https://doi.org/10.1016/j.rvsc.2021.03.011

Torres MA, Pedrosa AC, Novais FJ, Alkmin DV, Cooper BR, Yasui GS, Fukumasu H, Machaty Z, de Andrade AFC (2022) Metabolomic signature of spermatozoa established during holding time is responsible for differences in boar sperm freezability. Biol Reprod 106:213–226. https://doi.org/10.1093/biolre/ioab200

Torres P, Malcervelli DM, Cartelle JH, Fischman ML, Cisale HO (2024) Cold-induced damage in boar sperm cooled to 5°C in lactose egg yolk (LEY) medium with different glycerol concentrations. Anim Reprod Sci 264:107455. https://doi.org/10.1016/j.anireprosci.2024.107455

Waberski D, Luther A-M (2024) Boar semen storage at 5 °C for the reduction of antibiotic use in pig insemination: pathways from science into practice. Anim Reprod Sci 269:107486. https://doi.org/10.1016/j.anireprosci.2024.107486

Waberski D, Henning H, Petrunkina AM (2011) Assessment of storage effects in liquid preserved boar semen. Reprod Domest Anim 46(Suppl 2):45–48. https://doi.org/10.1111/j.1439-0531.2011.01836.x

Waberski D, Riesenbeck A, Schulze M, Weitze KF, Johnson L (2019) Application of preserved boar semen for artificial insemination: past, present and future challenges. Theriogenology 137:2. https://doi.org/10.1016/j.theriogenology.2019.05.030

Wang P, Wang Y-F, Wang C-W, Bu S-H, Hu J-H, Li Q-W, Pang W-J, Yang G-S (2014) Effects of low-density lipoproteins extracted from different avian yolks on boar spermatozoa quality following freezing–thawing. Zygote 22:175–181. https://doi.org/10.1017/S096719941200041X

Wang H, Guo M, Li T, Zhang H, Wang Q, Zhang Q, Song Y, Feng H, Li Y, Wei G, Li J (2023) Phosphocreatine addition to extender enhances the quality and antioxidant capacity of cryopreserved boar sperm. Reprod Domest Anim 58:1087–1096. https://doi.org/10.1111/rda.14404

Wang S, Wu Y, Chen M (2024) Predicting the cryotolerance of boar sperm through antioxidant stress. Reprod Domest Anim 59:e14554. https://doi.org/10.1111/rda.14554

Wiebke M, Hensel B, Nitsche-Melkus E, Jung M, Schulze M (2022) Cooled storage of semen from livestock animals (part I): boar, bull, and stallion. Anim Reprod Sci 246:106822. https://doi.org/10.1016/j.anireprosci.2021.106822

Winn E, Whitaker BD (2020) Quercetin supplementation to the thawing and incubation media of boar sperm improves post-thaw sperm characteristics and the *in vitro* production of pig embryos. Reprod Biol 20:315–320. https://doi.org/10.1016/j.repbio.2020.06.002

Wu T, Cheng F, Chen I, Yang C, Tsai M, Chang M, Wang J, Wu J (2013) The combinatorial effect of different Equex STM paste concentrations, cryoprotectants and the straw-freezing methods on the post-thaw boar semen quality. Reprod Domest Anim 48:53–58. https://doi.org/10.1111/j.1439-0531.2012.02022.x

Xu D, Wu L, Yang L, Liu D, Chen H, Geng G, Li Q (2020) Rutin protects boar sperm from cryodamage via enhancing the antioxidative defense. Anim Sci J 91:e13328. https://doi.org/10.1111/asj.13328

Xu Z, Xie Y, Wu C, Gu T, Zhang X, Yang J, Yang H, Zheng E, Huang S, Xu Z, Li Z, Cai G, Liu D, Hong L, Wu Z (2024) The effects of boar seminal plasma extracellular vesicles on sperm fertility. Theriogenology 213:79–89. https://doi.org/10.1016/j.theriogenology.2023.09.026

Yang C-H, Wu T-W, Cheng F-P, Wang J-H, Wu J-T (2016) Effects of different cryoprotectants and freezing methods on post-thaw boar semen quality. Reprod Biol 16:41–46. https://doi.org/10.1016/j.repbio.2015.12.008

Yeste M (2016) Sperm cryopreservation update: cryodamage, markers and factors affecting the sperm freezability in pigs. Theriogenology 85:47–64. https://doi.org/10.1016/j.theriogenology.2015.09.047

Yeste M, Estrada E, Rivera Del Álamo M-M, Bonet S, Rigau T, Rodríguez-Gil J-E (2014) The increase in phosphorylation levels of serine residues of protein HSP70 during holding time at 17°C is concomitant with a higher cryotolerance of boar spermatozoa. PLoS One 9:e90887. https://doi.org/10.1371/journal.pone.0090887

Yue S, Wang S, Liu X, Bian X, Ding C, Wu T, Li D, Zhou J (2023) Ameliorative effect of silymarin on the quality of frozen–thawed boar spermatozoa. Reprod Domest Anim 58:298–306. https://doi.org/10.1111/rda.14286

Zambrano F, Namuncura C, Uribe P, Schulz M, Pezo F, Burgos RA, Taubert A, Hermosilla C, Sanchez R (2021) Swine spermatozoa trigger aggregated neutrophil extracellular traps leading to adverse effects on sperm function. J Reprod Immunol 146:103339. https://doi.org/10.1016/j.jri.2021.103339

Zasiadczyk Ł, Kurpanik K, Fraser L, Kordan W (2024) Post-thaw quality of boar spermatozoa is affected by ejaculate fractions and extenders. Pol J Vet Sci 27:147–150. https://doi.org/10.24425/pjvs.2024.149345

Zeng C, Tang K, He L, Peng W, Ding L, Fang D, Zhang Y (2014) Effects of glycerol on apoptotic signaling pathways during boar spermatozoa cryopreservation. Cryobiology 68:395–404. https://doi.org/10.1016/j.cryobiol.2014.03.008

Zhang W, Yi K, Chen C, Hou X, Zhou X (2012) Application of antioxidants and centrifugation for cryopreservation of boar spermatozoa. Anim Reprod Sci 132:123–128. https://doi.org/10.1016/j.anireprosci.2012.05.009

Zhang B, Wang Y, Wu C, Qiu S, Chen X, Cai B, Xie H (2021) Freeze-thawing impairs the motility, plasma membrane integrity and mitochondria function of boar spermatozoa through generating excessive ROS. BMC Vet Res 17:127. https://doi.org/10.1186/s12917-021-02804-1

Zhang W, Li Y, Zhu Z (2022) Carboxylated ε-poly-L-lysine supplementation of the freezing extender improves the post-thawing boar sperm quality. Animals 12:1726. https://doi.org/10.3390/ani12131726

Zhang Y, Yuan W, Liu Y, Liu Y, Liang H, Xu Q, Liu Z, Weng X (2023) Plasma membrane lipid composition and metabolomics analysis of Yorkshire boar sperms with high and low resistance to cryopreservation. Theriogenology 206:28–39. https://doi.org/10.1016/j.theriogenology.2023.04.016

Zhang G, Wen F, Li Y, Sun P, Li Y, Hu Z, Wang H, Ma Y, Liang G, Chen L, Yang K, Hu J (2024a) Sulforaphane acts through the NFE2L2/AMPK signaling pathway to protect boar spermatozoa from cryoinjury by activating antioxidant defenses. Theriogenology 230:330–340. https://doi.org/10.1016/j.theriogenology.2024.09.030

Zhang S, Zhang H, Liu K, Xu X, Qin Y, Xiao L, Zhou C, Wu J, Liu Y, Bai J (2024b) Effect of cholesterol-loaded cyclodextrin treatment on boar sperm cryopreservation. Anim Biosci 37:1558–1567. https://doi.org/10.5713/ab.24.0030

Zhou J, Zhang K, Gao J, Xu J, Wu C, He M, Zhang S, Zhang D, Dai J, Sun L (2023) Effect of Poria cocos mushroom polysaccharides (PCPs) on the quality and DNA methylation of cryopreserved Shanghai white pig spermatozoa. Cells 12:1456. https://doi.org/10.3390/cells12111456

Nanoparticles Applied to Assisted Reproductive Technologies: Development, Applications, and Concerns

Gabriela Garrappa, María Jiménez Movilla, and Francisco Alberto García Vázquez

Abstract

Nanoparticle (NP) applications are growing worldwide, influencing diverse aspects of daily life. Nanotechnologies are presented as disruptive innovations with the potential to solve unsolvable situations. From self-cleaning construction materials and environmental remediation systems to advancements in the textile and cosmetics industries, the impact of these technologies is widespread. However, the most remarkable progress has been made in nanotechnological applications for biomedicine. Novel NP applications have been developed for both in vivo and in vitro diagnostics, drug delivery systems, and revolutionary oncological treatments. In the field of reproductive medicine, the development of NP-based applications has also seen significant growth, particularly in improving assisted reproductive technologies (ART). Systems based on NP functionalization with specific biomarkers for the selection of high-quality spermatozoa have gained considerable importance and are now routinely used in human fertility clinics. Despite these advances, the development and application of nanotech-

G. Garrappa
National Institute of Agricultural Technology, INTA, Rafaela, Argentina

M. J. Movilla
Department of Cell Biology and Histology, Faculty of Medicine, University of Murcia, Murcia, Spain

Pascual Parrilla Murcian Institute for Biosanitary Research (IMIB-Arrixaca), Campus of Health Sciences, Murcia, Spain

F. A. G. Vázquez (✉)
Pascual Parrilla Murcian Institute for Biosanitary Research (IMIB-Arrixaca), Campus of Health Sciences, Murcia, Spain

Department of Physiology, Faculty of Veterinary, Campus Mare Nostrum, University of Murcia, Murcia, Spain
e-mail: fragarcia@um.es

J. C. Gardón, K. Satué Ambrojo (eds.), *Assisted Reproductive Technologies in Animals Volume 2*, https://doi.org/10.1007/978-3-031-87198-6_4

nology for oocytes and embryos have been more limited. Some progress has been made in improving maturation rates by adding NP to in vitro culture media, and post-thaw development rates have shown improvement through the incorporation of NP during a vitrification process. Moreover, a system for labeling and identifying oocytes and embryos based on NP has been introduced. Recently, a novel NP-based tool has been developed for handling oocytes and embryos, showing promising results for its application in ART. This chapter covers the ultimate applications, developments, and concerns surrounding the use of nanotechnologies in ART.

Keywords

ART · Gametes · Nanotechnologies · Nanomaterial · Reproduction

1 Introduction

Assisted reproductive technologies (ART) consist of a series of sophisticated and highly complex techniques used to achieve reproductive improvements, whether for reproductive purposes in themselves or for productive purposes in the animal world (Zegers-Hochschild et al. 2017). In recent decades, the use of these technologies has undergone rapid development, mainly due to the increase in fertility problems in human couples (Chambers et al. 2021) but also because these technologies have become important tools in the search for improvements in productive results (efficiency and quality of livestock production), as well as an essential technology to help with the conservation of wild endangered species (Bolton et al. 2022; García-Vázquez et al. 2019; Hansen 2020). However, besides the great advances achieved in these technologies, there is still plenty of room for improvement. On the other hand, nanotechnology is a relatively new field of science in charge of studying molecules within the nanoscale, known as nanoparticles (NP) (Durfey et al. 2019; Hill and Li 2017). This new area of knowledge has recently arrived as a disruptive technology with the potential to address unresolved challenges across various sectors, especially in biomedicine. In this sense, nanotechnology not only helps overcome existing technological obstacles but also creates opportunities for the development of novel and groundbreaking innovations that once seemed impossible (Bilgili and Uysal 2019). Since nanotechnology allows to elaborate, characterize, and adapt specifically the functional properties of NP for many uses, numberless nano-applications have attracted followers in almost all aspects of life (Angelakeris 2017; Mehta 2017; Mohammed et al. 2017). The advantages of using nanotechnologies are being exploited in research, industrial communities, biomedicine, environmental processes and many other fields (Berry and Curtis 2003; Feugang 2017; Mehta 2017). In an overly optimistic view, it has been proposed that nanotechnologies have the potential to change the global economy, leading to changes of incalculable dimensions in human life over the next 15–20 years (Bilgili and Uysal 2019). In reproduction, nanotechnology has emerged as a tool to fill the existing lacunae in many unresolved limitations in ART. In concordance, the development of NP-based applications in reproduction also experienced significant growth, mainly with the development of many nanotechnologies for the selection and separation of sperm (Durfey et al. 2019; Feugang et al. 2019). Meanwhile, the development and application of nanotechnology for oocytes and embryos were scarce but serve a range of utilities, from improvements in in vitro culture media, vitrification processes, oocyte and embryo labels to a novel system for oocyte and embryo magnetic-assisted manipulation methods (Fynewever et al. 2007; García-Vázquez et al. 2024; Kim et al. 2018, 2021; Novo et al. 2013; Remião et al. 2018). However, the particularities that make NP so interesting for reproductive uses also represent a potential risk of toxicity and there is a general concern on its effects on reproductive cells and organs. Therefore, researchers have a special interest in studying the potential toxicity associated with NP application (Markides et al. 2012). This chapter delves into the intersection of nanotechnology and ART providing a summary of the key developments in this area. Additionally, we address the major concerns surrounding the use of this controversial technology in reproductive tissues.

2 Nanoparticles

Although the overall concept of NP defines them as particles between 1 and 1000 nm in size (Kreuter 1996), there is no consensus on the limits of the size of a particle to be considered a NP (Lu et al. 2007). Even though a large number of authors point out the importance of reduced size for their use on organisms (e.g., the recommended size for its uses in cancer treatment varies between 70 and 200 nm), size influences both the clearance and the biodistribution of NP (Gaumet et al. 2008). In this sense, exposure routes and duration, dose, size, shape, charge, composition, and surface coating of NP are some of the key factors in determining their impact (Silva et al. 2021). As a general rule, all NP should have an exceptionally high surface area/volume ratio; this is one of the reasons for some of its most impressive properties (Christian et al. 2008).

In general terms, a NP consists of a basic structure divided into two or three layers, a core material (which usually gives it its name), a protective layer that covers the core, and a functionalizable surface layer, responsible for the functionality of the NP (Christian et al. 2008; Feugang 2017). Different classifications have been proposed according to specific particularities of NP. For example, a widely used criterion is to classify the NP (1) by its shape as spherical, cylindrical, tubular, conical, hollow core, spiral, flat, etc.; (2) by its application or utility for therapeutic, diagnosis, nutrition, etc. (El-Sayed and Kamel 2020); or (3) by its composition, which is divided into two types: organic and inorganic. Organic nanomaterials such as proteins, lipids and polymers are the most widely used in the biomedical field for drug delivery, as they are highly biodegradable and non-toxic. In the inorganic NP group, many subcategories can be recognized: carbon-based (fullerenes, nanotubes), semiconductors (cadmium, selenium, etc.), metal-based (gold, silver, platinum), and oxides-based (magnetic and superparamagnetic iron oxides) (Anselmo and Mitragotri 2016; Barkalina et al. 2016; Ealias and Saravanakumar 2017).

Within the great variety of types of NP, magnetic nanoparticles (MNP) have gained great interest in the last decades, mainly due to their characteristics and properties. A number of high-impact applications have been developed with MNP, with special emphasis on the biomedical industry (Cardoso et al. 2017; Huang and Juang 2011; Mohammed et al. 2017; Reddy et al. 2012).

The MNP usually constituted two different parts: a magnetic core material (cobalt, nickel, and iron oxide being the most popular) and a functionalization coating (Cristea et al. 2017; Kozissnik et al. 2013) (Fig. 1). Their nucleus gives them the peculiarity of being manipulated and directed to a defined location, using external magnetic fields, while their coating (silica, phospholipids, fatty acids, polysaccharides, peptides, or other surfactants and polymers) allows researchers to functionalize them according to almost every interest (Cristea et al. 2017; Lu et al. 2007; Xiao and Du 2020).

Moreover, MNP have many special attributes, one of the most fundamental being their biotolerance. In general, all iron oxide MNP, also called superparamagnetic iron oxide NP (SPIONs), are well tolerated by the cells since this kind of iron is endogenous to the body. Furthermore, most of SPIONs are coated with materials that are highly biocompatible and also allow to functionalize the external cover with molecules such as drugs, proteins, plasmids, etc. (Neuberger et al. 2005). Other properties of SPIONs include their superparamagnetism, which prevents these MNP from retaining magnetism once the magnetic field is removed, a process known as hysteresis (Xiao and Du 2020). This feature is essential for the use of NP in vivo, as it prevents the formation of agglomerates that could cause embolisms (Berry and Curtis 2003; Feugang 2017; Kim et al. 2012b; Singh et al. 2010). Also, their limited size (nanometer) and huge surface area to mass ratio are benefits for biological novel applications (Lu et al. 2007; Singh et al. 2010). Medical imaging,

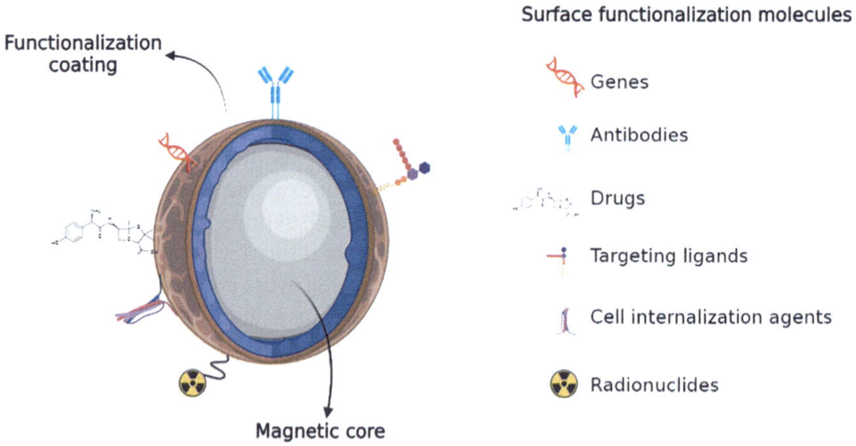

Fig. 1 Schematic image of a magnetic nanoparticle composition

magnetically guided vectors, targeted drug delivery, gene therapy, protein purification, biolabeling, bioseparation, magnetic resonance imaging, and many other applications give MNP a privileged place in biomedicine (Kim et al. 2012a, b; Laurent et al. 2014).

3 Nanotechnology Applied to Assisted Reproductive Technologies (ART)

The term ART was defined as all interventions that include the in vitro handling of oocytes and sperm or embryos for a reproductive purpose, representing a set of highly sophisticated techniques and improvements in order to bring reproduction (Zegers-Hochschild et al. 2017). In recent decades, ART has experienced a spectacular increase in demand, mainly due to the increase in fertility problems in human couples (Chambers et al. 2021) but also due to the productive improvements that their implementation in livestock entails and the possibility of their application in the conservation of the genome of endangered species (Bolton et al. 2022; García-Vázquez et al. 2019; Hansen 2020).

This increase in demand came along with a significant advancement in the development and effectiveness of these techniques, driven by the medical and scientific community's interest in further improving them (Sutcliffe and Ludwig 2007). Techniques like in vitro fertilization (IVF) have become essential today and they are now being implemented even in middle-income countries, with some governments offering them to their citizens (Adamson et al. 2006). Moreover, other technologies have also arrived and revolutionized ART as we know, intracytoplasmic sperm injection (ICSI), cryopreservation of gametes and embryos, or novel systems for embryo culture (El-Toukhy 2019; Hwang et al. 2010).

Even though there have been recent advances in this field, many challenges and limitations remain to be addressed (Mishra and Lohiya 2016). In this context, nanotechnology has found a more productive and broad niche for developing applications aimed at improving ART. For some time now, biomedicine has been focused on developing nanotechnological applications to enhance the diagnosis and treatment of fertility issues (Shandilya et al. 2020). The study and development of new uses of nanotechnology for ART could represent a significant advancement in improving these techniques (Table 1).

Applications of nanomaterials were developed mainly directed to the male gamete, as tools for high-quality spermatozoa selection (Durfey et al. 2019; Feugang et al. 2019). Many NP uses to improve sperm-related biotechnologies were reported, such as sex sorting (Rath et al. 2013), sperm quality selection (Durfey et al. 2019; Feugang et al. 2015, 2019; Odhiambo et al. 2014; Yousef et al. 2020), sperm-mediated gene transfer (Kim et al. 2009; Wang et al. 2017), spermatozoa imaging and detection (Vasquez et al. 2016), and the protection of high-quality sperm to extend their preservation (Jahanbin et al. 2021; Sánchez-Rubio et al. 2020). Recently, some approaches have even been made for the production of sperm microrobots, achieving control of sperm movement by MNP and producing

Table 1 NP development for its application in ART

Type of NP	Model/ species	Application	Reference
Sperm			
AuNP	General model	Sex sorting	Rath et al. (2013)
AuNP	Bovine	Sex sorting	Mancini et al. (2015)
MNP	Horse-donkey	Sex sorting	Domínguez et al. (2018), Ramírez Castex and Losinno (2018)
MNP + IGM	Bovine	Sex sorting	Jo et al. (2021)
MNP + anti-ubiquitin + PNA/ PSA	Bovine	Sperm quality/ sperm magnetic cell sorting	Odhiambo et al. (2014)
MNP+ lectins (PNA/PSA)	Porcine	Sperm quality/ sperm magnetic cell sorting	Feugang et al. (2015)
MNP + annexin V/PNA/LCA/ PSA	Porcine	Sperm quality/ sperm magnetic cell sorting	Durfey et al. (2019)
MNP + PNA	Donkey	Sperm quality/ sperm magnetic cell sorting	Yousef et al. (2020)
MNP + annexin V	Human	Sperm quality/ sperm magnetic cell sorting	Dirican et al. (2008), Romany et al. (2010), Khalid et al. (2011)
MPN + PNA + LCA + annexin V	Camel	Sperm quality/ sperm magnetic cell sorting	Rateb (2021)
MNP + anti-ubiquitin	Buffalo	Sperm quality/ sperm magnetic cell sorting	Bisla et al. (2021)
MNP + DNA	Porcine	Nano-SMGT (sperm-mediated gene transfer)	Kim et al. (2009)
MNP + DNA	Bovine	Nano-SMGT (sperm-mediated gene transfer)	Campos et al. (2011)
MNP + DNA	Mice	Nano-SMGT (sperm-mediated gene transfer)	Wang et al. (2017)
Zinc (Zn) NP	Bovine	MNP as antioxidant during semen processing and cryopreservation	Jahanbin et al. (2021)
Vitamin E nanoemulsions	Red deer	NE against oxidative stress	Sánchez-Rubio et al. (2020)

(continued)

Table 1 (continued)

Type of NP	Model/ species	Application	Reference
ZnO NP	Ram	Antioxidant effects	Zhang et al. (2015)
ZnO NP	Rats	Antioxidant effects	Afifi et al. (2015)
ZnO NP	Bovine	Antioxidant effects	Yazdanshenas et al. (2016)
ZnO NP	Bovine	Antioxidant effects	Farhadi et al. (2022)
ZnO NP	Human	Antioxidant effects	Isaac et al. (2017)
CeO_2 NP	Rats	Antioxidant effects	Moridi et al. (2018)
CeO_2 NP	Human	Antioxidant effects	Hosseinmardi et al. (2022)
MNP, positively and negatively charged	Porcine	Sperm movement control utilizing surface-charged magnetic nanoparticles	Chang et al. (2019a)
MNP	Bovine	IRONSperm: Sperm-templated soft magnetic microrobot	Magdanz et al. (2020)
Oocytes/embryos			
Polystyrene-based nanoparticles	Mice	NP applications for in vitro embryo tagging	Fynewever et al. (2007)
Polysilicon microparticles (barcodes)	Mice	Microinjection of encoded polysilicon microparticles (barcodes) into the perivitelline space of pronuclear (PN) embryos	Novo et al. (2011)
Polysilicon microparticles (barcodes)	Human	Tagging system for oocytes for ICSI and embryos	Novo et al. (2014b)
Polysilicon microparticles (barcodes)	Bovine	The targeting of bovine embryos	Novo et al. (2014a)
TRITC-Loaded PLGA NP	Mice	Safe and biocompatible nanocarrier inside oocytes and embryos	Kim et al. (2018); Kim et al. (2021)
Lipid-core NP + melatonin	Bovine	NP for improvement of IVM	Remião et al. (2016)

(continued)

Table 1 (continued)

Type of NP	Model/ species	Application	Reference
Se, cu, and ZnO NP	Bovine	NP for improvement of IVM	Abdel-Halim and Helmy (2017), Abdel-Halim et al. (2018)
CeO_2 NP	Ovine	NP for improvement of IVM	Ariu et al. (2017)
Chitosan NP	Porcine	NP for improvement of IVM	Roy et al. (2021)
HA-NP	Porcine	NP for improvement of oocyte vitrification results	Zhou et al. (2015)
NP (HA, SO_2, Al_2O_3, and TiO_2)	Porcine	NP for improvement of GV oocyte vitrification results	Li et al. (2016)
MNP	Mice	NP for improvement of GV oocyte vitrification results	Abbasi et al. (2021)
Fe_3O_4 NP + OVGP1	Porcine, rabbit	Magnetic-assisted control of egg and embryos	García-Vázquez et al. (2024)

microswimmers that emulate the movement of motile sperm cells (Chang et al. 2019a; Magdanz et al. 2020). On the other hand, the development and application of nanotechnology for oocytes and embryos are scarce, though this area is increasingly attracting the interest of researchers (Fynewever et al. 2007; Kim et al. 2018, 2021; Novo et al. 2013; Remião et al. 2018). Novel techniques for targeting and labeling oocytes and embryos based on NP adhesion to the external surface of the ZP were developed (Santaló Pedro et al. 2010; Novo et al. 2014a, b). Moreover, a revolutionary system for magnetic-assisted control of oocytes and embryos via ZP-linked NP was presented to the world recently (García-Vázquez et al. 2024).

3.1 Nanotechnology Applied to Sperm

3.1.1 Sperm Magnetic Sorting

Recently, novel techniques for the selection and separation of abnormal spermatozoa were developed based on nanotechnological tools due to the many advantages MNP applications offer (Feugang et al. 2015). Simplicity and low cost, along with high sensitivity and specificity, make nanotechnologies an excellent tool (Odhiambo et al. 2014; Said et al. 2008). Biomarker-based sperm selection has undergone a great deal of development and application since abnormal sperm have been shown to have a particular expression of surface determinant molecules (Feugang et al. 2019; García-Vázquez et al. 2020). The basis of the technology is simple; the functionalization of the MNP with these molecules allows the separation of damaged spermatozoa through their selective binding and subsequent separation by applying an external magnetic field (Bisla et al. 2021; Gil et al. 2013; Kang and Park 2005; Yousef et al. 2020).

For example, when acrosomal damage happens, glycans are exposed to the surface, and a variety of lectins have the ability to bind specifically to these glycoconjugates and can be excellent candidates for use in this technique (Graham et al. 1990; Nagy et al. 2003). Moreover, another molecule, ubiquitin, is exposed on the surface of abnormal sperm, so an anti-ubiquitin antibody could be employed to recognize defective sperm and remove them from the semen sample (Bisla et al. 2021).

Among the first published applications of this technology was its use for the separation of apoptotic spermatozoa in human ejaculates by functionalizing MNP with annexin V for its ability to target apoptotic sperm (Dirican et al. 2008). This method was proposed as a tool for the improvement of ART outcomes, especially in the conception rate of couples with unexplained infertility (Dirican et al. 2008; Khalid and Qureshi 2011; Romany et al. 2010). The integration of this method as part of sperm preparation techniques is currently used routinely in human reproduction laboratories (Falchi et al. 2018; Zhang et al. 2018).

On the other hand, in the animal world, one of the first approaches in this area was made by Odhiambo et al. in 2014 (2014) using bull semen. In this study, the authors achieve sustainable improvement in viability and fertilizing ability in vitro and in vivo by magnetic nanopurification of spermatozoa. About 30% of abnormal sperm were removed from the sample, using two different functionalized MNP: one coated with the lectin peanut agglutinin and another coated with an anti-ubiquitin antibody (Odhiambo et al. 2014). Other subsequent work has shown the efficiency of lectins (Feugang et al. 2015). The results showed an improvement in sperm motility after nanopurification, showing no detrimental effect either in sperm fertilization capacity or in piglet viability. Moreover, Durfey et al. (2019) found similar results, an upgrade in boar sperm motility on nanoselected spermatozoa, employing MNP functionalized with annexin V for apoptotic sperm elimination (Gil et al. 2013). Similarly, significant improvements have been achieved in studies involving donkey (Yousef et al. 2020), camel (Rateb 2021), and buffalo (Bisla et al. 2021) semen samples.

3.1.2 Sperm-Mediated Gene Transfer

The pioneering study of Brackett et al. in 1971 (1971) describes for the first time that sperm cells have the capacity to capture foreign DNA. This knowledge was the basis for the development of a method described by Lavitrano et al. (1989) based on this intrinsic ability of sperm cells to bind and internalize exogenous DNA and transfer it inside the oocyte at fertilization (Lavitrano et al. 1989, 1992) called sperm-mediated gene transfer (SMGT). SMGT represents a simple and low-cost alternative technology for producing transgenic animals (Katebi et al. 2016; Lavitrano et al. 1992). This promising technique has been widely applied in a wide variety of animal species (Gadea et al. 2012; García-Vázquez et al. 2009; Ma et al. 2011). Notwithstanding, the major limitation of this method is its low efficiency, mostly due to the low uptake of exogenous DNA by spermatozoa (García-Vázquez et al. 2011; Lavitrano et al. 2006; Spadafora 2007) but also because the spermatozoon can be severely damaged during the process, beyond the possibility of repair (Ma et al. 2011; Spadafora 2007).

Focusing on tools to improve sperm cells' DNA uptake and survival rate, many approaches have been proposed (Wang et al. 2017). In this sense, MNP exhibits several valuable properties that make it a well-suited tool for its use as a gene carrier for SMGT. MNP have the ability to carry long DNA strands, exhibit very low immunogenicity, are easily functionalized with desired molecules, possess high cell penetration capability, and are relatively low-cost (Kim et al. 2012a, b; Makhluf et al. 2006; Wang et al. 2017). Magnetofection is a technique based on the possibility of associating a transgene with MNP and introducing these complexes into a cell (sperm in this case) by applying a magnetic field and has the advantage of creating an improved expression of the transgene and minimizing the damage to the cell (Kim et al. 2009; Makhluf et al. 2006). Additionally, complexes DNA and MNP were demonstrated to be bound to spermatozoa at a higher efficiency when compared to traditional methods (just DNA incubation or lipofection). SMGT-NANO was successfully applied to produce transgenic pigs (Kim et al. 2009), mice (Wang et al. 2017), and cattle (Campos et al. 2011).

3.1.3 Sperm Antioxidant Activity

Cryopreservation and storage of semen at −196 °C in liquid nitrogen is the most widely used technique to maintain the fertilizing capacity of spermatozoa in cattle and humans for an unlimited time (Bisla et al. 2020a, b, 2021; Di Santo et al. 2012). Controversially, there is a large amount of evidence indicating that this procedure could cause damage to spermatozoa (DNA damage, aneuploidy, and chromosome fragmentation, among others) (Jahanbin et al. 2021; Ozkavukcu et al. 2008). These damages would result in a decrease in viable sperm cells and, consequently, worse reproductive rates. Oxidative damage is one of the major limitations on sperm quality conservation after cryopreservation (Bisla et al. 2020a), mostly due to changes in the sperm plasma membrane and nucleus (Bisla et al. 2020a, b). In any living cell, there is a balanced system that maintains the generation of reactive oxygen species (ROS) in correct equilibrium. When something alters this equilibrium, leading to an imbalance between the production of ROS and the possibility of clearing these molecules, a phenomenon known as oxidative stress begins (Cetica et al. 2001).

The use of numerous antioxidants for ROS elimination and/or neutralization has been proposed in order to reduce the negative effects of semen cryopreservation (Kumar et al. 2022, 2019). In this sense, metallic oxides of NP have been postulated as antioxidants with great effectiveness in reducing the damage produced by cryopreservation in sperm. For example, the use of zinc oxide NP (ZnO NP) was tested with a significant improvement in sperm quality in ram by oral administration

(Zhang et al. 2015) and in rats by intraperitoneal path (Afifi et al. 2015). Moreover, several studies have been testing the supplementation of semen samples with ZnO NP as antioxidants in order to achieve an improvement in sperm quality, viability, and fertilizing capacity. The addition of different volumes of ZnO NP on bull semen samples showed an improvement in mitochondrial sperm activity, a decrease in malondialdehyde (MDA) (one of the final products of fatty acid peroxidation in cells) (Yazdanshenas et al. 2016), an enhancement on the functionality of sperm plasma membrane (Jahanbin et al. 2015), and an increase in blastocyst rate with no adverse effect on further embryo development and pregnancy rate (Jahanbin et al. 2021). Also, the addition of these NP in the cryopreservation medium reduced the injury to human spermatozoa due to freeze and thaw techniques (Isaac et al. 2017).

The administration of cerium oxide NP (CeO_2 NP) has also been successfully tested intraperitoneally in rats (Moridi et al. 2018), and by supplementation in semen samples, achieving an improved sperm quality and sperm DNA integrity, along with a reduction in membrane lipid peroxidation (Falchi et al. 2018; Hosseinmardi et al. 2022).

3.1.4 Sperm Movement Control Based on Surface-Charged MNP

Over the past decade, the development of microrobots has emerged in the scientific community, with researchers from a wide variety of fields making significant contributions to this area (Giltinan et al. 2021; Sitti and Wiersma 2020). Moreover, the advances of nanotechnological applications have given even more impetus to the development of microrobots or better called nano-microrobots (Shen et al. 2023). The moving control of these tiny robots is realized by humans or computer programs, based on optical, acoustic, or magnetic fields (Sitti and Wiersma 2020). Between these sources, magnetic field control has undergone more impressive innovation and application in microrobot development. This is largely due to the advantages this technology offers for medical applications, including being harmless to organisms and allowing for easy, precise, and fast remote control (Chang et al. 2019a; Giltinan et al. 2021; Shen et al. 2023; Sitti and Wiersma 2020).

Among many limitations, asthenozoospermia (reduced percentage of motile sperm in the ejaculate below the lower reference limit) is a substantial issue related to male fertility alteration (Chang et al. 2019a; Singh et al. 2020). In this condition, the spermatozoon is normal and healthy but incapable of moving effectively in order to achieve oocyte fertilization. Although there are technologies capable of performing IVF, independent to a certain extent from the movement of the sperm, they are pretty complex, and their cost is often excessive (Singh et al. 2020). In this sense, several studies have been carried out in an attempt to provide the sperm with the ability to move quickly and effectively and to reach the oocytes based on a magnetic field.

One of the first approaches was made by Khalil et al. (2014), where they demonstrated the possibility of imitating the shape and movement of sperm in a self-propelled microrobot referred to as MagnetoSperm. These microrobots consist of a magnetic head and a flexible tail that can swim using a propulsion mechanism based on oscillating weak magnetic fields.

Some years later, Chang et al. (2019a) designed a mechanism to manipulate sperm movement by an external magnetic field using surface-charged MNP. This advance demonstrates that the simple fact of being the spermatozoa membrane is negatively charged allows it to be attracted by MNP with a positive charge and when this duo is subjected to a magnetic field, the NP are attracted, mobilizing the sperm. In addition, the authors postulate that the use of a magnetic field and iron oxide (Fe_3O_4) MNP has the advantage of being harmless to the organisms in which it must be used (Chang et al. 2019a). Some years later, in 2020, a new technique for the production of soft magnetic microswimmers that emulate sperm movement via electrostatic self-assembly propulsion was presented (Khalil et al. 2020; Magdanz et al. 2020). The authors proposed that applying a partial coating of the sperm head with NP through electrostatic self-assembly enables the propagation of motion waves in the flagellum under the influence of a magnetic field (Khalil et al. 2020; Magdanz et al. 2020). This technology was called IRONSperm, and the biocompatibility of these microrobots was successfully demonstrated.

3.2 Nanotechnology Applied to Oocytes and Embryos

3.2.1 Use of Nanoparticles and Microparticles Bound to ZP for Targeting and Drug Delivery (Within the Oocyte-Embryo)

The zona pellucida (ZP) is a relatively thick acellular glycoprotein matrix that fully surrounds and protects the oocyte's plasma membrane (Bleil and Wassarman 1980; Shabanowitz and O'Rand 1988). ZP has a vital role in mammalian oocytes and embryos up to the early blastocyst stage (Wassarman 2008). This external structure is responsible for the protection of the oocytes from the foreign environment but also has fundamental functions on fertilization (prevents polyspermy) and early development of the embryo prior to implantation (Bleil and Wassarman 1980; Kim et al. 2021; Shabanowitz and O'Rand 1988).

However, the presence of ZP is responsible for making it extremely difficult to achieve delivery of molecules such as drugs or genes to oocytes/embryos using traditional carriers (Kim et al. 2021). In this respect, several advances have been proposed regarding the addition of molecules (either microparticles or NP) to the external part of the ZP in oocytes and embryos (Fynewever et al. 2007; Novo et al. 2013). Additionally, there have been efforts to facilitate the entry of these molecules through the ZP (Kim et al. 2018, 2021) either for tagging purposes or for drug and factor delivery. In this sense, one of the earliest approaches was conducted by Fynewever et al. (2007), which aimed to compare different methods of NP applications for embryo tagging in mice. The study compared the intracytoplasmic injection of NP into embryos with the external application of NP to the ZP for tagging. The authors concluded that external tagging was more effective, but only up to the blastocyst stage, as the presence of NP beyond this stage inhibited embryo hatching.

Furthermore, a novel method for labeling oocytes and embryos using encoded microparticles was developed and patented by Santaló Pedro et al. (2010). In this method, micrometer-sized polysilicon barcodes were attached to the ZP for individual labeling of oocytes and embryos, with mouse samples used as the model. Three different biomolecules were tested for barcode binding to the ZP: two lectins, wheat germ agglutinin (WGA) and phytohemagglutinin-L, and an anti-ZP2 antibody. WGA demonstrated to be the most suitable biomolecule for the attachment to the ZP, as it remained bound to the ZP during and after the zona reaction and persisted up to the blastocyst stage in in vitro embryo culture (Novo et al. 2013). In another approach, these authors analyzed the possibility of binding these polysilicon barcodes to the ZP of human oocytes and embryos during ART. In this case, the tagging system was effective only for oocytes used in ICSI and for embryos, but not for oocytes destined for IVF or for sperm (Novo et al. 2014a, b). Furthermore, this technology was tested in bovine ART, where effective labeling and identification of zygotes from different donors were confirmed, allowing for successful collective embryo culture. Targeting bovine embryos with this biomolecule proved feasible, without affecting embryo development rates or vitrification/devitrification survival percentages (Novo et al. 2014a, b). This method promises a simple and secure manner to identify oocytes and embryos, ensuring safe traceability. It holds great potential application in assisted reproduction programs, both in humans and in animal production (Santaló Pedro et al. 2010).

On the other hand, Kim et al. (2018) have demonstrated the successful delivery of polylactic-co-glycolic acid NP (PLGA-NP) labeled with a fluorescence dye inside sperm, oocytes, and early embryos without harmful effects on embryo development. Further, the same group in 2021 studied the use of NP as a drug delivery carrier inside the oocytes, by passing through the ZP (Kim et al. 2021). Among three different types of NP tested, only tetramethylrhodamine-loaded PLGA-NP (TRNP) was able to completely pass through the ZP and can be proposed as a feasible and safe oocyte drug delivery carrier. Interestingly, in this work, it was found that size was not the main limiting factor for the entry of NP into the oocyte cytoplasm. In this sense, it was observed that the effect of the charge and the surface composition of the NP had a greater influence than the size on penetration into the ZP. NP with a neutral charge have shown greater acceptance to enter the oocyte.

3.2.2 NP for Enhancing ART and Developmental Competences of Oocyte in Vitro Maturation

Oocyte in vitro maturation (IVM) technique involves the collection and selection of immature oocytes (complex cumulus-oocytes) from non-stimulated females, followed by their exposure to an IVM medium to achieve gametes ready for fertilization by spermatozoa. This procedure is the first step in in vitro embryo production, and it is well-established that the quality of the embryo directly depends on the success of IVM in obtaining a high-quality mature oocyte (Ferris et al. 2016; Lonergan et al. 2003).

It has been shown that the IVM process could alter mRNA expression and DNA integrity, leading to oocyte damage that reduces its quality (Cetica et al. 2001; Lonergan et al. 2003). One of the main causes of damage to the oocyte during IVM is the increase in ROS production due to in vitro environment conditions (Cetica et al. 2001; Dalvit et al. 2005; Tamura et al. 2008; Wang et al. 2014). The supplementation of the IVM medium with antioxidant biomolecules was raised as a possible solution to oxidative stress while the oocytes are under in vitro conditions, but it was shown that these molecules are unstable under culture conditions (review by Remião et al. 2018).

In this sense, Remião et al. (2016) confirmed that the addition of nano-encapsulated melatonin significantly improves bovine embryonic development in vitro. The study demonstrated improvements in cleavage and blastocyst rate and embryo quality, reduced levels of ROS, a low number of apoptotic cells per blastocyst, and better gene expression regulation when melatonin was supplemented in nano-encapsulated form compared to its unencapsulated counterpart. The authors conclude that this effect is due to the fact that melatonin encapsulation improved its release throughout the in vitro embryo production process.

Other approaches to improving bovine oocyte IVM have involved supplementing the maturation medium with selenium (Se), copper (Cu), or zinc oxide (ZnO) NP (Abdel-Halim et al. 2018; Abdel-Halim and Helmy 2017). The addition of Se, Cu, and ZnO NP to the IVM medium resulted in better maturation rates. Additionally, a decrease in DNA damage and an increase in intracellular glutathione (GSH) concentrations in both oocytes and cumulus cells were observed.

Moreover, the addition of cerium dioxide NP (CeO_2 NP) during IVM of oocytes from prepubertal ovine was also analyzed. While these NP were internalized by cumulus cells, they were not found in the oocytes. The study observed a lower abundance of mRNA related to the apoptosis pathway and oxidative stress. It was confirmed that a low concentration of CeO_2 NP improved *in vitro* embryo production using oocytes from prepubertal ovine (Ariu et al. 2017). Additionally, another study tested the addition of chitosan NP to the IVM medium, achieving a significant reduction in oxidative damage (higher levels of GSH, lower ROS levels) and better developmental competence in porcine oocytes (Roy et al. 2021).

A recent study by our group has described the promising possibility of magnetic selection of mature oocytes using superparamagnetic nanoparticles functionalized with OVGP1 (NPOv) (García-Vázquez et al. 2024). In this study, it has been proposed that OVGP1 as a link between the NP and the ZP of the oocytes could provide these NP with the possibility of selecting mature oocytes, given their greater affinity for the ZP of effectively mature oocytes. Also, it was observed that more than 85% of the oocytes that remained unattracted by the magnetic field after a short co-incubation with NPOv were immature (García-Vázquez et al. 2024).

3.2.3 NP for ART's Improvements: Enhancing Oocytes' Survival After Vitrification

Similar to sperm cryopreservation, the cryopreservation of oocytes and embryos is a valuable tool for the conservation and storage of genetic material for future use (Bernard and Fuller 1996). Vitrification, an ultrarapid cooling technique, has emerged as one of the most effective strategies for oocyte and embryo cryopreservation (Amir and Roth 2008; Chang et al. 2019b). The main advantages of vitrification include preventing ice crystal formation within the cells and minimizing exposure to potentially toxic cryoprotective agents (Zhou et al. 2015). Despite being one of the most effective cryopreservation techniques for oocytes, vitrification still presents limitations, as fertility results and embryo development rates after warming are often inferior to those achieved with fresh gametes (Zhou et al. 2015). In this sense, the incorporation of NP (specially MNP) into cryoprotectant has been shown to enhance the survival and development rates of oocytes after vitrification (Abbasi et al. 2021; Li et al. 2016; Zhou et al. 2015).

One of the first approaches to improving porcine oocyte vitrification demonstrated that the survival rate of oocytes was significantly higher when hydroxyapatite NP (HA-NP) was added to the vitrification solution (Zhou et al. 2015). The authors hypothesized that the presence of HA-NP at an optimal concentration improved oocyte survival rates by preventing recrystallization of the solution during rewarming (Zhou et al. 2015). In a subsequent study, the vitrification of porcine germinal vesicle (GV) oocytes was analyzed using four different types of NP. HA-NP not only improved oocyte survival but also enhanced embryo development rates with minimal toxicity (Li et al. 2016).

Moreover, the effects of supplementing cryoprotectants with zinc oxide NP (ZnO NP) on vitrified bovine oocytes and embryos were also studied (Abdel-Halim and Helmy 2017). The study described a significantly higher percentage of normal oocytes after vitrification and warming when ZnO NP was added compared to the control group without NP. Likewise, blastocysts vitrified with ZnO NP showed a higher rate of embryo re-expansion rate after vitrification-devitrification than those in the control group studied (Abdel-Halim and Helmy 2017). Furthermore, the addition of MNP (Fe_3O_4 NP) into the vitrification medium was tested for the cryopreservation of mouse GV oocytes. The results indicated that Fe3O4 NP may protect immature oocytes from cryodamage and help address some of the major issues associated with the vitrification of mouse GV oocytes (Abbasi et al. 2018, 2021).

3.3 NP for ART Improvements: Enhancing Oocyte Manipulation and Culture Systems

3.3.1 Oocyte Culture Recreating the Physiological Follicular Environment

Conventional oocyte culture systems (defined culture media) do not accurately recreate the physiological conditions that occur in vivo (Rossi et al. 2019). This challenge becomes even more difficult when culturing preantral follicles in vitro, as it is essential to mimic the dynamics of the physiological environment to achieve correct cellular communication and interaction of factors necessary for producing a quality oocyte ready for fertilization (Green and Shikanov 2016). Potential applications of this technology include fertility restoration and/or preservation in women and girls undergoing aggressive chemotherapy treatments (De Vos et al. 2014; Jeruss and Woodruff 2009; Shea et al. 2014). Similarly, it could aid in genome conservation efforts for endangered species (Santos et al. 2010). In livestock, the ability to culture preantral follicles in vitro would facilitate the reproduction of prepubertal females with high genetic merit, reducing the generation interval (Baruselli et al. 2016).

Further work in this area has developed an easier and more cost-effective method for 3D culture using MNP (de Antonino 2017; de Antonino et al. 2019). This system consists of the assembly of MNP (gold and iron oxide) to the external surface of bovine follicles, followed by levitating the follicle-NP complex using an external magnetic field. This technique resulted in higher follicle viability and lower degeneration rates of follicles compared to the traditional culture systems. Moreover, the culture models used in this approach were able to sustain follicle development for up to 16 days, whereas in traditional systems, follicles typically begin to degenerate as early as day 4 (de Antonino et al. 2019).

3.3.2 Magnetic Force Application for Oocytes and Embryo Manipulation

The culture of a wide variety of cells is a routine activity in laboratories worldwide, often requiring continuous cell manipulation. In some cases, precise positioning of individual cells or groups of cells is necessary. The manipulation and positioning of cells have gained importance in recent years, driven by advances in targeted drug therapies, regenerative medicine, and reproductive technologies (Abdullah et al. 2022; Feng et al. 2021). Over the past decades, there has been a significant demand for improvements in cell handling techniques, alongside a growing need for automation in the manipulation of cells (Hu and Sun 2011). It has been suggested that cell manipulation can be carried out in two ways, with physical contact or without physical contact (Feng et al. 2021). The more traditional technique involves the use of micropipettes (Zhang et al. 2012). However, advances in nano- and microrobotics have led to numerous studies focused on achieving contactless cell manipulation which could ultimately enable the automation of cell culture processes (Feng et al. 2021).

In there production field, as the global demand and need for ART continue to increase, surpassing the current capacity of fertility clinics, there is a substantial requirement for automation in ART treatments which is crucial for enhancing the efficiency and accessibility of these procedures (Abdullah et al. 2022). Furthermore, to enhance outcomes and reduce costs, there is growing interest in minimizing reliance on highly qualified personnel. This can be achieved through the use of intelligent robots capable of performing standardized and automated tasks (Syed et al. 2020).

In recent decades, magnetic micro-/nanoparticle-based techniques have been developed to contribute to the automatization of ART (Boukallel et al. 2007). Novel systems designed for this purpose include devices for mechanical cell characterization, such as oocyte classification, and for noninvasive cell transport (Boukallel et al. 2007). The first system measures the mechanical rigidity of the oocytes for cell characterization and classification, using a micro-/nano-force sensor that levitates by a diamagnetic levitation principle. The second device, known as "micropusher," is a novel system for single-cell transportation. It consists of a ferromagnetic particle that can be mobilized by a magnetic field in a desired direction (Boukallel et al. 2007). These developments were the first approaches toward potential automation in ART.

Recently, a novel method for oocyte and embryo handling without physical contact has been developed based on the attaching magnetic NP to the ZP via the specific binding ability of the oviductin protein (NPOv) (García-Vázquez et al. 2024). This system allows the manipulation of oocytes and embryos, moving them in a desired direction by the application of an external magnetic field. This novel technology is extremely promising for its use in ART, as the absence of toxic effects of these NP on gametes and embryos has been extensively proven. Furthermore, the potential effects of this system were studied in gametes and embryos, gestational organs, and offspring, with no adverse effects in either in vivo or in vitro analysis, confirming the safety of this technology for its use in ART.

4 Concerns About the Use of Nanomaterials on Reproduction: Potential Harmful Effects and Toxicity

To ensure the safe use of NP-based applications, the absence of toxicity is one of the most critical requirements. Unfortunately, many general toxic effects of NP have been described (reviewed by Jiang et al. 2019). Toxicity from different nanomaterials has been observed in plants, fungi, vertebrates, and invertebrates (Levard et al. 2012). Moreover, the negative effects of their use were documented on the environment and human health (Klaine et al. 2008; Levard et al. 2012). As a result, a specialized field of toxicology, known as nanotoxicology, has emerged to study the adverse effects of NP on cells and organisms (Caldeira et al. 2017). NP properties that make NP so interesting for biomedical uses, such as size, shape, load, material, quantity, structure, and surface coating, are highly correlated to their potential risks (Feugang 2017; Hou and Zhu 2017; Singh et al. 2010). It has been demonstrated that NP biodistribution and accumulation in vivo are highly dependent on their size (Semmler-Behnke et al. 2014; Yang et al. 2015). A major limitation related to their small size is that the smallest NP can cross many physiological barriers, such as the blood-brain barrier (Hussain et al. 2009; Wang et al. 2010), potentially causing harmful effects on cells and organs (McAuliffe and Perry 2007).

In the field of reproduction, there has been a considerable increase in the development of NP-based applications aimed at improving reproductive technologies, which has consequently increased the need to study their potential toxicity. Research on the reproductive toxicity (reprotoxic effects) of NP remains controversial, with differing findings across studies. Some publications have specifically addressed their effects on gametes, highlighting the need for further investigation in this area (Falchi et al. 2018). Numerous NP-based applications for male gamete selection techniques have been developed (Feugang et al. 2019). Selected sperm viability was analyzed in ART protocols, and no significant toxic effects were observed. For example, studies reported no alterations in motility, functionality, or viability in bovine or porcine sperm (Caldeira et al. 2017; Durfey et al. 2019; Feugang et al. 2015). Even healthy offspring were obtained in some of these studies (Feugang et al. 2015; Makhluf et al. 2006; Odhiambo et al. 2014). Controversially, several studies have demonstrated the toxic effects of NP on sperm cells at different stages

(Ema et al. 2010; Falchi et al. 2018; Lan and Yang 2012). In vitro studies demonstrated cytotoxic effects of titanium dioxide (TiO_2) and carbon black (CB) (Komatsu et al. 2008) NP on mouse Leydig cells and alterations on mouse spermatogonia stem cells after exposure to silver (Ag), aluminum (Al), and molybdenum trioxide (MoO_3) NP (Braydich-Stolle et al. 2005). Likewise, toxic effects on human sperm motility have been observed caused by gold NP exposure (Wiwanitkit et al. 2009). In vivo studies have shown adverse effects of TiO_2 and CB NP (Hong et al. 2015; Takeda et al. 2009; Yoshida et al. 2009) on spermatogenesis in mice and of nickel (Ni) NP on sperm motility in rats (Kong et al. 2014). In contrast, no effects on sperm quality parameters were observed after exposure of boar sperm to OVGP1-functionalized iron oxide NP, neither through their presence in the IVF medium nor after their binding to the ZP surface of oocytes during the IVF protocol, without compromising sperm functionality or their ability to bind to the ZP (Garrappa et al. 2024). Also, no effects on sperm metabolisms and oxidative stress were observed after sperm contact with this NPOv (García-Vázquez et al. 2024).

Furthermore, female gametes are potentially at risk when exposed to nanoscale materials (Hou and Zhu 2017; Taylor et al. 2015; Tiedemann et al. 2014). Some studies have reported detrimental effects of different types of NP on oocyte maturation and related parameters. The addition of gold, Silver, and gold-silver alloy (Taylor et al. 2015; Huang et al. 2018), CeO_2 (Preaubert et al. 2015), and fullerenol NP (Lei et al. 2018) to in vitro maturation medium has shown an important reduction in oocyte quality, as evidenced by decreased meiosis resumption, and adverse effects on fertilization, cleavage, and blastocyst rate. On the contrary, the presence of iron oxide NP conjugated with OVGP1 bounded to the external part of ZP of matured oocytes has not shown any effects on oocyte viability, after an IVF and CE protocol (García-Vázquez et al. 2024). Also, no effects of these NP or the magnetic field were observed on embryo quality by a complete RNA-seq analysis of embryos obtained from oocytes conjugated with NPOv and their exposure to an external magnetic field (García-Vázquez et al. 2024).

In addition, evidence indicates that certain types of NP can cross reproductive barriers (Wang et al. 2018) such as the blood-testis barrier, placental barrier, and epithelial barrier, which protect reproductive tissues and the fetus. For example, the translocation of gold NP from maternal blood to the fetus was observed in mice when the particle size ranged from 1.4 to 18 nm, whereas gold particles of 80 nm could not cross the placental barrier (Kim et al. 2012b; Semmler-Behnke et al. 2014). It has also been postulated that silver NP may be more toxic to oocytes and embryos than gold NP, with toxicity linked to dosage and particular size (reviewed by Taylor et al. 2015). In this aspect, the negative effects of the presence of rabbit embryos conjugated with iron oxide NP-OVGP1 on the oviduct and uterus were evaluated. These results showed no effects on reproductive tissues, by histological analysis, neither on cell proliferation nor inflammation markers evaluated on these tissues (García-Vázquez et al. 2024).

There is currently insufficient data to fully understand the potential harmful effects of NP on reproductive organs and cells. The lack of studies, especially regarding NP toxicity on oocytes and embryos, combined with the rapid development of novel nanomaterial-based techniques to improve ART, underscores the urgent need for further research and knowledge generation (Shandilya et al. 2020; Bisla et al. 2022).

5 Concluding Remarks

This chapter provides a comprehensive overview of emerging nanotechnology applications for improving ART. In the field of reproduction, nanotechnology has found a productive and broad niche for the development of applications for ART improvement. The plasticity of nanomaterials has allowed the creation of many nanotechnology-based tools for the most varied applications, including the efficient selection of high-quality sperm and matured oocytes, improvements in culture media, protection of gametes for better cryopreservation rates, and the manipulation and movement of gametes and embryos without physical contact. Many of these applications are already successfully employed in assisted reproductive clinics. However, the exposure of gametes and embryos to NP remains a topic of debate, requiring further research on fully toxic effects.

References

Abbasi Y, Baniasadi F, Hajiaghalou S, Ghalamboran MR, Fathi R (2018) Vitrification of mouse GV-oocytes by the means of nanoparticles. Reprod Biomed Online 37:e17–e18. https://doi.org/10.1016/j.rbmo.2018.04.035

Abbasi Y, Hajiaghalou S, Baniasadi F, Mahabadi VP, Ghalamboran MR, Fathi R (2021) Fe_3O_4 magnetic nanoparticles improve the vitrification of mouse immature oocytes and modulate the pluripotent genes expression in derived pronuclear-stage embryos. Cryobiology 100:81–89. https://doi.org/10.1016/j.cryobiol.2021.03.006

Abdel-Halim BR, Helmy NA (2017) Effect of nano-selenium and nano-zinc particles during in vitro maturation on the developmental competence of bovine oocytes. Anim Prod Sci 50(11):2021–2028. https://doi.org/10.1071/AN17057

Abdel-Halim B, Moselhy W, Helmy N (2018) Developmental competence of bovine oocytes with increasing concentrations of nano-copper and nano-zinc particles during in vitro maturation. Asian Pac J Reprod 7(4):161–166. https://doi.org/10.4103/2305-0500.237053

Abdullah KAL, Atazhanova T, Chavez-Badiola A, Shivhare SB (2022) Automation in ART: paving the way for the future of infertility treatment. Reprod Sci 30(4):1006–1016. https://doi.org/10.1007/s43032-022-00941-y

Adamson GD, de Mouzon J, Lancaster P, Nygren KG, Sullivan E, Zegers-Hochschild F (2006) World collaborative report on in vitro fertilization, 2000. Fertil Steril 85(6):1586–1622. https://doi.org/10.1016/j.fertnstert.2006.01.011

Afifi M, Almaghrabi OA, Kadasa NM (2015) Ameliorative effect of zinc oxide nanoparticles on antioxidants and sperm characteristics in Streptozotocin-induced diabetic rat testes. Biomed Res Int 2015(153573):1–6. https://doi.org/10.1155/2015/153573

Amir A, Roth Z (2008) Do chilling injury and heat stress share the same mechanism of injury in oocytes? Mol Cell Endocrinol 282(1–2):150–152. https://doi.org/10.1016/j.mce.2007.11.026

Angelakeris M (2017) Magnetic nanoparticles: A multifunctional vehicle for modern theranostics. Biochim Biophys Acta Gen Subj 1861(6):1642–1651. https://doi.org/10.1016/j.bbagen.2017.02.022

Anselmo AC, Mitragotri S (2016) Nanoparticles in the clinic. Bioeng Transl Med 1(1):10–29. https://doi.org/10.1002/btm2.10003

Ariu FA, Bogliolo LA, Pinna AB, Malfatti LB, Innocenzi PB, Falchi LA, Bebbere D, Ledda S (2017) Cerium oxide nanoparticles (CeO_2 NPs) improve the developmental competence of in vitro-matured prepubertal ovine oocytes. Reprod Fertil Dev 29(5):1046–1056. https://doi.org/10.1071/RD15521

Barkalina N, Jones C, Coward K (2016) Nanomedicine and mammalian sperm: lessons from the porcine model. Theriogenology 85(1):74–82. https://doi.org/10.1016/j.theriogenology.2015.05.025

Baruselli PS, Batista EOS, Vieira LM, Ferreira RM, Guerreiro BG, Bayeux BM, Sales JNS (2016) Factors that interfere with oocyte quality for in vitro production of cattle embryos: effects of different developmental & reproductive stages. Anim Reprod 13(3):264–272. https://doi.org/10.21451/1984-3143-AR861

Bernard A, Fuller BJ (1996) Cryopreservation of human oocytes: A review of current problems and perspectives. Hum Reprod Update 2(3):193–207. https://doi.org/10.1093/humupd/2.3.193

Berry CC, Curtis ASG (2003) Functionalisation of magnetic nanoparticles for applications in biomedicine magnetic nanoparticle-based cancer therapy functionalisation of magnetic nanoparticles for applications in biomedicine. J Phys D Appl Phys 36(13):198–206. https://doi.org/10.1088/0022-3727/36/13/203

Bilgili A, Uysal MH (2019) Use areas of nanotechnology in veterinary medicine. J Food Sci Eng 9(6):210–216. https://doi.org/10.17265/2159-5828/2019.06.003

Bisla A, Rautela R, Yadav V, Saini G, Singh P, Ngou AA, Kumar A, Ghosh S, Kumar A, Bag S, Mahajan S, Srivastava N (2020a) Synthesis of iron oxide nanoparticles–antiubiquitin antibodies conjugates for depletion of dead/damaged spermatozoa from buffalo (Bubalus bubalis) semen. Biotechnol Appl Biochem 68(6):1453–1468. https://doi.org/10.1002/BAB.2066

Bisla A, Rautela R, Yadav V, Singh P, Kumar A, Ghosh S, Kumar A, Bag S, Kumar B, Srivastava N (2020b) Nano-purification of raw semen minimises oxidative stress with improvement in post-thaw quality of buffalo spermatozoa. Andrologia 52(9):1–17. https://doi.org/10.1111/and.13709

Bisla A, Rautela R, Yadav V, Alex Ngou A, Kumar A, Ghosh SK, Bag S, Srinivastava N (2021) Effect of cryopreservation on semen quality parameters in relation to lipid peroxidation and antioxidant profile in Indian Buffalo. CryoLetters 42(1):33–38

Bisla A, Honparkhe M, Srivastava N (2022) A review on applications and toxicities of metallic nanoparticles in mammalian semen biology. Andrologia 54(11): https://doi.org/10.1111/and.14589

Bleil JD, Wassarman PM (1980) Structure and function of the zona pellucida: identification and characterization of the proteins of the mouse oocyte's zona pellucida. Dev Biol 76(1):185–202. https://doi.org/10.1016/0012-1606(80)90371-1

Bolton RL, Mooney A, Pettit MT, Bolton AE, Morgan L, Drake GJ, Appeltant R, Walker SL, Gillis JD, Hvilsom C (2022) Resurrecting biodiversity: advanced assisted reproductive technologies and biobanking. Reprod Fertil 3(3):R121–R146. https://doi.org/10.1530/RAF-22-0005

Boukallel M, Gauthier M, Dauge M, Piat E, Abadie J (2007) Smart microrobots for mechanical cell characterization and cell convoying. IEEE Trans Biomed Eng 54(8):1536–1540. https://doi.org/10.1109/TBME.2007.891171

Brackett BG, Baranska W, Sawicki W, Koprowski H (1971) Uptake of heterologous genome by mammalian spermatozoa and its transfer to ova through fertilization. Proc Natl Acad Sci USA 68(2):353–357. https://doi.org/10.1073/PNAS.68.2.353

Braydich-Stolle L, Hussain S, Schlager JJ, Hofmann MC (2005) In vitro cytotoxicity of nanoparticles in mammalian germline stem cells. Toxicol Sci 88(2):412–419. https://doi.org/10.1093/toxsci/kfi256

Caldeira DF, Paulini F, Silva RC, de Azevedo RB, Lucci CM (2017) In vitro exposure of bull sperm cells to DMSA-coated maghemite nanoparticles does not affect cell functionality or structure. Int J Hyperth 34(4):415–422. https://doi.org/10.1080/02656736.2017.1341646

Campos VF, Komninou ER, Urtiaga G, de Leon PM, Seixas FK, Dellagostin OA, Deschamps JC, Collares T (2011) NanoSMGT: transfection of exogenous DNA on sex-sorted bovine sperm using nanopolymer. Theriogenology 75(8):1476–1481. https://doi.org/10.1016/j.theriogenology.2011.01.009

Cardoso VF, Francesko A, Ribeiro C, Bañobre-López M, Martins P, Lanceros-Mendez S (2017) Advances in magnetic nanoparticles for biomedical applications. Adv Healthc Mater 7(5):1700845. https://doi.org/10.1002/adhm.201700845

Cetica PD, Pintos LN, Dalvit GC, Beconi MT (2001) Antioxidant enzyme activity and oxidative stress in bovine oocyte in vitro maturation. IUBMB Life 51(1):57–64. https://doi.org/10.1080/15216540119253

Chambers GM, Dyer S, Zegers-Hochschild F, de Mouzon J, Ishihara O, Banker M, Mansour R, Kupka MS, Adamson GD (2021) International Committee for Monitoring Assisted Reproductive Technologies world report: assisted reproductive technology. Hum Reprod 36(11):2921–2934. https://doi.org/10.1093/humrep/deab198

Chang M, Chang Y-J, Wang T-Y, Yu Q (2019a) Sperm movement control utilizing surface charged magnetic nanoparticles. J Nanosci Nanotechnol 19(9):5713–5722. https://doi.org/10.1166/jnn.2019.16551

Chang H, Chen H, Zhang L, Wang Y, Xie X, Zhang Y, Quan F (2019b) Effect of oocyte vitrification on DNA damage in metaphase II oocytes and the resulting preimplantation embryos. Mol Reprod Dev 86(11):1603–1614. https://doi.org/10.1002/mrd.23247

Christian P, Von Der Kammer F, Baalousha M, Hofmann T (2008) Nanoparticles: structure, properties, preparation and behaviour in environmental media. Ecotoxicology 17(5):326–343. https://doi.org/10.1007/s10646-008-0213-1

Cristea C, Tertis M, Galatus R (2017) Magnetic nanoparticles for antibiotics detection. Nano 7(6):119. https://doi.org/10.3390/nano7060119

Dalvit GC, Cetica PD, Pintos LN, Beconi MT (2005) Reactive oxygen species in bovine embryo in vitro production. Biocell 29(2):209–212

Antonino DC de (2017) Cultivo in vitro de folículos pré-antrais bovinos asociado á nanopartículas e levitação magnética. Master Thesis-Sistema de Bibliotecas Da UFU, MG, Brasil, 72

de Antonino DC, Soares MM, de Melo Júnior J, de Alvarenga PB, de Mohallem RFF, Rocha CD, Viera LA, De Souza AG, Beletti ME, Alves BG, Jacomini JO, Goulart LR, Alves KA (2019) Three-dimensional levitation culture improves in-vitro growth of secondary follicles in bovine model. Reprod Biomed Online 38(3):300–311. https://doi.org/10.1016/j.rbmo.2018.11.013

De Vos M, Smitz J, Woodruff TK (2014) Fertility preservation in women with cancer. Lancet 384:1302–1310. https://doi.org/10.1016/S0140-6736(14)60834-5

Di Santo M, Tarozzi N, Nadalini M, Borini A (2012) Human sperm cryopreservation: update on techniques, effect on dna integrity , and implications for ART. Adv Urol 2012:854837. https://doi.org/10.1155/2012/854837

Dirican EK, Özgün OD, Akarsu S, Akin KO, Ercan Ö, Uğurlu M, Camsari C, Kanyilmaz O, Kaya A, Ünsal A (2008) Clinical outcome of magnetic activated cell sorting of non-apoptotic spermatozoa before density gradient centrifugation for assisted reproduction. J Assist Reprod Genet 25(8):375–381. https://doi.org/10.1007/s10815-008-9250-1

Domínguez E, Moreno-Irusta A, Castex HR, Bragulat AF, Ugaz C, Clemente H, Giojalas L, Losinno L (2018) Sperm sexing mediated by magnetic nanoparticles in donkeys, a preliminary in vitro study. J Equine Vet Sci 65:123–127. https://doi.org/10.1016/j.jevs.2018.04.005

Durfey CL, Swistek SE, Liao SF, Crenshaw MA, Clemente HJ, Rooban VK, Thirumalai G, Steadman CS, Ryan PL, Willard ST, Feugang JM (2019) Nanotechnology-based approach for safer enrichment of semen with best spermatozoa. J Anim Sci Biotechnol 10(1):1–12. https://doi.org/10.1186/s40104-018-0307-4

Ealias AM, Saravanakumar MP (2017) A review on the classification, characterisation, synthesis of nanoparticles and their application. IOP Conf Ser 263(3):032019. https://doi.org/10.1088/1757-899X/263/3/032019

El-Sayed A, Kamel M (2020) Advanced applications of nanotechnology in veterinary medicine. Environ Sci Pollut Res 27(16):19073–19086. https://doi.org/10.1007/s11356-018-3913-y

El-Toukhy T (2019) Reproductive medicine in 2020 and beyond. BJOG 126(2):133–134. https://doi.org/10.1111/1471-0528.15352
Ema M, Kobayashi N, Naya M, Hanai S, Nakanishi J (2010) Reproductive and developmental toxicity studies of manufactured nanomaterials. Reprod Toxicol 30(3):343–352. https://doi.org/10.1016/j.reprotox.2010.06.002
Falchi L, Khalil WA, Hassan M, Marei WFA (2018) Perspectives of nanotechnology in male fertility and sperm function. Int J Vet Sci Med 6:265–269. https://doi.org/10.1016/j.ijvsm.2018.09.001
Farhadi F, Towhidi A, Shakeri M (2022) Zinc oxide nanoparticles have beneficial effect on frozen – thawed spermatozoa of holstein bulls. Iran J Appl Ani Sci 12:49–55
Feng L, Bai X, Jia L, Zhang C, Chen Y, Chen H (2021) Field-controlled micro-nano manipulations and micro-nano robots. In: Micro and nano systems for biophysical studies of cells and small organisms. Academic Press, pp 201–225. https://doi.org/10.1016/B978-0-12-823990-2.00009-
Ferris J, Mahboubi K, MacLusky N, King WA, Favetta LA (2016) BPA exposure during in vitro oocyte maturation results in dose-dependent alterations to embryo development rates, apoptosis rate, sex ratio and gene expression. Reprod Toxicol 59:128–138. https://doi.org/10.1016/j.reprotox.2015.12.002
Feugang JM (2017) Novel agents for sperm purification, sorting, and imaging. Mol Reprod Dev 84(9):832–841. https://doi.org/10.1002/mrd.22831
Feugang JM, Liao SF, Crenshaw MA, Clemente H, Willard ST, Ryan PL (2015) Lectin-functionalized magnetic iron oxide nanoparticles for reproductive improvement. J Fertil Cell 03(02):2. https://doi.org/10.4172/2375-4508.1000145
Feugang JM, Rhoads CE, Mustapha PA, Tardif S, Parrish JJ, Willard ST, Ryan PL (2019) Treatment of boar sperm with nanoparticles for improved fertility. Theriogenology 137:75–81. https://doi.org/10.1016/j.theriogenology.2019.05.040
Fynewever TL, Agcaoili ES, Jacobson JD, Patton WC, Chan PJ (2007) In vitro tagging of embryos with nanoparticles. J Assist Reprod Genet 24(2–3):61–65. https://doi.org/10.1007/s10815-006-9084-7
Gadea J, Garcia-vazquez FA, Canovas S, Parrington J (2012) Sperm-mediated gene transfer in agricultural species. In: Sperm-mediated gene transfer: concepts and controversies. Bentham Books, pp 76–101
García-Vázquez FA, García-Roselló E, Gutiérrez-Adán A, Gadea J (2009) Effect of sperm treatment on efficiency of EGFP-expressing porcine embryos produced by ICSI-SMGT. Theriogenology 72(4):506–518. https://doi.org/10.1016/j.theriogenology.2009.04.006
García-Vázquez FA, Ruiz S, Grullón LA, de Ondiz A, Gutiérrez-Adán A, Gadea J (2011) Factors affecting porcine sperm mediated gene transfer. Res Vet Sci 91(3):446–453. https://doi.org/10.1016/j.rvsc.2010.09.015
García-Vázquez FA, Mellagi APG, Ulguim RR, Hernández-Caravaca I, Llamas-López PJ, Bortolozzo FP (2019) Post-cervical artificial insemination in porcine: the technique that came to stay. Theriogenology 129:37–45. https://doi.org/10.1016/j.theriogenology.2019.02.004
García-Vázquez FA, Luongo C, Garrappa G, Rodríguez Tobón E (2020) Reproductive biotechnologies applied to artificial insemination in swine. In: Biotechnologies applied to animal reproduction. Apple Academic Press, pp 283–323. https://doi.org/10.1201/9780367817527-12
García-Vázquez FA, Garrappa G, Luongo C, Hamze JG, Caballero M, Marco-Jiménez F, Vicente-Antón JS, Molina-Cuberos GJ, Jiménez-Movilla M (2024) Magnetic-assisted control of eggs and embryos via zona Pellucida-linked nanoparticles. Adv Sci 2306901:1–13. https://doi.org/10.1002/advs.202306901
Garrappa G, Martínez-López C, Jiménez-Movilla M, García-Vázquez FA (2024) In vitro exposure of porcine sperm to functionalized superparamagnetic nanoparticles. Reprod Domest Anim 59:e14654
Gaumet M, Vargas A, Gurny R, Delie F (2008) Nanoparticles for drug delivery: the need for precision in reporting particle size parameters. Eur J Pharm Biopharm 69(1):1–9. https://doi.org/10.1016/j.ejpb.2007.08.001
Gil M, Sar-Shalom V, Melendez Sivira Y, Carreras R, Checa MA (2013) Sperm selection using magnetic activated cell sorting (MACS) in assisted reproduction: a systematic review and meta-analysis. J Assist Reprod Genet 30(4):479–485. https://doi.org/10.1007/s10815-013-9962-8

Giltinan J, Sridhar V, Bozuyuk U, Sheehan D, Sitti M (2021) 3D microprinting of iron platinum nanoparticle-based magnetic Mobile microrobots. Adv Intell Syst 3(1):2000204. https://doi.org/10.1002/aisy.202000204
Graham JK, Kunze E, Hammerstedt RH (1990) Analysis of sperm Acrosomal flow mitochondrial function using flow cytometry. Biol Reprod 43:55–64. https://doi.org/10.1095/biolreprod43.1.55
Green LJ, Shikanov A (2016) In vitro culture methods of preantral follicles. Theriogenology 86:229–238. https://doi.org/10.1016/j.theriogenology.2016.04.036
Hansen PJ (2020) Implications of assisted reproductive technologies for pregnancy outcomes in mammals. Ann Rev Anim Biosci 8:395–413. https://doi.org/10.1146/annurev-animal-021419-084010
Hill EK, Li J (2017) Current and future prospects for nanotechnology in animal production. J Anim Sci Biotechnol 8(1):1–13. https://doi.org/10.1186/s40104-017-0157-5
Hong F, Si W, Zhao X, Wang L, Zhou Y, Chen M, Ge Y, Zhang Q, Wang Y, Zhang J (2015) TiO_2 nanoparticle exposure decreases spermatogenesis via biochemical dysfunctions in the testis of male mice. J Agric Food Chem 63(31):7084–7092. https://doi.org/10.1021/acs.jafc.5b02652
Hosseinmardi M, Siadat F, Sharafi M, Roodbari NH, Hezavehei M (2022) Protective effect of cerium oxide nanoparticles on human sperm function during cryopreservation. Biopreserv Biobank 20(1):24–30. https://doi.org/10.1089/bio.2021.0020
Hou CC, Zhu JQ (2017) Nanoparticles and female reproductive system: how do nanoparticles affect oogenesis and embryonic development. Oncotarget 8(65):109799–109817. https://doi.org/10.18632/oncotarget.19087
Hu S, Sun D (2011) Automatic transportation of biological cells with a robot-tweezer manipulation system. Int J Robot Res 30(14):1681–1694. https://doi.org/10.1177/0278364911413479
Huang SH, Juang RS (2011) Biochemical and biomedical applications of multifunctional magnetic nanoparticles: a review. J Nanopart Res 13(10):4411–4430. https://doi.org/10.1007/s11051-011-0551-4
Huang CH, Yeh JM, Chan WH (2018) Hazardous impacts of silver nanoparticles on mouse oocyte maturation and fertilization and fetal development through induction of apoptotic processes. Environ Toxicol 33(10):1039–1049. https://doi.org/10.1002/tox.22590
Hussain SM, Braydich-Stolle LK, Schrand AM, Murdock RC, Yu KO, Mattie DM, Schlager JJ (2009) Toxicity evaluation for safe use of nanomaterials: recent achievements and technical challenges. Adv Mater 21(16):1549–1559. https://doi.org/10.1002/adma.200801395
Hwang K, Yatsenko AN, Jorgez CJ, Mukherjee S, Nalam RL, Matzuk MM, Lamb DJ (2010) Mendelian genetics of male infertility. Ann N Y Acad Sci 1214:1–17. https://doi.org/10.1111/j.1749-6632.2010.05917.x
Isaac AV, Kumari S, Nair R, Raj D, Raj S, Kalthur G, Adiga SK, Manikkath J, Mutalik S, Sachdev D, Pasricha R (2017) Biochemical and biophysical research communications supplementing zinc oxide nanoparticles to cryopreservation medium minimizes the freeze-thaw-induced damage to spermatozoa. Biochem Biophys Res Commun 494(3–4):656–662. https://doi.org/10.1016/j.bbrc.2017.10.112
Jahanbin R, Yazdanshenas P, Amin Afshar M, Mohammadi Sangcheshmeh A, Varnaseri H, Chamani M, Nazaran MH, Bakhtiyarizadeh MR (2015) Effect of zinc nano-complex on bull semen quality after freeze-thawing process. Anim Prod 17(2):371–380. https://doi.org/10.22059/JAP.2015.54040
Jahanbin R, Yazdanshenas P, Rahimi M, Hajarizadeh A, Tvrda E, Nazari SA, Mohammadi-Sangcheshmeh A, Ghanem N (2021) In vivo and in vitro evaluation of bull semen processed with zinc (Zn) nanoparticles. Biol Trace Elem Res 199(1):126–135. https://doi.org/10.1007/s12011-020-02153-4
Jeruss JS, Woodruff TK (2009) Preservation of fertility in patients with cancer. N Engl J Med 360(9):902–911. https://doi.org/10.1056/nejmra0801454
Jiang Z, Shan K, Song J, Liu J, Rajendran S, Pugazhendhi A, Jacob JA, Chen B (2019) Toxic effects of magnetic nanoparticles on normal cells and organs. Life Sci 220(December 2018):156–161. https://doi.org/10.1016/j.lfs.2019.01.056

Jo S-Y, Hwangbo Y, Lee S-H, Cheong H-T, Kim D-K, Park C-K (2021) Effect of antibodies binding to Y chromosome-bearing sperm conjugated with magnetic nanoparticles on bull sperm characteristics. J Anim Reprod Biotechnol 36(4):239–246. https://doi.org/10.12750/JARB.36.4.239

Kang JH, Park J (2005) Cell separation technology. Asia-Pacific Biotech News 9(21):1135–1146. https://doi.org/10.1016/0923-1137(94)90034-5

Katebi S, Esmaeili A, Ghaedi K (2016) Static magnetic field reduced exogenous oligonucleotide uptake by spermatozoa using magnetic nanoparticle gene delivery system. J Magn Magn Mater 402:184–189. https://doi.org/10.1016/j.jmmm.2015.11.057

Khalid SN, Qureshi IZ (2011) Pregnancy rate improves in couples with unexplained infertility following intrauterine insemination (IUI) with magnetically selected non-apoptotic sperms. Fertil Steril 96(3):S25. https://doi.org/10.1016/j.fertnstert.2011.07.104

Khalil ISM, Dijkslag HC, Abelman L, Sarthak M (2014) MagnetoSperm: a microrobot that navigates using weak magnetic fields. Appl Phys Lett 104(223701):1–4. https://doi.org/10.1063/1.4880035

Khalil ISM, Magdanz V, Simmchen J, Klingner A, Sarthak M (2020) Resemblance between motile and magnetically actuated sperm cells. Appl Phys Lett 116(063702):1–5. https://doi.org/10.1063/1.5142470

Kim TS, Lee SH, Gang GT, Lee YS, Kim SU, Koo DB, Shin MY, Parck CK, Lee DS (2009) Exogenous DNA uptake of boar spermatozoa by a magnetic nanoparticle vector system. Reprod Domest Anim 206:201–206. https://doi.org/10.1111/j.1439-0531.2009.01516.x

Kim M, Shen M, Ngoy S, Karamanlidis G, Liao R, Tian R (2012a) Journal of molecular and cellular cardiology original article AMPK isoform expression in the normal and failing hearts. J Mol Cell Cardiol 52(5):1066–1073. https://doi.org/10.1016/j.yjmcc.2012.01.016

Kim JE, Shin JY, Cho MH (2012b) Magnetic nanoparticles: an update of application for drug delivery and possible toxic effects. Arch Toxicol 86(5):685–700. https://doi.org/10.1007/s00204-011-0773-3

Kim YS, Park JS, Park M, Ko MY, Yi SW, Yoon JA, Yang SC, Shim SH, Park KH, Song H (2018) PLGA nanoparticles with multiple modes are a biologically safe nanocarrier for mammalian development and their offspring. Biomaterials 183(May):43–53. https://doi.org/10.1016/j.biomaterials.2018.08.042

Kim HJ, Lee S, Lee JH, Park JM, Hong SJ, Lee OH, Park JS, Choi Y, Park KH (2021) TRITC-loaded PLGA nanoparticles as drug delivery carriers in mouse oocytes and embryos. ACS Appl Mater Interfaces 13(5):5975–5988. https://doi.org/10.1021/acsami.0c19792

Klaine SJ, Alvarez PJJ, Batley GE, Fernandes TF, Handy RD, Lyon DY, Mahendra S, McLaughlin MJ, Lead JR (2008) Nanomaterials in the environment: behavior, fate, bioavailability, and effects. Environ Toxicol Chem 27(9):1825–1851. https://doi.org/10.1897/08-090.1

Komatsu T, Tabata M, Kubo-Irie M, Shimizu T, Suzuki K, Nihei Y, Takeda K (2008) The effects of nanoparticles on mouse testis Leydig cells in vitro. Toxicol In Vitro 22(8):1825–1831. https://doi.org/10.1016/j.tiv.2008.08.009

Kong L, Tang M, Zhang T, Wang D, Hu K, Lu W, Wei C, Liang G, Pu Y (2014) Nickel nanoparticles exposure and reproductive toxicity in healthy adult rats. Int J Mol Sci 15(11):21253–21269. https://doi.org/10.3390/ijms151121253

Kozissnik B, Bohorquez AC, Dobson J, Rinaldi C (2013) Magnetic fluid hyperthermia: advances, challenges, and opportunity. Int J Hyperth 29(8):706–714. https://doi.org/10.3109/02656736.2013.837200

Kreuter J (1996) Nanoparticles and microparticles for drug and vaccine delivery. J Anat 189((Pt) 3(Ii)):503–505. https://doi.org/10.1002/bit

Kumar A, Prasad JK, Srivastava N, Ghosh SK (2019) Strategies to minimize various stress-related freeze-thaw damages during conventional cryopreservation of mammalian spermatozoa. Biopreserv Biobank 17(6):603–612. https://doi.org/10.1089/BIO.2019.0037

Kumar A, Ghosh SK, Katiyar R, Gemeda AE, Rautela R, Bisla A, Srivastava N, Bhure SK, Devi HL, Chandra V (2022) Supplementation of Mito TEMPO and acetovanillone in semen

extender improves freezability of buffalo spermatozoa. Andrology 10(4):775–788. https://doi.org/10.1111/andr.13158
Lan Z, Yang WX (2012) Nanoparticles and spermatogenesis: how do nanoparticles affect spermatogenesis and penetrate the blood-testis barrier. Nanomedicine 7(4):579–596. https://doi.org/10.2217/nnm.12.20
Laurent S, Saei AA, Behzadi S, Panahifar A, Mahmoudi M (2014) Superparamagnetic iron oxide nanoparticles for delivery of therapeutic agents: opportunities and challenges. Expert Opin Drug Deliv 11(9):1449–1470. https://doi.org/10.1517/17425247.2014.924501
Lavitrano M, Camaioni A, Fazio VM, Dolci S, Farace MG, Spadafora C (1989) Sperm cells as vectors for introducing foreign DNA into eggs: genetic transformation of mice. Cell 57(5):717–723. https://doi.org/10.1016/0092-8674(89)90787-3
Lavitrano M, French D, Zani M, Frati L, Spadafora C (1992) The interaction between exogenous DNA and sperm cells. Mol Reprod Dev 31(3):161–169. https://doi.org/10.1002/MRD.1080310302
Lavitrano M, Busnelli M, Cerrito MG, Giovannoni R, Manzini S, Vargiolu A (2006) Sperm-mediated gene transfer. Reprod Fertil Dev 18(1–2):19–23. https://doi.org/10.1071/RD05124
Lei R, Bai X, Chang Y, Li J, Qin Y, Chen K, Gu W, Xia S, Zhang J, Wang Z, Xing G (2018) Effects of fullerenol nanoparticles on rat oocyte meiosis resumption. Int J Mol Sci 19(3):699. https://doi.org/10.3390/ijms19030699
Levard C, Hotze EM, Lowry GV, Brown GE (2012) Environmental transformations of silver nanoparticles: impact on stability and toxicity. Environ Sci Technol 46(13):6900–6914. https://doi.org/10.1021/es2037405
Li W, Zhou X, Liu B, Dai J, Song P, Teng Y (2016) Effect of nanoparticles on the survival and development of vitrified porcine GV oocytes. Cryo-Letters 37(6):401–405
Lonergan P, Gutiérrez-Adán A, Rizos D, Pintado B, De la Fuente J, Boland MP (2003) Relative messenger RNA abundance in bovine oocytes collected in vitro or in vivo before and 20 hr after the preovulatory luteinizing hormone surge. Mol Reprod Dev 305:297–305. https://doi.org/10.1002/mrd.10357
Lu AH, Salabas EL, Schüth F (2007) Magnetic nanoparticles: synthesis, protection, functionalization, and application. Angew Chem Int Ed 46(8):1222–1244. https://doi.org/10.1002/anie.200602866
Ma H, Quan F, Chen D, Zheng Y, Zhang B, Wang Y, Zhang Y (2011) Protective function of alpha-lipoic acid on sperm motility and mitochondrial function during goat sperm-mediated gene transfer. Small Rumin Res 99(2–3):191–198. https://doi.org/10.1016/j.smallrumres.2011.04.006
Magdanz V, Khalil ISM, Simmchen J, Furtado GP, Mohanty S, Gebauer J, Xu H, Klingner A, Aziz A, Medina-Sánchez M, Schmidt OG, Misra S (2020) IRONSperm: sperm-templated soft magnetic microrobots. Sci Adv 6(28):1–15. https://doi.org/10.1126/sciadv.aba5855
Makhluf SBD, Qasem R, Rubinstein S, Gedanken A, Breitbart H (2006) Loading magnetic nanoparticles into sperm cells does not affect their functionality. Langmuir 22(23):9480–9482. https://doi.org/10.1021/la061988z
Mancini R, Gamrad L, Werner D, Tiedemann D, Taylor U, Rehbock C, Klein S, Kues W, Barchikowsky S, Rath D (2015) Specific nanotargeting of bovine Y – chromosome bearing sperm. 10th Pacific Symp Flow Vis Image Process 1(99):18–20
Markides H, Rotherham M, El Haj AJ (2012) Biocompatibility and toxicity of magnetic nanoparticles in regenerative medicine. J Nanomater 2012:13–15. https://doi.org/10.1155/2012/614094
McAuliffe ME, Perry MJ (2007) Are nanoparticles potential male reproductive toxicants? A literature review. Nanotoxicology 1(3):204–210. https://doi.org/10.1080/17435390701675914
Mehta RV (2017) Synthesis of magnetic nanoparticles and their dispersions with special reference to applications in biomedicine and biotechnology. Mater Sci Eng 79:901–916. https://doi.org/10.1016/j.msec.2017.05.135
Mishra PK, Lohiya NK (2016) Prioritizing reproductive health: can it be the real game changer for India? J Reprod Health Med 2(1):1–3. https://doi.org/10.1016/j.jrhm.2015.08.001

Mohammed L, Gomaa HG, Ragab D, Zhu J (2017) Magnetic nanoparticles for environmental and biomedical applications: a review. Particuology 30:1–14. https://doi.org/10.1016/j.partic.2016.06.001

Moridi H, Hosseini SA, Shateri H, Kheiripour N, Kaki A, Hatami M, Ranjbar A (2018) Protective effect of cerium oxide nanoparticle on sperm quality and oxidative damage in malathioninduced testicular toxicity in rats: an experimental study. Int J Reprod Biomedicine 16(4):261–266. https://doi.org/10.29252/ijrm.16.4.261

Nagy S, Jansen J, Topper EK, Gadella BM (2003) A triple-stain flow cytometric method to assess plasma-and acrosome-membrane integrity of cryopreserved bovine sperm immediately after thawing in presence of egg-yolk particles. Biol Reprod 68(5):1828–1835. https://doi.org/10.1095/biolreprod.102.011445

Neuberger T, Schöpf B, Hofmann H, Hofmann M, Von Rechenberg B (2005) Superparamagnetic nanoparticles for biomedical applications: possibilities and limitations of a new drug delivery system. J Magn Magn Mater 293(1):483–496. https://doi.org/10.1016/j.jmmm.2005.01.064

Novo S, Barrios L, Santaló J, Gómez-Martínez R, Duch M, Esteve J, Plaza JA, Nogués C, Ibáñez E (2011) A novel embryo identification system by direct tagging of mouse embryos using silicon-based barcodes. Hum Reprod 26(1):96–105. https://doi.org/10.1093/humrep/deq309

Novo S, Ibáñez E, Barrios L, Castell O, Nogués C (2013) Biomolecule screening for efficient attachment of biofunctionalized microparticles to the zona pellucida of mammalian oocytes and embryos. Biomed Microdevices 15(5):801–809. https://doi.org/10.1007/s10544-013-9766-8

Novo S, Morató R, Penon O, Duran S, Barrios L, Nogués C, Plaza JA, Pérez-García L, Mogas T, Ibáñez E (2014a) Identification of bovine embryos cultured in groups by attachment of barcodes to the zona pellucida. Reprod Fertil Dev 26(5):645–652. https://doi.org/10.1071/RD13066

Novo S, Nogués C, Penon O, Barrios L, Santaló J, Gómez-Martínez R, Esteve J, Errachid A, Plaza JA, Pérez-García L, Ibáñez E (2014b) Barcode tagging of human oocytes and embryos to prevent mix-ups in assisted reproduction technologies. Hum Reprod 29(1):18–28

Odhiambo JF, DeJarnette JM, Geary TW, Kennedy CE, Suarez SS, Sutovsky M, Sutovsky P (2014) Increased conception rates in beef cattle inseminated with nanopurified bull semen. Biol Reprod 91(4):1–10. https://doi.org/10.1095/biolreprod.114.121897

Ozkavukcu S, Erdemli E, Isik A (2008) Effects of cryopreservation on sperm parameters and ultrastructural morphology of human spermatozoa. J Assist Reprod Genet 25:403. https://doi.org/10.1007/s10815-008-9232-3

Preaubert L, Courbiere B, Achard V, Tassistro V, Greco F, Orsiere T, Bottero JY, Rose J, Auffan M, Perrin J (2015) Cerium dioxide nanoparticles affect in vitro fertilization in mice. Nanotoxicology 10(1):111–117. https://doi.org/10.3109/17435390.2015.1030792

Ramírez Castex H, Losinno L (2018) Magnetic nanoparticles for x sperm separation from equine semen. Field and commercial results - ProQuest. CES Med Vet y Zootec 2(13):268–275

Rateb SA (2021) Purification of cryopreserved camel spermatozoa following protease-based semen liquefaction by lectin-functionalized DNA-defrag magnetic nanoparticles. Reprod Domest Anim 56(1):183–192. https://doi.org/10.1111/rda.13863

Rath D, Barcikowski S, de Graaf S, Garrels W, Grossfeld R, Klein S, Knabe W, Knorr C, Kues W, Meyer H, Michl J, Moench-Tegeder G, Rehbock C, Taylor U, Washausen S (2013) Sex selection of sperm in farm animals: status report and developmental prospects. Reproduction 145(1):R15–R30. https://doi.org/10.1530/rep-12-0151

Reddy LH, Arias JL, Nicolas J, Couvreur P (2012) Magnetic nanoparticles: design and characterization, toxicity and biocompatibility, pharmaceutical and biomedical applications. Chem Rev 112(11):5818–5878. https://doi.org/10.1021/cr300068p

Remião MH, Lucas CG, Domingues WB, Silveira T, Barther NN, Komninou ER, Basso AC, Jornada DS, Beck RCR, Pohlmann AR, Varela AS Jr, Fabiana Kömmling Seixas F, Vinicius Farias Campos V, Staniscuaski Guterres S, Collares T (2016) Melatonin delivery by nanocapsules during in vitro bovine oocyte maturation decreased the reactive oxygen species of oocytes and embryos. Reprod Toxicol 63:70–81. https://doi.org/10.1016/j.reprotox.2016.05.016

Remião MH, Segatto NV, Pohlmann A, Guterres SS, Seixas FK, Collares T (2018) The potential of nanotechnology in medically assisted reproduction. Front Pharmacol 8(JAN):1–9. https://doi.org/10.3389/fphar.2017.00994

Romany L, Meseguer M, Garcia-Herrero S, Pellicer A, Garrido N (2010) Magnetic activated sorting selection (MACS) of non-apoptotic sperm (NAS) improves pregnancy rates in homologous intrauterine insemination (IUI). Preliminary data. Fertil Steril 94(4):S14. https://doi.org/10.1016/j.fertnstert.2010.07.055

Rossi G, Di Nisio V, Macchiarelli G, Nottola SA, Halvaei I, De Santis L, Cecconi S (2019) Technologies for the production of fertilizable mammalian oocytes. Appl Sci 9(1536):1–17. https://doi.org/10.3390/app9081536

Roy PK, Qamar AY, Fang X, Kim G, Bang S, De Zoysa M, Shin ST, Cho J (2021) Chitosan nanoparticles enhance developmental competence of in vitro-matured porcine oocytes. Reprod Domest Anim 56(2):342–350. https://doi.org/10.1111/rda.13871

Said TM, Agarwal A, Zborowski M, Grunewald S, Glander HJ, Paasch U (2008) Utility of magnetic cell separation as a molecular sperm preparation technique. J Androl 29(2):134–142. https://doi.org/10.2164/jandrol.107.003632

Sánchez-Rubio F, Soria-Meneses PJ, Jurado-Campos A, Bartolomé-García J, Gómez-Rubio V, Soler AJ, Arroyo-Jimenez MM, Santander-Ortega MJ, Plaza-Oliver M, Lozanod MV, Gardea JJ, Fernández-Santos MR (2020) Nanotechnology in reproduction: vitamin E nanoemulsions for reducing oxidative stress in sperm cells. Free Radic Biol Med 160(June):47–56. https://doi.org/10.1016/j.freeradbiomed.2020.07.024

Santaló Pedro J, Barrios Sanromà L, Ibáñez De Sans E, Nogués Sanmiguel C, Esteve Tinto J, Plaza Plaza JA (2010). European patent application: encoded microparticles. http://scholar.google.com/scholar?hl=en&btnG=Search&q=intitle:Designated+Extension+States+:#0

Santos RR, Amorim C, Cecconi S, Fassbender M, Imhof M, Lornage J, Paris M, Schoenfeldt V, Martinez-madrid B (2010) Cryopreservation of ovarian tissue: an emerging technology for female germline preservation of endangered species and breeds. Anim Reprod Sci 122(3–4):151–163. https://doi.org/10.1016/j.anireprosci.2010.08.010

Semmler-Behnke M, Lipka J, Wenk A, Hirn S, Schäffler M, Tian F, Schmid G, Oberdörster G, Kreyling WG (2014) Size dependent translocation and fetal accumulation of gold nanoparticles from maternal blood in the rat. Part Fibre Toxicol 11(1):1–12. https://doi.org/10.1186/s12989-014-0033-9

Shabanowitz RB, O'Rand MG (1988) Characterization of the human zona pellucida from fertilized and unfertilized eggs. J Reprod Fertil 82(1):151–161. https://doi.org/10.1530/jrf.0.0820151

Shandilya R, Mishra PK, Pathak N, Lohiya NK, Sharma RS (2020) Nanotechnology in reproductive medicine: opportunities for clinical translation. Clin Exp Reprod Med 47(4):245–262. https://doi.org/10.5653/cerm.2020.03650

Shea LD, Woodruff TK, Shikanov A (2014) Bioengineering the ovarian follicle microenvironment. Annu Rev Biomed Eng 16:29–52. https://doi.org/10.1146/annurev-bioeng-071813-105131

Shen H, Cai S, Wang Z, Ge Z, Yang W (2023) Materials & design magnetically driven microrobots: recent progress and future development. Mater Des 227(111735):111735. https://doi.org/10.1016/j.matdes.2023.111735

Silva JRV, Barroso PAA, Nascimento DR, Figueira CS, Azevedo VAN, Silva BR, dos Santos RP (2021) Benefits and challenges of nanomaterials in assisted reproductive technologies. Mol Reprod Dev 88(11):707–717. https://doi.org/10.1002/mrd.23536

Singh N, Jenkins GJS, Asadi R, Doak SH (2010) Potential toxicity of superparamagnetic iron oxide nanoparticles (SPION). Nano Rev 1(1):5358. https://doi.org/10.3402/nano.v1i0.5358

Singh AV, Hasan M, Ansari D, Mahajan M (2020) Sperm cell driven microrobots — emerging opportunities and challenges for biologically inspired robotic design. Micromachines 11(448):2–17. https://doi.org/10.3390/mi11040448

Sitti M, Wiersma DS (2020) Pros and cons: magnetic versus optical microrobots. Adv Mater 32(1906766):1–9. https://doi.org/10.1002/adma.201906766

Spadafora C (2007) Sperm-mediated gene transfer: mechanisms and implications. Soc Reprod Fertil Suppl 65:459–467
Sutcliffe AG, Ludwig M (2007) Outcome of assisted reproduction. Lancet 370(9584):351–359. https://doi.org/10.1016/S0140-6736(07)60456-5
Syed R, Suriadi S, Adams M, Bandara W, Leemans SJJ, Ouyang C, Ter Hofstede AHM, Van de Weerd I, Wynn MT, Reijers HA (2020) Robotic process automation: contemporary themes and challenges. Comput Ind 115:103162. https://doi.org/10.1016/j.compind.2019.103162
Takeda K, Suzuki KI, Ishihara A, Kubo-Irie M, Fujimoto R, Tabata M, Oshio S, Nihei Y, Ihara T, Sugamata M (2009) Nanoparticles transferred from pregnant mice to their offspring can damage the genital and cranial nerve systems. J Health Sci 55(1):95–102. https://doi.org/10.1248/jhs.55.95
Tamura H, Takasaki A, Miwa I, Taniguchi K, Maekawa R, Asada H, Matsuoka A, Yamagata Y, Shimamura K, Morioka H, Ishikawa H, Reiter RJ, Sugino N (2008) Oxidative stress impairs oocyte quality and melatonin protects oocytes from free radical damage and improves fertilization rate. J Pineal Res 44(3):280–287. https://doi.org/10.1111/j.1600-079X.2007.00524.x
Taylor U, Tiedemann D, Rehbock C, Kues WA, Barcikowski S, Rath D (2015) Influence of gold, silver and gold-silver alloy nanoparticles on germ cell function and embryo development. Beilstein J Nanotechnol 6(1):651–664. https://doi.org/10.3762/bjnano.6.66
Tiedemann D, Taylor U, Rehbock C, Jakobi J, Klein S, Kues WA, Barcikowski S, Rath D (2014) Reprotoxicity of gold, silver, and gold-silver alloy nanoparticles on mammalian gametes. Analyst 139(5):931–942. https://doi.org/10.1039/c3an01463k
Vasquez ES, Feugang JM, Willard ST, Ryan PL, Walters KB (2016) Bioluminescent magnetic nanoparticles as potential imaging agents for mammalian spermatozoa. J Nanobiotechnol 14(1):1–9. https://doi.org/10.1186/s12951-016-0168-y
Wang J, Chen Y, Chen B, Ding J, Xia G, Gao C, Cheng J, Jin N, Zhou Y, Li X, Tang M, Wang XM (2010) Pharmacokinetic parameters and tissue distribution of magnetic Fe3O4 nanoparticles in mice. Int J Nanomedicine 5(1):861–866. https://doi.org/10.2147/IJN.S13662
Wang F, Tian X, Zhang L, Gao C, He C, Fu Y, Ji P, Li Y, Li N, Liu G (2014) Beneficial effects of melatonin on in vitro bovine embryonic development are mediated by melatonin receptor 1. J Pineal Res 56(3):333–342. https://doi.org/10.1111/jpi.12126
Wang Y, Zhao X, Du W, Liu J, Chen W, Sun C, Cui B, Zeng Z, Shen Y, Gao F, Wang A, Liu G, Cui H (2017) Production of transgenic mice through sperm-mediated gene transfer using magnetic nano-carriers. J Biomed Nanotechnol 13(12):1673–1681. https://doi.org/10.1166/jbn.2017.2456
Wang R, Song B, Wu J, Zhang Y, Chen A, Shao L (2018) Potential adverse effects of nanoparticles on the reproductive system. Int J Nanomedicine 13:8487–8506. https://doi.org/10.2147/IJN.S170723
Wassarman PM (2008) Zona pellucida glycoproteins. J Biol Chem 283(36):24285–24289. https://doi.org/10.1074/jbc.R800027200
Wiwanitkit V, Sereemaspun A, Rojanathanes R (2009) Effect of gold nanoparticles on spermatozoa: the first world report. Fertil Steril 91(1):e7. https://doi.org/10.1016/j.fertnstert.2007.08.021
Xiao Y, Du J (2020) Superparamagnetic nanoparticles for biomedical applications. J Mater Chem 8(3):354–367. https://doi.org/10.1039/c9tb01955c
Yang L, Kuang H, Zhang W, Aguilar ZP, Xiong Y, Lai W, Xu H, Wei H (2015) Size dependent biodistribution and toxicokinetics of iron oxide magnetic nanoparticles in mice. Nanoscale 7(2):625–636. https://doi.org/10.1039/c4nr05061d
Yazdanshenas P, Jahanbin R, Mohammadi Sangcheshmeh A, Aminafshar M, Vaseghi Dodaran H, Varnaseri H, Chamani M, Nazaran MH (2016) Effect of zinc nano-complex on bull semen quality and pregnancy outcome. Anim Product 18(1):173–181. https://doi.org/10.22059/JAP.2016.54598
Yoshida S, Hiyoshi K, Ichinose T, Takano H, Oshio S, Sugawara I, Takeda K, Shibamoto T (2009) Effect of nanoparticles on the male reproductive system of mice. Int J Androl 32(4):337–342. https://doi.org/10.1111/j.1365-2605.2007.00865.x

Yousef MS, López-Lorente AI, Diaz-Jimenez M, Consuegra C, Dorado J, Pereira B, Ortiz I, Cárdenas S, Hidalgo M (2020) Nano-depletion of acrosome-damaged donkey sperm by using lectin peanut agglutinin (PNA)-magnetic nanoparticles. Theriogenology 151:103–111. https://doi.org/10.1016/j.theriogenology.2020.04.011

Zegers-Hochschild F, Adamson GD, Dyer S, Racowsky C, De Mouzon J, Sokol R, Rienzi L, Sunde A, Schmidt L, Cooke ID, Simpson JL, Van Der Poel S (2017) The international glossary on infertility and fertility care, 2017. Hum Reprod 32(9):1786–1801. https://doi.org/10.1093/humrep/dex234

Zhang XP, Leung C, Lu Z, Esfandiari N, Casper RF, Sun Y (2012) Controlled aspiration and positioning of biological cells in a micropipette. IEEE Trans Biomed Eng 59(4):1032–1040. https://doi.org/10.1109/TBME.2012.2182673

Zhang C, Qin X, Guo L, Zhang GL, Zhang JX, Ren YS (2015) Effect of different Nano-zinc levels in dietary on semen quality, activities of antioxidant enzyme and expression of copper zinc superoxide in epididymis of ram lambs. Sci Agric Sin 48(1):154–164. https://doi.org/10.3864/J.ISSN.0578-1752.2015.01.15

Zhang H, Xuan X, Yang S, Li X, Xu C, Gao X (2018) Selection of viable human spermatozoa with low levels of DNA fragmentation from an immotile population using density gradient centrifugation and magnetic-activated cell sorting. Andrologia 50(1):1–7. https://doi.org/10.1111/and.12821

Zhou X, Li W, Fang L, Zhang D, Dai J (2015) Hydroxyapatite nanoparticles improved survival rate of vitrified porcine oocytes and its mechanism. Cryo-Letters 36(1):45–50

Part II

Future Developments and Considerations

Reproductive Studies in the Native Breeds of the Region of Murcia. Current and Future Status of the IMIDA Animal Germplasm Bank

Laura Almela Veracruz, Ángel Poto Remacha, Sonia Galián Arnaldos, and Begoña Peinado Ramón

Abstract

In the Region of Murcia, there are native breeds in a critical situation such as the Chato Murciano pig, the Murcian Hen, or the Murciano-Levantina cow, but also other native breeds "in development" such as the Murciano-Granadina goat, which in recent years have been the subject of various studies within the Animal Genetic Improvement and Livestock Development Team belonging to the Department of Aquaculture and Animal Production Technology of the Murcian Institute of Agrarian and Environmental Research and Development (IMIDA), among them, those of the reproductive type. To safeguard the genetics of threatened breeds and to improve reproductive results in the case of native breeds "in development," a Germplasm Bank composed of male gametes has been created for each of them. Currently, these seminal doses continue to be maintained in liquid nitrogen containers, which in the case of threatened breeds means the protection of very valuable genetic material that may need to be used in the future. However, the maintenance of these doses is not free of problems that must be solved by studying other alternatives that allow their maintenance more sustainably and economically. Furthermore, the necessary and inevitable path towards the most advanced technologies leads to the use of these smart technologies also in the field of reproduction to achieve greater efficiency and better yields.

L. A. Veracruz · S. G. Arnaldos · B. P. Ramón (✉)
Animal Genetic Improvement and Livestock Development Team, Department of Aquaculture and Animal Production Technology, Murcian Institute of Agrarian and Environmental Research and Development (IMIDA), Murcia, Spain
e-mail: laura.almela@carm.es; begona.peinado@carm.es

Á. P. Remacha
Independent Scientific Advisor, Murcia, Spain

J. C. Gardón, K. Satué Ambrojo (eds.), *Assisted Reproductive Technologies in Animals Volume 2*, https://doi.org/10.1007/978-3-031-87198-6_5

Keywords

Germplasm bank · Chato Murciano pig · Murciano-Granadina goat · Murciano-Levantina cow · Spermatozoa cryoconservation

1 Introduction

The conservation of the biodiversity of domestic animals is a priority concern of Western societies because their great development and industrialization have led to the replacement of traditional production systems with other intensive ones based on the exploitation of a small number of animals (breeds, strains, lines, etc.) (Almela 2006).

The number of endangered livestock breeds in Spain is large (according to the Ministry of Agriculture, Fisheries and Food, there are 31 bovine breeds in danger of extinction, 7 pig breeds, 19 poultry breeds, 32 sheep breeds, and 15 goat breeds) (BOE 2254/2012, February 15, 2012). With some of them, on the other hand, work is being carried out that helps their study and recovery, either because they have some character of interest, or because they receive the support of regional or local organizations, specific associations, or even the support of some interested research group (Sierra 1998).

Thirty percent to 40% of total agri-food production is related to the direct or indirect contribution of domestic animals. This is due to their capacity to transform forage and by-products into food for man, in addition to providing and responding through work, fiber, fertilizer, fuel, income, and in general, to the required local needs (FAO 1998).

For those breeds that are in a critical situation or at risk of extinction, in-depth studies on their characteristics are necessary, as well as the implementation of recovery programs.

These livestock breeds present characteristics related to rusticity, work, resilience, or adaptation to low-nutritious foods, generating, despite this, high-quality productions. This is why the conservation of these genes is of great importance. Although we currently consider that these livestock breeds are not interesting, it may be necessary to recover them in the not-too-distant future to satisfy the increasingly growing demands for sustainability, quality, or differentiation (García-Menacho Osset 1999).

Germplasm Banks are, as defined by Royal Decree 45/2019, of February 8, collections of genetic material (sperm, oocytes, embryos, somatic cells, or DNA) officially recognized within the framework of the breeding program, whose purpose is ex-situ conservation or sustainable use of pure breeds of livestock.

The establishment of Germplasm Banks may be promoted by purebred animal breeders' associations, especially native ones, or official services recognized for the management of a breeding program, autonomously or in collaboration with other institutions. These collections must have the recognition of the competent authorities of the Autonomous Communities where they are located, except for the National Animal Germplasm Bank, which depends on the General State Administration.

Germplasm Banks provide a series of advantages that make them irreplaceable elements within the structure of our livestock breeding programs. The most obvious is their ability to preserve a breed or variety indefinitely, as it is the only tool that allows us to guarantee, without any doubt, its very long-term conservation (Galián et al. 2010).

To the above, we must add that conservation is carried out independently of the conditions that affect the population managed in situ, protecting the stored material from natural disasters and epizootics. Likewise, it is also protected from genetic changes (intentional or fortuitous) that it may suffer, allowing the cryopreserved material to reverse the effects of incorrect selection or genetic drift caused by low census sizes (Poto et al. 2000).

Thanks to all these attributes, Germplasm Banks can fulfill various functions, among which the most important is to be able to recreate an extinct breed or strain on farms. However, they can also perform other tasks, supporting in situ conservation programs, selection programs, or the creation of new lines or breeds.

Finally, the relevance of ex-situ conservation activities is also reflected in the fact that the stored reproductive material has been considered as one of the indicators to monitor the achievement of the UN Sustainable Development Goal that aims to "End hunger," achieve food security and improved nutrition, and promote sustainable agriculture.

The National Animal Germplasm Bank regulated and coordinated by the Ministry of Agriculture, Fisheries and Food (MAPA) is managed by the CENSYRA in Colmenar Viejo, a facility of the Madrid Institute of Rural, Agrarian and Food Research and Development of the Community of Madrid. Its objective is to constitute a permanent and closed collection of reproductive material of all the breeds in the Catalog of Livestock Breeds of Spain, which allows it to act as a security element in the event of a loss of their genetic variability, and even in the event of their extinction.

The censuses of the Chato Murciano pig, Murciano-Levantina cow, Murciana hen, Montesina sheep, and Blanca-Celtibérica goat breeds have decreased rapidly in a few years and they are on the verge of extinction or with a reduced number of specimens, sometimes caused by little economic interest and others because their possibilities of use are limited to spaces where ranchers do not find the appropriate socioeconomic circumstances for a decent life. However, the most coherent explanation is the lack of supporting studies required to understand the possible alternatives to the skills that they possessed in the past and continue to possess in the present and the absence of programs for their conservation, which until recent times have not been produced and when implemented today they encounter problems of scarcity of animals, lack of genetic variability, and high inbreeding (Poto et al. 2000a).

In 1997, the IMIDA Animal Genetic Improvement Team began sperm cryofreezing work with the Chato Murciano pig thanks to the approval of a project for the recovery of this breed at IMIDA. In subsequent years, sperm cryofreezing work has continued thanks to the approval of various projects in the rest of the species: Murciano-Levantina cow, Murciano hen, and Murciano-Granadina goat; currently having a Germplasm Bank of male and female gametes of these breeds stored in liquid nitrogen tanks (Peinado et al. 1998a, 2002, 2014; Duchi et al. 2008; Galián et al. 2021).

In the context of conserving domestic animal diversity, to avoid the serious erosion being suffered by most breeds, ex-situ conservation means storage. This entails the preservation of both animals from a sample of a breed in a situation far from their normal production environment or habitat (ex-situ and in vivo conservation) and/or the collection and cryopreservation of resources in the form of semen, oocytes, embryos or living tissues, which can be used to regenerate or recover animals (FAO/UNEP 1995).

Considering all animal species, not all ex-situ conservation techniques have the same results in achieving the desired objectives. In this way, effective semen cryopreservation techniques are available for practically all species, although the cost of maintenance or storage is very different depending on the final characteristics at the time of fertilization of the female. Likewise, all cryofreezing techniques cannot be extrapolated, from breeds highly selected for assisted reproduction to more environmental and/or rustic breeds (Peinado et al. 1998a).

Cryopreservation implies a high cost for obtaining the samples to be obtained and therefore must be justified. Cryopreservation priority should be for breeds that do not have viable in vivo conservation programs and in order of priority: a) they show a phenotypic adaptation to their environment or have new products, b) they have social or cultural importance, c) they are breeds that appear genetically distant from the other conserved breeds based on DNA analysis, and d) all other breeds. Cryopreservation allows us to preserve semen or embryos. As a general rule, the cost of embryo production is higher than the cost of semen production, even though the conservation costs are the same. Frozen semen is as effective or even more so than embryos, except when the objective is to recreate a breed that has disappeared, which is why the priority material to conserve is semen (Poto et al. 2000).

Even when samples of genetic material have been obtained and are preserved, a cryopreservation program presents risks. Samples should always be kept in liquid nitrogen and this involves appropriate storage facilities and a regular supply of liquid nitrogen. Any breach of this procedure will lead to the loss of all conserved material.

One of the most useful aspects of cryopreservation is the support of in vivo conservation. In this case, it is a complement to a broader conservation program, especially for in situ conservation projects, in which the cryopreservation of genetic material is initiated as security in case of catastrophe or if a genetic problem appears as a consequence of the accumulation of recessive genes harmful to the living population. However, it should be emphasized that cryopreservation is not necessary for the success of in situ conservation but is only a means of reducing risks when funding and facilities allow (Poto et al. 2000b).

Cryopreservation is the long-term preservation of living cells at very low temperatures such as the boiling point of liquid nitrogen, which is −196 °C. Such temperatures prevent any enzymatic and chemical reaction, and a sample can be preserved there indefinitely. Frozen semen has been used successfully after 30 years in an endangered breed preservation program. Tests carried out on mice have shown that there is no evidence of modifications in DNA due to freezing (Glenister et al. 1990). To verify the quality and effectiveness of seminal samples frozen and stored between 6 and 12 years in liquid nitrogen containers, in 2019 our team carried out a seminal quality study of frozen samples and all the parameters studied showed an average quality suitable to be used in AI, in addition to demonstrating the effectiveness of the cryofreezing and long-term storage technique (Galián et al. 2019).

Thus, the first requirement for cryopreservation is the availability of facilities that allow the storage of samples in liquid nitrogen (NL2), and the ability to regularly replenish the NL2 tanks. Although NL2 is produced industrially in many countries, prices are sometimes prohibitive.

In addition to the cryofreezing technique, more reproductive actions have been carried out in these species to solve or improve the situation of those breeds that are in a threatened situation, such as artificial insemination (Peinado et al. 1998; Almela et al. 2011; Galián et al. 2019). However, new times and the change in trend towards more technological and efficient management led to the use of new technologies and digitalization to become a model or reference of smart livestock farming for other livestock farms.

2 Chato Murciano Swine Breed

2.1 Introduction

Pig farming in the Region of Murcia, based in the past on the native Chato Murciano breed, has traditionally constituted high-quality production with an important socio-economic influence in the Region. On the other hand, the role played by this native breed in the birth and subsequent development of modern Murcian pig farming is indisputable, without underestimating the influence that the flourishing of the meat industry had since it was the basis of numerous genuine meat products (Peinado et al. 2001).

This breed is the last exponent that we have today of the ancient primitive Murcian pig breed, the fruit of the imagination and work of generations of Murcian livestock and agricultural farmers, on the other hand, it constitutes an authentic reserve of genes that enriches the biodiversity of Spanish livestock species and, therefore, the common heritage (Peinado et al. 2001).

In Murcia and throughout the Levant, in the nineteenth century, there existed a Black Mediterranean pig of the Iberian type, which has been called the Black Mediterranean pig, whose characteristics were slow growth, reaching slaughter weight at an age of approximately 2 years, thick subcutaneous fat, and a small litter size (Poto et al. 2002).

The current Chato Murciano pig comes from crosses made on the Primitive Murciano pig, of which there were two types or varieties known in the region as the

Gabana and Pintada varieties, differentiated only by the color of the coat. This breed of pig does not exist today and there are only references in the livestock literature of the time.

The formation of the Murcian breed or the Chato Murciano pig was the consequence of the development of the meat industries in the southeast of Spain, which caused the diffusion of a type of pig improved by crossing the primitive Murcian pig with imported foreign breeds, especially Yorkshire, Berkshire, Tamworth, Craonés and Alderney, and with Colorado Extremeño. These breeds genetically influenced to a greater or lesser extent the formation of the Chato Murciano; The Middle White and Berkshire breeds were the ones that contributed the greatest number of characters and the rest are cited or not by different authors, their contribution being merely testimonial (Panés 1916; Díaz 1953; Lobera 1998) (Fig. 1).

Consequently, an animal was obtained that gained in plasticity, performance, and proportion of leanness, as well as in precocity and prolificacy; it is also a magnificent transformer of a wide variety of foods, such as agricultural by-products and residues produced in the normal rotation of crops, as well as waste from family farms in agricultural and irrigated areas (Peinado et al. 2001).

The genetic situation of the breed in the mid-nineties raised fears of the impossibility of numerically recovering the animal group. There were six breeding females

Fig. 1 Chato Murciano breed boars

and two boars with a high degree of kinship and population consanguinity greater than 10.9% (Poto and Peinado 1998), presenting reproductive problems, poor growth, and poor food transformation, with litters observed with two types of animals: some with normal growth and development and others non-viable, which normally did not reach adulthood.

These measures have not managed to substantially reduce the degree of inbreeding of the population in general, but have allowed their numbers to increase (Peinado et al. 2001). Other complementary measures have been taken to ensure the conservation of the breed, such as boar sperm cryopreservation, currently having a Germplasm Bank with sufficient genetic material to carry out recovery through absorption crossing with other breeds when it is needed; the appearance of another group of Chato pigs in the Guadalentín region and the segregation in several farms and facilities at sufficient distance from the first line pigs, avoiding the health risks of mandatory slaughter due to epizootics (Poto et al. 2002).

Currently, this pig breed is considered endangered, according to the Official Catalog of Livestock Breeds of Spain (Annex 1 of RD 45/2019), with a total of 256 adult females and 15 boars of the breed (ARCA-National Breed Information System).

2.2 Reproductive Performances in the Chato Murciano Pig

In the case of pigs, ex-situ conservation is conditioned by economic and technical problems, and the use of all known methods is not always possible. Freezing semen is the most economical way to maintain the breed, providing frozen genetic material with a view to its recovery when required. The rest of the known techniques present serious drawbacks in this species: high maintenance cost, low fertility achieved in the case of freezing oocytes, not very encouraging results in the case of frozen embryos, and although the data published by Dobrinsky (2002) indicate a serious advance in this technique, we cannot yet recognize it as applicable for the conservation of pig genetic resources.

Spermatozoa cryopreservation was conceived mainly for the distribution of the genetic improvement achieved in different animal species over long distances. Today, the role of sperm cryopreservation is also to keep the male gametes of endangered breeds stored for long periods to ensure their survival (Poto and Peinado 1999). The technique has been widely used in practically all animal species, with pigs being an exception. In this case, the use of pig semen refrigerated at 15 °C, for more than a week, was shown to be highly effective in transferring the genetic improvement of highly productive breeds to distant areas of the planet, so semen cryopreservation studies were reduced to research centers. Furthermore, the number of sperm cells required to constitute a fertile seminal dose is much greater, sometimes more than double, for frozen semen than for refrigerated semen (Poto and Peinado 1999).

The starting point in the development of sperm cryopreservation techniques began in 1949 when Polge et al. discovered the protective action of glycerol.

Between the 40 s and 50 s, the semen freezing technique was perfected and some fertility data were obtained.

The technique of freezing pig semen has been a reality since 1969 the first birth was obtained from a sow inseminated with sperm subjected to a freezing process (Vicente 1972); sow deliveries were also obtained with thawed spermatozoa applied directly to the oviduct (Polge et al. 1970). Since then, many trials have been launched to improve swine semen freezing techniques, in an attempt to make it useful to the industry. This is the case of the methodology described by Westendorf et al. (1975), from which good results have been obtained, with modifications pointed out by various authors (Martín 1982).

The application of this technique to pig production ceased to have economic interest since the application of refrigerated semen became widespread in modern pig farming. In such a way, there are currently large production centers for swine semen, preserved at 15 °C, that satisfy all the productive needs of entrepreneurs in the sector. However, the trends for the conservation of animal genetic resources make cryopreservation techniques for gametes, particularly sperm, necessary, the need existing to resume initiatives from more than 30 years ago. Thus, numerous authors have used modifications of the Westendorf et al. technique (1975), but maintaining the general lines in seminal processing (Thilmant 1997; Peinado et al. 1998; Poto et al. 1999; Bussiére et al. 2000; Córdova et al. 2001; Peinado et al. 2002). Furthermore, freezing semen in itself provides benefits to pig companies for different reasons: it has an incalculable value for Artificial Insemination Centers in the event of any setback that could affect the reproductive males, in most cases of great economic value, a more rational use of labor, using semen from improving boars after their death, etc. The less than encouraging results obtained were partly attributed to the special morphology of the boar sperm head, which favored the formation of crystals and the subsequent damage to the sperm cell, with the loss of motility and alteration of the acrosome. The increasingly extensive knowledge we have about the metabolic processes that occur in the sperm cell during freezing and thawing provides new solutions to the problem.

In the specific case of the Chato Murciano pig and due to the past and current situation, with a reduced number of breeders, the conservation and recovery program of this pig breed began in 1997, launching at the same time all the trials and techniques for developing sperm cryopreservation. This allowed for the creation and maintenance of the Chato Murciano pig Animal Germplasm Bank at IMIDA.

However, the semen cryopreservation process, focused on pigs, involves the following phases:

1. Management of boars. Training boars to mount the dummy. Study of libido and behavior in the semen extraction room. Semen collection and evaluation of seminal quality.
2. Carrying out the semen freezing technique. Storage of seminal doses.
3. Thawing of seminal doses and their use in the artificial insemination of female pigs.

When the process of ex-situ conservation of pig breeds in danger of extinction begins, we frequently find ourselves faced with the impossibility of being able to choose or select the boars destined for seminal production. The low number of remaining male animals of some breeds entails all boars to be used to maintain and conserve the little genetic variability that still remains. In addition, we find male animals of different ages, and the responses to training can be very different.

When it comes to select breeds, all authors agree that animals must be chosen that provide semen of the best quality and suitable for freezing (Martín 1982; Thilmant 1997). The issue is complicated when breeds are used that have never been subjected to semen collection and that individual management is difficult due to their extensive exploitation system, as is the case of the different varieties of the Iberian pig.

Therefore, in these breeds, the training process for mounting the dummy usually takes longer than in the breeds usually used in pig artificial insemination, and for this reason it is necessary to know the behavior of the boars at different ages and in the different seasons of the year. In this regard, it has been found that boars that have recently reached puberty of the Chato Murciano breed have faster training and shorter courtship time in front of the dummy than adult boars (Peinado et al. 1998b). The behavior before the boar's mounting will be a good indicator of the quality of semen collection in terms of ejaculate speed, time spent, volume of the various fractions, sperm concentration, etc.

(a) Selection and management of boars.

* In pig breeds in danger of extinction, it is necessary to take into account that it is not always possible to make a good choice of the breeding animal due to the circumstances of the breed or how they are exploited (small number of specimens, aggressiveness, inbreeding, etc.). In any case, the breeding boar candidate must approximate the reproductive characteristics of the adult animal, and its body development at the time of puberty must be adequate, this requires to study of the standard body condition of the breed. Furthermore, it is advisable to study the genitals of the future boar and their harmonious relationship with the body as a whole, ruling out observable anomalies, both in the testicular area and in the entire area related to the penis.

The strength of its legs is essential for the choice of the boar, with good disposition of legs and hooves.

* All animals will continue to receive the appropriate diet for the breed, which allows them to optimize sexual behavior, sperm production, and the viability and fertilizing capacity of sperm cells.

* They will be subject to a hygienic-sanitary program, to avoid diseases that have negative effects on sperm production.

* The boars will be located in individual pens and with protection against extreme environmental phenomena, which do not alter animal well-being (being one of the key elements for training semen collection and the subsequent collection regime), and which cannot affect the sexual behavior of the animal and its sperm quality.

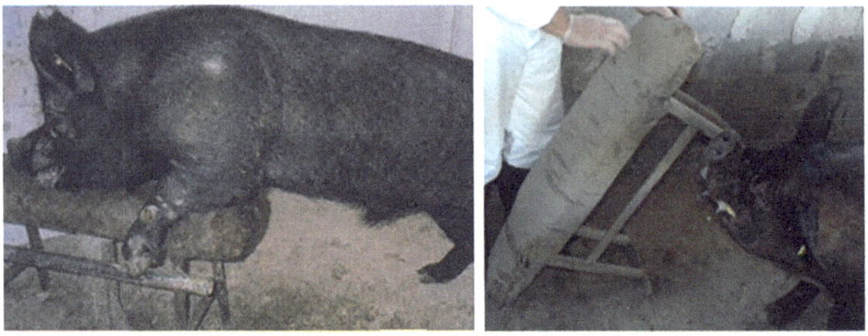

Fig. 2 Chato Murciano breed boars mounting the dummy

(b) Training the boars to jump onto (mount) the dummy. Study of libido and behavior in the semen extraction room (Fig. 2).

* It is interesting that the future breeder has the necessary sexual instinct so that when the time comes he shows sexual interest towards the female, and in our case towards the semen extraction dummy. This is of great importance for the success of training the boar to mount. It must also be taken into account that low sexual interest is usually correlated with poorer ejaculate quality.

We have observed that the time needed, once the boar enters the semen collection room, to mount the dummy and begin to ejaculate, is less in animals that are younger or closer to puberty, and that this time increases as the animals advance in age. This situation also occurs in wild suids, thus guaranteeing the genetic variability and renewal of the breed (Peinado et al. 1998b).

* The time it takes for the animals to mount the semen extraction dummy and begin to ejaculate depends on whether the animals perform sexual courtship before this or not and on its duration. However, we have not found influences on the quality of the seminal material depending on the time it takes to mount the dummy or whether or not they carry out the full sexual courtship.

Likewise, there are also differences due to the age of the animal on the duration of ejaculation. This time is shorter in younger animals and as age increases, the duration of ejaculation also increases. Likewise, the quality of the seminal material is not affected by this parameter, but it does affect the volume emitted (in studies carried out on the Chato Murciano breed boar, it has been found that the greatest difference is in the volume of the sperm-poor fraction.

* The period from training to semen collection depends on many circumstances, such as the age of initiation, the season of the year, temperatures and weather conditions, and sexual status before training (sometimes this training is not possible due to its use in natural mating), body development, health, and nutritional status.

Training has different guidelines depending on whether they are young animals or those that have previously been destined for natural breeding. Furthermore, we have observed a different adaptation behavior to mounting a dummy for young boars in various breeds. Thus, for Chato Murciano breed pigs, training from 8 months of age has consisted of one or two sessions, keeping the dummy impregnated with the saliva of other boars. However, for young Iberian pigs, we have not found this facility, so that by impregnating the room with remains of saliva and urine from other boars mounting has not been achieved until after several sessions, taking in some cases several months and, circumstantially, it has been necessary to use other training methods (Peinado et al. 1998b).

* The semen collection room will have enough space for any boar to carry out courtship before mounting. It is also necessary to have a height-adjustable dummy and powerful metal protection to avoid possible attacks and accidents. The extraction room must be located in a place away from powdery production and sources of dirt typical of pig farms and close to the semen preparation laboratory.

* All animals must be subjected to semen collection per week using the manual method of Hancock and Howell (1959) with the hand protected by a vinyl glove.

The first studies carried out on Chato Murciano boars were aimed at knowing the behavioral parameters, such as the time the pig remains in the semen extraction room, the time spent ejaculating, and the speed of semen emission (Peinado et al. 1998c). For McGlone (1992), there is a genetic influence on the sexual behavior of pigs that has not yet been well studied and naturally, there will be a breed component (Fig. 2).

In the Chato Murciano pig breed, significant decreases in semen production have been described, both in ejaculated volume and semen quality, not only due to the photoperiod but also to the temperatures reached in the region where these pigs are exploited (Peinado et al. 1998a). This environmental variable, like feeding (Mateos and Medel 1998), only produces alterations in the volume of the ejaculate when it decreases or increases abnormally, and may not affect the animals as a whole but rather the particularities of the individual, although both feeding and ambient temperature can be corrected through modern pig production systems (Peinado et al. 2001).

(c) Semen collection and evaluation of seminal quality.

* As we have mentioned, in this process there are differences between males in the libido they manifest, in the time it takes to mount the extraction dummy and begin to ejaculate, and in the duration of ejaculation depending on the age of the animal.

* The ejaculate-rich fraction is collected in a sterile beaker, warmed by a collection thermos that maintains body temperature during the process. A filter or sterile gauze is placed on said beaker to filter out the granules derived from Cowper's gland and any particles that could be contaminants (Fig. 3).

* Once the rich fraction has been collected, it is placed inside a thermostatic bath at 37 °C, where the first semen assessment is carried out as soon as possible to determine the macroscopic characteristics such as color, odor, the volume of the rich seminal fraction, and the presence or absence of foreign particles. Color: it will vary physiologically depending on whether it is the sperm fraction (milky whitish as it is rich in sperm) or the pre- and post-sperm fraction (transparent as it is poor in sperm). It gives us an idea of the state of the ejaculate obtained and also the animal's state of

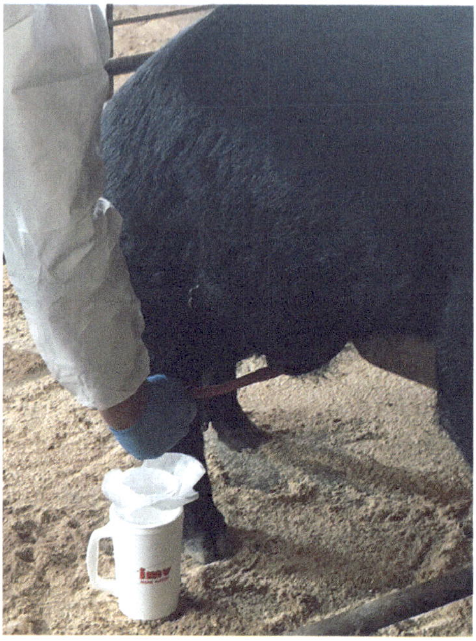

Fig. 3 Semen collection in a thermos heated to 37 °C

health (it should not be reddish, yellowish, or brown). Smell: any other smell that is not the typical semen smell will indicate the existence of alterations in the seminal material and the animal's health; it is often contamination with the boar's urine, which is usually due to failures of the operator who collects the semen. The volume of the rich seminal fraction: is a parameter that is taken into account when making seminal doses since it helps us know the concentration of sperm in the extracted semen. The volume study is subdivided into the volume of the sperm fraction and the volume of the pre- and post-sperm fraction. This parameter varies greatly depending on the age of the animal, time of year, and rate of semen collection. According to Peinado et al. (1998), the volume of the poor fraction is the one that tends to fluctuate the most. The presence or absence of foreign particles will tell us if the semen obtained is in good condition and if it should continue to be evaluated for subsequent use. Sometimes it is due to the analyst's failure.

* Next, we proceed to study, with an optical microscope, the microscopic characteristics of the ejaculate (Martín 1982): individual motility, quality of sperm movement, sperm concentration, percentage of sperm with intact membrane, state of the acrosome and morpho-anomalies.

* To know the individual motility, a drop of undiluted semen (diluted when carried out in the freezing protocol) is placed on a slide warmed to 37 °C, observing the set of several fields in an optical microscope at 100X magnification, assessing the semen according to a subjective scale from 0% to 100% to differentiate those that present motility from those that are immotile, and the presence or absence of agglutinations in the entire ejaculate. The quality of the movement is assessed at the same time using a subjective scale from 0 to 5, depending on whether the movement is

progressive and very fast (5) or absent (0) (du Mesnil du Buisson and Signoret 1971; Peinado et al. 1997).

* To study sperm concentration, a spectrophotometer was used, which by turbidity expresses the number of sperm in millions per milliliter of ejaculate. To do this, a drop of undiluted semen is absorbed by capillarity with a microcuvette, placed in the spectrophotometer, and the concentration is measured. Once the number of spermatozoa is obtained, it is multiplied by the volume obtained from the ejaculate-rich fraction, and we obtain the sperm concentration expressed as the number of spermatozoa per milliliter.

* The percentage of spermatozoa with an intact membrane was determined using vital eosin-nigrosin staining, the basis of which is based on the fact that spermatozoa with an altered membrane allow the dye to enter (Eliasson and Treichi 1971). The contrast of this test is carried out by mixing one drop of semen and another of the staining solution, and the extension is carried out, which once dry is observed under an optical microscope at 400X magnification. Sperm cells with a damaged membrane appear red-brown, while those with an intact membrane are not stained.

* To determine the percentage of spermatozoa with a normal acrosome, a 0.1% solution of formaldehyde in physiological saline is prepared, to observe the spermatozoa that are mixed with this solution in a phase contrast optical microscope at 1000X magnification. In the normal acrosome, it is observed at the apical edge of the sperm head, in the form of a perfectly defined black crescent. Any irregularity of the apical edge would correspond to sperm with damaged acrosomes. According to the previous considerations, the spermatozoa are divided into two groups, normal and altered to any degree (Pursel et al. 1973), counting the percentage of sperm with a normal acrosome.

* In the study of sperm morpho-anomalies, we evaluate with the same solution and in the same way as to know the percentage of spermatozoa with normal acrosomes, the number of spermatozoa that are morphologically abnormal, among which the following stand out, and according to Howard and Pace (1988) and Bonet et al. (2000):

Cytoplasmic gout: its presence has been taken as a way to assess sperm maturity. Depending on their location, they can be proximal, medial, or distal.

Abnormalities in the tail: this type of alteration implies an impediment to the movement of the spermatozoa. They can be bent or round tails.

Alterations of the head: they are the result of an alteration of spermatogenesis. They represent a very small percentage in the spermiogram. They can be macrocephaly, microcephaly, loose heads, pyriform heads, double heads, etc.

Other alterations: abnormal tail implantations, double tails, etc.

For sperm freezing, only the rich fraction of the ejaculate that shows good seminal quality at the time of collection is used: individual motility >3, live sperm >70%, normal acrosomes >90%, and total morphoanomalies <25%.

In studies carried out on the Chato Murciano pig breed, it was found that the youngest boars (10 and 20 months) had a shorter response time to the immobility of the dummy than the older boars (12 years): finding values of 3.57 s for the youngest boar, just reached puberty, and 64.62 s for the oldest. The time spent climbing onto the dummy increases in the months of high temperatures in adult boars, but not in younger ones (Peinado et al. 1998a).

The time spent climbing onto the dummy increases in the months of high temperatures in adult boars, but not in younger ones (Peinado et al. 1998a).

Concerning the time taken to ejaculate in boars, it has been observed that younger animals take longer due to a lower speed of ejaculate emission. Regarding the change in photoperiod, there do not seem to be differences in the time spent by each boar, although in all pigs a decrease could be observed during the summer months for the total ejaculation, due to a decrease in the volume of the fraction poor in spermatozoa, without variations for the other seminal parameters of this period (Peinado et al. 2001) (Fig. 3).

The speed or flow of semen emission of our native pig breed, deduced from the time spent ejaculating and the total volume emitted does not differ between the various seasons of the year, although it is influenced by the age of the animals. The maximum and minimum mean values found in boars of this breed are 0.69 and 0.29 ml/s, respectively (Peinado et al. 1998a) (Fig. 3).

The seminal characteristics of the Chato Murciano swine breed are not different from the results obtained for other swine breeds (Peinado et al. 1998b), and the suitability for the preparation of seminal doses follows the patterns referenced for the species (Table 1) (Peinado et al. 2001).

(d) Carrying out the semen freezing technique. Storage of seminal doses.

* The entire process of conditioning the sperm cell for freezing does not present great difficulties for it to be repeated successfully in any laboratory well-equipped with the material. From the study of the quality of the starting semen to the storage in liquid nitrogen containers, they are perfectly typified in the bibliographical references. There are many techniques and a large number of results for defrosting that indicate the possibility of success when each of the techniques is applied (Martín 1982; Thilmant 1997; Bussiére et al. 2000), although there are few studies on results that use semen from males not selected for reproduction, as is the case of breeds in a critical situation or endangered.

Table 1 Semen parameters of the Chato Murciano pig. February–December 1997

N° of boar	215	222	CH-Viejo
Rich fraction volume (ml)	74.2	87.5	90.6
Lean fraction volume (ml)	152.7	165.9	164.9
Movement quality (%)	76.8	58.2	70.9
Individual motility (×10)	38.7	34.1	38.4
Concentration (×109)	57.1	44.0	65.0
% Normal acrosomes	95.7	93.0	94.3

Peinado et al. (2001)

* The semen freezing technique requires control of critical points where the cell can be damaged. In this regard, we consider that the technique proposed by Thilmant (1997) is one of the most careful for the cryopreservation of porcine semen.

* Before this freezing technique (Thilmant 1997), the Westendorf et al. (1975) technique was used modified but with worse results in vitro and in vivo. Both methods begin to differ from the moment that the temperature drop curve is followed more rigorously in Thilmant's by the use of programmable cryofreezers. Likewise, there is a difference in the osmotic pressure suffered by the sperm throughout the entire process and, above all, in its speed of action.

From the moment the semen is collected, it must be taken from boar studs with high seminal quality and it is noted that they have good freezing capacity, always starting from the sperm-rich fraction of the ejaculate (Martín 1982; Thilmant 1997), and in this sense, it is indicated that the overall effectiveness of the method is what determines the minimum number of spermatozoa useful for thawing (Bussiére et al. 2000), this must be taken into account in the case of endangered pig breeds.

The dilution of the sperm-rich fraction is carried out at temperatures close to 30 °C (Martín 1982) or 25 °C (Thilmant 1997), using diluents for preserving semen refrigerated at 15 °C, MRA (trademark) or Beltsville thawing solution (BTS). It does not seem advisable to remove the seminal plasma by centrifugation (800 g) at 15 °C until 2 h after dilution (Thilmant 1997), since the action of the seminal lipoproteins has a certain beneficial effect on the sperm membrane (Martín 1982), and good results obtained with ejaculates collected the previous day have recently been reported (Bussiére et al. 2000).

Various preservative diluents have been used for freezing swine semen, adding after centrifugation and removal of the supernatant, those based on egg yolk accompanied by lactose (Westendorf et al. 1975), fructose (Thilmant 1997), or glucose (Polge et al. 1970) are the most used, improving sperm quality of the acrosome integrity and revival when *Orvus ES Paste*® (O.E.P.), a surfactant product whose main raw material is Sodium Lauryl Sulfate, is added (Pursel et al. 1978). The concentrations of O.E.P. vary from 0.3% to 1.6%, depending on the recommended technique (Martín 1982; Paquignon et al. 1980; Thilmant 1997; Bussiére et al. 2000). The type of diluent is one of the issues that must be improved in the future (Fig. 4).

The use of glycerol as a cryoprotectant is widespread and several tests have been used to know the response of the sperm plasma membrane and the integrity of the acrosome depending on the temperature at which it is added (Paquignon et al. 1980), the final concentration (Aimlid and Johnson 1988; Fiser and Fairfull 1990) and equilibration time. The addition of glycerol is accepted as having a greater effect on membrane integrity than on sperm metabolism at the level of the intermediate piece (Murdoch and Jones 1978). The final concentration considered optimal for glycerol is 3% (Fiser and Fairfull 1990), the results improving if it is added in three steps (Thilmant 1997). The purpose is to increase the osmotic pressure of the diluent to extract as much water as possible from the cell; if the glycerol concentration is very high, the osmotic pressure will increase and the water will escape very quickly, compromising the membrane integrity. This step of the technique can be easily

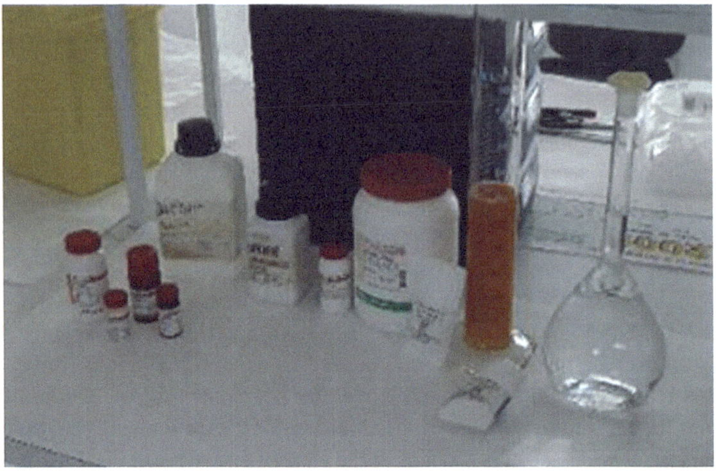

Fig. 4 Products used to prepare semen extender

solved with an osmotic gradient that increases the osmotic pressure in four steps, from 300 mOsm/kg to 1540 mOsm/kg (Poto et al. 1999).

Packaging geometry is another issue that must be studied, not only to be applied to a specific breed but also due to the economic problems that storage costs may present. The use of packaging geometries has undergone an evolutionary process towards larger container sizes to be able to store the total number of spermatozoa per dose. Initially, 5 ml macrotubes with a total concentration of 5×10^9 sperm seemed to give good results (Westendorf et al. 1975). The tests carried out on flattened containers of small volume (0.25, 0.5, or 1 ml) aim to disseminate identical cold or heat treatment for all the cells in the package as a whole, improving the results in sperm revival (Bwanga et al. 1991), but they present handling drawbacks. Other flattened containers with a capacity for the total number of spermatozoa and made of various materials have been tested, using Teflon (Bwanga et al. 1991) and aluminum bags (Kovachev et al. 1994; Barba et al. 1998) with similar results. Other authors recommended the use of frozen spherical "pellets" on dry ice (Nagase and Niwa 1964; Paquignon et al. 1974; Pursel and Johnson Ohnson 1975; Larsson et al. 1977) although the 0.5 ml straws have been more effective for pigs in comparative studies (freezing in other geometries not only does not improve the final fertility rate but also impairs the management of semen thawing). Packaging in 0.5 ml straws has been imposed since temperature changes are received at the same time by all cells, heat loss or gain during changes of state is substantially lower than in large containers and they are very manageable even in field conditions (Paquignon et al. 1986; Thilmant 1997; Poto et al. 1999; Labroue et al. 2000; Bussiére et al. 2000). Although we wanted to highlight the importance of the container geometry, given its impact on the economy and effectiveness of a swine germplasm bank, and taking into account the large number of doses/boar that need to be stored, it is necessary to take into account the incidence of other factors on the entire pig semen freezing technique (Fig. 5).

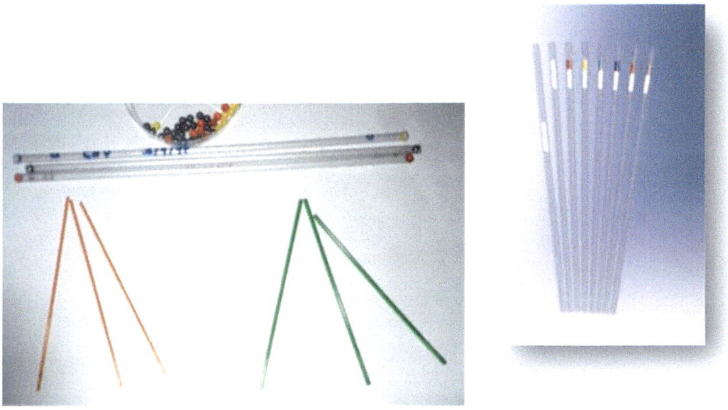

Fig. 5 Different geometry packaging for sperm cryopreservation

The cooling process until reaching the storage temperature in liquid nitrogen requires a slow process that occupies several different periods: starting with the temperature at which the rich fraction is obtained, the pig semen has to decrease the temperature to 5 °C in at least 5 h, at a speed of −0.08 °C/min, maintaining at 15 °C for two centrifugations at 800 g, for 20 min, to eliminate the supernatant contained in the seminal plasma and perform the first dilution without glycerol. When the temperature reaches 5 °C, it is time to add the part of the diluent that contains glycerol. In this circumstance, all authors who use various techniques agree, that obtaining fertility results by artificial insemination is close to natural mating (Martín 1982; Paquignon et al. 1986; Thilmant 1997; Bussiére et al. 2000). From this moment on, it is considered necessary to maintain the semen in a stabilization period of 90 min, although techniques that contain egg yolk-lactose recommend immediate freezing over liquid nitrogen vapors once glycerol has been added (Martín 1982; Martín et al. 1996).

The change of state process requires a drop in temperature at a rate of −1 °C/min to −4.5 °C, the temperature at which the fusion and seeding or crystallization process occurs (Thilmant 1997). Although using the same packaging geometry as this author, heat release temperatures due to a change of state close to −16 °C have been found (Poto et al. 1999). It is important to note that with geometries that contain greater mass, the amount of heat released is greater and the sperm will be subjected to thermal changes for a longer time, which will occur both in the freezing and thawing processes with the consequent cellular damage. The final temperature drop must be carried out at a speed of −30 °C/min. (Fiser and Fairfull 1990) to −196 °C (Fig. 6).

Sperm concentration is another factor that affects the freezing process. Concentrations of 10^9 spermatozoa/ml in 5 ml macrotubes have been proposed by several authors (Westendorf et al. 1975; Martín 1982; Martín et al. 1996), and currently, concentrations of 1.5×10^9 or 1.6×10^9 are being recommended for 0.5 ml containers (Thilmant 1997; Bussiére et al. 2000). Fertility outcomes and ease of handling seem to lean on the side of high spermatozoa concentrations (Fig. 7).

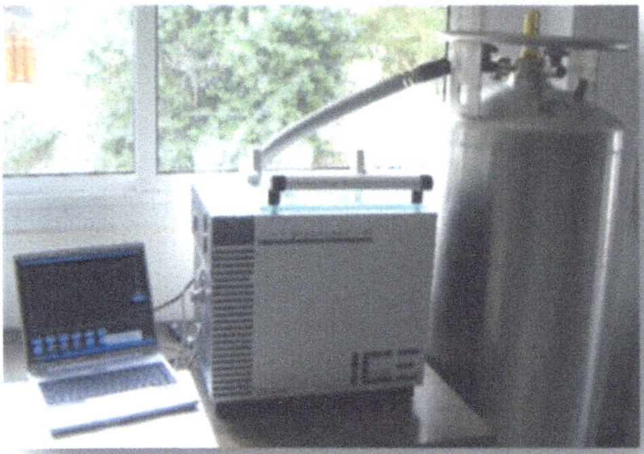

Fig. 6 Automatic cryofreezer

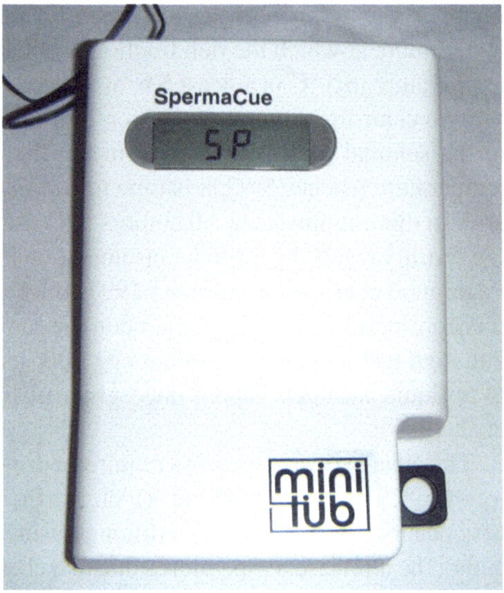

Fig. 7 A spectrophotometer used to study sperm concentration

To obtain good results, any technique must be followed with the greatest accuracy, and this can only be achieved with good equipment, with a refrigeration capacity that maintains all elements at the same temperature, and the use of automatic cryofreezers being essential.

Therefore, the critical points in the freezing of swine semen can be summarized as follows: (a) elimination of seminal plasma, (b) diluent preparation time, (c) type of diluent, with the addition of easily metabolized sugars that are very assimilable by the cell, (d) the percentage of cryoprotectant, with slow evolution in the cryoprotectant addition system, with a stepwise increase in osmotic pressure, (e) precision in the temperature drop curve, using automatic cryofreezers, (f) the packaging geometry (microtubes of 0.25 or 0.50 ml, macrotubes of 2–5 ml, bags of various materials: Teflon, aluminum), pellets, which allow the loss or entry of heat precisely to all the cells it contains, (g) adequate thawing with very rapid solid-liquid state changes.

From 1997 and early 1998, a total of 953 semen doses from five Chato Murciano boars were obtained after the following actions were frozen, reflected in Table 2 (Peinado et al. 2001).

The number of frozen seminal doses is also influenced by the age of the boar, as can be seen in Table 2 (Peinado et al. 2001).

* Storage of frozen seminal doses

- The frozen semen doses are stored in containers/tanks of liquid nitrogen, each boar in a different canister and in containers where the freezing date and the number of the boar whose semen is cryofrozen are identified, waiting to be evaluated in vitro and in vivo by using it for the artificial insemination of some female pigs (Fig. 8).
- Storage is one of the issues that require special attention, because the present and future costs of liquid nitrogen can make an ex-situ conservation program unfeasible, and therefore it is essential to calculate the number of minimum doses that must be stored per boar. It is necessary to clarify that the concept of dose refers to the minimum number of frozen/thawed spermatozoa capable of fertilizing a female pig. There is a certain tendency to confuse the dose with the container; certain containers can contain the complete required dose 5–6000 million total spermatozoa, case of macrotubes of 5 ml capacity (Westendorf et al. 1975; Martín 1982), aluminum bags (Kovachev et al. 1994), and Teflon bags (Bwanga et al. 1991), while currently there is a trend towards the use of smaller containers, 0.5 ml microtubes with a capacity for 750 or 800 million spermatozoa (Thilmant 1997; Barba et al. 1998; Bussiére et al. 2000).
- The number of doses to be stored for each breed is calculated as proposed by Ollivier and Renald (1995), depending on the scheme applied and mainly on the number of generations, as well as the reproductive parameters of the species. In the case of swine, assuming only one litter per generation and one generation per year, the number of minimum doses to be stored is given by the formula:

$$\mathrm{D} = dN\left(\mathrm{r}^{\mathrm{n}} - 1\right) / \left(\mathrm{r} - 1\right)$$

Table 2 Freezing of porcine semen doses

Ch-M Pig Age	N° collected	N° cryoconservations	Total doses	Average frozen doses
11 months	2	1	5	5
1.8 years	12	11	96	8.72
3 years	38	31	320	10.31
4.5 years	36	23	242	10.54
12 years	31	24	290	12.09

Peinado et al. (2001)

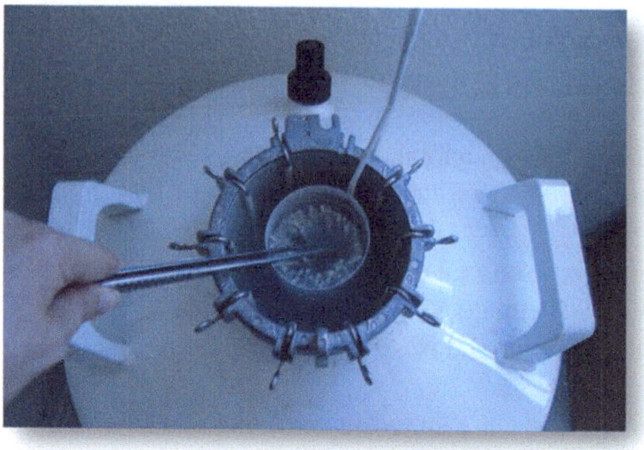

Fig. 8 Liquid nitrogen tank where stored seminal cryofrozen doses of Chato Murciano boar

- Where **d** is the number of doses per insemination, **N** is the number of reproductive females to be obtained, **r** is the inverse of the expected number of fertile females obtained by insemination with the d doses, and **n** is the number of generations.
- If it is necessary to store 80 doses/boar and the endangered breed has at least 8 semen donors, the total number of 5 ml containers would be 860. When the technique uses microtubes, the total number of containers to be stored will be at least 7000. The recommendations received from the European Union GENRES012 project establish a minimum of 25 donor males with a total storage of 2000 doses per breed, located in two different places to avoid possible accidents. In this way, the seminal doses of the injured bank can be replaced.
- An important factor to take into account in the construction of a Germplasm Bank is the minimum number of males required to guarantee the genetic variability of the cryopreserved population. According to Smith (1984), the minimum number of cryopreserved males should be 25 to guarantee maintenance of the 2% limits of inbreeding at the time of breed reconstruction. In this sense, Ollivier and Renald (1995) and Poto and Peinado (1999) indicate that from a genetic point of view, a minimum number of 80 fertile doses per boar is recommended, stored in liquid nitrogen with a view to the future recovery of a boar breed in danger of extinction. Furthermore, for optimal use of the germplasm bank resources, consisting of maximally adjusting the number of males and the number of doses necessary, taking into account the storage costs of cryofrozen seminal doses, it is important to monitor the critical points of the cryofreezing technique, already mentioned.
- An important aspect of the storage of frozen seminal doses is their identification. The growth of international exchanges of frozen swine semen requires, for several reasons, reliable identification of each dose of semen. In 1992, a European group of A.I. specialists approved, at the International Congress on Animal Reproduction and Artificial Insemination in The Hague (Netherlands), a com-

mon procedure for identifying semen dosage. This identification can easily be adapted to other species by adding two letters indicating the species at the beginning of the identification number.
- Another important aspect of conservation programs is to reduce risk by storing duplicate samples in two geographically distant locations. This serves to protect the frozen breed from disasters such as fires or earthquakes, accidents, as well as any errors or accidents in the management and operation of the Germplasm Bank. Once the doses have been collected and identified they can be divided into two identical subsets for long-term storage.
- About the location of the samples, after freezing, the semen dose is stored in canisters inside the liquid nitrogen thermos. Each canister (there are different sizes on the market) contains only the doses of the same male and the same ejaculate. An appropriate registration system must be developed that allows rapid and reliable identification of the ejaculate of a given stud and a certain date, inside the thermos.
- Liquid nitrogen thermoses. Samples can be transported in special double-walled thermoses with a content of 10–50 L. Semen storage flasks come in sizes ranging from 2 L to several thousand liters. For practical reasons, conservation programs should store material in 30–60 L thermoses. This is due to the small number of straws per donor or doses of semen per male and ejaculate, as well as the health classification of the samples. These thermoses have very low evaporation rates (as long as they are not opened frequently) and thus ensure the duration of the appropriate level for several months without the need to replenish liquid nitrogen. Furthermore, in the event of an unexpected breakage of a thermos, the damage will be less by using small thermoses instead of large ones.
- Stock maintenance and level replacement. Once in liquid nitrogen, the semen doses should not be thawed before the time of use and, then, the appropriate procedures must be followed, which we will see below. If accidental thawing occurs, the semen will not be viable.
- Access to the Germplasm Bank will consume the stock. It is essential to take into account the replenishment of stock levels whenever possible. It can be foreseen that all the breeds conserved in the Germplasm Bank will have advanced towards extinction and that replacement will no longer be possible (at least of identical quality). For this reason, users must replenish the stock level to the extent possible. Given the objectives of gene banks, there appears to be no justification for throwing away stored material because it has not been requested for long periods.
- Consequences of the health classification of semen on storage.

The compromise between the need to conserve the genes of an endangered breed and the health status of these latter representatives and/or the availability of appropriate facilities in the vicinity has led to recommending the introduction of a semen scoring system according to the quality of the health control carried out and the health status of the donors.

For the classification of mammalian semen, this system grades the health status of males into three categories A (the highest), B, and C (the lowest). On the straws, it will be identified as GA, GB, and GC, respectively. This will facilitate the storage in the germplasm bank of animals with similar health status but from different origins and will also facilitate their access and future use.

The semen doses, once prepared, are hermetically sealed and there is very little risk of disease spreading directly from the semen, although there is limited risk from the thermoses and during handling. Therefore, it is advisable to store semen doses of different grades in separate rooms. Each room must have its equipment (tweezers, containers, etc.) that must be marked and must not be passed from one room to another. Thermoses, canisters, and plastic tubes must be carefully washed and disinfected before being introduced into one of the rooms. Samples from different countries but of the same sanitary grade can be kept in the same room, but always in different thermoses.

The issue becomes more complicated regarding the storage of duplicates. No difficulties are expected for grade A since this type of semen can move freely. Grade B semen must be transferred to a country that has the same animal health conditions as those prevailing in the country of origin. More difficulties are expected with grade C doses, for which it will be necessary to find a country that has the same general (poor) animal health conditions, but good, safe, and reliable storage facilities and that accepts samples of this low quality.

The grading system should allow maximum flexibility by permitting the collection of any animal identified by its genetic singularity, providing maximum information to establish the necessary security measures to avoid the spread of diseases when the samples are used.

Currently, there are 4064 straws of 0.5 ml stored in the IMIDA Animal Germplasm Bank, belonging to 24 Chato Murciano breed boars. This Bank began in 1999, and the intention is to increase it as new donors of seminal material become available for cryofreezing.

In 2018, the quality of semen from different Chato Murciano pig specimens, frozen for more than 10 years, was evaluated and compared with the quality they presented at the time they were obtained and cryofrozen (Galián et al. 2018). To thaw the 0.5 ml straws in a water bath at 56 °C for 12 seconds, 4.5 ml/straw of BTS medium warmed to 37 °C was used. The parameters analyzed were individual motility, percentage of motile sperm, percentage of viable sperm (eosin-nigrosin stain), and percentage of damaged acrosomes (with phase contrast microscope). The analysis was carried out immediately after thawing and every hour for 4 h (thermal resistance test) (Table 3).

In conclusion, the result obtained was that after 10 years or more of conservation in liquid nitrogen tanks, the seminal doses of Chato Murciano pigs presented optimal quality after thawing at zero time. This quality decreased to a greater extent than those of the doses that remained frozen for only 24 h throughout the 24 h of the analysis.

In the coming years, these analyses will be repeated depending on the incorporation of new semen donors into the Chato Murciano pig Germplasm Bank.

Table 3 Resistance test for 5 h of seminal doses cryopreserved in liquid nitrogen

	Individual Motility	% Motile sperm	Vital stain (% spz. Motiles)	Intac Acrosomes (%)
Time 0	3.83 (3.5–4.5*)	55.42 (40–80%*)	76.8	81.8
Time 1	3.89 (3.5–4.5*)	48.75 (35–75%*)	68.2	74.2
Time 2	3.70 (3.5–4.5*)	42.08 (30–60%*)	61.1	68.7
Time 3	3.44 (2–4.5*)	32.50 (30–50%*)	53.05	63.6
Time 4	3.00 (0–4*)	21.81 (0–40%*)	44.1	60.4

Peinado et al. (2001). * Minimum and maximum value obtained

3 Murciano-Granadina Goat

3.1 Introduction

The Murciano-Granadina (MG) goat breed comes phylogenetically from the *Capra aegagus*, whose secondary form is the *Capra pirenaica*. Both the Granadina and Murciano breeds were different breeds until 1975 when the Herd Book of the MG breed was approved (cabrandalucia.com). The first time the MG breed is mentioned in legal texts is in the Resolution of 03/28/1979, of the General Directorate of Agricultural Production, which approves the Genetic-Functional Assessment Scheme of Reproductive Males of the MG Breed. The Ministry of Agriculture joined by decree the destinies of the then Granadina and Murciano breeds, two perfectly defined breeds in those times (Galián 2022).

Over the years of joint management of both breeds and after the circumstances that have affected the Spanish goat in the last 40 years (transition to democracy, federalization of the state, accession to the EU, globalization, etc.), the breed's current situation is reflected in the existence of animals that respond to the two original morphotypes and a third MG morphotype resulting from the crossing between the original varieties that is in great expansion and recognition (murcigran.es).

Currently, according to the Official Catalog of Livestock Breeds of the Ministry of Agriculture, Fisheries and Food, the MG breed is included in the group of Native Breeds in Development and is distributed mainly in the Autonomous Communities of Andalucia, Murcia, Valencia, Castilla-La Mancha, Baleares Islands, Cataluña and Extremadura (feagas.com). It is presently the most cosmopolitan goat breed in Spain with an estimated census of 500,000 animals. In addition to being distributed throughout Spain, it is also present in different countries on the European, American, and African continents. This great diffusion is possible due to its great capacity for adaptation and the great cheese yield of its milk, which is why the production of the MG breed is oriented toward obtaining milk (Galián 2022).

All these characteristics make it a breed of enormous economic value, thanks to its resistance to arid climates, its continuous reproduction throughout the year, and its large milk production, which provides benefits to livestock farms. The genetic improvement of this breed through assisted reproduction techniques such as AI can help make it even more productive, always respecting animal welfare. Hence the interest in the AI technique in this breed is being optimized and achieving the highest possible fertility and birth rates (Galián 2022).

3.2 Reproductive Performances in the Murciano-Granadina Goat

The creation of a Goat Germplasm Bank requires, logically, the extraction of semen from the male goat (buck). Ideally, this collection should provide semen of optimal quality (sufficient to allow freezing and post-thawing use), not be excessively complicated to perform, and not compromise animal welfare.

The first methods of collecting semen in goats consisted of introducing cloths and/or sponges into the vagina of females in heat, which were wrung out after copulation with the male, collecting the semen deposited there. Moving forward in time, at the beginning of the twentieth century, semen collection was carried out by introducing a glass collector, Dubois collector, into the female's vagina, in which the male's semen was deposited after mating (Poto et al. 2021a, b). These traditional forms of collection had a very important advantage in that animal welfare was not compromised. The approach between the male and the female occurred naturally, as did courtship and penetration. The main problem with this method was the removal of the collector from the vagina, which was complicated due to the movements of the goat, and normally a part of the ejaculate was lost.

Another way to collect seminal fluid is through the use of an electroejaculator. This method requires the application of an electric current through the rectal mucosa to suddenly stimulate the nerve pairs that innervate the spinal cord with the male genital apparatus at the lower part of the sacral region and the lumbar nerve pairs, triggering the entire process of stimulation, erection, and ejaculation. Sudden muscle contractions also occur throughout the posterior third, mainly in the lumbar area and in the posterior extremities, which means the unphysiological stimulation of the muscle plates, nerve endings, and blood vessels, causing the relaxation of certain sphincters such as those of the urethra (Veru Montano) and the rectum, with the possibility of allowing the discharge of bladder and intestinal contents. It is not a method frequently used in goats, mainly due to the commitment it entails for the well-being of the donor male. In addition, it can affect seminal quality by reducing sperm concentration due to retrograde ejaculation (Tecirlioglu et al. 2002), obtaining a lower concentration of semen than when collecting with an artificial vagina (AV) (Memon et al. 1986) and often the semen is mixed with urine or has lower quality than that collected through AV (Cueto et al. 2016).

Finally, some methods that have remained for use in the specific research field are those that extract semen by puncture in the urethral canal, the vas deferens, or directly from the epididymis once the animal has been sacrificed.

Currently, these technologies are not used and have been replaced by semen collection with AV (Poto et al. 2021a, b). For semen collection through AV, a well-known device is used; it consists of an external box (a rigid cylindrical body) and an internal latex or rubber jacket that fits tightly at the ends (Fig. 9). The space between these two elements can be filled with air and hot water (about 45 °C), in such a way that the interior of the AV simulates the environment that the penis receives during intercourse in terms of temperature, texture, pressure, and humidity conditions (the interior of the flexible element is usually lubricated with neutral Vaseline) (Poto et al. 1995; Cortés 2003). The ejaculate is collected in graduated glass tubes placed at the end of the vagina and fitted with a flexible funnel (Fig. 9).

Using this technique, we can obtain an ejaculate with the same characteristics as natural mating, while respecting the male's courtship. This courtship before penetration, by both the male and the female in heat, is more important than it may seem at first, since its performance helps the emission of gametes and protective and nutritional fluids, improving their quantity and quality (Poto et al. 2021a, b; Galián 2022).

One of the main disadvantages of the AV method is that it requires a training period (Maxwell and Evans 1990; Wulster-Radclife et al. 2001), so the buck becomes accustomed to mounting and ejaculating in human presence. Bucks intended to be used as semen donors must become accustomed to mounting a female in human presence and allow the technician to approach to withdraw the penis at the

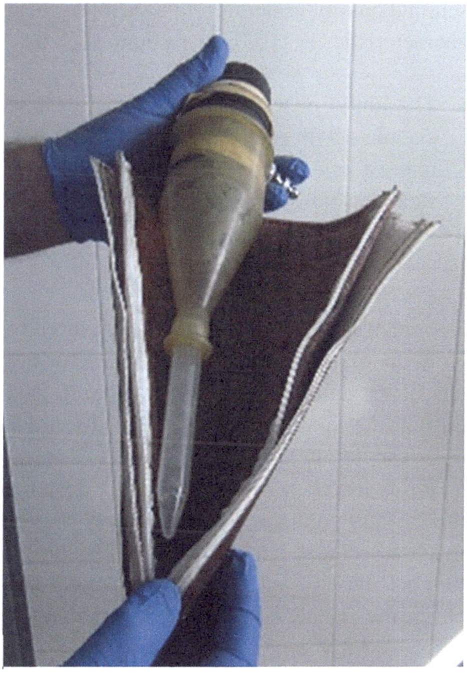

Fig. 9 Artificial vagina

moment before penetration and divert it towards the AV. Traditionally, this process has been carried out by coupling the male with an estrogenized female or a female in natural heat attached to a dummy. In this way, after carrying out a previous courtship to know the willingness of the female goat (doe) to be mounted, the male makes an attempt at copulation by mounting her, a moment that is used to divert the penis towards the artificial vagina where he will ejaculate and the semen will be collected to carry out the following steps according to its subsequent use. This procedure repeated over several periods finally gets the males used to donating semen outside the caprine vagina.

3.3 Choice of Male Semen Donors

Bucks must meet a series of requirements to be used as gamete donors to maintain a germplasm bank. Their seminal quality is essential since males whose semen does not meet minimum seminal quality parameters in fresh ejaculate will be excluded as donors. It must be taken into account that the semen obtained will be frozen to be preserved over time and, after thawing, it must maintain optimal quality to be used in artificial insemination. Freezing semen will cause the death of a percentage of sperm cells, so starting from semen with high sperm concentration, viability, and motility is essential.

Semen quality will depend, in turn, on a series of factors, such as breed, age, season of the year, diet, and the intrinsic characteristics of each male. Not only seminal quality is an important factor for using a male as a donor, but there are other characteristics to take into account, such as:

- The libido of a given male. It will facilitate semen collection if certain males ejaculate with each extraction attempt compared to those males who do not show interest in copulation.
- Reaction time (RT). RT is defined as the time in seconds that passes from the moment the male makes contact with the female until he ejaculates. Short RTs speed up semen collection, especially if several males are used.

Semen quality, libido, and RT are directly related to the age of the specimen.

In the Murciano-Granadina (MG) goat breed, seminal quality will increase as the age of the animal increases, stabilizing from 1 year of age, and remaining stable during the male's reproductive life (quality parameters are maintained until 6–7 years of age).

RT is lower in MG males under 1 year old, compared to that shown by older males. Very young males barely carry out courtship, they sniff the female or not and in a few seconds, between four and twenty, they mount her. This process changes as they advance in age and around 1 year old they begin to carry out courtship to check the female's acceptance of mating, sniffing her sexual organs, sniffing the environment raising his upper lip, licking the animal's genitals and flanks, striking its hooves on the female's flanks, checking for immobility by approaching head and neck and making sounds together with pseudo-sneezes, normally directed at other males, who respond in the same way to these sounds.

Some authors attribute the lack of guidelines in sexual behavior before mating to the inexperience of young males and competition with adult males (Ungerfeld and González 2009; Orihuela 2014). Dominant males, of greater corpulence and strength than the young ones, can attack the latter if they dedicate considerable time to courtship before mounting, with the young males losing the possibility of mounting the females of the entire herd, so the strategy of these males would be the speed of copulation, limiting or even ignoring courtship time to a minimum (Galián 2022).

Finally, libido decreases as goats' age increases, which can be verified by counting the number of mounts and ejaculations that they perform in the same morning. Males of the MG breed under 1 year of age perform between four and seven ejaculations without difficulty. When they reach 4 years of age, they ejaculate on average about four times in one morning, while 6- and 7-year-old males usually do not ejaculate more than two in a row, with numerous days when they do not ejaculate even once. This characteristic may compromise their use as studs (Galián 2022).

Both semen quality, RT, and libido have great variability between specimens, so each male must always be tested to decide whether or not to include him as a sperm donor. A suitable male to act as a semen donor will be one who combines optimal semen quality when fresh and post-thawing, a high libido, and a short RT.

3.4 Semen Obtaining Process in Goats

As mentioned previously, semen collection through an artificial vagina is the method most frequently used in the goat species. To obtain semen in this way, a female goat in heat is usually used as a sexual lure to excite the donor males. A female that is not in heat can also be used, often in which case she has been hormoned using an injectable estrogen, normally estradiol benzoate or valerate. The goat is usually immobilized and attached to a collection dummy (Figs. 10, 11 and 12).

It is important that the collection is always made in the same place, which will be recognized by the males as such and will facilitate its completion.

Once we have the female prepared, we introduce the male into the enclosure where the technician who is going to carry out the collection with the prepared AV is already located. The AV will have a funnel and a graduated glass tube attached and warmed to 37 °C to collect the semen, avoiding thermal shock.

When the male mounts the female, externalizing the penis, it will be manually diverted by the technician (avoiding penetration) towards the AV where ejaculation will occur, collecting the semen in the glass tube (Fig. 15).

During the process, the RT can be measured with a stopwatch and it can be observed whether courtship is carried out on the female or not.

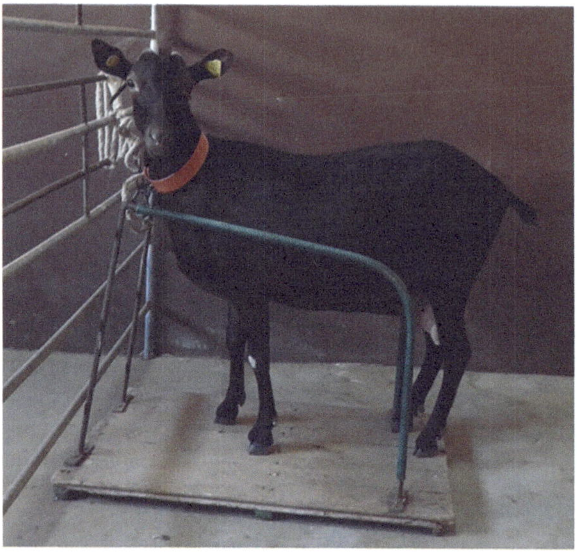

Fig. 10 Immobilized goat attached to a collection dummy

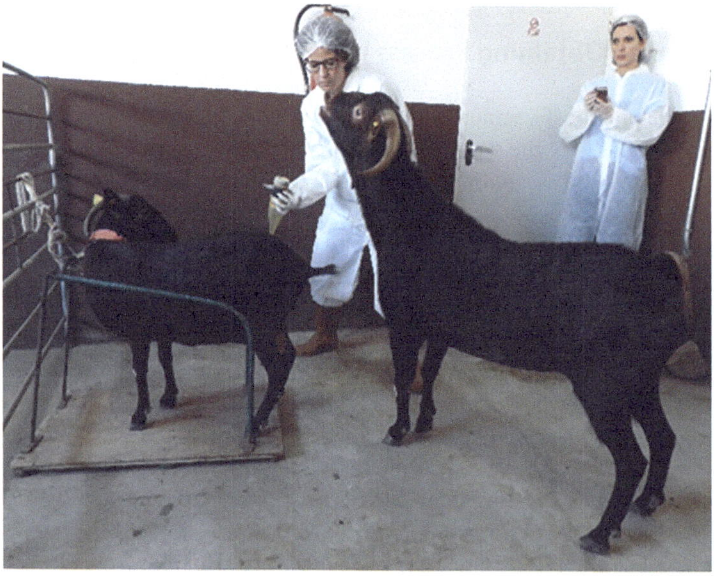

Fig. 11 Buck carrying out courtship on an immobilized goat

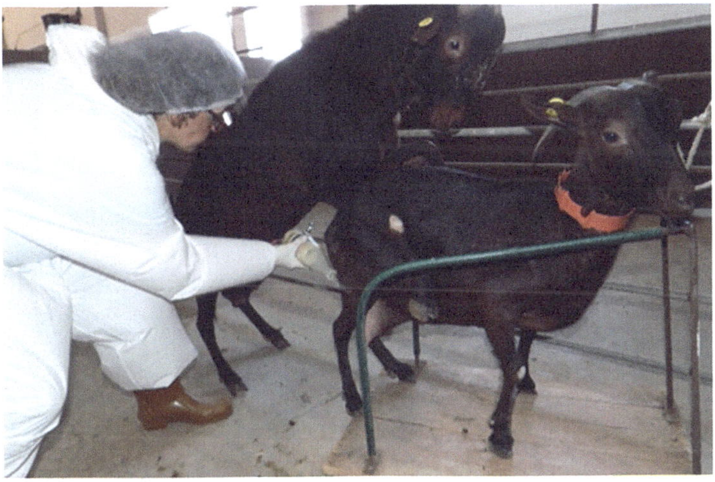

Fig. 12 Semen collection using an artificial vagina with an immobilized goat

3.5 Use of Another Goat or a Dummy as a Sexual Stimulus to Improve Animal Welfare in Semen Extraction

As previously mentioned, semen is collected using a female in natural or artificial heat, obtained by injecting hormones, to avoid the male's refusal to mate. This implies that the female used as a sexual stimulus and, therefore, subjected to estrogenization, must receive several injections throughout the entire collection period, so it is usually the same animal that is used in all collections at the center or farm where goat semen is prepared. This compromises her well-being, both due to hormonal overdose, which can cause the accumulation of metabolic derivatives in the detoxification process (Silva and Loaiza 2007; García 2018), and because this female will have to endure the actions of typical male courtship goat that includes kicking, sniffing, hitting the female's flanks and mounting attempts, of any males that are going to be used as donors. To avoid animal suffering, at IMIDA, semen extraction trials have been carried out on goats using, instead of a female, another buck or even an inanimate dummy in the shape of a doe as a decoy, and the behavior that has occurred has been evaluated show the donor males regarding the change in sexual stimulus and the seminal quality of the ejaculates thus obtained (Figs. 13 and 14).

The results of these studies have shown that the use of another buck as a sexual stimulus does not affect the sexual behavior of the donor males (they had been trained using a female) since there were no significant differences in reaction time or the performance of courtship nor in the number of ejaculates they had in one morning (Galián 2022; Galián et al. 2024). There were also no significant differences in the seminal quality parameters for the averages obtained when using a female as a decoy (Tables 4 and 5). Therefore, the use of another male, stronger and larger than a goat, and therefore capable of withstanding the usual courtship of this species, would be a good method of obtaining goat semen that would reduce animal suffering.

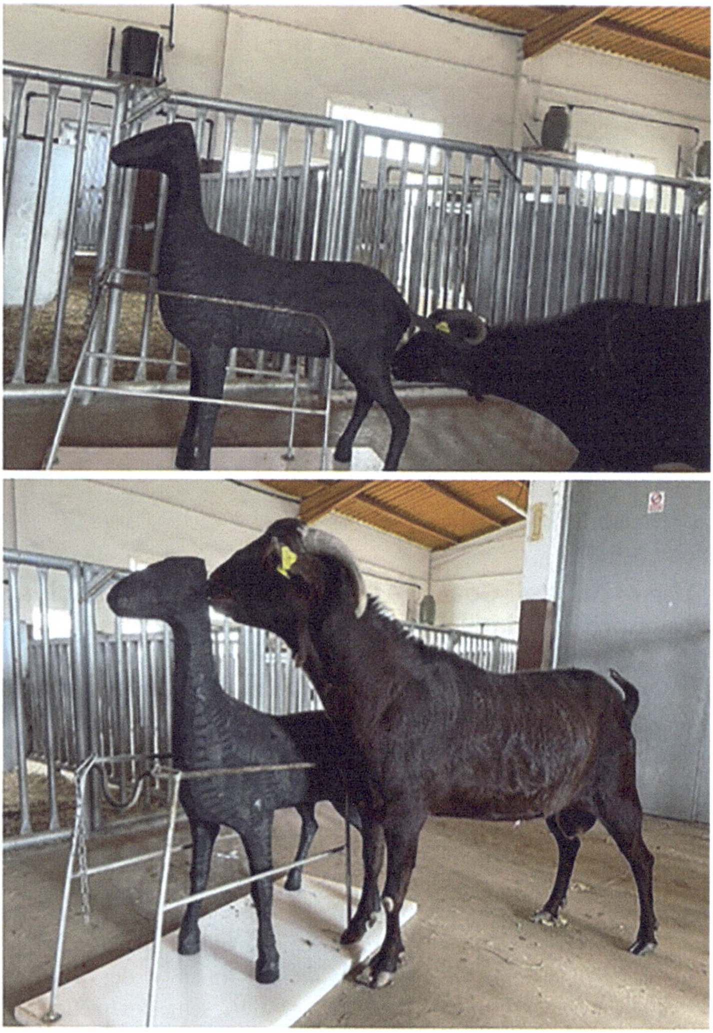

Fig. 13 Buck sniffs the dummy as part of sexual behavior before mating

However, the use of an inanimate dummy in the shape of a doe failed to excite the donor bucks in the same way as a female or another male, causing only the males with the highest libido to ejaculate, reducing the number of matings they had one morning and decreasing the seminal quality obtained in the parameters of ejaculate volume and sperm concentration (Table 5) (Galián 2022; Galián et al. 2024).

A more in-depth study of the dummy as a sexual stimulus is required, introducing it during the stud males' training period and accompanying it with some organic fluid, such as urine, follicular fluid, or semen from another male, to see if these first results are improved. The advantages that this method would provide would be extremely interesting. Of course, achieving maximum animal welfare is the main objective, but it would have other advantages, such as being able to attach an AV to

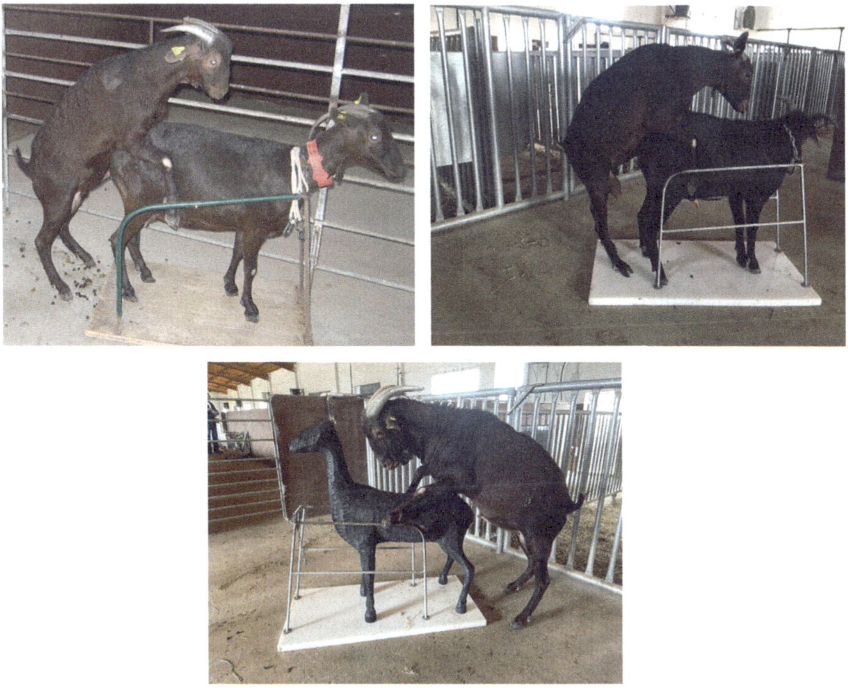

Fig. 14 Buck mounting on the goat, another buck, and a dummy, respectively

Table 4 Mean values of reaction time (RT), ejaculate volume, total sperm concentration, IM and % MOT when using a female or a male as sexual stimulus (Galián 2022)

	RT (s)	Ejaculate volume (ml)	[Esp] (x10^6)	IM	MOT (%)
FEMALE	19.2^a	0.77^a	3604^a	4.18^a	76.2^a
(N = 35)	± 10.2	± 0.07	± 273	± 0.07	± 2.3
MALE	22.1^a	0.66^a	2966^a	4.20^a	73.4^a
(N = 43)	± 7.6	± 0.06	± 193	± 0.06	± 1.9

Mean ± standard error. Different letters in the same column indicate significant differences. N = number of ejaculates analyzed

Table 5 Mean values of seminal quality parameters evaluated when comparing the use of a female and a dummy as a sexual stimulus (only males that ejaculate with both stimuli) (Galián 2022)

	Reaction time (s)	Ejaculate volume (ml)	[Esp] (x10^6)	IM	MOT (%)
FEMALE	17.6^a	1.15^a	2552^a	4.11^a	69.5^a
(N = 58)	± 2.23	± 0.06	± 168.4	± 0.05	± 2.11
DUMMY	**51.6^b**	**0.86^b**	**1967^b**	**4.11^a**	**67.1^a**
(N = 58)	**± 9.12**	**± 0.06**	**± 128.4**	**± 0.06**	**± 2.23**

Mean ± standard error. [Esp] = total sperm concentration. Different letters in the same column indicate significant differences. N = number of ejaculates analyzed the dummy that allows the males to ejaculate into it without having the technician nearby. This would be especially useful with those males of genetic interest that do not ejaculate in human presence.

3.6 Evaluation of Goat Semen Before Its Cryopreservation

The semen collected in the glass tube attached to the AV must be evaluated to verify that it has sufficient quality to be frozen. The first step is to keep it at a temperature of 37 °C, for which it is immersed in a thermostatic bath or a heater at that temperature.

From this sample the following is determined:

- Volume of undiluted ejaculate: by direct observation in the graduated glass tube, the milliliters of semen obtained from the male are observed. The mean volume in the buck is 1 ml, although ejaculates can be obtained between 0.3 and more than 1.5 ml, depending on the age, season of the year, general condition of the male, and rate of collection (Delgadillo 2004) (Fig. 15).
- Mass motility: consists of observing the waves produced by a moving sperm mass. It should be performed immediately after obtaining the ejaculate and without diluting it (Evans and Maxwell 1990; Garde 1993). To measure it, a drop of semen is placed on a slide warmed to 37 °C, and without covering, the characteristic waves are observed on the edge of the drop. It is assessed subjectively, being rated from 0 to 5 (Aisen and Venturino 2004).
- After these evaluations, the sample is diluted 1:10, adding the corresponding volume of an appropriate diluent, such as medium Krebs-Ringer-Phosphate (KRP) (Corteel 1974), which must also be at 37 °C. Once diluted, the following can be determined:
- Sperm concentration: usually expressed in the number of spermatozoa/ml (spermatozoa/ml), and normal values range between 3000 and 7000 × 10^6 spermatozoa/ml (Aisen and Venturino 2004). It can be determined using microscopic counting chambers, photocolorimeters, and turbidimeters. Dilution of the sample

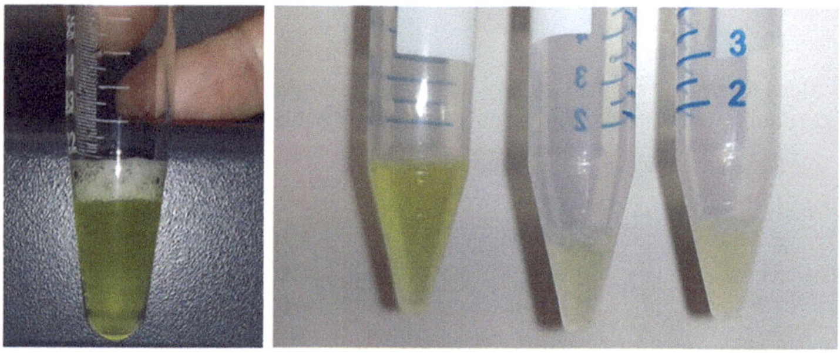

Fig. 15 Goat semen recently ejaculated, in which we can observe different volumes and colorations

is essential, since without it the sperm cells could not be visualized independently, due to their high concentration in undiluted semen.

- Individual Motility (IM): consists of evaluating the quality of sperm movement on a scale from 0 to 5 according to the progression of that movement, with the following grades 0: immotile spermatozoa, 1: non-progressive motile spermatozoa, 2: slow progressive motile, with stops in movement, 3: slow continuous progressive movements, 4: fast progressive movement, and 5: very fast progressive movements (it is difficult to follow them with your eyes) (Aisen and Venturino 2004). A drop of this dilution is taken, placed on a slide, and covered, both warmed to 37 °C. It is normally observed at 40X or 100X, evaluating several different fields.
- Percentage of motile spermatozoa (%MOT): Using the previous preparation, this parameter can also be estimated, assigning a subjective value of between 0 and 100, corresponding to the fraction of the number of spermatozoa presenting movement with respect to the total amount visualized (Aisen and Venturino 2004).

These are the essential tests to evaluate semen quality and accept or reject the sample for freezing. An ejaculate volume ≥0.5 ml, with IM of 3.5 or more, and a percentage of motile sperm of at least 60% are recommended to accept a sample for freezing (Galián 2022).

The following can be added to the tests mentioned above:

Vitality: evaluate the integrity of the head plasma membrane using dyes that stain damaged spermatozoa, whose membrane has lost its integrity, and, therefore, is considered dead. These spermatozoa allow the dyes to penetrate and appear colored in the stains, compared to the live ones, which remain unstained. The stains most frequently used to determine a sample's vitality are eosin-nigrosin (Fig. 16) (Colas 1980), trypan blue (Suttiyotin and Thwaiter 1991), and triple stain (Talbot

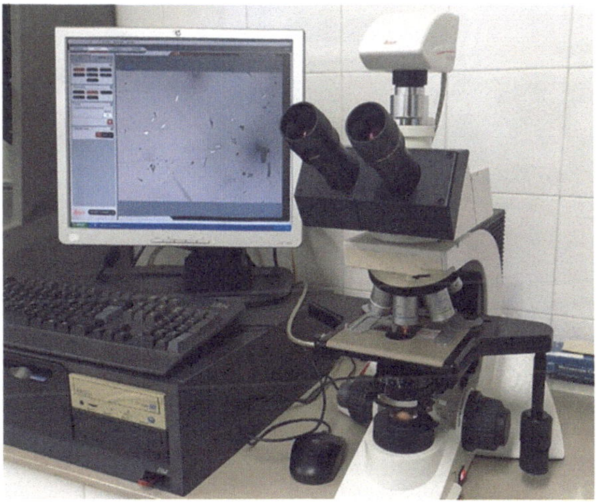

Fig. 16 Microscopic evaluation of sperm vitality using eosin-nigrosin stain

and Chacon 1981) which allows the observation of the live/dead spermatozoa, their acrosomal state, and their morphology. It is expressed as a percentage of live spermatozoa.

Morphoanomalies: The determination of abnormal spermatozoa can be done in stained smears for bright field microscopy, the most frequently used being eosin-nigrosin (Colas 1980; Mickensen et al. 1981). The degree of morphoanomalies can also be established with phase contrast microscopy, after including the samples in a formolated saline solution (Memon and Ott 1981) or in a glutaraldehyde solution (Voglmayr et al. 1983). It is recommended not to exceed 15–20% of abnormal shapes, although it is not a parameter related to the fertility of the sample (Garde 1993).

Acrosome integrity: the state of the acrosome can be assessed with different techniques such as bright field microscopy for samples stained with a 6% Giemsa solution (Watson and Martin 1972); by phase contrast microscopy, using a glutaraldehyde solution to fix the samples, where the state of the apical edge of the acrosome is observed (Pontbrandt et al. 1989) or by fluorescence microscopy, staining the sperm cells with a combination of fluorophores (Propidium iodide and PNA lectin) that distinguishes spermatozoa with reacted and unreacted acrosomes, in addition to their vitality (Silva and Gadella 2006; Hernández et al. 2012).

The aforementioned tests have a high degree of subjectivity. If you want to objectively determine the values of the parameters determining the quality of an ejaculate, the most frequently used evaluation system is the CASA system (Computer Assisted Sperm Analysis), through which computer software coupled to a microscope determines a series of parameters that show the semen quality of the evaluated sample (Fig. 17). These parameters are:

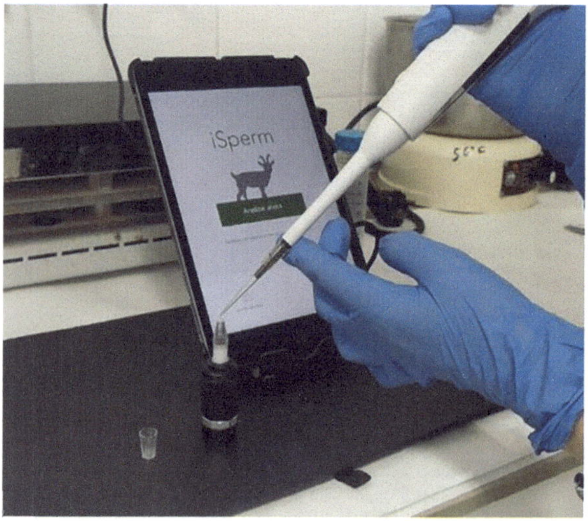

Fig. 17 Preparation of the seminal sample to be evaluated using the CASA system

- Curvilinear velocity (VCL)
- Straight-line velocity (VSL)
- Average path velocity (VAP)
- Beat cross frequency (BCF)
- Lateral head displacement (ALH)
- Mean angular displacement of the head (MAD)
- Linearity index (LIN = VSL/VCL × 100)
- Straightness index (STR = VSL/VAP × 100)
- Oscillation index (WOB = VAP/VCL × 100)

An optimal semen sample should have high displacement velocities as well as ALH. To penetrate the cervical mucus, the minimum sperm characteristics required are: VSL >25 μm/s, STR >90%, ALH >2.5 μm/s. Muiño et al. (2006) determined in bovine sperm that to consider a spermatozoon to be fast its VCL had to be greater than 110 μm/s, while Bravo et al. (2011), in ram spermatozoa, those with a VCL >75 μm/s were classified as fast spermatozoa.

It must be taken into account that elevated VCL and ALH show very active spermatozoa, but they do not have to be progressive since to determine progressivity the values of LIN and STR must be considered. For LIN we consider a good value if it is greater than 50% (Cox et al. 2006), and for STR if it is greater than 80% (Tomás 2007; Bravo et al. 2011).

3.7 Semen Freezing in Goats

Once a semen sample of optimal quality is obtained, it can be used immediately for artificial insemination, refrigerated for use in the next 24 h, or frozen to allow it to last for longer periods.

Freezing semen samples will allow the creation of germplasm banks of breeds of interest, and also presents numerous advantages, such as:

- Prolonged preservation of semen, even after the death of the specimen of genetic interest.
- Easy transport of genetic material over long distances of frozen semen samples, compared to the transport of a male of interest.
- Increased reproductive efficiency: several doses of frozen semen can be obtained from a single ejaculate.

However, it has an important disadvantage compared to the use of fresh semen, which is that freezing semen causes ultrastructural, biochemical, and functional damage to the spermatozoa, causing a decrease in viability and motility (Leboeuf et al. 2000; Watson 2000) and their fertilizing capacity, which leads to a lower fertility rate (31.2%, Ferrari et al. 1998; 40–45%, Gibbons 2002; 18%, Vera et al. 2004) if we compare it with the results provided by the use of fresh semen (60–80%, Lorenzo et al. 1997; 65–75%, Greyling and Van der Nest 2000). These damages occur mainly during the freezing-thawing process (between 15 and 60 °C) and not during storage in N_2L, and mainly affect the plasma membrane and the acrosome (Aisen 2004).

To try to reduce the damage caused to the sperm cells during the freezing process, we can act on two factors that will directly influence the generation of this damage: the diluents/extenders used and the speed of temperature decrease, without forgetting the form of thawing the straw, through which cellular metabolism will be restored (Galián et al. 2021).

3.8 Extenders for Freezing Goat Semen

The use of a good semen extender can make the difference between the success and failure of the freezing technique. A good diluent should contain:

- Sugars, which serve as a source of energy, such as fructose, glucose, and sucrose... and also act as cryoprotectants by maintaining or increasing the extracellular osmotic pressure in the spermatozoa, which helps maintain the integrity of the membrane when stored for long periods.
- Salts, which allow the maintenance of adequate osmotic pressure (Barbas and Mascarenhas 2009). The extender must be isotonic with the semen so as not to damage the plasma membrane.
- A buffer, which ameliorates changes in the pH of the medium, neutralizing the acids produced by the spermatozoa metabolism (Salamon and Maxwell 2000).
- Cryoprotective agents, which lower the eutectic point (the minimum temperature at which the state of a solution is liquid) to avoid damage to the sperm plasma membrane, as well as the formation of intracellular ice, since they induce dehydration of the cell (Ávila-Portillo et al. 2006; Rojas 2014).
- Antibiotics, to prevent the proliferation of microorganisms (Purdy 2006).

Generally, the means for refrigerating and freezing the semen of many species include egg yolk in a proportion of 15–20%. Extenders based on skimmed milk have also been widely used in goats. Both types of diluents present the drawback of the interaction between secretions from the bulbourethral glands of males with some egg components such as lecithin, which is hydrolyzed to fatty acids and lysolecithin (Iritani and Nishikawa 1963), which are toxic to spermatozoa. In the case of skimmed milk, a fraction of a protein called SBUIII, also produced by the bulbourethral glands, is capable of hydrolyzing the triglycerides in milk (Pellicer-Rubio and Combarnous 1998), resulting in products that are toxic to spermatozoa. For the use of both media, seminal washing by centrifugation is necessary (Cabrera et al. 2005).

Detergents are capable of solubilizing and altering cell membranes and their components. Sodium dodecyl sulfate (SDS) is an alkyl-ionic group detergent capable of denaturing the native structure of proteins. Added to the freezing diluent in small quantities, it has a beneficial effect on sperm motility and acrosomal integrity during the freezing-thawing processes (Pursel et al. 1978; Arriola and Foote 1987) and even prevents premature sperm capacitation into spermatozoa frozen and thawed (Peña and Forsberg 2000). This beneficial effect would be due to the modification it causes in the tertiary structure of the lipoproteins of the egg yolk of the diluent (Arriola and Foote 1987; Peña and Forsberg 2000), which exert their protective action on the cell membranes (Watson 1976). The addition of SDS alone or contained in commercial detergents such as Equex STM® Paste (Orvus ES Paste®) to an egg yolk-based extender improves sperm survival after thawing (Rota et al. 1997; Peña and Forsberg 2000) and also the ability of the spermatozoa to bind to the zona pellucida of homologous oocytes (Stornelli et al. 2001, 2002). It is, therefore, a common component in diluents for semen freezing of different species.

3.9 Goat Semen Freezing Protocol

Just as an appropriate diluent must be available, the freezing protocol used must damage the sperm cells as little as possible. The addition of the internal cryoprotectant, whose concentration will determine the speed of water passage through the cell membranes, and the speed of temperature drop, can cause damage to the cellular structure due to the formation of ice crystals and the way of thawing of the straw, through which cellular metabolism will be restored, are points that can be improved to avoid sperm damage.

The semen freezing protocol proposed by Thilmant for the swine species in 1997 provides optimal post-thaw quality results. This protocol has been adapted for the goat species, achieving high seminal quality after thawing semen samples from Murciano-Granadina goats (Figs. 18, 19 and Table 6).

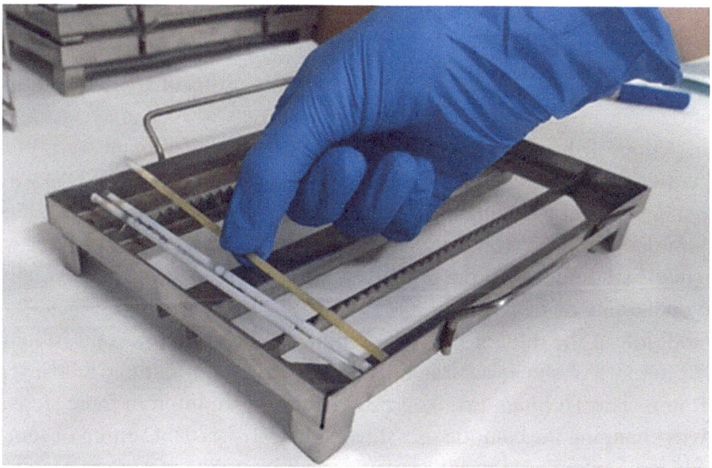

Fig. 18 Semen straws, before being placed inside cryo-freezer

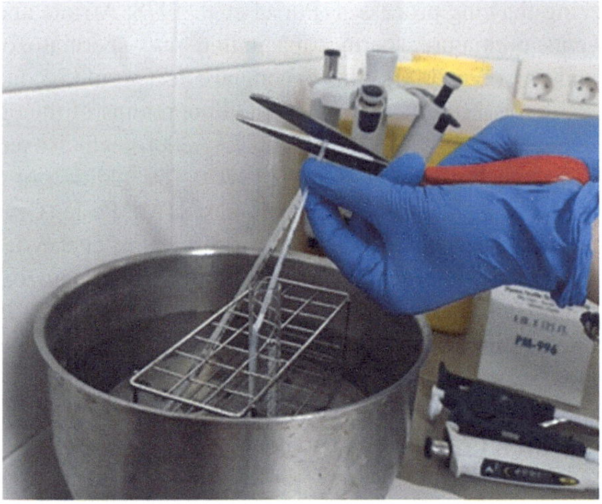

Fig. 19 Thawing of straws into graduated glass tube at 56 °C

Table 6 Composition of the diluent for sperm freezing (fraction A) for a total volume of 1 l (distilled H_2O + egg yolk) (Galián 2022)

Fructose (*Sigma-Aldrich, Steinheim, Germany*)	56.82 g
Sodium bicarbonate (Na HCO_{3-}) (*Sigma-Aldrich, Steinheim, Germany*)	1.0 g
Cysteine (L-cysteine $C_3H_7NO_2S$) (*Sigma-Aldrich, Steinheim, Germany*)	0.099 g
Egg yolk (farm fresh eggs)	227.27 ml
H_2O (Milli-Q® direct)	772.7 ml
Sodium dodecyl sulfate ($C_{12}H_{25}NaO_4S$) (*Sigma-Aldrich, Steinheim, Germany*)	6 g diluted in 10 ml of distilled H_2O

The main differences with respect to the original protocol are detailed below:

- Composition of the diluent: the diluent described by Thilmant for porcine was used with a modification; the detergent sodium dodecyl sulfate (SDS) was added independently and not contained in Orvus ES Paste®. In goats, extenders based on skimmed milk are frequently used. In this book, we will not refer to them, since in the studies carried out at IMIDA they provided lower semen quality than those containing an egg yolk base.
- The addition of the diluent fraction containing the cryoprotectant (diluent B) is carried out in five steps, separated by 5 min, instead of in three steps, separated by 10 min. The Thilmant protocol establishes the addition in three steps, which involves changing the osmotic pressure from 450 to 470 mOsm/kg of semen with non-glycerol extender to progressively higher osmotic pressures after each addition (650 mOsm/kg, 900 mOsm/kg, 1500 mOsm/kg). By adding two steps in the

Table 7 Mean semen quality parameters evaluated using the CASA system of thawed samples of MG goat semen (Galián 2022)

% MOT	% MP	VCL (µm/s)	VSL (µm/s)	LIN	STR	ALH
71.8 ± 4.2	56.1 ± 3.4	138.9 ± 3.0	121.3 ± 2.7	82.9 ± 1.3	89.5 ± 1.0	2.4 ± 0.05

Mean ± standard error. Number of ejaculates analyzed = 8

Table 8 Mean semen quality parameters through subjective evaluation of thawed MG goat semen samples (Galián 2022)

IM	% MOT	VS (% life)	% NAR
4.1 ± 0.4	66.2 ± 3.0	61.5 ± 6.3	72.5 ± 4.5

Mean ± standard error. Number of ejaculates analyzed = 8

addition, we achieve that this increase in osmotic pressure is even more gradual, and therefore, that the release of water from the cell interior occurs more slowly, causing less damage to the membranes due to the passage of water through them.
- The drop in temperatures until reaching freezing temperatures is carried out in a controlled manner using a computer system, with the straws being introduced into a cryofreezer coupled to a computer. The original protocol curve is followed, instead of carrying out the freezing process using liquid N_2 vapors, which is the most frequently used form of semen freezing in goats.
- The concentration of spermatozoa per straw is also adjusted according to the species because the sperm concentration of goat ejaculate is lower than that of porcine ejaculate.

Tables 7 and 8 show some results of seminal quality upon thawing of cryopreserved semen doses from Murciano-Granadina goats, through two assessments: subjective with an optical microscope and objective using the CASA system. These trials will be repeated and expanded to a greater number of semen donor studs during 2024 and 2025.

4 Murciano-Levantina Bovine Breed

4.1 Introduction

The Murciano-Levantina (ML) bovine breed (Fig. 20) is part of the genetic resources of the Region of Murcia, having been highly esteemed in the past for its aptitude as a traction animal in agricultural work, dragging fishing nets, demolishing houses, transport of goods and construction materials, since it consists of a very muscular front quarter, a strong head and a very powerful neck; castrated males or oxen are also used on some occasions (Sánchez 1984, 2002; Poto et al. 2021a, b).

There used to be four varieties of this breed: the Huertana, which was the smallest of the four; the Lorquina, used for work in fields and orchards; the Almazoreña, for the hauling of aggregates from the mines located in the region of Cuevas de

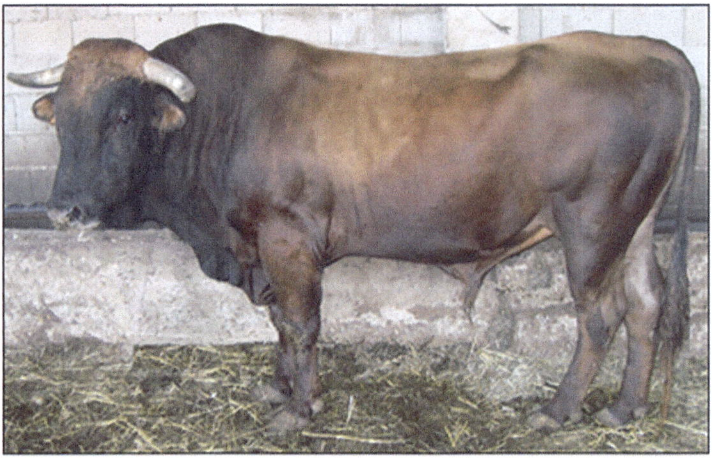

Fig. 20 Murciano-Levantino bull

Table 9 Evolution of the number of animals of the ML breed (Sánchez 1984, 2002)

Year	Number of animals
1916	60,793
1955	15,600
1980	125
1988	60
1991	70
1997	36
2002	34
2009	36
2010	56

Almanzora (Almería); and the Calasparreña, which was used for work in the rice fields near the municipality of Calasparra, also being the variety that exists today in breed purity (Poto et al. 2021a, b).

Furthermore, these animals gave birth to one calf a year, which they fed with their mother's milk at the beginning of their life, and at 15 months, they fed it with garden products and cereal, which they finally sold, because they were in high demand in the slaughterhouses of Madrid or Valencia (Poto et al. 2021a, b).

They produced around 4 L of milk (not counting what the calf consumed), using both this and colostrum as food for families, with the highest production being about 7–8 L (Poto et al. 2021a, b).

In 1916, the census of this breed was 66,000 specimens, but with the arrival, in the second half of the twentieth century, of advanced agricultural technology, its use was abandoned and the number of animals decreased, becoming disregarded as a genetic resource since other possible aptitudes were not competitive with the large milk and meat producing breeds of the bovine sector (Table 9).

Fig. 21 Murciano-Levantina cows pulling a cart in popular folk and in a religious festival

As we have mentioned previously, this breed is currently useless due to its unprofitable characteristics, since the parts that are considered most valuable, such as the back and leg, are also the most depressed. This is without taking into account the low carcass yield, which is 45% when the yield demanded in bovine breeds is 55–60%. No interesting yield has been found in terms of cheese production, nor do the current regulations in force help, which makes it difficult to open new farms or farmyards in the agricultural farming environment. All this without taking into account the productive and reproductive problems derived from high inbreeding (Poto et al. 2011).

For all of the above, and because, currently, it has little potential to offer, only is used in popular folk and/or religious festivals (Fig. 21), although attempts have been made to give them other uses by crossbreeding with meat-producing breeds such as Limousin to obtain a greater weight at slaughter in less time, or also castrating some specimens after semen extractions for the Germplasm Bank, to obtain oxen that will be marketed when they reach the requested weight. However, and as long as this situation continues, it has been included in the Official Catalog of Livestock Breeds of Spain (ARCA), of the Ministry of Agriculture, Food and Environment, as a breed at a critical point of extinction, with a current census of 51 specimens (Poto et al. 2011).

In 2007, the Association for the Conservation and Recovery of the Murciano-Levantina Cow (COREMUR) was created and managed by IMIDA, within which breeders and farmers who own some specimens of this breed at different levels of breed purity are associated, and through which they can access the transfer of semen doses for the AI of their females (Fig. 22), whenever they request them, selecting the cryo-frozen semen from the IMIDA Germplasm Bank, belonging to bulls the most genetically distant from the specimens that will be inseminated, to avoid future problems derived from inbreeding. To know the degree of inbreeding of each of the specimens, and given that at that time there were only a few specimens distributed, not only in Murcia but also in Almería, Granada, and Alicante (Peinado et al. 2013), interviews were carried out with each of the ranchers and breeders who owned these specimens to clarify their genealogical origins. Photographs were also taken of those specimens alive at that time, as well as ethnozoometric measurements or

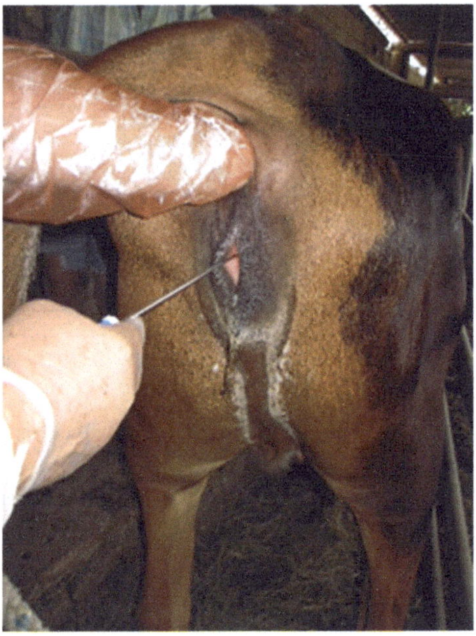

Fig. 22 Artificial insemination of a Murciano-Levantina cow

genetic-molecular characterization. All these data, along with the REGA number, owner, or percentage of inbreeding, were gathered and recorded in what would become a library of Genealogical Charts for each of the specimens, in which it was possible to consult which animal would be the most suitable for use in AI. This Association was in force until 2011, and it was not until 2023 that a new Association called AVAMUR (Murciano-Levantina Cow Breeders Association) was established, this time independent of IMIDA, although both were united by a collaboration contract, to which the Genealogical Charts of the registered animals were transferred, as well as the continuation of the keeping of the Herd Books and the realization of the AI, with IMIDA providing the genetic material from the Germplasm Banks of this breed. Currently, a census of 51 animals (32 females and 19 males) belonging to 7 different farms or owners is maintained, with maintenance and breed recovery work continuing (MAPA).

4.2 Germplasm Banks and Reproductive Performances in the ML Bull

In 2002, the first successful ML bull semen freezing (Fig. 23) was carried out by the IMIDA Animal Genetic Improvement Team, storing the resulting semen doses in liquid nitrogen tanks. It continued to be carried out during the following years, until 2018 when the sperm freezing work of this breed ceased, forming a Germplasm Bank composed of 9652 straws of 0.5 ml belonging to 16 ML bulls, extemporaneous to each other.

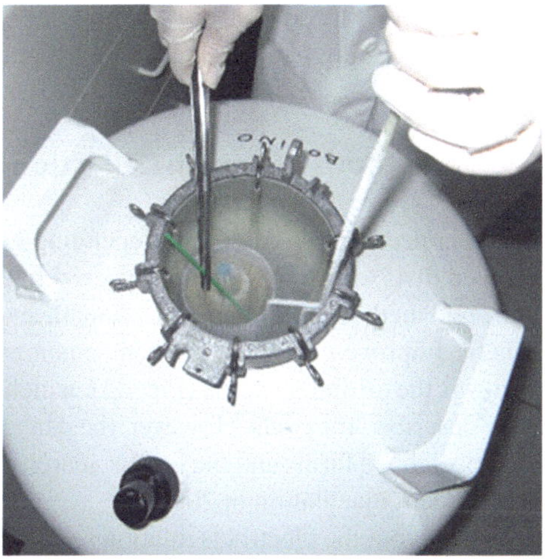

Fig. 23 Liquid nitrogen tank where seminal doses of Murciano-Levantino bull are stored

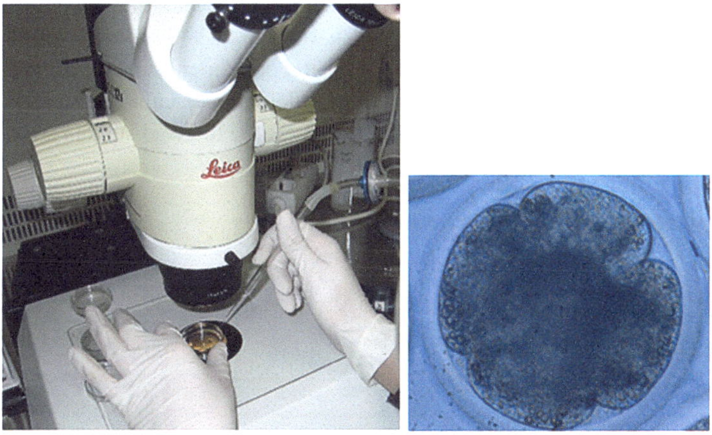

Fig. 24 Oocyte selection for In Vitro Fertilization (IVF) and embryo development

Obtaining these semen doses contributes, both to in vivo conservation, by transferring them to farmers to obtain new breeders, and to ex situ in vitro conservation for the future recovery of the breed if necessary, although the number of bulls is insufficient, since semen from 25 bulls preserved in vitro is required (Almela 2014).

On the other hand, from 2009 to 2010, work on In Vitro Maturation (IVM), In Vitro Culture (IVC), and In Vitro Fertilization (IVF) was undertaken at IMIDA using ovaries from the slaughterhouse to obtain oocytes and thawed semen from male ML bulls from the Animal Germplasm Bank (Fig. 24) (Galián et al. 2010). These works were carried out thanks to the research project that is currently underway and called "Improvement of Reproductive techniques regarding the manipulation of cryopreserved gametes and embryos destined for the Germplasm Bank of animal breeds of economic interest and social of Murcia (VITROGEN)," and that after the work carried out, provided a total of 170 embryos that are also currently stored and forming part of the IMIDA Animal Germplasm Bank.

4.3 Methods of Semen Collection from the ML Bull

Semen collection methods vary depending on the species we are dealing with and are even diverse within the same species. In the case of semen collection from ML bulls, two methods of semen collection have been carried out by the IMIDA Animal Genetic Improvement Team. When semen extractions began in 2001, the method used was the artificial vagina (Fig. 25), which follows the same methodology previously described for goats. However, the dangerous handling conditions of the bull at non-specialized farms and the lack or scarcity of libido in some specimens led to the use of electroejaculation in 2007.

To carry out the electroejaculation semen extraction method (Fig. 26), the bull is previously sedated and introduced into a holding chute. Next, a probe is introduced into the animal's rectum, which will apply small electrical stimuli incident on the lumbosacral area with a voltage of between 10 and 24 V depending on the bull's response, controlling it manually until protrusion of the penis is induced and subsequent ejaculation. Once the bull begins to ejaculate, the ejaculate is received using a semen collector, and it is deposited in a warmed tube to avoid sudden temperature

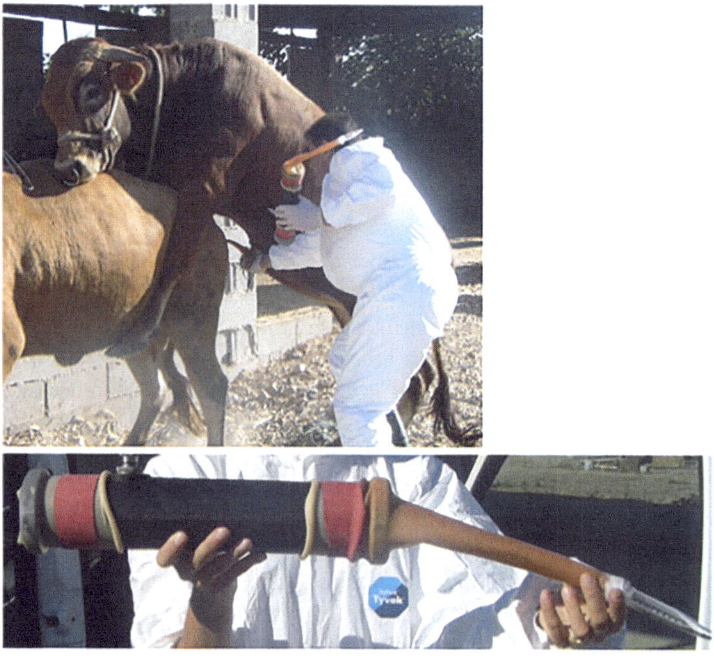

Fig. 25 Semen collection through the artificial vagina

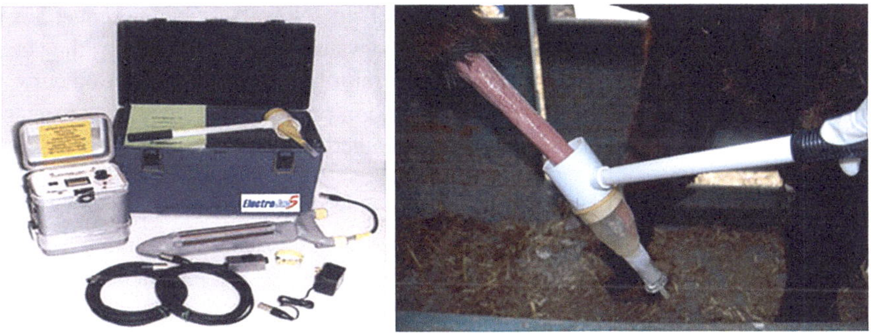

Fig. 26 Bovine electroejaculator and semen collection

changes that could affect sperm quality. Next, ejaculate quality studies are carried out taking into account the physicochemical parameters of color, odor, volume, and pH, in addition to other parameters such as concentration, motility, and vitality. If these parameters are above the minimum required quality, the ejaculate will be processed, frozen, and stored in liquid nitrogen tanks to form part of this breed's Germplasm Bank (Poto et al. 2013).

4.4 Semen Freezing in ML Bulls

Once the ejaculate is obtained using the techniques described above, it is processed following the protocol described by Watson (1976, 1981) and published by MAPA (1985).

The first step that must be followed after extraction is the assessment of seminal quality, since an ejaculate of poor quality or that has a strange color or odor will have to be eliminated. In addition, a minimum is established according to which, below 50% live spermatozoa and 3.5 motility, the ejaculate will be discarded and will no longer be processed for freezing.

Assuming that the ejaculate presents optimal conditions for processing, and once the calculations have been made to find out the number of seminal doses to obtain for a concentration of 40 million live spermatozoa per dose (Number of doses = Volume × concentration/40 million), The first dilution is added, which contains, among other compounds, egg yolk, a protector of the sperm plasma membrane, and an antioxidant, in an amount equivalent to the volume of semen collected. This dilution is added in three parts, separated by 10–15 min each, thus allowing a progressive increase in osmotic pressure that does not damage the sperm cells. Next, glycerolization is carried out, or the addition of the second dilution, which will contain glycerol, a membrane cryoprotectant against the formation of ice crystals produced during freezing. It will also be added progressively, in five parts, every 10 min.

Until now, this entire process has been carried out at 20 °C, but now it is time to check the semen quality, and if it continues to maintain the starting quality, begin to reduce the temperature from 20 °C to 5 °C in 30 min. Now the semen is kept at 5 °C for 4–6 h. Then the semen is put into 0.5 ml straws and sealed, and finally, they are frozen in an automatic biofreezer with a temperature-programmed reduction curve.

The straws obtained are deposited and stored in properly labeled liquid nitrogen tanks, and their seminal quality will be maintained by controlling the appropriate levels of liquid nitrogen.

To verify the quality and effectiveness of seminal samples frozen and stored for years in liquid nitrogen containers, in 2019 our team carried out a seminal quality study of samples frozen between 6 and 12 years ago. All the parameters studied, individual motility, percentage of motile sperm, acrosomes, viability using eosin-nigrosin staining, and fluorescence with propidium iodide and PNA lectin, were evaluated every hour until completing 4 h, thus checking their resistance over time. The results showed an optimal average quality to be used in artificial insemination, in addition to demonstrating that the freezing and storage technique is effective in the long term (Galián et al. 2019).

4.5 Artificial Insemination (AI) in the ML Cow

To carry out AI, the animals must meet certain health conditions according to current regulations, or certain characteristics related to body condition, such as secondary characteristics or genital organs, that correspond to the standards required to designate a quality breeder. However, in the case of breeds in danger of extinction or at a critical point of extinction such as the ML cow, these minimum selection conditions cannot be taken into account when choosing a good breeder, since the number of specimens available is limited, having to accept the few existing donors (Poto et al. 2013).

Taking into account the genealogical charts prepared by our team, when carrying out the inseminations, it was possible to select the semen belonging to the most suitable males, that is, those that were the most genetically distant from the females that were going to be inseminated, thus trying to reduce the problems derived from the high inbreeding that results from these breeds in danger of extinction with so few specimens. Thus, at least, they must have 75% of the genes of the breed, that is, a grandfather of the breed, 50% of maternal genes, and 100% of paternal genes (Poto et al. 2011).

4.6 Inbreeding in ML Cattle

Until now, to carry out the recovery, crossbreeding was carried out through natural mating and/or AI with the gametes of those breeders registered in the breed's Stud Book. For an animal to belong to the Stud Book, it must have at least 92% of the breed's genes, which is achieved through successive crossings over generations. In addition to keeping live purebred animals.

To find out if the actions regarding the conservation and recovery of this breed were being effective, in 2013 our team carried out work to determine the degree of inbreeding of this population by calculating the individual and population inbreeding quotient according to Cavalli-Sforza and Bodmer (1981) with all the copies available at that time. The individual and population inbreeding indices of this breed are $F = 0.125$ as the minimum value, $F = 0.453$ as the maximum value and the average population inbreeding coefficient is 0.233 (Peinado et al. 2013).

Furthermore, a work was published in which the individual inbreeding coefficients were calculated and compared with the number of births per year of each of the breeders that had had offspring. The results showed a decrease in calving in those cows with the highest inbreeding coefficient, although some specimens calved every year, despite the high inbreeding. On the other hand, the coefficients of those specimens of very low inbreeding were compared with those of higher inbreeding, resulting in significant differences between the averages (0.8 births/year without influence of inbreeding versus 0.5 births/year for those that do), so a decrease in the number of births per year was observed the higher the inbreeding coefficient was.

Furthermore, it was concluded that males of the breed used for artificial insemination with high inbreeding resulted in a lower number of births per year, as well as part of the offspring, presenting reproductive and metabolic problems (Almela et al. 2013).

The animals that remain today have a very high inbreeding coefficient and, therefore, reproductive problems are frequent. This results in weak newborn calves, unable to get up or drink their mother's milk without help, requiring an operator to administer it (Poto et al. 2021a, b).

4.7 Birth of an ML Calf Through Reproductive Biotechnology Procedures

One of the works that had the most impact at a scientific and media level was the birth of "Fuensanta" (Fig. 27), the first calf born through reproductive biotechnology procedures. The IMIDA Animal Genetic Improvement Team participated in this work, together with the Department of Physiology of the Faculty of Veterinary Medicine of the University of Murcia and the Department of Animal Reproduction of the National Institute of Agricultural and Food Research and Technology (INIA) of Madrid (Astiz et al. 2012).

Within the framework of a research project, financed by the Ministry of Economy and Competitiveness, within the Animal Genetic Resources Program managed by the INIA (FEDER-INIA funds) for the recovery of animal breeds that are in danger of extinction.

To make this birth possible, the following biotechnological procedures were carried out (Astiz et al. 2012):

Fig. 27 Fuensanta, the first calf born through reproductive biotechnology procedures

- Seminal cryofreezing: cryofrozen semen from ML bulls was used and carried out by the IMIDA Animal Genetic Improvement Team. To obtain it, the semen freezing technique described by MAPA (1985) was applied.
- Transrectal ultrasound-guided follicle aspiration (*Ovum Pick Up*; OPU) (Kruip et al. 1994; Ruiz and Astiz 2010).
- In vitro maturation (IVM) of the recovered oocytes (Ruiz and Astiz 2010; Aragonés et al. 2022).
- In vitro fertilization (IVF) of oocytes obtained by OPU with cryo-frozen semen from an ML bull (Ruiz and Astiz 2010; Aragonés et al. 2022).
- In vitro culture (IVC): for 7 days (Ruiz and Astiz 2010).
- Vitrification: This technique uses the *Cryotop* device (Kitazato Fujinomiya, Japan) and the vitrification and devitrification solutions defined by Kuwayama et al. (2005) with the protocol described by Morató et al. (2010).

Before the embryo transfer, the selection and synchronization of a herd of Holstein dairy heifers, belonging to the commercial farm Agropecuaria Puente Alto de Los Dolores (Murcia), was carried out.

Once the embryo was transferred, the ultrasound diagnosis occurred after 30 days post-transfer through transrectal palpation and ultrasound, as well as its confirmation and follow-up.

The pregnancy and birth took place without incidents, with the birth of the "Fuensanta" heifer of the ML breed being a "normal" birth, although unique, since it was developed and born from a Holstein breed heifer, and through complex reproductive biotechnological procedures, which have proven to be more than established techniques.

4.8 Sperm Refreezing in ML Bull

In the last 10 years, several conservation projects of the ML breed have been carried out, both in vivo and in vitro, in an attempt to combine the activities of animals kept on small farms, with obtaining samples of gametes for the ML breed Germplasm Bank. In addition, these projects have been implementing novel techniques for their application in the effective in vitro maintenance of spermatozoa for their use in the in vivo conservation program through artificial insemination (AI) and the study of new alternatives in the use of these gametes that may be required in the future, such as the reduction of the number of useful spermatozoa per dose to use them for in vitro fertilization (IVF), the refreezing of semen doses that allow the application of sex separation techniques in male gametes and, as a prediction of fertility from the doses stored in the Germplasm Bank of this bovine breed (Poto et al. 2011), studies whose results were materialized and presented with the doctoral thesis presented in 2014 entitled "Contributions to the cryopreservation of gametes males in the ML bovine breed: sperm refreezing."

One of the drawbacks of storing seminal doses is their maintenance cost, due to the continuous increase in N_2L prices. Sperm sexing would reduce these costs by half, although later refreezing of the gametes would be necessary.

The objectives of this doctoral thesis were to achieve effectiveness and efficiency in refrozen semen, thus reducing the number of stored seminal doses stored, which would be thawed from the Germplasm Bank, and to which a treatment similar to that required by the technology of sperm sexing would be applied; a derived objective being the study with conventional techniques of refrozen semen and relating the seminal quality parameters of refrozen semen with those bulls that achieved the best fertility values with only frozen semen. Another objective was to develop a refreezing technique that would allow seminal doses to be stored more efficiently (Almela 2014).

To carry out this work, processed data on semen doses from 12 bulls of this breed and frozen sperm from 6 of those bulls were used. The methods for studying ejaculate and seminal doses were conventional for the bovine species. The freezing and refreezing techniques used were the same as for the Germplasm Bank (MAPA 1985). In the cold application experiments, the thawed semen was subjected to refreezing after having undergone a treatment (CT), or refrozen without treatment (ST). To know the value of the parameters studied, optical microscopy and fluorescence techniques, resistance in a thermostatic water bath, E-N, PI, lectins, toluidine blue, and AI stains were used with refrozen semen from seven Friesian breed heifers (Almela 2014).

From the results of the study, it was concluded that in ML cattle, there was variability in the mean values of seminal quality between the bulls, although not due to age. Furthermore, there were no notable differences in the mean values of the semen parameters from the first and second ejaculation of each bull (Table 10) (Almela 2014).

The analysis of the results allowed us to know the deterioration of the sperm during refreezing, with the seminal quality in the thawed semen being higher than that

Table 10 Mean and statistical values of the parameters individual motility and percentage (%) of live "Esp" in frozen (C) and refrozen semen with treatment (CT) and without treatment (ST)

	Individual motility (n = 6)			% live sperm(n = 6)		
	ST	CT	C	ST	CT	C
Mean	3.58	2.93	4.08	39.29	21	60
SD	0.65	1.21	0.58	14.58	16.75	5.48
CV (%)	18.12	41.39	14.30	37.11	79.76	9.12

of the refrozen one, with wide variability between bulls. One of the bulls presented better values in seminal parameters after freezing and refreezing, although in some parameters there were no differences in any bull. The sperm anomalies frequently found were "whip tails," but slightly lower in CT refrozen semen (Almela 2014).

The insemination of 7 Friesian heifers provided a fertility percentage of ST refrozen semen of 57.1%, so it was concluded that sperm refreezing could be used as a fertility prediction technique and be used for in vitro conservation (Almela 2014).

Since 2001 to the present day, various research projects have been carried out in which this bovine breed so prized by the most nostalgic and those who revere popular festivals have had a place, resulting in the conservation of the genes of several specimens of ML bull, which have resulted in the durability and continuity over time of this breed, latently awaiting its possible future usefulness.

References

Aimlid T, Johnson LA (1988) Effects of glycerol concentration, equilibration time and temperature of glycerol addition on post-thaw viability of boar spermatozoa frozen in straws. J Anim Sci 66:2899–2905. https://doi.org/10.2527/jas1988.66112899x

Aisen EG (2004) Procesamiento y conservación del material seminal. In: Reproducción Ovina y Caprina. Intermédica, Buenos Aires, pp 73–83. ISBN: 950-555-278-5

Aisen EG, Venturino A (2004) Recolección y evaluación del semen. En: Reproducción Ovina y Caprina. Ed. Intermédica. Buenos Aires, Argentina, 55–69. ISBN: 950-555-278-5

Almela L (2006) Estudio sobre la situación actual de cuatro especies en peligro de extinción en la Región de Murcia: cerdo Chato Murciano, vaca Murciana, gallina Murciana y cabra Blanca Celtibérica. Trabajo fin de carrera. Universidad Miguel Hernández. Escuela Politécnica Superior de Orihuela

Almela L (2014) Aportaciones a la crioconservación de gametos masculinos en la raza bovina Murciano-Levantina: recongelación de espermatozoides. Tesis Doctoral. Facultad de Veterinaria. Universidad de Murcia

Almela L, Poto A, Galián S, Ruiz S, Romero J, Peinado B (2011) Murcia se esfuerza por la supervivencia de su raza bovina autóctona. Albéitar 143:18–20. https://doi.org/10.1016/j.livsci.2019.04.003.231

Almela L, Peinado B, Poto A (2013) Relación entre la consanguinidad y la reproducción en la vaca Murciano-Levantina. AICA 3:41–45

Anexo I del Real Decreto 45/2019. De 8 de febrero, por el que se establecen las normas zootécnicas aplicables a los animales reproductores de raza pura, porcinos reproductores híbridos y su material reproductivo, y se actualiza el Programa Nacional de Conservación, Mejora y Fomento de las Razas Ganaderas

Aragonés P, Bermúdez I, Ortíz I, Sánchez J, Madrigal M (2022) Aspectos relacionados con la fertilización *in vitro* (FIV) en bovinos, una revisión bibliográfica. Revista Agroinnovación en el Trópico Húmedo, 52–61. ISSN: 2215-5368 http://revistas.tec.ac.cr/index.php/agroinn/index
Arriola J, Foote RH (1987) Glycerolation and thawing effects on bull spermatozoa frozen in detergent-treated egg yolk and whole egg extenders. J Dairy Sci 70(8):1664–1670. https://doi.org/10.3168/jds.S0022-0302(87)80194-7
Astiz S, Romero J, Poto A, Almela L, Peinado B, Ruíz S (2012) Caso clínico. Primera ternera Murciano-Levantina nacida, obtenida por biotecnologías reproductivas (OPU, Fecundación *In Vitro*, Cultivo *In vitro* y Vitrificación). Boletín ANEMBE 98:19–20. https://www.researchgate.net/publication/302589616
Ávila-Portillo LM, Madero JI, López C, León MF, Acosta L, Gómez C, Delgado LG, Gómez C, Lozano JM, Reguero MT (2006) Fundamentos de criopreservación. Revista Colombiana de Obstetricia y. Ginecologia 57:291–300. https://doi.org/10.18597/rcog.468
Barba C, Peinado B, Martínez M, María Dolores E, Lobera JB, Delgado JV, Poto A (1998) Influence of the geometry of the frozen semen doses container in the conservation of the Chato Murciano pig. In The 49th annual meeting of the European Association for Animal Production (E.A.A.P). Warsaw (Poland)
Barbas JP, Mascarenhas RD (2009) Cryopreservation of domestic animal sperm cells. Cell Tissue Bank 10(1):49–62. https://doi.org/10.1007/s10561-008-9081-4
Bonet S, Briz MD, Pinart B, Sancho S, García-Gil N, Badia E (2000) Morfología espermática en porcino. Institut d'Estudis Catalans, Barcelona. https://doi.org/10.15517/ma.v29i2.30613
Bravo JA, Montanero J, Calero R, Roy TJ (2011) Relación entre variables subjetivas e informatizadas del movimiento espermático del morueco. Archivos de Zootecnia 60(232):1087–1094. https://doi.org/10.4321/S0004-05922011000400024
Bussiére JF, Bertaud G, Guillouet P (2000) Conservation de la semence congelèe de verrat. Résultats *in vitro* et aprés insémination. Journées de Recherche Porcine en France 32:429–432
Bwanga CO, Einarsson S, Rodriguez-Martinez H (1991) Deep freezing of boar semen in plastic bags and straws. Reprod Domest Anim 26(117–1):25. https://doi.org/10.1111/j.1439-0531.1991.tb01528.x
Cabrera F, Gonzalez F, Batista M, Calero P, Medrano A, Gracia A (2005) The effect of removal of seminal plasma, egg yolk level and season on sperm freezability of canary buck (Capra hircus). Reprod Domest Anim 40:191–195. https://doi.org/10.1111/j.1439-0531.2005.00544.x
Cavalli-Sforza L, Bodmer W (1981) Genética de las poblaciones humanas. Genética y Evolución. Omega S.A., ediciones. ISBN: 8428206600, 9788428206600
Colas G (1980) Variations Saisonniéres de la qualité du sperme chez le bélier Ille de France. I. Etude de la morphologie cellulaire et de la motilité massale. Reprod Nutr Dev 20(6):1789–1799. https://doi.org/10.1051/rnd:19801005
Córdova A, Perez J, Lleo B, Artiga CG, Rillo SM (2001) *In vitro* fertilizing capacity of deep frozen boar semen packaged in 0.5 and 5 ml straws. Reprod Domest Anim 36:199–202. https://doi.org/10.1046/j.1439-0531.2001.00296.x
Corteel JM (1974) Viabilité des spermatozoïdes de bouc conservés et congelés avec ou sans leur plasma séminal: effet du glucose. Annales de Biologie Animale Biochimie Biophysique 14:741–745. https://doi.org/10.1051/rnd:19740502
Cortés S (2003) Efecto de la conservación sobre la fisiología espermática de semen caprino. Tesis Doctoral. Universidad Complutense de Madrid, Madrid, Spain
Cox F, Alfaro V, Montenegro V, Rodriguez-Martinez H (2006) Computer assisted analysis of sperm motion in goats and its relationship with sperm migration in cervical mucus. Theriogenology 66:860–867. https://doi.org/10.1016/j.theriogenology.2006.01.062
Cueto MI, Gibbons AE, Bruno MM, Fernandez J (2016) Manual de obtención, procesamiento y conservación del semen ovino. Ed. INTA, 1–21. http://hdl.handle.net/11336/106147
Delgadillo JA (2004) Características anatómicas y funcionales del sistema reproductor del macho. In: Reproducción Ovina y Caprina. Intermédica, Buenos Aires, pp 1–9. ISBN: 950-555-278-5
Díaz R (1953) Ganado porcino. Ed. Salvat Editores. Madrid, 68

Dobrinsky J (2002) Advancements in cryopreservation of domestic animal embryos. Theriogenology 57:285–602. https://doi.org/10.1016/S0093-691X(01)00672-0
Duchi N, Almela L, Peinado B, Poto A (2008) Criopreservación de semen de gallo: una alternativa para la recuperación y conservación de la gallina de raza Murciana. VIII Congreso de la Sociedad Española de Agricultura Ecológica. Bullas, Murcia
Eliasson R, Treichi L (1971) Supravital staining of human spermatozoa. Fertil Steril 22:134–137. https://doi.org/10.1016/S0015-0282(16)42927-4
Evans G, Maxwell WM (1990) Conservación de semen durante corto tiempo. In: Evans G, Maxwell WM (eds) Inseminación Artificial en Ovejas y Cabras. Acribia, Zaragoza, pp 119–122. ISBN: 978-84-200-0675-8
FAO (1998) Primer documento de líneas directrices para la elaboración de planes nacionales de gestión de los recursos genéticos de animales de granja
FAO/UNEP (1995) World watch list for domestic animal diversity. 1st edn. Ed. B. Scherf. FAO, Rome, 369 pp. https://doi.org/10.1079/WPS20020031
Ferrari S, Leinz F, Bernanbe VH (1998) Inseminaçao Artificial em cabras com sêmen congelado: Resultados preliminares. Braz J Vet Res Anim Sci 35(5):223–224. https://doi.org/10.1590/S1413-95961998000500007
Fiser PS, Fairfull RW (1990) Combined effect of glycerol concentration and cooling velocity on motility and acrosomal integrity of boar spermatozoa frozen in 0.5 ml straws. Mol Reprod Dev 25:123–129. https://doi.org/10.1016/0093-691x(86)90057-9
Galián S (2022) Optimización y mejora de la inseminación artificial en la raza caprina Murciano-Granadina. Tesis Doctoral. Universidad de Murcia. Murcia, Spain
Galián S, Almela L, Peinado B, Azevedo J, Ruiz S, Romero J, Poto A (2010) Actuaciones en el Banco de Germoplasma de la raza bovina Murciano-Levantina. Feagas 36:131–135. ISSN: 1887-4177
Galián S, Almela L, Peinado B, Barceló MD, Poto A (2018) Evaluación de la calidad seminal porcina post descongelación en dosis seminales almacenadas durante más de diez años. Resultados preliminares. XI Congreso Ibérico sobre Recursos Genéticos Animales. Murcia, Spain. 27–28 September
Galián S, Peinado B, Poto A, Almela L (2019) Estudio de la calidad seminal posdescongelación de dosis de toros de raza Murciano-Levantina, con almacenamiento de larga duración. Boletín de ANEMBE 125:20–27
Galián S, Peinado B, Almela L, Poto A (2021) Calidad del semen caprino post-descongelación al utilizar un protocolo de congelación convencional para la especie porcina. Actas Iberoamericanas de Conservación Animal 16:13–20. ISSN: 2253-9727
Galián S, Peinado B, Ruiz S, Poto A, Almela L (2024) Influencia del estímulo sexual utilizado en la libido y la calidad seminal del macho cabrío de raza Murciano-Granadina. Información Técnica Agraria-ITEA, publication accepted
García Y (2018) Efectos adversos de los estrógenos endógenos en hembras caninas. Tesina de Grado. Facultad de Ciencias Veterinarias UNCPBA. Tandil, Argentina. https://www.ridaa.unicen.edu.ar/handle/123456789/2017
García-Menacho Osset V (1999) Recursos genéticos ganaderos tradicionales de la Comunidad Valenciana. Ruralia, 5. http://hdl.handle.net/10396/8064
Garde J (1993) Congelación de semen en la especie ovina: características biológicas de las dosis descongeladas. Tesis Doctoral. Universidad Complutense de Madrid. Madrid, Spain
Gibbons A (2002) Inseminación artificial con semen congelado en cabras de raza Angora. Taurus 16:24–32
Glenister BF, Baker C, Furnish WM, Thomas GA (1990) Additional Early Permian ammonoid cephalopods from Western Australia. J Paleontol 64:392–399. https://doi.org/10.1017/S0022336000018618
Greyling JPC, Van der Nest M (2000) Synchronization of oestrus in goats: dose effect of progestagen. Small Rumin Res 36(2):201–207. https://doi.org/10.1016/S0921-4488(99)00165-0
Hancock JL, Howell GJR (1959) The collection of boar semen. Vet Rec 71:664

Hernández PJ, Fernández RF, Rodríguex SJ, Negrete RM, Soto MY, García RA (2012) Efecto de la criopreservación de semen de conejo Nueva Zelanda (Oryctolagus cuniculus) sobre su viabilidad y estado acrosomal. Revista Salud Animal 34:188–191. Corpus ID: 42977633

Howard TH, Pace MM (1988) Seminal evaluation and artificial insemination. In: Laing JA, Morgan WJ, Wagner WC (eds) Fertility and infertility in veterinary practice. Baillière Tindall, London, pp 39–51

Iritani A, Nishikawa Y (1963) Studies on the egg-coagulating enzyme in goat semen: IV. On the position of yolk constituents attacked by the coagulating enzyme. Japan J Anim Reprod 8(4):113–117

Kovachev KD, Zagorski DI, Ivanova MG, Babadov ND (1994) Cryogenic damage of boar spermatozoa frozen in pellets and in tubes. Theriogenology 42(8):1369–1379. https://doi.org/10.1016/0093-691X(94)90257-J

Kruip TA, Boni R, Wurth YA, Roelofsen MW, Pieterse MC (1994) Potential use of ovum pick-up for embryo production and breeding in cattle. Theriogenology 42:675–684. https://doi.org/10.1016/0093-691X(94)90384-U

Kuwayama M, Vajta G, Ieda S, Kato O (2005) Comparison of open and closed methods for vitrification of human embryos and the elimination of potential contamination. Reprod Biomed Online 11(5):608–614. https://doi.org/10.1016/s1472-6483(10)61169-8

Labroue F, Luquet M, Guillouet P, Bussière JF, Glodek P, Wemheuer W, Gandini G, Pizzi F, Delgado JV, Poto A, Ollivier L (2000) La cryoconservation des races porcines menacées de disparition La situation en France, en Allemagne, en Italie et en Espagne. Journées Recherche Porcine en France 32:419–427

Larsson K, Einarsson S, Sensor T (1977) The development of a practicable method for deep freezing of boar spermatozoa. Nord Vet Med 29:113–118

Leboeuf B, Restall B, Salamon S (2000) Production and storage of goat semen for artificial insemination. Anim Reprod Sci 62:113–141. https://doi.org/10.1016/S0378-4320(00)00156-1

Lobera JB (1998) El cerdo Chato Murciano. Orígenes e historia. Comunidad Autónoma de la Región de Murcia. Consejería de Medio Ambiente, Agricultura y Agua

Lorenzo M, Fresno M, Darmain N, Molina A, Ramos R (1997) Estudio del comportamiento reproductivo y calidad seminal de los machos y de la fertilidad mediante inseminación artificial en la Agrupación caprina Canaria. XXII Jornadas Científicas SEOC. Tenerife, Spain, 307–318

Martín S (1982) Reproducción e inseminación artificial porcina. Biblioteca Agrícola Aedos. Editorial Aedos

Martín S, Martínez C, García C, De Alaba C (1996) Boar semen evaluation in practice. Reprod Domest Anim 35:519–526

Mateos G, Medel P (1998) Necesidades nutricionales del verraco de alta selección. Alimentación: Reproductores. Porci 44:67–89

Maxwell WMC, Evans G (1990) Inseminación Artificial de Ovejas y Cabras. Ed. Acribia. Madrid, Spain, 19–32

McGlone JJ (1992) Técnicas para la evaluación y cuantificación del comportamiento sexual, ingestivo y social del cerdo. Anaporc 112:36–52

Memon MA, Ott RS (1981) Methods of semen preservation and Artificial Insemination in sheep and goats. Anim Prod 17(1):19–25

Memon MA, Bretzlaff KN, Ott RS (1986) Comparison of semen collection techniques in goats. Theriogenology 26:823–827. https://doi.org/10.1016/0093-691X(86)90011-7

du Mesnil du Buisson F, Signoret JP (1971) Utilisation du sperme de verrat pour l'insemination artificielle. Zagreb Symposium, 41

Mickensen WD, Paisley LG, Dahmen JJ (1981) The effect of scrotal circumference, sperm motility and morphology in the ram on conception rates and lambing percentage in the ewe. Theriogenology 16:53–57. https://doi.org/10.1016/0093-691x(81)90112-6

MAPA (1985) Ministerio de Agricultura, Pesca y Alimentación. Dirección General de la Producción Agraria. Subdirección General de la Producción Animal. Manual de Normas Técnicas para la Congelación de semen en la especie bovina.

Morató R, Izquierdo D, Paramio MT, Mogas T (2010) Survival and apoptosis rates after vitrification in cryotop devices of *in vitro*-produced calf and cow blastocysts at different developmental stages. Reprod Fertil Dev 22(7):1141–1147. https://doi.org/10.1071/RD10013

Muiño R, Fernández M, Peña AI (2006) Parámetros cinéticos en eyaculados bovinos de toros de raza frisona y rubia gallega. Información Técnica Agraria-ITEA 102(1):55–66

Murdoch RN, Jones RC (1978) The effects of glycerol of the metabolism and ultrastructure of boar spermatozoa. J Reprod Fertil 54:419–422. https://doi.org/10.1530/jrf.0.0540419

Nagase H, Niwa T (1964) Deep freezing bull semen in concentrated pellet form. I. Factors affecting survival of spermatozoa. Proceedings of Vth international congress on animal reproduction and artificial insemination, Trento. Vol. IV, 410

Ollivier L, Renald JP (1995) EAAP 46th annual congress. Prague, 7 pp 4–7 September

Orihuela A (2014) La conducta sexual del carnero. Revisión Revista Mexicana de Ciencias Pecuarias 5(1):49–90

Panés A (1916) Ganadería Murciana. Asociación General de Ganaderos. Artes Gráficas Mateu, Madrid

Paquignon M, Mergounis D, Courot M, du Mesnil du Buisson F (1974) Technologie de congélation de la semence de verrat: étude « in vitro ». Journées de la Recherche Porcine en France 6:71–74. https://doi.org/10.1051/animres:19740320

Paquignon M, Bussiere J, Bariteau F, Courot M (1980) Sperm conservation semen freezing pig artificial insemination ungulata mammalia reproduction. Theriogenology 14(3):217–226. https://doi.org/10.1016/0093-691X(80)90006-0

Paquignon M, Quellier P, Dacheux JL (1986) Congélation du sperme de verrat: comparaison de différents dilueurs, techniques de préparation de la semence, modes de conditionnement et températures de décongélation. Ann Zootech 35(2):173–184

Peinado B, Poto A, Lobera JB, Martín J, Fernández A (1997) Calidad seminal de los eyaculados de verraco de raza Chato Murciano. Primer Congreso Nacional de la Sociedad Española para los Recursos Genéticos Animales (SERGA). Córdoba, Spain

Peinado B, Poto A, Lobera JB, Martín J, Fernández A (1998) Calidad seminal de los eyaculados de verraco de raza Chato Murciano. Arch Zoot 47(178–179):311–317

Peinado B, Poto A, Gadea J, Ruiz S (1998a) Estudios preliminares en la crioconservación de espermatozoides porcinos de raza Chato Murciano. Arch Zoot 47:305–310

Peinado B, Barba C, Rosique M, Martínez M (1998b) Comportamiento del cerdo Chato Murciano frente al maniquí de extracción de semen. Estudio preliminar de la líbido. II Congreso Nacional de la Sociedad Española para los Recursos Genéticos Animales. Palma de Mallorca

Peinado B, Poto A, Lobera JB, Martín J, Fernández A (1998c) Calidad seminal de los eyaculados de verraco de raza Chato Murciano. Arch Zoot 47:311–317

Peinado B, Poto A, Marín M, Lobera JB (2001) Raza porcina Chato Murciano. Porci 61:39–55

Peinado B, Poto A, Marín M, De la Fuente J (2002) Banco de Germoplasma del Centro de Investigación y Desarrollo Agroalimentario de La Alberca. Técnicas utilizadas. V Congreso de la Sociedad Española para los Recursos Genéticos Animales. III Congreso Ibérico sobre Recursos Genéticos Animales. Madrid

Peinado B, Almela L, Poto A (2013) Situación genética actual de la raza bovina Murciano-Levantina. AICA:46–50

Peinado B, Poto A, Almela L (2014) Los Bancos de Germoplasma aseguran el mantenimiento de los recursos genéticos. Mundo Ganadero 261:6–13

Pellicer-Rubio MT, Combarnous Y (1998) Deterioration of goat spermatozoa in skimmed milk-based extenders as a result of oleic acid released by the bulbourethral lipase BUSgp60. Reproduction 112(1):95–105. https://doi.org/10.1530/jrf.0.1120095

Peña A, Forsberg LC (2000) Effects of Equex, one-or two-step dilution, and two freezing and thawing rates on post-thaw survival of dog spermatozoa. Theriogenology 54(6):859–875. https://doi.org/10.1016/S0093-691X(00)00397-6

Polge C, Smith AU, Parks AS (1949) Revival of spermatozoa after vitrification and dehydration at low temperatures. Nature 164:666. https://doi.org/10.1038/164666a0

Polge C, Salamon S, Wilmut I (1970) Fertilizing capacity of boar semen following surgical insemination. Vet Rec 87:424–428. https://doi.org/10.1136/vr.87.15.424

Pontbrandt D, Howard JC, Schiewe ML, Stuart LD, Wildt DE (1989) Effect of cryoprotective diluent and method of freeze-thawing on survival and acrosomal integrity of ram spermatozoa. Cryobiology 26:341–354. https://doi.org/10.1016/0011-2240(89)90058-8

Poto A, Peinado B (1998) Razas autóctonas de Murcia. Estimación del censo y aptitudes. Comunicación oral. II Congreso Nacional de la Sociedad Española para los Recursos Genéticos Animales (SERGA). Palma de Mallorca

Poto A, Peinado B (1999) Diferentes aplicaciones de las técnicas de conservación "*Ex situ*" en las razas autóctonas. Horticult Ganad Trop 21:4–9

Poto A, Peinado B, Lorenzo M, Domínguez MC, Gergatz E, Göker E, Bali Pap A (1995) Inseminación artificial en caprino. Aportaciones en ganado de raza Murciano-Granadina. Mundo Ganadero 12:46–50

Poto A, Peinado B, Marín M (1999) Estudio comparativo entre dos técnicas de preparación de semen congelado porcino. III Congreso de la Sociedad Española para los Recursos Genéticos Animales (CIRGA). Lugo. 11–13 de noviembre

Poto A, Peinado B, Barba C, Delgado JV (2000) Congelación de semen porcino de razas autóctonas en peligro de extinción. Influencia de la metodología en Banco de Germoplasma para pequeñas poblaciones. Arch Zoot 49:493–495

Poto A, Lobera JB, Peinado B (2000a) Razas autóctonas de Murcia. Estimación del censo y aptitudes. Arch Zoot 49(185–186):107–114

Poto A, Peinado B, Sereno JRB, Cabello A, Barba C (2000b) Programas de conservación *ex situ* en España. Porci 60:49–59

Poto A, Peinado B, Marín M, Lobera JB (2002) El cerdo Chato Murciano. EdiPorc 54:7–17

Poto A, Aguirregomezcorta J, Peinado B, Ruiz S, Almela L (2011) La conservación de la raza bovina Murciano-Levantina. Coag-IR Murcia 41:16–19

Poto A, Almela L, Peinado B, Ruiz S (2013) Respuesta a las técnicas reproductivas de las razas en peligro de extinción. AICA 3:29–40

Poto A, Almela L, Galián S, Peinado B (2021a) Diversas metodologías en la extracción de semen caprino y su influencia en el bienestar animal. Tierras Caprino 33:50–53

Poto A, Galián S, Almela L, Peinado B (2021b) La vaca Murciano-Levantina. RumiNews, 62–68

Purdy PH (2006) A review on goat sperm cryopreservation. Small Rumin Res 63(3):215–225. https://doi.org/10.1016/j.smallrumres.2005.02.015

Pursel VG, Johnson Ohnson LA (1975) Freezing of boar spermatozoa. Fertilizing capacity with concentrated semen and a new thawing procédure. J Anim Sci 40:99–102. https://doi.org/10.2527/jas1975.40199x

Pursel VG, Johnson LA, Schulman LL (1973) Effect of dilution, seminal plasma and incubation period on cold shock susceptibility of boar spermatozoa. J Anim Sci 37:528–531. https://doi.org/10.2527/jas1973.372528x

Pursel VG, Schulman LL, Johnson LA (1978) Effect of Orvus ES Paste on acrosome morphology, motility and fertilizing capacity of frozen-thawed boar sperm. J Anim Sci 47(1):198–202. https://doi.org/10.2527/jas1978.471198x

Real Decreto 45/2019 (2019) De 8 de febrero, por el que se establecen las normas zootécnicas aplicables a los animales reproductores de raza pura, porcinos reproductores híbridos y su material reproductivo, se actualiza el Programa Nacional de Conservación, Mejora y Fomento de las Razas Ganaderas y se modifican los Reales Decretos 558/2001, de 25 de mayo; 1316/1992, de 30 de octubre; 1438/1992, de 27 de noviembre; y 1625/2011, de 14 de noviembre

Rojas A (2014) Optimización del protocolo de crioconservación de semen caprino de la raza autóctona en peligro de extinción Blanca de Rasquera. Tesis Doctoral. Universidad Autónoma de Barcelona, Barcelona, Spain, 25–264

Rota A, Ström B, Linde-Forsberg C, Rodriguez-Martinez H (1997) Effects of Equex STM paste on viability of frozen-thawed dog spermatozoa during *in vitro* incubation at 38 C. Theriogenology 47(5):1093–1101. https://doi.org/10.1016/S0093-691X(97)00066-6

Ruiz S, Astiz S (2010) Producción *in vitro* de embriones (PIV) en biotecnología de la reproducción bovina. Boletín ANEMBE 88:256–232

Salamon S, Maxwell WMC (2000) Storage of ram semen. Anim Reprod Sci 62(1–3):77–111. https://doi.org/10.1016/S0378-4320(00)00155-X

Sánchez A (1984) Razas Bovinas Españolas. Extensión Agraria. MAPA. Madrid, Spain
Sánchez A (2002) Razas Ganaderas Españolas Bovinas. FEAGAS and MAPA. Madrid, Spain
Sierra I (1998) Interés del estudio de algunas razas en peligro con peculiares características. Arch Zoot 47:411–416
Silva P, Gadella B (2006) Detection of damage in mammalian sperm cells. Theriogenology 65:958–978. https://doi.org/10.1016/j.theriogenology.2005.09.010
Silva RF, Loaiza AM (2007) Piómetra en pequeños animales. Rev Vet Zoot 1(2):71–86. https://revistasojs.ucaldas.edu.co/index.php/vetzootec/article/view/5759
Smith RL (1984) Human sperm competition. In: Smith RL (ed) Sperm competition and the evolution of animal mating systems. Academic Press, New York, pp 601–660. eBook ISBN: 9780323143134
Stornelli MA, Stornelli MC, Arauz MS, Savignone CA, García M, De La Sota RL (2001) Estudio comparativo del efecto de tres diluyentes sobre la supervivencia de semen canino almacenado refrigerado a 4°C. Rev Bras Reprod Anim 25(3):468–470
Stornelli MA, Stornelli MC, Arauz MS, Savignone CA, García M, de la Sota RL (2002) Effects of two different temperatures and three different extenders on survival and longevity of chilled canine semen. Theriogenology 57:483
Suttiyotin D, Thwaiter CJ (1991) The ability of trypan blue to differentiate live and dead ram spermatozoa. Anim Reprod Sci 25:209–224. https://doi.org/10.1016/0378-4320(91)90016-S
Talbot P, Chacon RS (1981) A triple-stain technique for evaluating normal acrosome reactions of human sperm. J Exp Zool 215:201–208. https://doi.org/10.1002/jez.1402150210
Tecirlioglu RT, Hayes ES, Trounson AO (2002) Semen collection from mice: electroejaculation. Reprod Fertil Dev 14(6):363–371. https://doi.org/10.1071/RD02015
Thilmant P (1997) Congélation du Sperme de Verrat en Paillete de 0.5 ml. Résultats sur le Terrain. Ann Med Vet 141:457–462
Tomás C (2007) Nuevos protocolos para la crioconservación de espermatozoides de macho cabrío. Tesis Master. Universidad Politécnica de Valencia. Valencia, Spain
Ungerfeld R, González SP (2009) Social dominance and courtship and mating behaviour in rams in non-competitive and competitive pen tests. Reprod Domest Anim 44(1):44–47. https://doi.org/10.1111/j.1439-0531.2007.00987.x
Vera TA, Chagra-Dib EP, Leguiza HD, Brizuela R, Valdivia CL (2004) Incidencia del tipo de servicio sobre parámetros reproductivos de cabras criollas en Los Llanos de La Rioja. Revista Argentina de Producción. Animal 24(1):273–274. https://doi.org/10.13140/RG.2.2.21855.30886
Vicente A (1972) Technology of freezing boar semen. In VIIth international congress on animal reproduction and artificial insemination, Munich, 1624–1626
Voglmayr OJ, Chartier D, Sawyer R (1983) Viability of ram testicular spermatozoa following cryopreservation in rete testis fluid. Cryobiology 20:421–431. https://doi.org/10.1016/0011-2240(83)90032-9
Watson PF (1976) The protection of ram and bull spermatozoa by the low density lipoprotein fraction of egg yolk during storage at 5°C and deep-freezing. J Therm Biol 1:137–141. https://doi.org/10.1016/0306-4565(76)90003-6
Watson PF (1981) The roles of lipids and protein in the protection of ram spermatozoa at 5°C by egg-yolk lipoprotein. J Reprod Fertil 62:483–492. https://doi.org/10.1530/jrf.0.0620483
Watson PF (2000) The causes of reduced fertility with cryopreserved semen. Anim Reprod Sci 60:481–492. https://doi.org/10.1016/s0378-4320(00)00099-3
Watson PF, Martin ICA (1972) A comparison of changes in the acrosomes of deep-frozen ram and bull spermatozoa. J Reprod Fertil 28:99–101. https://doi.org/10.1530/jrf.0.0280099
Westendorf P, Richter L, Treu H (1975) Zur Tiefgefrierung von Ebersperma Labor-und Besamungsergebnisse mit dem Hulsenberger Pailleten-Verfahren. Dtsch Tierärztl Wschr 82:261–267
Wulster-Radclife MC, Williams MA, Stellflug JN, Lewis GS (2001) Artificial vagina vs a vaginal collection vial for collecting semen from rams. J Anim Sci 7:2964–2967. https://doi.org/10.2527/2001.79122964x

Decellularization of Reproductive Organs: Techniques, Applications, and Challenges

Cristina Martínez López, Mª José Izquierdo Rico, and Francisco Alberto García Vázquez

Abstract

Scaffolds are used in the field of tissue engineering for the reconstruction of organs and tissues. These can be synthetic, made from materials such as polymers, metals, ceramics, or graphene, or natural, using macromolecules like collagen or gelatin. While useful, synthetic scaffolds often face biocompatibility issues, whereas traditional natural scaffolds can elicit insufficient cellular responses. To overcome these limitations, recent research has focused on the development of scaffolds derived from decellularized extracellular matrices (dECM). These natural-origin scaffolds are produced through a process known

C. M. López
Department of Physiology, Faculty of Veterinary, Campus Mare Nostrum, University of Murcia, Murcia, Spain

Department of Cellular Biology and Histology, Faculty of Medicine, Universidad of Murcia, Murcia, Spain
e-mail: cristina.martinez67@um.es

M. J. Izquierdo Rico
Department of Cellular Biology and Histology, Faculty of Medicine, Universidad of Murcia, Murcia, Spain

Pascual Parrilla Murcian Institute for Biosanitary Research (IMIB-Arrixaca), Campus of Health Science, Murcia, Spain
e-mail: mjoseir@um.es

F. A. G. Vázquez
Department of Physiology, Faculty of Veterinary, Campus Mare Nostrum, Universidad of Murcia, Murcia, Spain

Pascual Parrilla Murcian Institute for Biosanitary Research (IMIB-Arrixaca), Campus of Health Science, Murcia, Spain
e-mail: fragarcia@um.es

J. C. Gardón, K. Satué Ambrojo (eds.), *Assisted Reproductive Technologies in Animals Volume 2*, https://doi.org/10.1007/978-3-031-87198-6_6

as decellularization, which aims to remove all cells from an organ or tissue while preserving the extracellular matrix components. Decellularization can be achieved through physical, chemical, and/or enzymatic methods, resulting in biocompatible, hypoimmunogenic scaffolds that replicate the functionality, structure, and biochemical properties of native tissue. These characteristics have led to the use of dECM in reproduction to treat infertility and enhance assisted reproductive technologies (ARTs). However, the decellularization technique is still in its early stages of development, with ongoing studies aimed at optimizing decellularization methods and clinical applications of derived materials. This chapter summarizes the basic principles of this technique and its main applications in reproduction, emphasizing the advances made to date and the challenges that remain.

Keywords

Artificial organs · 3D bioprinting · Hydrogel · Testicular scaffold · Uterine scaffold

6.1 Introduction

The female and male reproductive systems are comprised of organs whose integrity and functionality are crucial for adequate fertility. Consequently, any disease or dysfunction in these reproductive organs can lead to infertility. Current options for treating infertility include hormonal therapy, medications, or the use of assisted reproductive technologies (ARTs) such as artificial insemination (AI), in vitro fertilization (IVF), or intracytoplasmic sperm injection (ICSI) (Verberckmoes et al. 2004; O'Neill et al. 2018; Hansen 2020). However, these treatments do not always yield optimal results. Moreover, when infertility is due to ovarian insufficiency, absence of a functional uterus, or disorders in the process of spermatogenesis, a different approach is required. For instance, until recently, the only option for treating absolute uterine factor infertility (AUFI) was uterine transplantation, which has significant limitations such as the scarcity of donor organs and the risk of immunological rejection (Hellström et al. 2017; Brännström and Dahm-Kähler 2019). In response to these limitations, research has turned towards alternative treatments capable of regenerating tissues and restoring reproductive organ function. In this context, tissue engineering, and particularly decellularized extracellular matrix (dECM) scaffolds, is emerging as a promising alternative (Sittadjody et al. 2021). This technique employs physical, chemical, and enzymatic methods to remove all cells from an organ or tissue, preserving the components of the extracellular matrix (ECM) and mimicking an in vivo environment (Badylak et al. 2009). This capability to replicate the natural tissue environment and facilitate recellularization has led to the adoption of dECM scaffolds for treating reproductive pathologies and as three-dimensional (3D) substrates for culturing cells or embryos, aiming to improve existing ARTs (Miyazaki and Maruyama 2014; Laronda et al. 2015; Campo et al. 2019). However, despite the advances achieved in the technique of decellularization, tissue engineering still faces challenges and is in an early stage of development, with ongoing studies to optimize the methods of decellularization and the clinical applications of the materials obtained. The subsequent sections will delve into the fundamentals of decellularization, the employed methods and techniques, and their specific applications in reproduction. We will also discuss the persistent limitations and challenges faced by this technology, providing a comprehensive overview of its current status and future potential.

6.2 The Extracellular Matrix: Structure, Composition, and Biological Relevance

All tissues contain a mixture of cells and non-cellular components, which form well-organized networks called extracellular matrices. To better understand the decellularization process and why dECM is used as a biomaterial in bioengineering, it is necessary to know the composition and function of the ECM in tissues. The ECM is a complex, acellular 3D network that may be in contact with cells as part of the basal membrane (pericellular matrix) or around cells forming what is known as the interstitial matrix. The ECM is composed of different types of macromolecules, among which proteins are prominent. The constituent proteins of the ECM form the matrisome, representing between 1% and 1.5% of the total proteins in mammals. Some of these proteins are organized into fibers. Generally, we can distinguish three types of fibers that are part of the ECM: collagen fibers, elastic fibers, and reticular fibers (Naranjo et al. 2009; Theocharis et al. 2016).

Collagen is the most abundant structural protein of the ECM, synthesized and deposited in the ECM by fibroblasts. Its primary function is to provide structural support and mechanical resistance to tissues, especially against tensile forces. Additionally, collagen plays a crucial role in cell-matrix interactions, with some types acting as cell adhesion receptors. It is also fundamental in regulating cell migration and plays an essential role in tissue healing and regeneration processes. There are 28 types of collagen, which are classified into different groups based on their molecular structure and function in tissues. Among these, type III collagen forms small-diameter fibers known as reticular fibers. These reticular fibers cross-link to form a fine network. This network acts as a support mesh in soft tissues such as the liver, bone marrow, and tissues of the lymphatic system (Theocharis et al. 2016).

Elastic fibers, whose main component is elastin, are also part of the ECM. Elastin is responsible for the elastic properties of tissues and is therefore abundant in tissues that undergo regular contractions such as the lungs, heart, urinary bladder, uterus, and ligaments, among others (Rajab et al. 2020).

In addition to fiber-forming proteins, the ECM is composed of other proteinaceous macromolecules such as glycoproteins. Within this group, proteoglycans are a specific type of glycoproteins that are covalently bonded to carbohydrates called glycosaminoglycans (GAGs). This bond gives proteoglycans a high negative charge, attracting positively charged electrolytes and allowing them to bind to growth factors, cytokines, and chemokines. These features enable them to act as mediators of cellular activity, participating in cell proliferation, migration, adhesion, and apoptosis. Additionally, they provide resistance to compressive forces and are responsible for the viscoelastic properties of the tissue by attracting water molecules through osmosis. Lastly, glycoproteins such as fibronectin and laminin play a role in signal transduction and in regulating cellular activities through their binding to collagen and integrins. Both laminin and fibronectin are essential in tissue regeneration processes, providing the substrate that cells need to move and cover the wounded area. Additionally, their involvement in angiogenesis is crucial, supporting the formation of new blood vessels needed for nutrient supply and waste removal in repairing tissues (Singh et al. 2010; Theocharis et al. 2016; Rajab et al. 2020; Karamanos et al. 2021).

In addition to the proteins described, elements such as growth factors and cytokines are associated with the ECM playing a crucial role in its functionality (Theocharis et al. 2016).

The complex interaction among the components of the ECM creates a dynamic environment in which the matrix is continuously being formed and remodeled to meet the organism's needs, thus contributing to homeostasis. Fibroblasts and immune system cells, deeply integrated into the ECM, play key roles in its formation. These cells not only deposit essential matrix components but also actively participate in their remodeling by secreting specific enzymes such as matrix metalloproteinases (MMPs). MMPs are proteolytic enzymes responsible for the degradation of certain ECM components like collagen and elastin. This degradation process facilitates the continuous remodeling of the ECM, allowing it to constantly adapt to the changing needs of the organism (Cui et al. 2017).

In reproductive tissues, the ECM plays a key role in the processes of fertilization, embryonic implantation, and placental invasion. In the ovaries, during oocyte maturation, cumulus cells secrete hyaluronic acid and glycoproteins that form the zona pellucida, essential for the protection of the oocyte and its effective fertilization. Subsequently, the endometrium adapts and remodels to receive the embryo. This process is influenced by steroid hormones, which cause a reduction in antiadhesive glycoproteins and an increase in integrin expression, facilitating embryonic adhesion. After implantation, extravillous trophoblasts invade the endometrium, actively modifying the ECM through degradation and the deposition of new proteins. They employ enzymes such as the previously described MMPs to break down the native matrix and induce angiogenesis, ensuring adequate blood flow for embryonic development. During pregnancy and childbirth, the uterine ECM transforms to allow for the growth and expansion of the uterus, maintaining uterine pressure unchanged. This is achieved through an increase in elastin expression, allowing the uterus to stretch and expand up to 500 times its original size. Simultaneously, collagen fibers reorganize into thicker bundles, increasing the mechanical resistance of the tissue. This entire process is also regulated by hormones, such as estrogens and progesterone, which control the degradation of this additional matrix at the time of childbirth to facilitate the coupling of uterine myocytes and the initiation of contractions (O'Connor et al. 2020).

Therefore, the ECM not only acts as a structural support for cells but also functions as a dynamic structure that actively participates in signal transduction and the regulation of cellular behavior. These unique characteristics make the ECM not only vital for natural biological processes but also a focus of research in the field of bioengineering. Currently, the use of ECM-based scaffolds obtained through the decellularization process is being explored to enhance tissue regeneration, leveraging its ability to promote processes of angiogenesis, migration, proliferation, and cellular differentiation. Research in this area promises to revolutionize strategies for repairing and regenerating tissues, opening new avenues for advanced treatments in reproductive medicine that could solve infertility problems or revolutionize ARTs.

6.3 Decellularization and dECM Scaffolds

Scaffolds, whether of natural or synthetic origin, provide the necessary structural support for cell growth and proliferation, as well as for the delivery of growth factors essential for tissue regeneration. Within synthetic scaffolds, four main groups are described: polymers, ceramics, metals, and graphene. Polymers such as polylactic acid, polycaprolactone, polyethylene glycol, and polyvinyl alcohol, among others, are the most commonly used synthetic scaffolds. Despite their good mechanical properties, synthetic scaffolds often present disadvantages such as poor biocompatibility, deficient resorption, and limited properties for cell adhesion and signal transmission regulating cell activity (Neishabouri et al. 2022). Additionally, cells of the innate immune system often detect synthetic biomaterials as foreign bodies, leading to fibrous encapsulation of the graft. This phenomenon does not allow the biomaterial to carry out its regenerative function (Wong et al. 2016).

To overcome these issues, there has been a shift towards the use of natural biomaterials, such as collagen, gelatin, fibrin, alginate, or hyaluronic acid hydrogels. These materials, derived from macromolecules of the ECM, offer better biocompatibility and biodegradability. Nevertheless, they also face challenges, such as a lack of homogeneity between batches and, occasionally, an insufficient cellular response.

Therefore, recent research has focused on the use of scaffolds derived from dECM (Neishabouri et al. 2022). These scaffolds are obtained through a process known as decellularization. Decellularization aims to completely remove the cells from an organ or tissue while preserving the components of the ECM. This can be achieved through physical, chemical, and/or enzymatic procedures (Badylak et al. 2009). The main characteristic of dECM scaffolds is biomimicry, which denotes their capacity to replicate the functionality, structure, and biochemical properties of native tissue. The decellularization process preserves the 3D structure of the organ or tissue from which the scaffold is derived as well as all the components of the ECM. This enables dECM scaffolds to effectively simulate in vivo conditions. The preservation of these components, which include fibers of collagen and elastin, is essential for maintaining the elasticity and rigidity of the dECM scaffolds. This is especially important when these scaffolds are to be used in reproductive tissues such as the uterus, an organ that undergoes significant changes in weight, size, and volume during the gestation process (Kim et al. 2021).

These scaffolds not only maintain structure and mechanical properties but also retain bioactive molecules that promote cellular processes such as angiogenesis, migration, proliferation, and differentiation. Moreover, as they are derived from organic tissues, these scaffolds are biodegradable and present low immunogenicity, as the decellularization process removes potentially reactive antigens, thereby minimizing inflammatory responses and immunological rejection (Gadre et al. 2024). Furthermore, as they are composed of ECM, these scaffolds are well tolerated by xenogeneic recipients provided that antigens are effectively removed during the decellularization process. This is because the ECM is well-conserved across species (Wong et al. 2016). For this reason, dECM scaffolds can be obtained from various sources.

The selection of tissue or organ sources for obtaining dECM is an important aspect of both research and clinical applications. This choice not only affects the efficacy and safety of the scaffold but also its clinical acceptance and ethical considerations. In this context, it is essential to understand the advantages and limitations of various tissue sources, including xenogenic and human tissues, in order to develop dECM scaffolds that are safe and effective for medical applications (Zhang et al. 2021).

Due to the scarcity of human tissues, both from autologous and allogenic sources, current research is focusing on the use of dECM derived from xenogeneic sources. A variety of animal tissues, including those from pigs, cattle, rabbits, and rodents, have been explored for the creation of dECM. Porcine tissues have been particularly prevalent due to their availability and similarities to human tissues. However, the use of animal tissues carries risks such as disease transmission or variations in the composition and mechanical properties of the dECM depending on the tissue origin (Miyazaki and Maruyama 2014; López-Martínez et al. 2021; Padma et al. 2021).

On the other hand, the use of dECM derived from human tissues or organs avoids risks associated with the transmission of animal diseases and the potential contamination and immunogenicity of xenogeneic dECM. These can be sourced from a variety of sources, including cadavers, damaged or diseased tissues from patients, and donations stored in biobanks (Daryabari et al. 2022).

6.4 Phases of the Decellularization Process

The decellularization process, used to prepare dECM scaffolds from tissues or organs, is divided into three distinct phases, regardless of the type of tissue or organ involved. The first phase is cell lysis, which is designed to eliminate all cells from the organ or tissue. This stage is conducted using chemical, physical, or enzymatic methods that are specifically designed to break down and dissolve the cells and their components. This is followed by the rinsing phase, where washing solutions are used to remove cellular debris and any residue from the previously used chemical agents. This step is crucial to ensure the purity of the scaffold. Finally, the third phase is sterilization, carried out to eliminate microbial contamination. This last step ensures that the resulting scaffold is sterile and safe for use in clinical applications (Choudhury et al. 2020).

6.4.1 Cell Lysis Phase

The cell lysis phase is the first stage of the decellularization process. This initial phase aims to remove cells from the organ or tissue. To achieve this, chemical, physical, or enzymatic methods can be employed.

6.4.1.1 Chemical Methods

The chemical methods used in decellularization encompass a wide range of substances, including detergents, hypertonic and hypotonic solutions, acids, bases, alcohols, organic solvents, and chelating agents.

6.4.1.1.1 Detergents

Detergents are the most commonly used chemical agents in decellularization protocols due to their effectiveness in solubilizing cell membranes and dissociating DNA from proteins (Abaci and Guvendiren 2020). Detergents can be categorized into three groups based on their charge: ionic detergents, non-ionic detergents, and zwitterionic detergents (Guruswamy Damodaran and Vermette 2018). Ionic detergents are characterized by having a charge in their molecular structure, which can be negative (anionic detergents) or positive (cationic detergents). They are very effective in removing cells and genetic material from tissues as they solubilize DNA and disrupt cell membranes by increasing their curvature and tension. However, they are quite aggressive detergents capable of solubilizing proteins, altering the collagen of the matrix, and removing the content of GAGs and growth factors. This results in an alteration of the ECM structure and a loss of mechanical properties. Additionally, their high polarity makes their removal from the tissue difficult, requiring several washing phases subsequently. The most commonly used ionic detergents are sodium dodecyl sulfate (SDS), sodium deoxycholate (SDC), and Triton X-200 (Guruswamy Damodaran and Vermette 2018; Wang et al. 2022; Gadre et al. 2024).

Non-ionic detergents are characterized by having no charge in their molecular structure. They are less effective than ionic detergents at removing tissue cells but better preserve the ECM structure since their action focuses on disrupting lipid-lipid and lipid-protein bonds while leaving protein-protein bonds intact. Among non-ionic detergents, Triton X-100 is the most used (Abaci and Guvendiren 2020; Neishabouri et al. 2022).

Finally, we have zwitterionic detergents, which combine properties of both ionic and non-ionic detergents, featuring negative and positive charges within the same molecule. However, their net charge is neutral, so they protect proteins during the decellularization process and are less harmful to ECM components compared to ionic detergents (Crapo et al. 2011). The most commonly used zwitterionic detergents as decellularizing agents are 3-[(3-cholamidopropyl) dimethylammonio]-1-propanesulfonate (CHAPS) (O'Neill et al. 2013), sulfobetaine-10, and sulfobetaine-16 (Hudson et al. 2004). Although zwitterionic detergents are less harmful, they have not yet been used to decellularize reproductive organs.

In decellularization protocols, it is more common to use combinations of several detergents to improve the efficiency of the process. One of the most frequent combinations includes ionic detergents such as SDS and non-ionic detergents like Triton X-100. While the use of SDS alone can significantly alter the structure of the ECM, its combination with Triton X-100 has proven to be very effective. This synergy reduces the exposure time needed for each detergent separately, achieving more efficient DNA removal thanks to SDS, and maintaining a balance between cell removal and ECM preservation. Additionally, the combined use facilitates the removal of SDS residues, thereby reducing cytotoxicity in the treated tissue (Guruswamy Damodaran and Vermette 2018; Neishabouri et al. 2022). This strategy has been successfully applied in the decellularization of a wide variety of organs, including reproductive tissues such as the testis (Naeemi et al. 2021), ovary (Almeida et al. 2023), oviduct (Francés-Herrero et al. 2021a, b), and uterus (Campo et al. 2017).

6.4.1.1.2 Acids and Bases

Acids and bases can be used as decellularizing chemical agents due to their ability to generate extreme pH levels. These pH extremes facilitate the hydrolysis of cell membranes and cytoplasmic components. Acids, for example, can break molecular bonds and incorporate hydrogen ions, while bases tend to degrade nucleic acids (Paulo Zambon et al. 2020). Among the most commonly used acids are acetic acid, peracetic acid, hydrochloric acid, and sulfuric acid. On the other hand, commonly used bases include calcium hydroxide, sodium sulfide, sodium hydroxide, and ammonium hydroxide (Dong et al. 2009; Gadre et al. 2024). However, the effect of these chemical agents has not yet been explored in reproductive tissues, with the exception of peracetic acid. Although this acid is included in various decellularization protocols for reproductive tissues, it is generally used in the final phase of the process as a disinfecting agent (Hellström et al. 2016; Liu et al. 2017; Vermeulen et al. 2018).

6.4.1.1.3 Hypertonic and Hypotonic Solutions

Hypertonic and hypotonic solutions induce cellular lysis through osmotic shock. However, they do not act on the cytoplasmic or nuclear components that are released, so they are often used in combination with other methods to increase their efficacy. Commonly, they are used alternately, combining cycles of immersion in hypertonic solutions with cycles in hypotonic solutions. Sodium chloride (NaCl) is the most commonly used hypertonic solution, while Tris-HCl is commonly used as a hypotonic solution (Crapo et al. 2011; Kim et al. 2016; Mendibil et al. 2020). Although the use of these substances is not very common for decellularizing reproductive tissues, the use of NaCl to decellularize testicles (Bashiri et al. 2021) and uterus (Arezoo et al. 2021) has been reported.

6.4.1.1.4 Alcohols

Alcohols, including glycerol, ethanol, and methanol, are also used as decellularizing agents. Their mechanism of action is based on replacing intracellular water, causing cell dehydration and consequently, lysis (Neishabouri et al. 2022). In addition to lysing cells, alcohols are capable of solubilizing lipids, which facilitates the removal of these molecules from cell membranes. In fact, alcohols have been described as more effective than lipase in removing adipose tissue (Brown et al. 2011; Crapo et al. 2011). For this reason, alcohols can be used as a pretreatment to enhance the efficacy of other decellularizing agents used in later stages. However, the main drawback of using alcohols lies in their potential to alter the mechanical properties of the dECM scaffolds due to their ability to fix structures. This effect can compromise the functionality of the scaffold in subsequent clinical applications (Guruswamy Damodaran and Vermette 2018).

6.4.1.1.5 Organic Solvents

Organic solvents, such as acetone, can also be used as decellularizing agents. Like alcohols, these solvents are effective at removing lipids from cellular structures. However, their use can result in increased stiffness in the resulting scaffolds. For this reason, solvents are particularly useful in the preparation of decellularized ligaments or tendons, where specific mechanical properties different from the original tissue are sought. In this context, they act as crosslinking agents, contributing and helping to modify the mechanical properties of the scaffold (Crapo et al. 2011; Gadre et al. 2024). Other organic solvents used in decellularization include dimethyl sulfoxide (DMSO), an aprotic polar solvent that facilitates the penetration of other chemical agents due to its ability to dissolve a wide range of organic and inorganic materials (Daryabari et al. 2019), and tributyl phosphate (Deeken et al. 2011). The use of DMSO has been reported in decellularization protocols used in uterus (Hellström et al. 2014; Hellström et al. 2016).

6.4.1.1.6 Chelating Agents

Chelating agents act by disrupting cell adhesion by binding to divalent metal cations such as calcium and magnesium. This binding is critical because these cations often stabilize interactions between cells and the ECM, as well as cell membranes. By chelating these ions, chelating agents dissociate cells from the ECM and contribute to destabilizing cell membranes, facilitating cell lysis. Among the most notable chelating agents are ethylene glycol tetraacetic acid (EGTA) and ethylenediaminetetraacetic acid (EDTA) (Neishabouri et al. 2022). The latter has been used to decellularize the ovary (Park et al. 2023) and testicle (Vermeulen et al. 2018).

6.4.1.2 Physical Methods

Physical methods employ changes in pressure or temperature to induce cellular damage and are often used in combination with chemical and/or enzymatic methods to enhance decellularization processes. One example of physical treatment is cutting the organ into small pieces when it is not necessary to preserve the 3D structure of the organ in question. This chopping or slicing can be used as a pretreatment to facilitate decellularization by increasing the surface exposure of the tissue to decellularizing agents such as detergents (Skardal et al. 2015; Nikniaz et al. 2021).

Another commonly used physical method in decellularization protocols is freeze-thaw cycles. These cycles involve freezing the tissue at −80 °C or −20 °C and subsequently thawing at room temperature (Hellström et al. 2014; Campo et al. 2017). During the freezing process, ice crystals form within the cells, which upon thawing, cause cellular lysis. Freeze-thaw cycles have been shown to be beneficial as they preserve the ultrastructure of the ECM, contribute to more uniform decellularization, and prevent microbial growth (Burk et al. 2014; Abaci and Guvendiren 2020; Mendibil et al. 2020). Additionally, by inducing cellular lysis themselves, they reduce the exposure time to other decellularizing agents that may alter the structure of the ECM.

Applying hydrostatic pressures above 600 MPa can also induce cellular lysis. However, the cytoplasmic content and nucleic acids released after lysis often remain in the scaffold, making it necessary to combine this method with other treatments (Abaci and Guvendiren 2020).

Among the more innovative physical methods are sonication or ultrasonication, the use of supercritical CO_2, non-thermal irreversible electroporation, and vacuum-assisted decellularization. Among these methods, ultrasonication has been used to decellularize the ovary (Almeida et al. 2023), and vacuum-assisted decellularization has been employed for the testicle (Khazaei et al. 2023). Vacuum-assisted decellularization is typically used in combination with detergents as it is not effective on its own, but it does facilitate the penetration of other agents into the tissues (Butler et al. 2017; Khazaei et al. 2023). Supercritical CO_2 and non-thermal irreversible electroporation have not yet been used in reproductive tissues, although their properties are of interest. Supercritical CO_2 exhibits properties of both a gas and a liquid, reaching this supercritical state when subjected to temperatures and pressures above its critical point (31.1 °C and 7.38 MPa). Supercritical CO_2 can be used as a decellularizing agent due to its dual action: it removes lipids and alters cell membranes by acting as a solvent for nonpolar molecules while preserving polar components of the ECM, such as proteins. Furthermore, the high pressures used in this method also induce cellular lysis, adding efficiency to the process (Gafarova et al. 2020). Additionally, supercritical CO_2 acts as a sterilizing agent and, due to its high permeability, can be easily removed from tissues without the need for additional washes, thus preventing ECM degradation. However, this method requires equipment that can withstand high pressures, which is not available in all laboratories (Neishabouri et al. 2022). Non-thermal irreversible electroporation creates micropores in cell membranes, disrupting homeostasis and inducing lysis. Like other physical methods, it does not directly affect the components released after lysis, so it is generally combined with enzymes or chemical agents (Crapo et al. 2011).

6.4.1.3 Enzymatic Methods

Enzymes are not typically used to induce cellular lysis but rather to remove specific cellular or nuclear components, as well as unwanted elements from the ECM (Crapo et al. 2011). Enzymes commonly used in decellularization include endonucleases, such as DNases and RNases, which act on nucleic acids reducing DNA and RNA concentrations in the tissue (Campo et al. 2019; Francés-Herrero et al. 2021a, b; López-Martínez et al. 2021). This helps to reduce adverse immune responses and, therefore, enhances the biocompatibility of the scaffold. Proteases such as trypsin and pepsin can also be used in decellularization protocols. Their action is based on peptide cleavage, which can modify crucial proteins for future clinical applications. Moreover, it has been observed that prolonged exposure to these enzymes can significantly alter laminin, as well as cleave collagen and elastin, affecting the ultrastructure of the ECM (Gadre et al. 2024). In cases where the goal is to degrade lipid macromolecules, enzymes such as lipase can be used, and in cases where the goal is to degrade collagen, collagenase can be employed (Crapo et al. 2011). Additionally, when the tissue comes from a xenogeneic source, the enzyme α-galactosidase can be used to minimize immune responses to the cellular antigen galactose-α-(1,3)-galactose (Gal epitope) (Xu et al. 2008; Xu et al. 2009).

6.4.1.4 Other Decellularization Methods

In addition to chemical, physical, and enzymatic methods, other approaches such as apoptosis or the use of toxins, like latrunculin, are also described. Latrunculin B, in particular, has been used to effectively decellularize skeletal muscle without the need for detergents or proteases. The results obtained using this toxin showed a reduction in GAGs, but cell removal was effective, and the final mechanical properties were very similar to those of the native tissue (Gillies et al. 2011). On the other hand, apoptosis, also known as programmed cell death, has been described in the literature as a method of decellularization. This process occurs naturally in tissues and organs but can be induced in the laboratory by altering cellular homeostasis. Two pathways have been identified for the activation of apoptosis: intrinsic and extrinsic. The intrinsic pathway, also known as the mitochondrial pathway, involves the permeabilization of the outer mitochondrial membrane due to changes in environmental conditions (environmental stress), leading to the release of proteins that activate caspases, the enzymes responsible for cell death. The extrinsic pathway, on the other hand, is based on the use of specific ligands that bind to cell death receptors, such as tumor necrosis factor receptor 1. This ligand-receptor binding activates a signaling pathway that also results in the release of caspases. The main advantage of this technique is that it keeps the ECM intact. However, decellularization by apoptosis is not commonly used due to its complexity and because the molecular mechanisms involved are not yet fully understood (Bourgine et al. 2013).

6.4.2 Rinsing Phase

The rinsing phase is the second stage of the decellularization process. In this phase, washing solutions are used to remove cellular debris and residues of chemical or enzymatic agents that remain in the scaffold after the lysis phase. Washing solutions employed include phosphate-buffered saline (PBS), distilled water (dH_2O), and Milli-Q water (Choudhury et al. 2020). This phase is typically alternated with the lysis phase to progressively remove cells that are being lysed. It is also carried out once the decellularization protocol is completed to eliminate residues of the decellularizing agents used. The presence of these residues in the scaffold can cause cytotoxicity and inhibit cellular activity during subsequent recellularization. For this reason, it is crucial to perform a thorough wash once the decellularization protocol is completed and before the sterilization phase. To ensure the effectiveness of this phase, multiple washes are required, even in thin tissues such as the oviduct (Sanguansook et al. 2024). Typically, the washing period to effectively remove SDS from scaffolds obtained from reproductive organs ranges from 4 to 48 h (Campo et al. 2019; Bashiri et al. 2021; Padma et al. 2021).

6.4.3 Sterilization Phase

Sterilization is the final stage of the decellularization process. Sterilization and disinfection of dECM scaffolds are essential before their use in vitro or in vivo. Sterilization eliminates all microorganisms, including bacterial spores, unlike disinfection, which only removes vegetative microorganisms. A variety of sterilization methods can be applied to dECM scaffolds, and the selection depends on the specific properties of the scaffold, as well as the available time, accessibility, and purpose of the application (Tao et al. 2021; Neishabouri et al. 2022). Generally, these methods can be classified into two groups described in the following sections.

6.4.3.1 Physical Sterilization Methods

Physical sterilization methods use heat or radiation to achieve sterility in scaffolds. Heat sterilization can be carried out in ovens or by using equipment that raises the temperature above 140 °C, known as dry heat sterilization, or through high-temperature steam, called moist heat sterilization. These methods are accessible and common in most laboratories and are effective for synthetic scaffolds. However, they are not recommended for dECM scaffolds as they are thermosensitive, and the application of high temperatures could alter the ECM and denature important proteins (Mendibil et al. 2020; Fuentes et al. 2022).

An alternative to heat is radiation, such as gamma rays or ultraviolet light (UV). Gamma irradiation is an effective and non-toxic method that does not require heat. Additionally, the high penetrability of these radiations allows for effectively eradicating all microbial forms present in the scaffold. However, if the radiation dose is too high, it can damage the collagen in the matrix, altering the mechanical properties of the scaffolds. Gamma radiation at 10 Gy has been shown to effectively sterilize acellular heart valves without compromising their ultrastructure (Somers et al. 2009; Harrell et al. 2018). In the reproductive field, gamma radiation with cobalt-60 has been used to sterilize decellularized amniotic membranes, providing good results without altering the tissue structure (Chen et al. 2019). Nevertheless, the radiation dose and exposure time should be researched and adjusted for each type of scaffold. UV rays are useful for sterilizing thin tissues but are ineffective in dense decellularized tissues due to their weak penetrability. Their use is common for sterilizing reproductive organs such as the uterus or testis due to their easy availability in laboratories (Bashiri et al. 2021; Khazaei et al. 2023; Arezoo et al. 2021).

Supercritical CO_2 is also included among physical sterilization methods, although this method has not yet been employed in the sterilization of reproductive tissues (Tao et al. 2021; Neishabouri et al. 2022).

6.4.3.2 Chemical Sterilization Methods

Ethylene oxide, peroxides, and alcohols are widely used as chemical methods of disinfection and sterilization. Ethylene oxide has been used as a sterilizing agent on a wide variety of decellularized organs, providing good results (Dempsey et al. 2019). However, if used, it must be subsequently ensured that it is thoroughly removed by washing the scaffold, as ethylene oxide residues are toxic and can cause adverse reactions during clinical use. Its use is not recommended on lyophilized dECM scaffolds and on scaffolds containing water, as ethylene oxide in contact with water can form ethylene glycol, which is also toxic (Thier and Bolt 2000; Tao et al. 2021).

Peroxides used as sterilizing agents include peracetic acid and hydrogen peroxide. These compounds can eliminate bacterial spores if applied for an adequate time, approximately 30 min. A significant advantage is that their decomposition products are not toxic. For this reason, they are frequently used to sterilize decellularized reproductive tissues. For example, the use of 0.1% peracetic acid for periods of 30 min to 1 h has been reported to sterilize decellularized rat and sheep uteri, as well as porcine vaginas (Hellström et al. 2016; Greco et al. 2018; Padma et al. 2021).

Alcohols are also commonly included in sterilization protocols, acting mainly as disinfectants. Although they are effective in denaturing proteins and enzymes from microorganisms, alcohols do not eliminate bacterial spores (Tao et al. 2021). For this reason, they are often used in combination with other chemical agents, such as peracetic acid, to achieve complete sterilization. For example, a documented protocol includes immersing testicular scaffolds in a solution of 0.1% peracetic acid and 4% ethanol for 2 h (Naeemi et al. 2021). Another method involves immersing decellularized porcine ovaries in a solution of 0.1% peracetic acid and 20% ethanol at 4 °C also for a period of 2 h (Zheng et al. 2022).

Regardless of the sterilization method used, it is necessary to verify the effectiveness of the employed technique. Sterility tests must be conducted to confirm the efficacy of the method. Additionally, cytotoxicity tests are crucial, particularly when chemical methods of sterilization are used. These tests ensure that the scaffold does not contain toxic residues from the sterilizing substances. Mechanical tests are also necessary to guarantee that the method employed has not altered the properties of the ECM (Tao et al. 2021).

6.5 Decellularization Techniques

Decellularization techniques are those that allow decellularizing agents to come into contact with the organ or tissue. Although some physical methods such as non-thermal irreversible electroporation or supercritical CO_2 can be employed alone to decellularize organs or tissues, most protocols use chemical methods to achieve effective decellularization. Among the chemical methods, the use of detergents is particularly common. In fact, it is usual to employ two or more detergents, seeking to leverage synergistic effects that minimize the alteration of the biomechanical properties of the ECM. For this same reason, detergents are often combined with physical methods such as freeze-thaw cycles and enzymatic methods like the use of DNase, which removes DNA residues from the scaffolds. To bring these chemical and enzymatic agents into contact with the organs or tissues, two techniques can be employed: perfusion technique and immersion-agitation technique (Choudhury et al. 2020).

6.5.1 Perfusion Technique

The perfusion technique involves the introduction of decellularizing agents through the organ's vasculature, utilizing its natural vascular network to ensure uniform distribution. This technique is ideal for large or thick organs, as it ensures that the agents reach even the deepest parts. However, it requires complex equipment such as peristaltic or infusion pumps to pump the agents, systems to control the flow and rate of perfusion, and devices to monitor pressure. Furthermore, for the technique to be effective, it is crucial that the organ's vasculature remains intact. Otherwise, the decellularization may not be complete (Campo et al. 2017; Campo et al. 2019; Wang et al. 2022).

In reproduction, the perfusion technique is the most used method for uterine decellularization. Its application has been described for the decellularization of the uterus of rats (Hellström et al. 2016), rabbits (Campo et al. 2019), sheep (Padma et al. 2021), and porcine (Campo et al. 2017). The use of this technique has also been reported in other reproductive organs such as porcine ovaries (Almeida et al. 2023) and rat testes (Kiani et al. 2021), although its use in these organs is not very common.

6.5.2 Immersion-Agitation Technique

On the other hand, the immersion-agitation technique involves submerging the organs in a decellularizing solution and combining it with mechanical agitation to facilitate contact between the agents and the tissue. This technique is suitable for small organs, thin tissues, or when vascular access is complicated, or the vasculature is damaged. It requires a container for the organs and reagents and an orbital shaker that provides the necessary agitation. It is a simple method that can be implemented in any laboratory. Most decellularization protocols take several days to complete. Therefore, conducting this procedure at a low temperature, rather than at room temperature, favors the preservation of the organ and the maintenance of ECM components such as collagen (Negishi et al. 2011; Francés-Herrero et al. 2021a, b; Sanguansook et al. 2024).

In reproduction, the immersion-agitation technique is the most used method for the decellularization of the oviduct (Francés-Herrero et al. 2021a, b), ovary (Laronda et al. 2015), and testis (Vermeulen et al. 2018). For solid organs, such as the ovary and testis, a preliminary physical treatment involving cutting is employed. This means that the organs are sectioned into smaller pieces, thereby enhancing the exposure of the tissues to the decellularizing agents (Liu et al. 2017; Ashouri Movassagh et al. 2020; Nikniaz et al. 2021).

6.6 Decellularization Protocols Used in Reproductive Organs

Once the phases of the decellularization process, the agents used in each of these phases, and the available techniques for bringing the agents into contact with the organs are known, a protocol must be designed. This protocol must consider multiple factors. Among these factors, the equipment available in the laboratory stands out, as it will be a limiting factor in developing one technique or another. Another important factor is the type of reproductive organ. For instance, in the case of solid organs, it is recommended to use a perfusion technique that ensures the decellularizing agents reach all the cells of the organ, or alternatively, cut the tissue into small pieces and use the immersion-agitation technique. The size of the organ, which will depend on the species from which it originates, must also be considered. In larger organs, decellularization protocols must be more extensive, use a higher concentration of decellularizing agents, or combine different decellularization methods during the lysis phase to ensure that the process is effective. The ultimate goal is to obtain an acellular scaffold with a well-preserved ECM (Choudhury et al. 2020).

Below, some published protocols for decellularizing organs of the ovary (Table 1), oviduct (Table 2), uterus (Table 3), and testis (Table 4) are detailed. Each protocol specifies the animal species, the decellularization technique employed, the decellularizing agents used, and the corresponding application of the obtained dECM scaffolds.

6.7 Characterization of dECM Scaffolds

Once the decellularization protocol is completed, it is essential to perform a series of studies to characterize the dECM scaffolds. This characterization verifies the effectiveness of the protocol, ensuring that all cells and nucleic acids released after nuclear membrane rupture have been eliminated or reduced to acceptable levels. It is crucial to check that the 3D structure of the decellularized organ has been preserved and that its mechanical properties are comparable to those of the native tissue. This is important because the scaffold will act as support and must replicate the functions of the original tissue. Moreover, it is necessary to verify that the proteins and other key components of the ECM remain intact, as these bioactive molecules are fundamental for the future functionality of the scaffold. Finally, it must be

Table 1 Protocols for ovarian decellularization in different species, specifying agents used in each phase of the decellularization process and applications of the derived dECM scaffolds. "N/A" indicates "not applicable" when a protocol does not include a rising phase or a sterilization phase

Species	Decellularization technique	Lysis phase	Rinsing phase	Sterilization phase	Application	Reference
Bovine	Immersion-agitation	SDS 0.1% (24 h for 500μm ovarian sections or 3 weeks for whole ovary).	N/A	Ethanol followed by PBS washes.	Creation of an artificial ovary.	Laronda et al. (2015)
Porcine	Immersion-agitation (100 rpm, RT)	3 cycles of freeze-thaw (−80 °C/37 °C). Ovary sectioning (1.5 mm). Triton X-100 1% (9 h). SDS 0.5% (3 h). DNase at 37 °C (12 h).	Final rinse in deionized water (4 h)	Peracetic acid 3%	Medicinal regenerative (creation of dECM scaffolds for ovarian regeneration).	Liu et al. (2017)
Porcine	Immersion-agitation (200 rpm, RT)	1 cycle of freeze-thaw (−80 °C/37 °C). SDS 0.5% (3 h). Triton X-100 1% (overnight). SDC 2% (12 h).	Interspersed rinses with deionized water during lysis phase. Final rinse in deionized water (6 h).	Ethanol 70%	Research and optimization of protocols.	Pennarossa et al. (2020)
Mouse	Immersion-agitation	**Protocol 1:** SDS 0.5% (10 h) + DNase at 37 °C (30 min). **Protocol 2**: SDC 2% (16 h) + DNase at 37 °C (30 min). **Protocol 3:** SDS 0.5% (5 h) followed by SDC 2% (8 h) + DNase at 37 °C (30 min).	Final rinse with water for 24 h.	Peracetic acid 0.1% for 30 min + rinse for 24 h with sterile PBS supplemented with antibiotic-antimycotic.	Research and optimization of protocols.	Alshaikh et al. (2020)

Bovine	Immersion-agitation	Ovarian cortex cut into 2 mm pieces 3 cycles of freezing-thawing (−80 °C/37 °C). **Protocol 1:** SDS 0.1% (24 h). **Protocol 2:** SDS 0.1% (18 h). **Protocol 3:** SDS 0.5% (3 h) and triton X-100 1% (9 h). **Protocol 4:** SDS 0.5% (2 h) followed by treatment with triton X-100 1% and ammonium hydroxide 0.1% (22 h).	Final wash with deionized water for 24 h.	N/A	Research and optimization of protocols (protocol 4 most effective). Improvement of ARTs (development of ovarian dECM scaffolds for 3D culture of preantral follicles).	Nikniaz et al. (2021)
Porcine	Immersion-agitation	Ovarian cortex cutting into small pieces Trypsin-EDTA 0.05% at 37 °C (1 h). Tris (hydroxymethyl) aminomethane 0.69% + DS 2.5 mM (24 h).	Multiple rinses with deionized water during lysis phase and at the end of decellularization.	Ethyl alcohol 4% for 2 h. Incubation with PBS supplemented with 1% penicillin-streptomycin (overnight).	Improvement of ARTs (development of hydrogels mimicking the ovarian environment for 3D culture of preantral follicles).	Park et al. (2023)
Porcine	Perfusion	SDS 0.5% (48 h). Triton X-100 1% (5 h) follicle rupture with hypodermic needle. Ultrasound bath (15 min).	Initial organ wash with PBS and deionized water. Washes with PBS alternated with lysis phase. Final rinse with deionized water in ultrasound bath.	N/A	Research and optimization of protocols.	Almeida et al. (2023)

Table 2 Protocols for oviduct decellularization in different species, specifying agents used in each phase of the decellularization process and applications of the derived dECM scaffolds. "N/A" indicates "not applicable" when a protocol does not include a rising phase or a sterilization phase

Species	Decellularization technique	Lysis phase	Rinsing phase	Sterilization phase	Application	Reference
Rabbit	Immersion-agitation (300 rpm)	Cutting of oviducts into 5 mm pieces. SDS 0.1% (22 h). Triton X-100 0.5% (2 h) DNase 30µg/ml (4 h).	Multiple washes with PBS (30 min each) during the lysis phase. Final wash with PBS (3 washes of 30 min each).	N/A	Improvement of ARTs (development of oviductal hydrogels as a substrate for in vitro culture of rabbit embryos).	Francés-Herrero et al. (2021a, b)
Bovine	Immersion-agitation (4 °C)	SDC 4% (24 h). DNase 5µg/ml (3 h, RT).	Washes in sterile ultrapure water after detergent application and at the end of the process (5 min 3 times).	N/A	Research and optimization of protocols (investigation of the mechanical properties of decellularized oviducts).	Franko et al. (2024)
Cat	Immersion-agitation (150 rpm, 4 °C)	**Protocol 1**: 2 identical cycles with immersion in SDS 0.1% (18 h each cycle) and triton X-100 1% (30 min each cycle). **Protocol 2**: 3 identical cycles using the same detergents. **Protocol 3**: 3 cycles and treatment with DNase 100 U (24 h, 37 °C).	Washes with dH_2O (30 min) and PBS (5 h) during the lysis phase. Additional 24 h wash in PBS.	N/A	Research and optimization of protocols (protocol 3 more effective for cell removal but less effective for preserving ECM components).	Sanguansook et al. (2024)

Table 3 Protocols for uterus decellularization in different species, specifying agents used in each phase of the decellularization process and applications of the derived dECM scaffolds. "N/A" indicates "not applicable" when a protocol does not include a rising phase or a sterilization phase

Species	Decellularization technique (velocity of perfusion)	Lysis phase	Rinsing phase	Sterilization phase	Application	Reference
Rat	Perfusion (50 ml/h)	SDS (72 h): First SDS 0.01% (24 h). Second SDS 0.1% (24 h, 4 °C). Third SDS 1% (24 h, RT). Triton X-100 1% (30 min)	Wash in dH_2O during the lysis phase. Several final washes in PBS.	N/A	Regenerative medicine (development of a dECM scaffold with recellularization capacity and regeneration of partial uterine lesions, promoting normal gestation).	Miyazaki and Maruyama (2014)
Rat	Perfusion (2 ml/min)	**Protocol 1 and 2**: DMSO 4% (4 h) followed by triton-X100 1% (4 h). **Protocol 3**: SDC 2% (6 h).	**Protocol 1**: PBS (16 h) **Protocol 2:** dH_2O (16 h) **Protocol 3**: dH_2O (18 h)	Peracetic acid (0.1%) for 30 minutes. Multiple washes with PBS.	Regenerative medicine (protocols 1 and 2 yielded dECM scaffolds capable of recellularization and regeneration of full-thickness uterine lesions, promoting normal gestation).	Hellström et al. (2016)

(continued)

Table 3 (continued)

Species	Decellularization technique (velocity of perfusion)	Lysis phase	Rinsing phase	Sterilization phase	Application	Reference
Porcine	Perfusion (15 ml/min)	**Protocol 1:** 1 cycle of freeze-thaw followed by 2 identical cycles: SDS 0.1% (18 h) Triton X-100 1% (30 min) **Protocol 2**: Same as protocol 1, but with fresh organs.	Washes with PBS (5 h) and dH_2O (30 min) alternated with the lysis phase. Final wash in PBS (5 h).	N/A	Research and optimization of protocols (investigating the recellularization capacity of uterine dECM scaffolds for potential use in regenerative medicine or as an alternative to transplantation).	Campo et al. (2017)
Rabbit	Perfusion (8 ml/min)	2 identical cycles: SDS 0.1% (18 h) Triton X-100 1% (30 min) After the second cycle: DNase 2μg/ml (1 h).	Washes with PBS (5 h) and dH_2O (30 min) alternated with the lysis phase. Final wash in PBS (3–4 h).	N/A	Improvement ARTs (development of endometrial hydrogel as a medium for in vitro culture of rabbit embryos).	Campo et al. (2019)

Sheep	Perfusion (2–2,5 ml/min)	Pretreatment with EDTA. **Protocol 1**: Two cycles with SDS 0.5% (26 h in the first cycle and 8 h in the second cycle) and DNase (1 h; 37 °C; 8000 IU). **Protocol 2**: Two cycles with SDC 2% (26 h in the first cycle and 8 h in the second cycle) and DNase (1 h; 37 °C; 8000 IU). **Protocol 3**: Two cycles with SDC (4 h) and triton X-100 (12 h). After the second cycle: DNase (1 h; 37 °C; 8000 IU).	Washes with PBS and deionized water during the lysis phase. Final wash in deionized water (48 h).	Peracetic acid 0.1% (1 h). Wash for 48 h in sterile PBS.	Research and optimization of protocols (protocol 2 yielded better results). Regenerative medicine (investigation of the recellularization capacity of the dECM scaffold using fetal sheep bone marrow stem cells).	Tiemann et al. (2020)
Porcine	Perfusion (15 ml/min)	2 identical cycles: SDS 0.1% (18 h) Triton X-100 1% (30 min) After the second cycle: DNase 5µg/ml (1 h)	Washes with PBS (5 h) and dH_2O (30 min) alternated with the lysis phase. Final wash in PBS (6 washes of 30 min each).	N/A	Substrate medium for three-dimensional culture of human endometrial cells.	López-Martínez et al. (2021)
Sheep	Perfusion	**Protocol 1:** SDS 0.5% (8 h) and DNase (1 h; 37 °C; 8000 UI). **Protocol 2:** SDC 2% (8 h) and DNase (1 h; 37 °C; 8000 UI). **Protocol 3:** SDC 2% (4 h), triton X-100 1% (12 h) and DNase (1 h; 37 °C; 8000 UI).	Washes with deionized water and PBS during the lysis phase. Final wash with deionized water for 48 h.	Peracetic acid 0.1% (1 h). Wash for 48 h in sterile PBS.	Research and optimization of protocols for the use of dECM scaffolds in regenerative medicine.	Padma et al. (2021)

(continued)

Table 3 (continued)

Species	Decellularization technique (velocity of perfusion)	Lysis phase	Rinsing phase	Sterilization phase	Application	Reference
Mouse	Immersion	Cutting of uterine horns into 1 cm pieces. **Protocol 1**: SDS and triton X-100 using different concentrations: 0.1%, 0.3%, 0.6% (immersion for 24 h in each of these concentrations). **Protocol 2**: 1 cycle of freeze-thaw (0 °C/37 °C). Then immersion in SDS 0.5% (48 h) and triton X-100 0.1% (48 h). **Protocol 3**: Immersions in NaCl at concentrations of 1.5 M, 3 M, and 5 M (24 h each). Then immersion in SDS 0.5% (48 h) and triton X-100 0.1% (48 h).	N/A	Peracetic acid 0.1% (10 min) followed by washing with sterile PBS. Exposure to UV radiation (20 min).	Research and optimization of protocols (protocol 1 yielded better results). Regenerative medicine (creation of recellularized dECM scaffolds with human menstrual stem cells).	Arezoo et al. (2021)

Table 4 Protocols for testis decellularization in different species, specifying agents used in each phase of the decellularization process and applications of the derived dECM scaffolds. "N/A" indicates "not applicable" when a protocol does not include a rising phase or a sterilization phase

Species	Decellularization technique	Lysis phase	Rinsing phase	Sterilization phase	Application	Reference
Porcine	Immersion-agitation	1 freeze-thaw cycle (−80 °C/37 °C). **Protocol 1, 2, and 3**: SDS (0.01, 0.1, or 1%) and triton X-100 1% (8 h). **Protocol 4, 5, and 6:** Sequential treatment with triton 3%-SDS (0.01, 0.1, or 1%) - triton 1% (8 h). **Protocol 7 and 8:**Treatment with trypsin, EDTA, and triton X-100 (1 or 3%) (8 h).	Washes with deionized water (5 min) and PBS (15 min) during the lysis phase. Final wash in PBS (15 min).	Peracetic acid 0.1% in 4% ethanol, followed by three PBS washes.	Research and optimization of protocols (protocol 1, where 0.01% SDS along with triton X-100 was used, was the most effective and showed the best cytocompatibility).	Vermeulen et al. (2018)
Rat	Perfusion (0.5–0.9 cc/min, RT)	1 freeze-thaw cycle (−80 °C/RT). Triton X-100 1% (4 h). SDS 1% (24 h). DNase 1% (2 h).	Final wash in dH_2O (4 h)	3% peracetic acid in 4% ethanol followed by PBS washes.	Development of testicular organoids.	Kargar-Abarghouei et al. (2018)

(continued)

Table 4 (continued)

Species	Decellularization technique	Lysis phase	Rinsing phase	Sterilization phase	Application	Reference
Sheep	Perfusion (10 ml/min)	**First step**: First, the most effective decellularizing agent was tested (SDS, triton X-100, or EDTA). **Second step**: Using the most effective decellularizing agent (SDS 1%), the ideal perfusion time was determined (4, 6, 8, 10, or 12 hours).	Final wash with dH_2O (24 h)	N/A	Research and optimization of protocols.	Akbarzadeh et al. (2019)
Sheep	Immersion-agitation	Sectioning of the testicle into 100μm slices SDS 1% (24 h).	Washing in PBS (2 h)	70% ethanol (1 h), followed by washing with water and sterile PBS (2 h).	Improvement of ARTs (development of testicular scaffolds to promote human spermatogenesis in vitro).	Ashouri Movassagh et al. (2020)
Mouse	Immersion-agitation (50 rpm, 25 °C)	SDS 0.5% (18 h). Triton X-100 0.5% (18 h).	Washing in PBS (24 h)	Peracetic acid 0.1% in 4% ethanol (2 h). Three sterile PBS washes (4 h each).	Creation of an artificial testicle for in vitro sperm production.	Majidi Gharenaz et al. (2020)

Sheep	Immersion-agitation	Testicle cut into small pieces (1 cm^3). One freeze-thaw cycle (−80 °C/4 °C). **Protocol 1:** NaCl 3.4 M (30 min). **Protocol 2:** NaCl (30 min) followed by triton X-100 0.1% (15 min). **Protocol 3:** SDS 1% (30 min). **Protocol 4:** SDS 1% (30 min) followed by triton X-100 0.1% (15 min).	Washing in PBS (24 h)	UV light	Research and optimization of protocols (protocol 2 was the most efficient). Development of artificial testicles through 3D bioprinting.	Bashiri et al. (2021)
Rat	Perfusion (0.5 cc/min, RT)	SDS 0.5% (48 h). Triton X-100 al 1% (6 h). DNase 1% (1 h).	N/A	Washing in PBS supplemented with penicillin-streptomycin 0.1%.	Obtaining organoids from testicular dECM scaffold.	Kiani et al. (2021)
Mouse	Immersion-agitation (50 rpm)	SDS 0.5% (18 h). Triton X-100 0.5% (18 h).	Final wash with PBS (24 h)	Peracetic acid 0.1% in 4% ethanol (2 h). Three sterile PBS washes (4 h each).	Research and optimization of protocols (development of a conservation method for the prolonged storage of testicular scaffolds).	Gharenaz et al. (2021)

(continued)

Table 4 (continued)

Species	Decellularization technique	Lysis phase	Rinsing phase	Sterilization phase	Application	Reference
Bovine	Immersion-agitation (90 rpm)	Testicle cut into small pieces (1 cm^3). **Protocol 1:** 5 cycles of freeze-thaw (liquid N_2-R/T). **Protocol 2:** SD 1% (24 h). **Protocol 3:** SDS 0,1% (24 h). **Protocol 4:** SDS 1% (24 h). **Protocol 5:** SDS 0.1% (24 h) + vacuum (5 h). **Protocol 6:** SDS al 1% (24 h) + vacuum (5 h). **Protocol 7:** Triton X-100 (24 h).	Protocols that used detergents were washed with PBS (12 h)	PBS with 1% penicillin-streptomycin for 1 hour, followed by 70% ethanol for 10 min, and finally exposure to UV radiation for 15 min.	Research and optimization of protocols (protocol 6 achieves effective decellularization of the testicle, and the use of vacuum contributes to increased tissue porosity, which prevents the overlapping of the seminiferous tubules).	Khazaei et al. (2023)

confirmed, both in vitro and in vivo, that the scaffolds are non-cytotoxic and completely biocompatible.

The characterization of decellularized scaffolds begins with macroscopic and morphometric inspection to determine their appearance and similarity to the native tissue. Discoloration is a characteristic sign of the decellularization process. Therefore, a color change from pink to white and an opaque or translucent appearance are indicative of successful decellularization (Fig. 1a–c) (Campo et al. 2017). After the visual inspection, morphometric evaluation allows for assessing the physical changes (size and weight) that occurred in the decellularized organ or tissue. For instance, an increase in weight has been described after decellularization of the female cat reproductive tract. This weight gain can be attributed to the hydration of the decellularized tissues, as the spaces left in the tissue after cell removal are occupied by the liquid in which the tissue is immersed (Sanguansook et al. 2024).

Following the macroscopic evaluation, microscopic evaluation proceeds to confirm the effectiveness of the process. Complete elimination of antigens, nucleic acids, and other cellular components must be verified, as the presence of these elements can compromise the biocompatibility of the scaffold and produce immune responses or rejection if the scaffold is implanted. For this purpose, common histological techniques such as hematoxylin and eosin staining, and immunofluorescence (DAPI staining) are used. The hematoxylin-eosin staining technique employs two dyes: eosin, which stains the cytoplasm, muscle fibers, and connective tissue pink; and hematoxylin, which stains cell nuclei purple (Fig. 1d–f). Additionally, DAPI is a dye that can penetrate both cellular and nuclear membranes, binding to DNA and highlighting nuclei with fluorescent blue. These techniques allow us to assess the morphology of the tissues, the structural integrity of the cells, and the presence or absence of nuclear material (Almeida et al. 2023).

In addition to hematoxylin-eosin and DAPI, other histological stains can be used to evaluate ECM components. Masson's trichrome staining is generally used to evaluate or generally quantify collagen fibers (Campo et al. 2017). For more specific evaluation, stains such as Picrosirius Red, which discriminates between type I collagen fibers, which stain yellow or orange, and type III collagen fibers, which stain green-yellow, can be used (Daryabari et al. 2019). Azan staining is another option for evaluating collagen fibers (Kargar-Abarghouei et al. 2018). Weigert's resorcin-fuchsin staining (Almeida et al. 2023), orcein staining (Sanguansook et al. 2024), or fuchsin aldehyde staining can be used to highlight elastic fibers (Kargar-Abarghouei et al. 2018). Alcian blue staining can be used to assess the presence of GAGs as it dyes regions containing these components in bluish tones (Francés-Herrero et al. 2021a, b).

Although DAPI staining provides visual confirmation of nuclear removal, to obtain more objective and quantifiable results, it is necessary to measure the amount of residual DNA in the scaffold. Specific kits that extract and purify DNA are used for this purpose. Once isolated and purified, the DNA concentration (ng/mg of tissue) can be quantified using a NanoDrop spectrophotometer (Padma et al. 2021). According to the literature, acceptable standards indicate that DNA content should be less than 50 ng per mg of tissue, and the size of the DNA fragment should be less than 200 bp (Crapo et al. 2011). To confirm the size of these fragments, gel

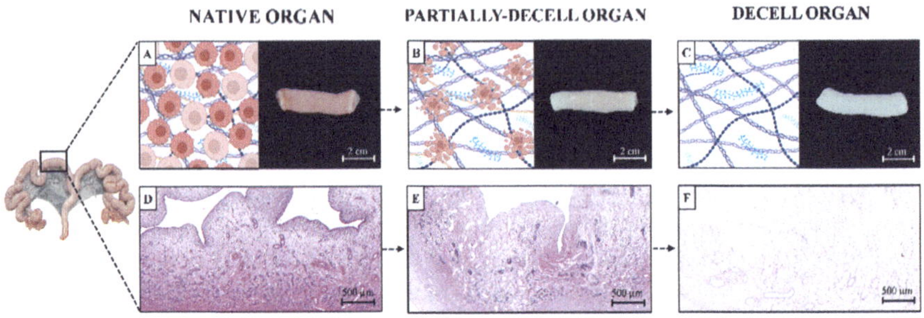

Fig. 1 Macroscopic and microscopic evolution of a porcine uterus segment during the decellularization process. (**a–c**) Graphical representation of the color change of the organ during the decellularization process. The native organ has a pink color, but as the decellularization process progresses, the organ becomes discolored, resulting in a completely white organ at the end of the process. To the left of the image of the uterine segment, a graphical representation shows what happens to the ECM during the decellularization process. In the native organ, the ECM is occupied by multiple cells. During the decellularization process, decellularizing agents cause cell lysis. By the end of the decellularization process, no cells are observed, but the ECM remains intact. (**d–f**) Histological images of native, partially decellularized, and fully decellularized uterine segments stained with hematoxylin-eosin. A decrease in cell nuclei is observed as the decellularization process progresses, resulting in completely acellular tissue at the end of the process

electrophoresis can be performed, by running the extracted DNA samples through an agarose gel (Jamaluddin et al. 2022).

ECM components can also be detected and quantified through biochemical assays. Among these techniques are immunohistochemical methods, which use specific antibodies to identify key components such as different types of collagen, elastin, hyaluronic acid, fibronectin, and laminin (Daryabari et al. 2019). A more comprehensive analysis of the matrix components is possible through proteomics, which allows for the evaluation of the protein profile of the decellularized samples. Additionally, through gene ontology analysis, this technique identifies the molecular function, biological processes involved, and the origin of the proteins detected in the samples (López-Martínez et al. 2021; Jamaluddin et al. 2022). Advanced techniques such as Fourier-transform infrared spectroscopy and Raman spectrometry can also be employed to analyze the molecular composition of the ECM, providing a more detailed understanding of its structure and functionality (Almeida et al. 2023).

Scanning electron microscopy and transmission electron microscopy are used to evaluate the external and internal ultrastructure of dECM scaffolds. These techniques confirm the preservation of the ECM's 3D architecture and allow us to observe the arrangement of the fibers and the porosity of the obtained scaffolds (Campo et al. 2017; Sanguansook et al. 2024).

Once the ECM structure has been verified, both macroscopically and microscopically after the decellularization process, it is crucial to check whether the mechanical properties of the tissue have remained intact or have been altered. For this, mechanical tests are used to assess the strength and elasticity of the dECM scaffolds. Among the mechanical tests, tensile tests stand out, which are performed using a dynamic servo-hydraulic machine. This machine measures the displacement and force applied during the test, which allows for determining the resistance to deformation, the modulus of elasticity, and the elongation to break of the tissue (Daryabari et al. 2019, 2022). Another important technique is atomic force microscopy (AFM), a type of scanning probe microscopy that, unlike traditional optical methods, does not use lenses or beam irradiation. This allows it to achieve a resolution up to 1000 times greater than conventional optical systems. AFM uses a micrometric or nanometric-sized tip (2–100 nm) attached to a movable cantilever. As the tip approaches the sample at a constant speed vertically, the tip contacts the surface, causing the cantilever to deflect, which is measured by a detector through a laser beam. This interaction not only generates high-resolution images of the surface topography but also allows measuring the local elastic stiffness or Young's modulus, thus providing a precise evaluation of the mechanical properties of the tissue at micro and nanometric scales (Alcaraz et al. 2018; Asgari et al. 2019).

In organs decellularized using the perfusion technique, where decellularizing agents are administered through the organ's vascular network, angiography or vascular corrosion tests can be resorted to. After the decellularization process, it is interesting to see if the blood vessels remain intact or have been damaged during the process. This is interesting because in this type of organ, the vascular tree can be subsequently used for recellularization processes (Campo et al. 2017; Daryabari et al. 2019).

Once the scaffold's structure, the absence of cellular and nuclear material, the ECM components, and the mechanical properties have been evaluated, the analysis of cytotoxicity and biocompatibility proceeds. Before in vivo use, the biocompatibility and cytotoxicity of the decellularized scaffold are evaluated through in vitro tests. The 3-[4,5-dimethylthiazol-2-yl]-2,5-diphenyltetrazolium bromide (MTT) assay is frequently used to determine the cytotoxicity of scaffolds. The MTT salt is added to cultured cells. Live and metabolically active cells convert MTT into formazan crystals, which are dark purple and insoluble in aqueous solutions. The amount of formazan produced can be measured using a spectrophotometer. The recorded absorbance is directly proportional to the number of metabolically active cells. Therefore, higher absorbance indicates greater cell viability. Conversely, lower absorbance suggests that the scaffold may be cytotoxic, implying that it has induced cell death (Daryabari et al. 2019). In vitro cytotoxicity can also be evaluated by seeding cells on the scaffolds. An example is the seeding of fibroblasts for 7 days on decellularized scaffolds to evaluate whether the cells are able to anchor, survive, and migrate through the scaffold (Almeida et al. 2023).

In specific contexts, such as in reproductive medicine, the cytotoxicity of dECM scaffolds can also be evaluated by incubating the scaffolds with sperm. This analysis not only studies the viability of sperm after coming into contact with the scaffold, but also studies how scaffolds affect sperm quality. This way, multiple sperm parameters can be determined to evaluate the interaction between spermatozoa and dECM scaffolds. Some of the parameters that can be assessed include effects on motility, mitochondrial activity, and acrosomal integrity can be determined (Sanguansook et al. 2024).

When scaffolds are intended to be implanted in living animals, it is important to evaluate hemocompatibility. When scaffolds are implanted in an organism, they will come into contact with blood and if the scaffold is not hemocompatible, hemolysis and damage to red blood cells will occur. To test hemocompatibility, an assay is conducted in which the scaffold is exposed to fresh blood containing anticoagulants. Subsequently, absorbance is measured to determine the degree of hemolysis caused. A scaffold is considered non-hemolytic if the percentage of hemolysis is less than 2% (Khazaei et al. 2023).

Once the cytotoxicity and biocompatibility of the scaffolds have been verified in vitro, in vivo testing can be performed. In vivo tests often employ rodents and involve the implantation of decellularized tissue patches at the site of interest. For instance, the implantation of decellularized ovine uterine tissue patches in adult fertile rats has been described to investigate biocompatibility and the ability for in vivo recellularization (Daryabari et al. 2019).

This series of evaluations ensures that dECM scaffolds are not only structurally and mechanically suitable but also safe and effective for their clinical applications.

6.8 dECM Scaffold Formats

Once the decellularization process is completed, it is necessary to decide on the final format of the dECM scaffold, which will depend on the desired applications or research focus.

6.8.1 Unmodified dECM Scaffolds

In this case, the dECM scaffolds undergo no further processing after the decellularization treatment. These scaffolds are commonly used to investigate the recellularization properties of the decellularized tissue, both in vivo and in vitro. They can be small patches of decellularized tissue that are implanted in animals or used for in vitro cell seeding to explore their recellularization potential (Yi et al. 2022). Additionally, this format can also include whole decellularized organs. Over the past decade, the technique of whole organ decellularization has been explored as a method to create scaffolds that preserve an intact vascular structure and a 3D configuration similar to that of the original tissues. Preserving the architecture of the organ offers several advantages: it allows the use of the existing vascular network for recellularization and ensures the biomechanical properties necessary for organ regeneration, which is crucial in applications such as uterine regeneration, given its potential for expansion during pregnancy (Campo et al. 2017).

6.8.2 Processed dECM Scaffolds

dECM scaffolds can be processed after obtaining them to form dECM hydrogels. Hydrogels are polymeric materials composed of synthetic or natural polymers that organize into a stable 3D network. These polymers contain numerous hydrophilic groups, allowing the hydrogel to retain water without being soluble in it. dECM hydrogels are a type of natural hydrogel formed through the self-assembly ability of collagen (Zhang et al. 2021).

The production of dECM hydrogels is achieved through enzymatic digestion of decellularized tissue. For effective enzymatic digestion, the decellularized tissue must be turned into powder. This is achieved by lyophilizing the tissue for a period ranging from 24 to 96 h. The lyophilization time will depend on the size of the decellularized tissue/organ and the retained moisture. After lyophilization, the lyophilized tissue is ground using a mortar and pestle or a cryogenic mill (Fig. 2). Once the dECM scaffold is pulverized, it is advisable to use a 40μm mesh sieve to filter the resulting powder and select fine particles, as larger particles may hinder complete digestion. Pepsin is commonly used for enzymatic digestion along with 0.01 M HCl. Higher concentrations of HCl (0.1 M) can also be used for faster digestion. The components are mixed in the following ratio: 1 g powdered dECM +100 mg pepsin +100 ml HCl. Once digestion is complete, PBS 10x (1:9 ratio) and 0.1 M NaOH (1:10 ratio) are used to adjust salt concentration and decrease the pH to physiological conditions. The digested and neutralized dECM is placed on a plate and incubated. The incubation should last for 30 min to 1 h at 37 °C for gelation to occur, transforming the digest from a liquid state to a semi-liquid state forming the hydrogel (Brown et al. 2017; McInnes et al. 2022).

Hydrogels have multiple advantages, making them particularly interesting for use in tissue engineering. First, their ability to remain in a liquid state at room temperature (RT) allows for direct and minimally invasive application in an animal via injection (Zhang et al. 2021). Second, it is possible to adjust the mechanical properties of hydrogels by altering their concentration or through crosslinking processes, allowing for precise control over their stiffness and elasticity. In this way, the hydrogel can be combined with other compounds, such as sodium alginate, to stabilize its structure, or with solubilized collagen to increase its strength. Furthermore, the structural stability of the hydrogel can be reinforced through crosslinking. This process facilitates the crosslinking of proteins and the formation of covalent bonds

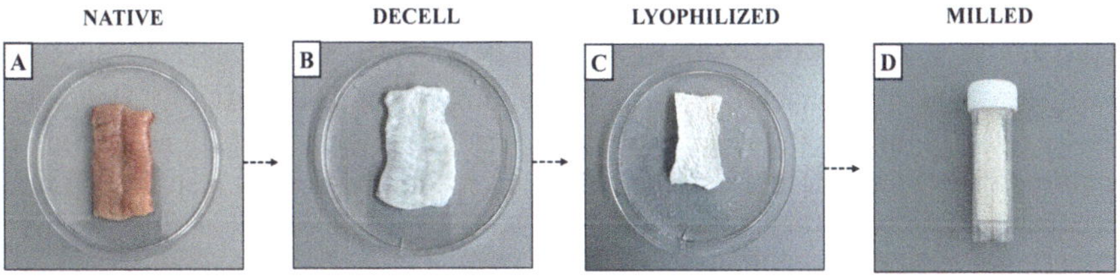

Fig. 2 Processing of a uterine dECM scaffold for hydrogel production. (**a**) Image of a uterine segment before the decellularization process. (**b**) Uterine segment at the end of the decellularization process. (**c**) Decellularized uterine segment after being subjected to a lyophilization process. (**d**) After lyophilization, the segment is milled using a mortar or a cryogenic mill. Subsequently, an enzymatic digestion is performed with pepsin and HCl. Once digestion is complete, pH and salts are neutralized using NaOH and PBS 10x, respectively. The result is the formation of a pre-gel solution that is liquid at room temperature. The final step is gelation, where the pre-gel solution is incubated at 37 °C for 1 h to induce gelation and form the hydrogel

between polymer chains. Crosslinking can be achieved physically (using light or heat), chemically (employing compounds such as carbodiimide, genipin, transglutaminase, or glutaraldehyde), or through photo-crosslinking (McInnes et al. 2022; Gadre et al. 2024). The third advantage is versatility, allowing for the incorporation of living cells, drugs, and other molecules such as extracellular vesicles to enhance their functionality. The 3D and porous structure allows hydrogels to be loaded with specific compounds, enabling controlled release of them (Ju et al. 2022). Both hydrogels alone (inks) and hydrogels loaded with cells (bioinks) can be used in 3D printers to create structures with complex and customized geometries, further expanding their applicability in the field of regenerative medicine (Kabirian and Mozafari 2020; Zhang et al. 2021).

6.9 Main Applications of dECM Scaffolds in Reproduction

The biomimetic capabilities of dECM scaffolds to replicate the functionality, structure, and biochemical properties of native tissues make them suitable as natural biomaterials for a wide range of applications. In reproductive medicine, these scaffolds are used to treat reproductive disorders through their regenerative abilities or to create artificial organs that can restore fertility. Additionally, the ability of these scaffolds to provide structural and biochemical support, promoting cell proliferation and differentiation, has led to their use as substrates for the 3D culture of gametes, embryos, and other reproductive cells. This application seeks to improve existing ARTs, offering new hopes and more effective methods for treating infertility. The subsequent sections will detail the principal applications of these scaffolds in reproductive medicine.

6.9.1 Regenerative Medicine and Treatment of Pathologies

In the field of regenerative medicine, dECM scaffolds have shown significant potential for treating various pathologies and restoring fertility, especially in the female reproductive system and more specifically in the uterus. The uterus is essential for pregnancy, and any congenital or acquired anomaly can lead to AUFI. Congenital causes of AUFI include the absence of the uterus or uterine malformations. Common acquired causes include intrauterine adhesions or uterine defects resulting from a cesarean section or hysterectomy to remove a tumor (Hellström et al. 2017). These defects can naturally heal through a process that involves three phases: the inflammatory phase, a tissue formation phase where fibroblasts begin to produce type III collagen to repair the wound, and a remodeling phase where the final organization of the collagen fibers occurs. The formation of fibrous scar tissue acts as a barrier to the complete recovery of uterine musculature, potentially leading to future infertility issues, placental abruption, or the risk of uterine rupture (Hanuman et al. 2023). Currently, the only solution for treating AUFI is uterine transplantation. Although there have been reports of successful births following uterine transplants (Brännström et al. 2015), it is important to consider that transplantation carries significant risks, including ischemic damage and the side effects of immunosuppressive therapies in recipients. In addition to the recipient-related risks, there is also a shortage of donor organs (Brännström and Dahm-Kähler 2019). Bioengineering can address these issues through the use of biomaterials that fully or partially replace a damaged uterus (Yoshimasa and Maruyama 2021). Among biomaterials, dECM scaffolds are increasingly used in regenerative medicine for their ability to stimulate the reparation of tissues and organs. Indeed, it has been demonstrated that these scaffolds can regenerate partial or full-thickness uterine injuries and restore fertility, resulting in a completely normal gestation. For this purpose, a previous study developed uterine dECM scaffolds from rats, which were then recellularized using uterine and mesenchymal stem cells. To evaluate the regenerative capacity in vivo, these recellularized dECM scaffold were implanted in rats with induced partial uterine injuries. These animals were compared with two control groups: one with an intact uterus and another with a uterine incision that was not covered with the dECM scaffold. The results showed that the dECM scaffold was able to regenerate the defective area and promote normal pregnancy, similar to the control group with an intact uterus. In contrast, in the group where the uterine incision was not covered with the dECM scaffold, no fetuses developed over the injured area, and the uteri developed alterations such as stenosis and hydrometra (Miyazaki and Maruyama 2014).

The regenerative capacity of these scaffolds has also been tested in full-thickness uterine injuries. For example, another similar study developed dECM scaffolds from decellularized rat uteri. These scaffolds were implanted as patches in rats that had sustained a full-thickness injury to one of the uterine horns. Six weeks after the surgery, the animals were mated. There were no signs of degeneration in the patches, and pregnancy rates were not affected compared to the control group. The number of fetuses was similar to that of the opposite non-patched uterine horn. The only limitation found was the absence of placental formation directly over the patched area. However, the morphology of the graft resembled that of a normal pregnant rat uterus, with well-defined myometrial and epithelial structures. Moreover, immunohistochemical analyses confirmed that most of the cells in the patch were host uterine cells. This suggests that the patch was able to promote regenerative processes, and the bioactive molecules retained in the decellularized patch fostered angiogenesis, migration, and cell proliferation, resulting in the complete regeneration of the uterine horn capable of supporting a normal pregnancy (Hellström et al. 2016).

The regenerative potential of dECM scaffolds has also been explored in other reproductive organs such as the ovary. In 2017, researchers developed scaffolds using porcine ovarian tissue. First, they confirmed that these scaffolds were biocompatible and non-toxic to cells through in vitro testing and in vivo implantation in rats. Subsequently, they investigated whether ovarian cells could adhere, migrate, and proliferate within these scaffolds. For this purpose, they implanted pieces of rat ovary into the scaffolds. The results revealed that granulosa cells not only infiltrated and grew within the scaffolds but also maintained their critical function in producing estradiol, an essential hormone for the female reproductive cycle. This finding suggests that dECM scaffolds could be a promising tool for regenerating ovarian lesions and restoring ovarian function (Liu et al. 2017).

6.9.2 Creation of Artificial Organs to Restore Fertility

The creation of artificial organs has been primarily studied in the context of the ovary. Ovarian tissue can deteriorate due to conditions such as polycystic ovary syndrome, systemic metabolic issues like diabetes and hypertension, and exposure to external factors such as chemotherapy treatments that induce tissue degeneration (Berek et al. 2021). These conditions can lead to challenges in achieving a successful gestation. Cryopreservation of oocytes or ovarian tissue may be a solution for patients with these issues. However, there are limitations to cryopreservation, including the high cost of the procedure, the toxic effects of cryoprotectants, and the deterioration of oocytes associated with freezing, thawing, or vitrification processes (Pai et al. 2021). Moreover, the transplantation of cryopreserved ovarian tissue post-tumor recovery also poses a risk of reintroducing malignant cells through the transplanted tissues (Kim et al. 2018). The creation of artificial ovaries through decellularization may offer a solution to this concern, as they would restore ovarian functions without the associated risk of tumor recurrence (Park et al. 2023).

In 2015, researchers developed an artificial ovary using decellularized bovine ovarian tissue and mouse ovarian cells. The artificial ovary was implanted in young mice from which ovaries and uterus had been previously removed. The results showed that the artificial ovary was capable of restoring hormonal function in the mice, resulting in elevated blood levels of estradiol and inhibin A. Additionally, it induced physical changes characteristic of sexual maturity, such as vaginal opening (Laronda et al. 2015).

However, one of the challenges associated with seeding cells onto dECM scaffolds is that cell distribution is primarily restricted to the surface, with only a small percentage of cells managing to infiltrate deeper into the scaffold. This limitation was observed in Laronda's studies, where ovarian cells cultured on the dECM scaffolds colonized the entire surface, but few cells migrated into the interior of the scaffold (Laronda et al. 2015). Overcoming this limitation is possible through 3D bioprinting techniques, in which cell-enriched bioinks allow for a more homogeneous and optimal cell distribution throughout the resulting structure (Rajab et al. 2020). Bioinks enable the creation of 3D structures that can be used for biomedical purposes, thanks to their printability, biocompatibility, mechanical stability, and biodegradability (Gadre et al. 2024). Combining 3D printing technology with decellularization techniques enables the creation of artificial organs. These artificial organs can be a solution to male and female infertility issues (Simopoulou et al. 2019).

In this context, a study successfully developed a bioink from decellularized porcine ovary dECM, which was used to fabricate 3D structures loaded with primary ovarian cells using 3D bioprinting technology. These structures significantly improved cell survival both in vitro and in vivo and were capable of emulating ovarian function in ovariectomized mice, inducing puberty and the estrous cycle (Zheng et al. 2022).

Another study employed a similar technique to simulate an artificial testis. For this purpose, a bioink composed of decellularized ram testicular ECM mixed with alginate and gelatin was used to enhance its mechanical properties. This scaffold showed a surface morphology that promotes cell adhesion and it was compatible with spermatogonial stem cells in both in vitro and in vivo assays, suggesting its potential as an effective biomimetic platform for sperm production (Bashiri et al. 2021).

6.9.3 Improvement of ARTs

ARTs have become increasingly significant in contemporary human and animal reproduction. These techniques encompass all in vitro manipulations of gametes and embryos for reproductive purposes. Examples include AI, in vitro oocyte maturation (IVM), IVF, ICSI, and in vitro embryo culture for subsequent transfer (De Geyter 2019). Although these techniques have been refined over time, a current challenge is to achieve in vitro conditions that closely mimic in vivo environments. In this regard, dECM scaffolds could offer a solution. The scaffolds provide not only structural support but also retain bioactive molecules that foster cellular processes such as migration, proliferation, and differentiation. Among these molecules are growth factors, which remain intact following the decellularization process. Growth factors such as vascular endothelial growth factor (VEGF) have significant implications for reproduction. Studies have shown that VEGF can enhance the maturation of porcine and bovine oocytes isolated from small follicles, as well as fertilization rates and blastocyst formation (Yang and Fortune 2007; Bui et al. 2017). Therefore, the use of dECM scaffolds that retain these growth factors can mimic in vivo processes and enhance ARTs.

Building on these findings, some researchers have explored the potential use of dECM scaffolds as a culture medium in these ARTs. For instance, a previous study reported the use of dECM hydrogels derived from sheep omentum. In this study, the oocyte maturation media was supplemented with this hydrogel to investigate its effect on rabbit oocyte maturation. The results showed that the supplemented media accelerated IVM, yielding mature oocytes after just 12 h of incubation (Fazelian-Dehkordi et al. 2022).

In another study, dECM scaffolds were developed from the bovine ovary, which proved effective for culturing murine preantral follicles. The injection of the follicles into the scaffold along with alginate improved estrogen production and oocyte maturation, thereby demonstrating the scaffold's utility in supporting follicular development (Nikniaz et al. 2021).

Additionally, the use of dECM hydrogels for in vitro embryo culture has also been explored. One example is the use of hydrogels obtained from decellularized rabbit endometrium. These hydrogels have been shown to provide an environment that promotes embryo growth without the need for supplementation with fetal bovine serum. This effect is attributed to the preservation of ECM components and the gradual release of growth factors retained in the scaffold after decellularization (Campo et al. 2019). Decellularized oviduct hydrogels have also been used for in vitro embryo culture, demonstrating embryonic development similar to or better than standard techniques (Francés-Herrero et al. 2021a, b). This shows that these materials can improve in vitro culture conditions by providing essential bioactive elements.

While traditional ARTs offer effective solutions for many infertility cases, alternative approaches are required in instances where spermatogenesis is impaired, resulting in the inability of an individual to develop sperm due to failures in germ cells. In such cases, the generation of sperm from autologous stem cells using 3D cultures represents a promising avenue for further research. Various biomaterials have been explored in these cultures, including synthetic materials like polylactic acid (Eslahi et al. 2013) and natural biomaterials like collagen (Khajavi et al. 2014), which have shown encouraging results. Moreover, dECM scaffolds, which more effectively simulate the in vivo environment, can offer significant improvements in the treatment of male infertility. For example, dECM scaffolds from sheep testicles to culture testicular stem cells and explore their differentiation in a 3D environment have been used (Ashouri Movassagh et al. 2020). The results showed that gene expression related to spermatogenesis was significantly higher compared to 2D cultures, indicating that dECM scaffolds can enhance existing techniques for producing sperm in vitro.

6.9.4 Creation of Organoids for Disease Research and Personalized Medicine

Another promising application of dECM scaffolds is their use as a 3D structure for organoid culture. Organoids are small 3D structures developed from stem cells that are capable of mimicking both the structure and function of a tissue or organ. Organoids have gained important roles in research, helping to better understand disease development. Organoids can also be created from a patient's cells, allowing for the development of personalized therapies. In the case of the reproduction field, organoids have enabled more detailed studies of processes such as embryonic development and implantation (Heremans et al. 2021).

Currently, organoids are cultured in Matrigel, which is derived from the Engelbreth-Holm-Swarm (EHS) sarcoma, a tumor that develops in mice. Cells from EHS sarcoma produce a large amount of ECM, which is purified to create Matrigel. This biomaterial is used to recreate a 3D environment that supports the culture of certain cells. However, Matrigel has limitations as its composition can significantly differ from that of normal tissue, prompting the search for more representative alternatives such as dECM scaffolds or hydrogels (Habanjar et al. 2021). In response to these limitations, researchers have explored the use of dECM scaffolds for organoid culture. For example, dECM scaffolds derived from porcine uterus have been used to culture human endometrial stromal and epithelial cells. After 3–4 days under hypoxic conditions, the seeded scaffolds rolled and contracted, and by day 9, organoid-like 3D structures had formed (Campo et al. 2017).

Furthermore, other authors have developed dECM hydrogels from porcine endometrium. These hydrogels were used as a culture medium for human endometrial organoids, resulting in significantly higher proliferation rates compared to standard media, especially between days 2 and 7 of culture (Francés-Herrero et al. 2021a, b). Moreover, other studies have highlighted the scaffolds' potential to facilitate the formation of seminiferous tubule-like structures and the differentiation of cells with phenotypes characteristic of Leydig cells (Kargar-Abarghouei et al. 2018; Kiani et al. 2021).

These examples illustrate how dECM scaffolds not only overcome some of the limitations of Matrigel but also provide a more natural and effective environment for organoid culture.

6.10 Limitations and Challenges of the Decellularization Technique

Decellularization began to be researched in the late twentieth century (Wilson et al. 1990). Since then, it has successfully decellularized a wide range of organs and tissues. Within the field of reproductive medicine, numerous studies have achieved the decellularization of both male and female reproductive organs (Tables 1, 2, 3 and 4). These studies utilize and examine various decellularization protocols, which are essential for identifying the optimal protocol that is most suitable for each case. It is important to consider that the protocols used directly influence the physical characteristics and properties of the resulting scaffold. Thus, a scaffold created using two distinct protocols will exhibit different characteristics, which can impact its final functionality (Hellström et al. 2016; Pennarossa et al. 2020; Almeida et al. 2023). Therefore, one of the ongoing challenges is to establish a standard protocol for a specific organ and species, considering that characteristics such as size and cell thickness vary significantly between species. These variations require adjustments in the timing and methods of decellularization to achieve complete cell removal.

To date, most decellularization protocols described for reproductive tissues use detergents. While detergents are effective, the cytotoxicity of these agents and their effects on the components of the ECM can compromise the final functionality of the scaffold and potentially produce adverse immune responses (Wong et al. 2016). Innovative techniques such as irreversible electroporation decellularization, the use of supercritical CO_2, and induced programmed cell death can circumvent these issues while preserving the integrity of the ECM without detergents. However, these techniques have not yet been applied to reproductive organs, underscoring the need for continued exploration and testing of new methods.

Research should not only focus on developing specific and standardized protocols for each type of organ and species but also on refining techniques that have already proven effective. An example of this is the achievement of successful pregnancies in rats using decellularized uterine material. However, despite the successful establishment of pregnancy, adequate placental formation on the graft was not achieved (Hellström et al. 2016). This issue could be due to insufficient epithelial reconstruction or the infiltration of inflammatory cells. These aspects must be improved if the ultimate goal is to create a fully functional uterus through bioengineering techniques (Hellström et al. 2017).

Most in vivo biocompatibility or regenerative capacity studies have been conducted in small rodents. For example, the creation of an artificial ovary capable of inducing puberty in ovariectomized mice (Laronda et al. 2015). Thus, one of the pending challenges is to expand this research to larger animals and even humans.

In fact, the use of these scaffolds in human medicine is the main challenge of tissue engineering and regenerative medicine, offering a promising alternative to the current limitations of organ donation. However, the diversity and immunogenicity of human tissues complicate the standardization of decellularization methods. Therefore, it is important to continue investigating the immune response triggered after the implantation of these scaffolds. Although dECM scaffolds are considered hypoimmunogenic because the decellularization process eliminates antigenic epitopes that could cause rejection, the question remains how the human body responds to their implantation and whether immunosuppressive therapy might be necessary. It has been observed that the immune response to scaffolds may include the activation of M2 macrophages, which promote angiogenesis and tissue remodeling, beneficial for scaffold integration. However, the activation of M1 macrophages, induced by the classical pathway, can trigger inflammatory response and tissue destruction. This inflammatory response could contribute to scaffold rejection and the development of autoimmune and chronic diseases (Hellström et al. 2017).

Despite these limitations and challenges, the commercialization of these scaffolds by companies like Biohorizon and Axogen, and the use of bioprinted hydrogels derived from dECM in drug detection and delivery, illustrate the current practical applications of these advances. Products like AlloDerm, Strattice, and OaSIS demonstrate the clinical use of acellular scaffolds in breast reconstruction, skin flap coverage, and the treatment of ulcers and wounds (Neishabouri et al. 2022). Some clinical products on the market have been obtained from reproductive tissue or are used in reproductive medicine, such as Matrion, a decellularized placenta produced by LifeNet Health, which is used for wound healing, or Axi by Coloplast, made from human dermis and used in utero (Gadre et al. 2024).

Although there is still a long way to go, dECM scaffolds represent a promising solution to overcome the limitations of traditional treatments in reproductive tissue regeneration. These scaffolds could become a viable alternative to transplantation in the future. Additionally, their ability to mimic in vivo conditions makes them an exceptional platform for cell or embryo culture, with the potential to revolutionize ARTs.

6.11 Concluding Remarks

This chapter explore the fundamental aspects of decellularization technology, an innovative technique in the field of tissue engineering that focuses on the development of scaffolds from ECM. These scaffolds, which are noted for their biomimicry, biocompatibility, and hypoimmunogenicity, have begun to be successfully used in reproductive medicine. Their capacity to replicate the natural environment of tissues and facilitate recellularization has enabled significant advancements, such as improvements in oocyte and embryo culture techniques, regeneration of uterine lesions, as well as the creation of artificial ovaries and testes that emulate essential biological functions. However, despite these advances, the decellularization technique still presents significant challenges. The lack of standardized protocols, cytotoxicity associated with the use of detergents, limited knowledge of some decellularization methods, and restricted research in large animals underscore the need for ongoing research to optimize these technologies and extend their application to a clinical setting.FundingThis research was supported by the Ministry of Science and Innovation (PID2019-106380RB-I00/AEI/10.13039/501100011033), and Fundación Séneca-Agencia de Ciencia y Tecnología de la Región de Murcia "Ayudas a la realización de proyectos para el desarrollo de investigación científica y técnica por grupos competitivos" (22001/PI/22). Additionally, Cristina Martínez-López is the beneficiary of a predoctoral contract from the Research Promotion Plan of the University of Murcia for 2022 (R-496/2022).

References

Abaci A, Guvendiren M (2020) Designing decellularized extracellular matrix-based bioinks for 3D bioprinting. Adv Healthc Mater 9(24):e2000734. https://doi.org/10.1002/adhm.202000734

Akbarzadeh A, Kianmanesh M, Fendereski K, Ebadi M, Daryabari SS, Masoomi A, Ghazisaeedi F, Seyyed Hossein Beigi R, Sheikh R, Kajbafzadeh AM (2019) Decellularised whole ovine testis as a potential bio-scaffold for tissue engineering. Reprod Fertil Dev 31(11):1665–1673. https://doi.org/10.1071/RD19070

Alcaraz J, Otero J, Jorba I, Navajas D (2018) Bidirectional mechanobiology between cells and their local extracellular matrix probed by atomic force microscopy. Semin Cell Dev Biol 73:71–81. https://doi.org/10.1016/j.semcdb.2017.07.020

Almeida GHDR, da Silva-Júnior LN, Gibin MS, Dos Santos H, de Oliveira Horvath-Pereira B, Pinho LBM, Baesso ML, Sato F, Hernandes L, Long CR, Relly L, Miglino MA, Carreira ACO (2023) Perfusion and ultrasonication produce a decellularized porcine whole-ovary scaffold with a preserved microarchitecture. Cells 12(14):1864. https://doi.org/10.3390/cells12141864

Alshaikh AB, Padma AM, Dehlin M, Akouri R, Song MJ, Brännström M, Hellström M (2020) Decellularization and recellularization of the ovary for bioengineering applications; studies in the mouse. Reprod Biol Endocrinol 18(1):75. https://doi.org/10.1186/s12958-020-00630-y

Arezoo N, Mohammad H, Malihezaman M (2021) Tissue engineering of mouse uterus using menstrual blood stem cells (MenSCs) and decellularized uterine scaffold. Stem Cell Res Ther 12(1):475. https://doi.org/10.1186/s13287-021-02543-y

Asgari M, Abi-Rafeh J, Hendy GN, Pasini D (2019) Material anisotropy and elasticity of cortical and trabecular bone in the adult mouse femur via AFM indentation. J Mech Behav Biomed Mater 93:81–92. https://doi.org/10.1016/j.jmbbm.2019.01.024

Ashouri Movassagh S, Ashouri Movassagh S, Banitalebi Dehkordi M, Pourmand G, Gholami K, Talebi A, Esfandyari S, Jabari A, Samadian A, Abbasi M (2020) Isolation, identification and differentiation of human spermatogonial cells on three-dimensional decellularized sheep testis. Acta Histochem 122(8):151623. https://doi.org/10.1016/j.acthis.2020.151623

Badylak SF, Freytes DO, Gilbert TW (2009) Extracellular matrix as a biological scaffold material: structure and function. Acta Biomater 5(1):1–13. https://doi.org/10.1016/j.actbio.2008.09.013

Bashiri Z, Amiri I, Gholipourmalekabadi M, Falak R, Asgari H, Maki CB, Moghaddaszadeh A, Koruji M (2021) Artificial testis: a testicular tissue extracellular matrix as a potential bio-ink for 3D printing. Biomater Sci 9(9):3465–3484. https://doi.org/10.1039/d0bm02209h

Berek JS, Renz M, Kehoe S, Kumar L, Friedlander M (2021) Cancer of the ovary, fallopian tube, and peritoneum: 2021 update. Int J Gynaecol Obstet 155(Suppl 1):61–85. https://doi.org/10.1002/ijgo.13878

Bourgine PE, Pippenger BE, Todorov A Jr, Tchang L, Martin I (2013) Tissue decellularization by activation of programmed cell death. Biomaterials 34(26):6099–6108. https://doi.org/10.1016/j.biomaterials.2013.04.058

Brännström M, Dahm-Kähler P (2019) Uterus transplantation and fertility preservation. Best practice & research. Clin Obstet Gynaecol 55:109–116. https://doi.org/10.1016/j.bpobgyn.2018.12.006

Brännström M, Johannesson L, Bokström H, Kvarnström N, Mölne J, Dahm-Kähler P, Enskog A, Milenkovic M, Ekberg J, Diaz-Garcia C, Gäbel M, Hanafy A, Hagberg H, Olausson M, Nilsson L (2015) Livebirth after uterus transplantation. Lancet 385(9968):607–616. https://doi.org/10.1016/S0140-6736(14)61728-1

Brown BN, Freund JM, Han L, Rubin JP, Reing JE, Jeffries EM, Wolf MT, Tottey S, Barnes CA, Ratner BD, Badylak SF (2011) Comparison of three methods for the derivation of a biologic scaffold composed of adipose tissue extracellular matrix. Tissue Eng Part C Methods 17(4):411–421. https://doi.org/10.1089/ten.TEC.2010.0342

Brown BN, Buckenmeyer MJ, Prest TA (2017) Preparation of decellularized biological scaffolds for 3D cell culture. Methods Mol Biol 1612:15–27. https://doi.org/10.1007/978-1-4939-7021-6_2

Bui TMT, Nguyễn KX, Karata A, Ferré P, Trần MT, Wakai T, Funahashi H (2017) Presence of vascular endothelial growth factor during the first half of IVM improves the meiotic and developmental competence of porcine oocytes from small follicles. Reprod Fertil Dev 29(10):1902–1909. https://doi.org/10.1071/RD16321

Burk J, Erbe I, Berner D, Kacza J, Kasper C, Pfeiffer B, Winter K, Brehm W (2014) Freeze-thaw cycles enhance decellularization of large tendons. Tissue Eng Part C Methods 20(4):276–284. https://doi.org/10.1089/ten.TEC.2012.0760

Butler CR, Hynds RE, Crowley C, Gowers KH, Partington L, Hamilton NJ, Carvalho C, Platé M, Samuel ER, Burns AJ, Urbani L, Birchall MA, Lowdell MW, De Coppi P, Janes SM (2017) Vacuum-assisted decellularization: an accelerated protocol to generate tissue-engineered human tracheal scaffolds. Biomaterials 124:95–105. https://doi.org/10.1016/j.biomaterials.2017.02.001

Campo H, Baptista PM, López-Pérez N, Faus A, Cervelló I, Simón C (2017) De- and recellularization of the pig uterus: a bioengineering pilot study. Biol Reprod 96(1):34–45. https://doi.org/10.1095/biolreprod.116.143396

Campo H, García-Domínguez X, López-Martínez S, Faus A, Vicente Antón JS, Marco-Jiménez F, Cervelló I (2019) Tissue-specific decellularized endometrial substratum mimicking different physiological conditions influences in vitro embryo development in a rabbit model. Acta Biomater 89:126–138. https://doi.org/10.1016/j.actbio.2019.03.004

Chen X, Sun J, Li X, Mao L, Cui L, Bai W (2019) Transplantation of oral mucosal epithelial cells seeded on decellularized and lyophilized amniotic membrane for the regeneration of injured endometrium. Stem Cell Res Ther 10(1):107. https://doi.org/10.1186/s13287-019-1179-z

Choudhury D, Yee M, Sheng ZLJ, Amirul A, Naing MW (2020) Decellularization systems and devices: state-of-the-art. Acta Biomater 115:51–59. https://doi.org/10.1016/j.actbio.2020.07.060

Crapo PM, Gilbert TW, Badylak SF (2011) An overview of tissue and whole organ decellularization processes. Biomaterials 32(12):3233–3243. https://doi.org/10.1016/j.biomaterials.2011.01.057

Cui N, Hu M, Khalil RA (2017) Biochemical and biological attributes of matrix metalloproteinases. Prog Mol Biol Transl Sci 147:1–73. https://doi.org/10.1016/bs.pmbts.2017.02.005

Daryabari SS, Kajbafzadeh AM, Fendereski K, Ghorbani F, Dehnavi M, Rostami M, Garajegayeh BA, Tavangar SM (2019) Development of an efficient perfusion-based protocol for whole-organ decellularization of the ovine uterus as a human-sized model and in vivo application of the bioscaffolds. J Assist Reprod Genet 36(6):1211–1223. https://doi.org/10.1007/s10815-019-01463-4

Daryabari SS, Fendereski K, Ghorbani F, Dehnavi M, Shafikhani Y, Omranipour A, Zeraatian-Nejad Davani S, Majidi Zolbin M, Tavangar SM, Kajbafzadeh AM (2022) Whole-organ decellularization of the human uterus and in vivo application of the bio-scaffolds in animal models. J Assist Reprod Genet 39(6):1237–1247. https://doi.org/10.1007/s10815-022-02492-2

De Geyter C (2019) Assisted reproductive technology: impact on society and need for surveillance. Best Pract Res Clin Endocrinol Metab 33(1):3–8. https://doi.org/10.1016/j.beem.2019.01.004

Deeken CR, White AK, Bachman SL, Ramshaw BJ, Cleveland DS, Loy TS, Grant SA (2011) Method of preparing a decellularized porcine tendon using tributyl phosphate. J Biomed Mater Res B Appl Biomater 96(2):199–206. https://doi.org/10.1002/jbm.b.31753

Dempsey SG, Miller CH, Hill RC, Hansen KC, May BCH (2019) Functional insights from the proteomic inventory of ovine forestomach matrix. J Proteome Res 18(4):1657–1668. https://doi.org/10.1021/acs.jproteome.8b00908

Dong X, Wei X, Yi W, Gu C, Kang X, Liu Y, Li Q, Yi D (2009) RGD-modified acellular bovine pericardium as a bioprosthetic scaffold for tissue engineering. J Mater Sci Mater Med 20(11):2327–2336. https://doi.org/10.1007/s10856-009-3791-4

Eslahi N, Hadjighassem MR, Joghataei MT, Mirzapour T, Bakhtiyari M, Shakeri M, Pirhajati V, Shirinbayan P, Koruji M (2013) The effects of poly L-lactic acid nanofiber scaffold on mouse spermatogonial stem cell culture. Int J Nanomedicine 8:4563–4576. https://doi.org/10.2147/IJN.S45535

Fazelian-Dehkordi K, Talaei-Khozani T, Fakhroddin Mesbah AS (2022) Three-dimensional in vitro maturation of rabbit oocytes enriched with sheep decellularized greater omentum. Vet Med Sci 8(5):2092–2103. https://doi.org/10.1002/vms3.891

Francés-Herrero E, De Miguel-Gómez L, López-Martínez S, Campo H, Garcia-Dominguez X, Diretto G, Faus A, Vicente JS, Marco-Jiménez F, Cervelló I (2021a) Development of Decellularized Oviductal hydrogels as a support for rabbit embryo culture. Reprod Sci 28(6):1644–1658. https://doi.org/10.1007/s43032-020-00446-6

Francés-Herrero E, Juárez-Barber E, Campo H, López-Martínez S, de Miguel-Gómez L, Faus A, Pellicer A, Ferrero H, Cervelló I (2021b) Improved models of human endometrial organoids based on hydrogels from decellularized endometrium. J Pers Med 11(6):504. https://doi.org/10.3390/jpm11060504

Franko R, Franko Y, Ribes Martinez E, Ferronato GA, Heinzelmann I, Grechi N, Devkota S, Fontes PK, Coeti R, Oshiro TSI, Ferraz MAMM (2024) Mechanical properties of native and decellularized reproductive tissues: insights for tissue engineering strategies. Sci Rep 14(1):7347. https://doi.org/10.1038/s41598-024-57867-5

Fuentes JM, Arrieta MP, Boronat T, Ferrándiz S (2022) Effects of steam heat and dry heat sterilization processes on 3D printed commercial polymers printed by fused deposition modeling. Polymers 14(5):855. https://doi.org/10.3390/polym14050855

Gadre M, Kasturi M, Agarwal P, Vasanthan KS (2024) Decellularization and their significance for tissue regeneration in the era of 3D bioprinting. ACS Omega 9(7):7375–7392. https://doi.org/10.1021/acsomega.3c08930

Gafarova ER, Grebenik EA, Lazhko AE, Frolova AA, Kuryanova AS, Kurkov AV, Bazhanov IA, Kapomba BS, Kosheleva NV, Novikov IA, Shekhter AB, Golubeva EN, Soloviova AB, Timashev PS (2020) Evaluation of supercritical CO2-assisted protocols in a model of ovine aortic root decellularization. Molecules 25(17):3923. https://doi.org/10.3390/molecules25173923

Gharenaz NM, Movahedin M, Mazaheri Z (2021) Comparison of two methods for prolong storage of decellularized mouse whole testis for tissue engineering application: an experimental study. Int J Reprod Biomed 19(4):321–332. https://doi.org/10.18502/ijrm.v19i4.9058

Gillies AR, Smith LR, Lieber RL, Varghese S (2011) Method for decellularizing skeletal muscle without detergents or proteolytic enzymes. Tissue Eng Part C Methods 17(4):383–389. https://doi.org/10.1089/ten.TEC.2010.0438

Greco KV, Jones LG, Obiri-Yeboa I, Ansari T (2018) Creation of an acellular vaginal matrix for potential vaginal augmentation and cloacal repair. J Pediatr Adolesc Gynecol 31(5):473–479. https://doi.org/10.1016/j.jpag.2018.05.003

Guruswamy Damodaran R, Vermette P (2018) Tissue and organ decellularization in regenerative medicine. Biotechnol Prog 34(6):1494–1505. https://doi.org/10.1002/btpr.2699

Habanjar O, Diab-Assaf M, Caldefie-Chezet F, Delort L (2021) 3D cell culture systems: tumor application, advantages, and disadvantages. Int J Mol Sci 22(22):12200. https://doi.org/10.3390/ijms222212200

Hansen PJ (2020) Implications of assisted reproductive technologies for pregnancy outcomes in mammals. Annu Rev Anim Biosci 8:395–413. https://doi.org/10.1146/annurev-animal-021419-084010

Hanuman S, Pande G, Nune M (2023) Current status and challenges in uterine myometrial tissue engineering. Bioengineered 14(1):2251847. https://doi.org/10.1080/21655979.2023.2251847

Harrell CR, Djonov V, Fellabaum C, Volarevic V (2018) Risks of using sterilization by gamma radiation: the other side of the coin. Int J Med Sci 15(3):274–279. https://doi.org/10.7150/ijms.22644

Hellström M, El-Akouri RR, Sihlbom C, Olsson BM, Lengqvist J, Bäckdahl H, Johansson BR, Olausson M, Sumitran-Holgersson S, Brännström M (2014) Towards the development of a bioengineered uterus: comparison of different protocols for rat uterus decellularization. Acta Biomater 10(12):5034–5042. https://doi.org/10.1016/j.actbio.2014.08.018

Hellström M, Moreno-Moya JM, Bandstein S, Bom E, Akouri RR, Miyazaki K, Maruyama T, Brännström M (2016) Bioengineered uterine tissue supports pregnancy in a rat model. Fertil Steril 106(2):487–496.e1. https://doi.org/10.1016/j.fertnstert.2016.03.048

Hellström M, Bandstein S, Brännström M (2017) Uterine tissue engineering and the future of uterus transplantation. Ann Biomed Eng 45(7):1718–1730. https://doi.org/10.1007/s10439-016-1776-2

Heremans R, Jan Z, Timmerman D, Vankelecom H (2021) Organoids of the female reproductive tract: innovative tools to study desired to unwelcome processes. Front Cell Dev Biol 9:661472. https://doi.org/10.3389/fcell.2021.661472

Hudson TW, Liu SY, Schmidt CE (2004) Engineering an improved acellular nerve graft via optimized chemical processing. Tissue Eng 10(9–10):1346–1358. https://doi.org/10.1089/ten.2004.10.1641

Jamaluddin MFB, Ghosh A, Ingle A, Mohammed R, Ali A, Bahrami M, Kaiko G, Gibb Z, Filipe EC, Cox TR, Boulton A, O'Sullivan R, Ius Y, Karakoti A, Vinu A, Nahar P, Jaaback K, Bansal V, Tanwar PS (2022) Bovine and human endometrium-derived hydrogels support organoid culture from healthy and cancerous tissues. Proc Natl Acad Sci USA 119(44):e2208040119. https://doi.org/10.1073/pnas.2208040119

Ju Y, Hu Y, Yang P, Xie X, Fang B (2022) Extracellular vesicle-loaded hydrogels for tissue repair and regeneration. Mater Today Bio 18:100522. https://doi.org/10.1016/j.mtbio.2022.100522

Kabirian F, Mozafari M (2020) Decellularized ECM-derived bioinks: prospects for the future. Methods 171:108–118. https://doi.org/10.1016/j.ymeth.2019.04.019

Karamanos NK, Theocharis AD, Piperigkou Z, Manou D, Passi A, Skandalis SS, Vynios DH, Orian-Rousseau V, Ricard-Blum S, Schmelzer CEH, Duca L, Durbeej M, Afratis NA, Troeberg L, Franchi M, Masola V, Onisto M (2021) A guide to the composition and functions of the extracellular matrix. FEBS J 288(24):6850–6912. https://doi.org/10.1111/febs.15776

Kargar-Abarghouei E, Vojdani Z, Hassanpour A, Alaee S, Talaei-Khozani T (2018) Characterization, recellularization, and transplantation of rat decellularized testis scaffold with bone marrow-derived mesenchymal stem cells. Stem Cell Res Ther 9(1):324. https://doi.org/10.1186/s13287-018-1062-3

Khajavi N, Akbari M, Abdolsamadi HR, Abolhassani F, Dehpour AR, Koruji M, Habibi Roudkenar M (2014) Role of somatic testicular cells during mouse spermatogenesis in three-dimensional collagen gel culture system. Cell J 16(1):79–90

Khazaei MR, Ami Z, Khazaei M, Rezakhani L (2023) The decellularized calf testis: introducing suitable scaffolds for spermatogenesis studies. Int J Fertil Steril 18(1):32–39. https://doi.org/10.22074/ijfs.2023.1989173.1433

Kiani M, Movahedin M, Halvaei I, Soleimani M (2021) Formation of organoid-like structures in the decellularized rat testis. Iran J Basic Med Sci 24(11):1523–1528. https://doi.org/10.22038/IJBMS.2021.58294.12948

Kim JK, Koh YD, Kim JO, Seo DH (2016) Development of a decellularization method to produce nerve allografts using less invasive detergents and hyper/hypotonic solutions. J Plast Reconstr Aesthet Surg 69(12):1690–1696. https://doi.org/10.1016/j.bjps.2016.08.016

Kim S, Lee Y, Lee S, Kim T (2018) Ovarian tissue cryopreservation and transplantation in patients with cancer. Obstet Gynecol Sci 61(4):431–442. https://doi.org/10.5468/ogs.2018.61.4.431

Kim SW, Kim YY, Kim H, Ku SY (2021) Recent advancements in engineered biomaterials for the regeneration of female reproductive organs. Reprod Sci 28(6):1612–1625. https://doi.org/10.1007/s43032-021-00553-y

Laronda MM, Jakus AE, Whelan KA, Wertheim JA, Shah RN, Woodruff TK (2015) Initiation of puberty in mice following decellularized ovary transplant. Biomaterials 50:20–29. https://doi.org/10.1016/j.biomaterials.2015.01.051

Liu WY, Lin SG, Zhuo RY, Xie YY, Pan W, Lin XF, Shen FX (2017) Xenogeneic decellularized scaffold: A novel platform for ovary regeneration. Tissue Eng Part C Methods 23(2):61–71. https://doi.org/10.1089/ten.TEC.2016.0410

López-Martínez S, Rodríguez-Eguren A, de Miguel-Gómez L, Francés-Herrero E, Faus A, Díaz A, Pellicer A, Ferrero H, Cervelló I (2021) Bioengineered endometrial hydrogels with growth factors promote tissue regeneration and restore fertility in murine models. Acta Biomater 135:113–125. https://doi.org/10.1016/j.actbio.2021.08.025

Majidi Gharenaz N, Movahedin M, Mazaheri Z (2020) Three-dimensional culture of mouse Spermatogonial stem cells using A Decellularised testicular scaffold. Cell J 21(4):410–418. https://doi.org/10.22074/cellj.2020.6304

McInnes AD, Moser MAJ, Chen X (2022) Preparation and use of decellularized extracellular matrix for tissue engineering. J Funct Biomater 13(4):240. https://doi.org/10.3390/jfb13040240

Mendibil U, Ruiz-Hernandez R, Retegi-Carrion S, Garcia-Urquia N, Olalde-Graells B, Abarrategi A (2020) Tissue-specific decellularization methods: rationale and strategies to achieve regenerative compounds. Int J Mol Sci 21(15):5447. https://doi.org/10.3390/ijms21155447

Miyazaki K, Maruyama T (2014) Partial regeneration and reconstruction of the rat uterus through recellularization of a decellularized uterine matrix. Biomaterials 35(31):8791–8800. https://doi.org/10.1016/j.biomaterials.2014.06.052

Naeemi S, Eidi A, Khanbabaee R, Sadri-Ardekani H, Kajbafzadeh AM (2021) Differentiation and proliferation of spermatogonial stem cells using a three-dimensional decellularized testicular scaffold: a new method to study the testicular microenvironment in vitro. Int Urol Nephrol 53(8):1543–1550. https://doi.org/10.1007/s11255-021-02877-9

Naranjo TA, Noguera-Salvá R, Guerrero FF (2009) La matriz extracelular: morfología, función y biotensegridad (parte I). Rev Esp Patol 42(4):249–261

Negishi J, Funamoto S, Kimura T, Nam K, Higami T, Kishida A (2011) Effect of treatment temperature on collagen structures of the decellularized carotid artery using high hydrostatic pressure. J Artif Organs 14(3):223–231. https://doi.org/10.1007/s10047-011-0570-z

Neishabouri A, Soltani Khaboushan A, Daghigh F, Kajbafzadeh AM, Majidi Zolbin M (2022) Decellularization in tissue engineering and regenerative medicine: evaluation, modification, and application methods. Front Bioeng Biotechnol 10:805299. https://doi.org/10.3389/fbioe.2022.805299

Nikniaz H, Zandieh Z, Nouri M, Daei-Farshbaf N, Aflatoonian R, Gholipourmalekabadi M, Jameie SB (2021) Comparing various protocols of human and bovine ovarian tissue decellularization to prepare extracellular matrix-alginate scaffold for better follicle development in vitro. BMC Biotechnol 21(1):8. https://doi.org/10.1186/s12896-020-00658-3

O'Connor BB, Pope BD, Peters MM, Ris-Stalpers C, Parker KK (2020) The role of extracellular matrix in normal and pathological pregnancy: future applications of microphysiological

systems in reproductive medicine. Exp Biol Med (Maywood) 245(13):1163–1174. https://doi.org/10.1177/1535370220938741
O'Neill JD, Anfang R, Anandappa A, Costa J, Javidfar J, Wobma HM, Singh G, Freytes DO, Bacchetta MD, Sonett JR, Vunjak-Novakovic G (2013) Decellularization of human and porcine lung tissues for pulmonary tissue engineering. Ann Thorac Surg 96(3):1046–1056. https://doi.org/10.1016/j.athoracsur.2013.04.022
O'Neill CL, Chow S, Rosenwaks Z, Palermo GD (2018) Development of ICSI. Reproduction 156(1):F51–F58. https://doi.org/10.1530/REP-18-0011
Padma AM, Carrière L, Krokström Karlsson F, Sehic E, Bandstein S, Tiemann TT, Oltean M, Song MJ, Brännström M, Hellström M (2021) Towards a bioengineered uterus: bioactive sheep uterus scaffolds are effectively recellularized by enzymatic preconditioning. NPJ Regen Med 6(1):26. https://doi.org/10.1038/s41536-021-00136-0
Pai HD, Baid R, Palshetkar NP, Pai A, Pai RD, Palshetkar R (2021) Oocyte cryopreservation - current scenario and future perspectives: a narrative review. J Hum Reprod Sci 14(4):340–349. https://doi.org/10.4103/jhrs.jhrs_173_21
Park EY, Park JH, Mai NTQ, Moon BS, Choi JK (2023) Control of the growth and development of murine preantral follicles in a biomimetic ovary using a decellularized porcine scaffold. Mater Today Bio 23:100824. https://doi.org/10.1016/j.mtbio.2023.100824
Paulo Zambon J, Atala A, Yoo JJ (2020) Methods to generate tissue-derived constructs for regenerative medicine applications. Methods 171:3–10. https://doi.org/10.1016/j.ymeth.2019.09.016
Pennarossa G, Ghiringhelli M, Gandolfi F, Brevini TAL (2020) Whole-ovary decellularization generates an effective 3D bioscaffold for ovarian bioengineering. J Assist Reprod Genet 37(6):1329–1339. https://doi.org/10.1007/s10815-020-01784-9
Rajab TK, O'Malley TJ, Tchantchaleishvili V (2020) Decellularized scaffolds for tissue engineering: current status and future perspective. Artif Organs 44(10):1031–1043. https://doi.org/10.1111/aor.13701
Sanguansook P, Martínez-López C, Izquierdo-Rico MJ, Martínez-Cáceres C, López-Orozco M, Chatdarong K, García-Vázquez FA (2024) Development of decellularization protocols for female cat reproductive organs. Res Vet Sci 173:105257. https://doi.org/10.1016/j.rvsc.2024.105257
Simopoulou M, Sfakianoudis K, Tsioulou P, Rapani A, Giannelou P, Kiriakopoulos N, Pantou A, Vlahos N, Anifandis G, Bolaris S, Pantos K, Koutsilieris M (2019) What will the future hold for artificial organs in the service of assisted reproduction: prospects and considerations. Front Med 13(6):627–638. https://doi.org/10.1007/s11684-019-0697-5
Singh P, Carraher C, Schwarzbauer JE (2010) Assembly of fibronectin extracellular matrix. Annu Rev Cell Dev Biol 26:397–419. https://doi.org/10.1146/annurev-cellbio-100109-104020
Sittadjody S, Criswell T, Jackson JD, Atala A, Yoo JJ (2021) Regenerative medicine approaches in bioengineering female reproductive tissues. Reprod Sci 28(6):1573–1595. https://doi.org/10.1007/s43032-021-00548-9
Skardal A, Devarasetty M, Kang HW, Mead I, Bishop C, Shupe T, Lee SJ, Jackson J, Yoo J, Soker S, Atala A (2015) A hydrogel bioink toolkit for mimicking native tissue biochemical and mechanical properties in bioprinted tissue constructs. Acta Biomater 25:24–34. https://doi.org/10.1016/j.actbio.2015.07.030
Somers P, Cuvelier CA, Somer FD, Cornelissen M, Cox E, Verloo M, Chiers K, van Nooten G (2009) Gamma radiation alters the ultrastructure in tissue-engineered heart valve scaffolds. Tissue Eng Part A 15(11):3597–3604. https://doi.org/10.1089/ten.TEA.2008.0690
Tao M, Ao T, Mao X, Yan X, Javed R, Hou W, Wang Y, Sun C, Lin S, Yu T, Ao Q (2021) Sterilization and disinfection methods for decellularized matrix materials: review, consideration and proposal. Bioact Mater 6(9):2927–2945. https://doi.org/10.1016/j.bioactmat.2021.02.010
Theocharis AD, Skandalis SS, Gialeli C, Karamanos NK (2016) Extracellular matrix structure. Adv Drug Deliv Rev 97:4–27. https://doi.org/10.1016/j.addr.2015.11.001
Thier R, Bolt HM (2000) Carcinogenicity and genotoxicity of ethylene oxide: new aspects and recent advances. Crit Rev Toxicol 30(5):595–608. https://doi.org/10.1080/10408440008951121

Tiemann TT, Padma AM, Sehic E, Bäckdahl H, Oltean M, Song MJ, Brännström M, Hellström M (2020) Towards uterus tissue engineering: a comparative study of sheep uterus decellularisation. Mol Hum Reprod 26(3):167–178. https://doi.org/10.1093/molehr/gaaa009

Verberckmoes S, Van Soom A, de Kruif A (2004) Intra-uterine insemination in farm animals and humans. Reprod Domest Anim 39(3):195–204. https://doi.org/10.1111/j.1439-0531.2004.00512.x

Vermeulen M, Del Vento F, de Michele F, Poels J, Wyns C (2018) Development of a Cytocompatible scaffold from pig immature testicular tissue allowing human Sertoli cell attachment, proliferation and functionality. Int J Mol Sci 19(1):227. https://doi.org/10.3390/ijms19010227

Wang B, Qinglai T, Yang Q, Li M, Zeng S, Yang X, Xiao Z, Tong X, Lei L, Li S (2022) Functional acellular matrix for tissue repair. Mater Today Bio 18:100530. https://doi.org/10.1016/j.mtbio.2022.100530

Wilson GJ, Yeger H, Klement P, Lee JM, Courtman DW (1990) Acellular matrix allograft small caliber vascular prostheses. ASAIO Trans 36(3):M340–M343

Wong ML, Wong JL, Vapniarsky N, Griffiths LG (2016) In vivo xenogeneic scaffold fate is determined by residual antigenicity and extracellular matrix preservation. Biomaterials 92:1–12. https://doi.org/10.1016/j.biomaterials.2016.03.024

Xu H, Wan H, Sandor M, Qi S, Ervin F, Harper JR, Silverman RP, McQuillan DJ (2008) Host response to human acellular dermal matrix transplantation in a primate model of abdominal wall repair. Tissue Eng Part A 14(12):2009–2019. https://doi.org/10.1089/ten.tea.2007.0316

Xu H, Wan H, Zuo W, Sun W, Owens RT, Harper JR, Ayares DL, McQuillan DJ (2009) A porcine-derived acellular dermal scaffold that supports soft tissue regeneration: removal of terminal galactose-alpha-(1,3)-galactose and retention of matrix structure. Tissue Eng Part A 15(7):1807–1819. https://doi.org/10.1089/ten.tea.2008.0384

Yang MY, Fortune JE (2007) Vascular endothelial growth factor stimulates the primary to secondary follicle transition in bovine follicles in vitro. Mol Reprod Dev 74(9):1095–1104. https://doi.org/10.1002/mrd.20633

Yi X, Liu F, Gao K, Chen F, Wang Y, Li H, Wang X, Huang Y, Fu H, Zhou W, Fan JB, Wang S, Gao Y (2022) Reconstructable uterus-derived materials for uterus recovery toward efficient live births. Adv Mater 34(8):e2106510. https://doi.org/10.1002/adma.202106510

Yoshimasa Y, Maruyama T (2021) Bioengineering of the uterus. Reprod Sci 28(6):1596–1611. https://doi.org/10.1007/s43032-021-00503-8

Zhang W, Du A, Liu S, Lv M, Chen S (2021) Research progress in decellularized extracellular matrix-derived hydrogels. Regen Ther 18:88–96. https://doi.org/10.1016/j.reth.2021.04.002

Zheng J, Liu Y, Hou C, Li Z, Yang S, Liang X, Zhou L, Guo J, Zhang J, Huang X (2022) Ovary-derived Decellularized extracellular matrix-based bioink for fabricating 3D primary ovarian cells-laden structures for mouse ovarian failure correction. Int J Bioprint 8(3):597. https://doi.org/10.18063/ijb.v8i3.597

Abnormal Offspring Syndrome

Héctor Nava-Trujillo, Callum G. Donnelly, and Rocío Melissa Rivera

Abstract

Abnormal offspring syndrome (AOS), also known as large offspring syndrome, is characterized by overgrowth, limb hyperflexion, abdominal wall defects, umbilical hernia, hardened umbilicus, open urachus, gelatinous material in the peritoneal cavity, hepatic cysts, abdominal ascites, hemorrhage, enlarged heart, kidney, lung, pancreas, and tongue, organ and skull asymmetry, and increase in the length of long bone. Published results suggest that the effect of IVP on birth weight is independent of the culture medium, the bull, or the breed, suggesting that molecular mechanisms beyond those at the DNA sequence level are at play in this phenomenon. In this review, we present the current knowledge about AOS, from its putative inception prior to embryo transfer through pregnancy and parturition, as well as the clinical findings relating to this condition. We and others have demonstrated that AOS is an epigenetic syndrome, and in this chapter, we summarize those findings. Lastly, we discuss the possibilities of early identification during pregnancy.

Keywords

IVP · Overgrowth · Epigenetic · IGF2R · Genomic imprinting

H. Nava-Trujillo · R. M. Rivera (✉)
Division of Animal Sciences, University of Missouri, Columbia, MO, USA
e-mail: hn4qn@umsystem.edu; riverarm@missouri.edu

C. G. Donnelly
Department of Clinical Sciences, College of Veterinary Medicine, Cornell University, Ithaca, NY, USA
e-mail: cgd43@cornell.edu

J. C. Gardón, K. Satué Ambrojo (eds.), *Assisted Reproductive Technologies in Animals Volume 2*, https://doi.org/10.1007/978-3-031-87198-6_7

1 Introduction/Background

The first calves produced by in vitro fertilization (IVP) were born in the 1980s (Brackett et al. 1982; Goto et al. 1988). At the time, it was estimated that with IVP, the impact of a bull on generating offspring could increase by 1000 times compared to artificial insemination (AI) (Brackett et al. 1982). Currently, the worldwide production of in vitro bovine embryos has surpassed 1.6 million (Viana 2023), and this represents an increase of 1160% from 2000 (139,372) (Thibier 2001). This progress reflects the impact of IVP in terms of genetic improvement as well as a possible way to circumvent the adverse consequences of heat stress on cow fertility (Lohuis 1995; Stewart et al. 2011; Baruselli et al. 2020). However, in spite of the measurable genetic improvement of desirable traits and similar pregnancy rates compared to AI (Silva et al. 2022; Crowe et al. 2024), IVP embryos exhibit alterations in the expression of genes and increased rate of programmed cell death (i.e. apoptosis). The aforementioned altered development, together with other shortfalls of the in vitro system, leads to reduced blastocyst rates, reduced developmental competence as well as increased sensitivity to cryopreservation when compared to their in vivo counterparts (Rizos et al. 2002; Sudano et al. 2011). Further, longer gestation, higher risk of abortion or stillbirth, increased rate of dystocia and congenital malformation, and reduced offspring survival have been associated with the use of IVP (Behboodi et al. 1995; Kruip and den Daas 1997; Stacchezzini et al. 1997; van Wagtendonk-de Leeuw et al. 1998; Hasler 2000; Pugh et al. 2003; Lopes et al. 2020; Crowe et al. 2022; Demetrio et al. 2023).

Another recognized consequence of IVP is increased birth weight of offspring (Hasler et al. 1995; Holm et al. 1996; Stacchezzini et al. 1997; Sinclair et al. 1997; Crowe et al. 2022; Beilby et al. 2023). For example, in a population of 16,343 Jersey calves, a recent study reported a 4.16% (~1.1 kg) increase in birth weight in those derived from IVP (Demetrio et al. 2023), while in Brahman, another report recorded that IVP calves were 7.89% heavier than AI calves (Pimenta-Oliveira et al. 2011) and yet others have observed birth weights >33% than expected (Behboodi et al. 1995; Bertolini et al. 2002). The promoting effects of IVP on birth weight have also been reported in sheep (Walker et al. 1992; Holm et al. 1996; Sinclair et al. 1997, 1999). In some cases, the increase in birth weight observed in IVP calves or lambs can reach an extreme/abnormal phenotype characterized by excessive birth weight. For example, Walker et al. 1992 reported a birth weight of 11 kg in a lamb derived from in vitro culture, which was twice the average expected weight. An extreme case of a lamb weighing more than 20 kg was reported (Brown and Radziewic 1998). Because of the extreme overgrowth/overweight, this condition has received such names as "large offspring phenomenon" or "large offspring syndrome" (Walker et al. 1996; Young et al. 1997). Although this syndrome has frequently been associated with the use of IVP or somatic cell nuclear transfer (SCNT), we have reported its spontaneous occurrence in offspring derived from natural service or AI and coined this phenomenon "spontaneous large offspring syndrome" (SLOS) (Li et al. 2019a, 2022b; Rivera et al. 2021).

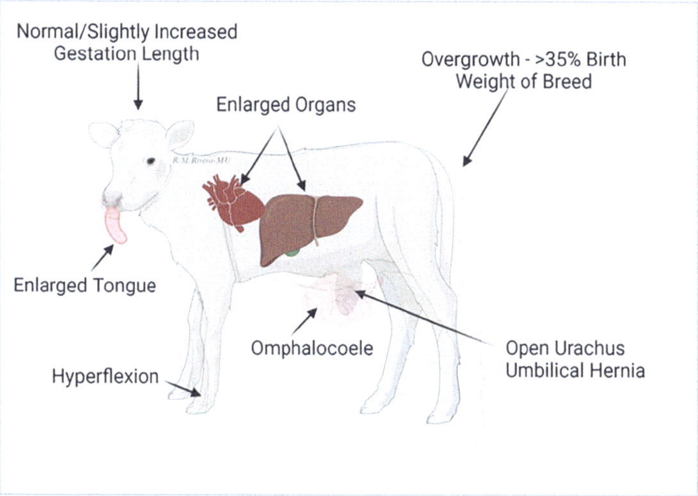

Fig. 1 Abnormal Offspring Syndrome. Depicted are the most common phenotypes observed in this syndrome in calves produced in vitro as well as in vivo. Calves with AOS may present with one or more of these phenotypes. Stillbirth and dystocia due to overgrowth are more severe consequences of AOS

Of note, pathological overgrowth is sometimes accompanied by other abnormalities, including tendon hyperflexion, enlarged tongue, and umbilical hernia (Fig. 1), and as a result, the condition was renamed "abnormal offspring syndrome" (AOS) (Farin et al. 2006). Therefore, AOS is characterized by various phenotypes that may or may not include overgrowth. Importantly, published results suggest that the effect of IVP on birth weight is independent of the culture medium (Lazzari et al. 2002; Siqueira et al. 2017), the bull (Hasler et al. 1995; van Wagtendonk-de Leeuw et al. 1998; Numabe et al. 2000), or the breed (Bertolini et al. 2002; Camargo et al. 2010), suggesting that molecular mechanisms beyond those at the DNA sequence level (i.e., those directly resulting from genetic selection for trait improvement) are at play in this phenomenon. In this review, we present the current knowledge about AOS, from its putative inception prior to embryo transfer through pregnancy and parturition, as well as the clinical interventions required for the treatment of dam and calf because of overgrowth or other abnormalities. Lastly, we discuss the possibilities of early identification during pregnancy.

2 Economic Implications of Fetal Overgrowth for the Cattle Industry

Birth weight is positively correlated with the incidence of dystocia and cesarean section. Calves born following assisted calving have a higher risk of perinatal mortality (Johanson and Berger 2003; Hohnholz et al. 2019), and calves with high birth weight have a higher incidence of diseases and mortality (de Rezende et al. 2009; Buczinski et al. 2011; Weiller et al. 2024). In the USA, dystocia is estimated to be responsible for losses of 220 million USD/year and 185 million USD/year in the dairy and beef industry, respectively (Bellows et al. 2002). In Iran, the cost of one stillbirth varies between 767 and 1189 USD (Mahnani et al. 2018), while in the US dairy industry, the cost of stillbirths is estimated at 125.3 million USD per year (Meyer et al. 2001). In cow-calf operations in Canada, it was reported that heifers and cows with calving assistance have negative profits per calving (−$76.11 and −$120.12 CAD, respectively) (Lucio et al. 2024).

Abnormal offspring syndrome has been associated with a high incidence of abortion and dystocia, stillbirth, longer gestation, difficulty standing/suckling, and reduced perinatal survival (Rérat et al. 2005; Nava-Trujillo et al. 2005; van Wagtendonk-de Leeuw 2006; Li et al. 2019a, 2022b). However, it is important to note that the increase in the duration of gestation observed in IVP pregnancies does not fully explain the excessive increase in birth weight (Walker et al. 1992; van Wagtendonk-de Leeuw et al. 1998). In a case reported in Canada, an IVP-derived female fetus weighing 56 kg was removed from a heifer at day 244 of gestation, approximately 35 days before the due day (MacDonald 2011). The fetus presented subcutaneous edema and enlarged liver and heart; the placenta had only 59 cotyledons, it was thick and edematous, and 85 L of allantoid fluid and 25 kg amniotic fluid were collected. This indicates that overgrowth is a consequence of abnormal conceptus development and not the result of a lengthened pregnancy or number of placentomes. In fact, at 7 months of pregnancy, fetuses derived from AI and IVP have been reported to weigh 15.4 ± 0.8 kg and 18.6 ± 1.1 kg ($p = 0.02$), respectively, with no difference in the number of placentomes (113 ± 7 and 110 ± 11 placentomes, respectively) (Farin and Farin 1995). Moreover, the divergence in the weight observed in AOS can be evident early in the pregnancy (days 56–105) (Chen et al. 2013; Siqueira et al. 2017; Rivera et al. 2022). At day 86 of gestation, Siqueira et al. 2017 reported an AOS fetus weighing 354.2 g, which is 2.28 times the weight of the other IVP fetuses obtained (average weight = 155.1 ± 15.5 g). For these abovementioned reasons, planning the pharmacological advancement of calving is a managerial solution that has been implemented in cows impregnated with IVP embryos in order to avoid the often fatal consequences of excessive birth weight to the calf and the dam (Lopes et al. 2020).

3 Phenotypes of AOS

Although overgrowth is how AOS was first recognized as an adverse consequence of IVP, other abnormal phenotypes may be present in the presence or absence of overgrowth. Enlarged heart, kidney, lung, pancreas, and tongue, hepatic cysts, abdominal ascites, organ and skull asymmetry, increase in the length of long- bone, ears malformations, abdominal wall defects, umbilical hernia, gelatinous material in the peritoneal cavity and hemorrhage have also been reported (Farin and Farin 1995; Young et al. 1999; Sinclair et al. 1999; Chen et al. 2013; Rivera et al. 2022).

All these phenotypes, as well as asymmetry of the pinna, joint deformities, microtia, hardened umbilicus, enlarged head, and limb hyperflexion, have been observed in the spontaneous cases of LOS (in this chapter referred to as SAOS) (Li et al. 2019a, 2022b). In addition, during the AOS open discussion forum at the International Embryo Technology Society meeting in Denver Colorado in January of 2024, other phenotypes such as an open urachus and female reproductive tract abnormalities were reported by embryo transfer practitioners/veterinarians.

4 Incidence and Risk Factors

While the overgrowth phenomenon has been known to occur because of invasive embryo manipulations in cattle for more than 30 years (i.e., in 1991 Willadsen reported that several cloned calves weighed more than 70 kg), the actual current incidence of AOS, especially those resulting from non-invasive IVP, is not known. Further compounding the problem is the fact that different criteria have been used to define the cases, and it is common to ignore the size of the offspring if the calving is unassisted; therefore some cases could be dismissed or underestimated. For example. Bebbere et al. 2013, defined a case as a fetus weighing more than four standard deviations above the control fetuses, and this resulted in 50% (2/4) of AOS at day 80 of pregnancy. In our laboratory, a fetus is categorized as AOS when its weight is above the 97th percentile of the weight of AI fetuses, and the incidence was 21% (9/43) on days 56 of gestation and 26% (7/27) and 37% (17/46) on day 105 (Chen et al. 2013; Rivera et al. 2022, respectively). Using the same criterion, Siqueira et al. 2017 observed at day 86 of gestation, 30% (4/12) of AOS in control IVP fetuses, 50% (4/8) in IVP fetuses treated with Colony-stimulating factor 2 during the embryo culture, and 11% (1/9) in AI-derived fetuses. It should be noted that using the 97th percentile of AI control weight criterion always results in at least one control fetus being categorized as AOS (Chen et al. 2013; Rivera et al. 2022).

In sheep, the percent of oversized fetuses at day 61 of gestation, used as a criterion to define oversize as a weight higher than three standard deviations of the mean of weight in the control group, varied between 0 and 20% depending on the type of media used during the embryo culture. In addition, oversized fetuses were 41.51% heavier than the controls (Sinclair et al. 1999). In the same study, but at day 125 of gestation, the incidence of oversized fetuses varied between 0% and 69.2%, and the oversized fetuses were 41.13% heavier than the controls (Sinclair et al. 1999). Additionally, this study reported that oversized fetuses at days 61 and 125 of gestation presented enlargement of the liver, heart, kidneys, and muscle. In another study, 25% (12/48) of fetuses (day 125 of gestation) were reported as AOS, and those fetuses weighed at least 5.5 kg, 22% more than the largest control fetus (Young et al. 2001), while (Carolan et al. 1998) observed an incidence of overgrowth of 44.73% (17/38). Moreover, the incidences of AOS in cloned sheep and cattle have been reported to be 31.3% (5/16) (Rutigliano et al. 2017) and 58.33% (7/12) (Mangiavacchi et al. 2022), respectively.

Wilson et al. (1995), observed that birth weight and its variation were greater in clones than IVP calves which in turn was greater than AI calves suggesting that there is a cumulative effect on birth weight outcome based on the number and invasiveness of embryo manipulation prior to embryo transfer. In fact, a recent study where IVP embryos were biopsied and cryopreserved (Agerholm et al. 2023) showed this to be the case. In that study, the variance of birth weight was higher ($p = 0.037$) in those calves derived from IVP biopsied and cryopreserved (range: 39–54 kg) when compared to those derived from embryos without biopsy and transferred fresh (range: 31–44 kg). Additionally, in the group of calves derived from biopsied and cryopreserved- embryos, the incidence of calves weighing more than 50 kg at birth was higher than those derived from biopsied and no cryopreserved or no biopsied no cryopreserved embryos (60%, 9.52%, and 0%, respectively) (Agerholm et al. 2023). This is in accordance with an early report that showed that calves derived from IVP frozen embryos were 1.2 kg heavier than those derived from fresh embryos (Merton et al. 1998). Similarly, in mice, it was observed that trophectoderm biopsy and vitrification have a summative negative effect on placental and fetal development and the induction of epigenetic alterations (discussed below), some of these changes seem to be sex- and procedure-specific and to have long-term repercussions during adult life (Ecker et al. 2004; Rhon-Calderon et al. 2024).

As we have seen, most reports of AOS are from the scientific literature (some of them are summarized in Table 1), and there is very little information regarding the incidence of AOS in the commercial embryo transfer industry nor of the occurrence and type of phenotypes observed besides high birth weight. In a 1997 report on calves derived from IVP produced for Holland Genetics and transferred on 120 farms, the percent of calves with birth weight exceeding 50 kg and 60 kg was 34.3% and 5.7%, respectively, and these percents were higher when compared to those calves from the same farms but derived from AI (7.4% and 0.3%) or MOET (multiple ovulation embryo transfer; 10.02% and 4.10%) (Kruip and den Daas 1997). However, much has changed in the media composition and in vitro procedures since 1997, which have improved IVP outcomes (Chronopoulou and Harper 2015; Hansen 2023; Keane and Ealy 2024), and it may be expected that with such improvements, there has been a concomitant elimination of the overgrowth phenotype; anecdotally, this is the case.

However, in 2023, Dr. Eric Mullart presented an abstract at The Association of Embryo Technology in Europe's annual meeting in which data from ten farms in The Netherlands confirmed that IVP-associated overweight continues to be a concern. In that abstract, using the 97th percentile of AI weight criterion to define AOS cases, he reported an incidence of 10% for males and 24% for females calves (Mullaart et al. 2023). It is important to mention that without the information from the commercial embryo transfer industry, it will be impossible to estimate the real impact of IVP and address alternative approaches to improve them and minimize the adverse outcomes in the short and long term. This is a critical aspect because the costs associated with AOS are high and could harm the economic performance of

Table 1 Scientific reports of AOS/overgrown calves derived from IVP

Reference	Criteria to define AOS/ overgrowth, or birth weight mean, or range observed (kg)	AOS/overgrowth cases	AOS/overgrowth frequency in calves from AI-MOET*
Behboodi et al. (1995)	>55 (61–74)	4/4 1/6 (SOVD)**	NA
Hasler et al. (1995)	>68	7%	NA
Sinclair et al. (1995)	>75	2/54	0/28
Schmidt et al. (1996)	58	1/21	NA
Kruip and den Daas (1997)	>50	Worldwide: 31.7%	Worldwide: 10%
	>60	Worldwide: 14.4%	Worldwide: 1%
Agca et al. (1998)	105–142	3/92	NA
McEvoy et al. (1998)	≥60	7/12	0/6
Numabe et al. (2000)	>40	32/376	8/586
Behboodi et al. (2001)	>55	6/164 (in vivo matured/fertilized) 6/25 (in vitro matured/fertilized)	NA
Lazzari et al. (2002)	>50	59.1% (BSA) 50% (serum) 11.8% (SOVD)	AI: 0% MOET: 1.6%
	>60	22.7% (BSA) 50% (serum) 2.9% (SOVD)	AI: 0% MOET: 0%
Heyman et al. (2002)	>59.5	9.5%	NA
Breukelman et al. (2004)	>50 (52–61)	2/25 (MOET) 2/22 (IVP-SOF) 2/14 (IVP-co-culture)	NA
Rérat et al. (2005)	63	1/11	0/11
Gómez et al. (2017))	>50	Fresh: 1/22 Frozen/thawed:3/22 Vitrified/warmed: 6/29 Total:10/73	NA

(continued)

Table 1 (continued)

Reference	Criteria to define AOS/overgrowth, or birth weight mean, or range observed (kg)	AOS/overgrowth cases	AOS/overgrowth frequency in calves from AI-MOET*
Agerholm et al. (2023)	>50	0/17 (no biopsied fresh embryos) 2/21 (biopsied fresh embryos) 3/5 (biopsied and cryopreserved embryos) Total: 5/43	NA
Hattori et al. (2024)	Females: >35.9 Males: >45.2	7/13 (serum) 3/13 (BSA) Total: 10/13	0/5

*AI: artificial insemination; MOET: superovulation; **SOVD: embryos cultured in sheep oviduct. Note: the AOS frequency in the AI or MOET group reported in this table is based on the cutoff for birth weight used for each specific study and represents only body weight and not any other abnormality associated with the syndrome

affected farms (Rivera et al. 2021), and the condition presents animal welfare concerns.

Notably, some species do not appear to be susceptible to fetal overgrowth induced by assisted reproductive technologies (ARTs). This seems to be true for water buffalos and goats (Hufana-Duran et al. 2004; Rodríguez-Dorta et al. 2007; Saliba et al. 2020). Goats may even be resistant to the adverse effects of SCNT (Rutigliano et al. 2017), as only one case has been reported in this species following SCNT. The overgrown kid from a twin pregnancy weighed 7 kg and presented anasarca, cardiac insufficiency, and respiratory arrest, dying soon after birth, while the twin kid weighed 4 kg and was normal (Martins et al. 2016). Alternatively, it is possible that, at least in goat, the severity of the abnormalities does not allow the fetus to develop to term, although this would need to be determined empirically. Together, these findings suggest that the incidence of AOS is affected by the interaction between the technology used to produce offspring (e.g., SCNT, IVP) and the species.

As mentioned above, it should be noted that using the 97th percentile of AI control weight criterion always results in at least one control fetus being categorized as AOS (Chen et al. 2013; Rivera et al. 2022). Therefore, additional methods or criteria, such as molecular analyses, must be developed to accurately identify cases, as has been achieved to some degree for Beckwith-Wiedemann Syndrome (BWS) in humans (Brioude et al. 2018). Moreover, in those farms where the transfer of IVP embryos is the only method of conception, and no AI-derived calves are produced, the criteria that must be used to define AOS cases is a conundrum but could include the average birth weight of the breed. Further, as we will show evidence of late pregnancy losses in our studies, the actual losses due to IVP-induced AOS may never be known as only those that manage to survive their abnormal development are the ones being noticed and/or recorded.

5 Conditions that Promote AOS

The occurrence of AOS has been associated with factors such as in vitro procedures (including cloning), supplementation of the culture medium with serum or BSA, coculture with somatic cells (with the confounding that serum is used as an additive to the culture medium), high levels of rumen-degradable nitrogen in the maternal diet, high levels of progesterone during early embryo development, and asynchrony with the uterine environment (Young et al. 1998; Sinclair et al. 1999; Farin et al. 2014; Rivera et al. 2021; Lazzari et al. 2002).

Oocyte maturation is a crucial step in embryo development, with in-vivo-matured oocytes exhibiting higher quality than their in vitro-matured counterparts (Rizos et al. 2002). In vitro maturation and in vitro fertilization can promote increased body weight (Behboodi et al. 1995, 2001). This was demonstrated in a study by Holm in which lambs from embryos derived from in vitro maturation and fertilization but cultured in vivo were found to have similar weights to those derived from IVP but were heavier than those derived from superovulation (Holm et al. 1996). In another study, calves from in vivo-derived embryos cultured in vitro were significantly lighter than their in vitro counterparts (42.3 ± 0.48 kg vs. 47.0 ± 0.79 kg, $p < 0.05$). These bovine embryos were also associated with a reduced rate of cesarean section (10% vs. 92%, $p < 0.05$) and a lower incidence of birth weight > 45 kg (19 vs. 48%, $p < 0.05$) (Behboodi et al. 2001).

The effects of supplementation with serum and BSA on bodyweight appear to affect all stages of early preimplantation development. For example, exposure to serum during in vitro culture of in vivo-derived zygotes before embryo compaction (i.e., in the first 48 h of culture) causes an increase in the weight of the gravid uterus, placenta, fetus, and cotyledons as well as the crown-rump length of ovine fetuses by day 125 of gestation (Rooke et al. 2007). Work performed in our laboratory showed that supplementing the culture medium with 10% serum from day 5–8 increased the diameter, cell number, rate of development, and number of embryos (Soto-Moreno et al. 2021). Others have shown that serum supplementation during embryo culture (day 0–7) increased the number of blastomeres by day 7 of culture, increased the length and width of day-12 conceptuses, and increased birth weight in calves (Lazzari et al. 2002). It should be noted, however, that AOS has been reported in embryo culture systems that lack BSA and serum (Nava-Trujillo et al. 2005; Siqueira et al. 2017); thus, in vitro culture itself, subtle alterations to the composition of serum-free embryo culture media, the level of nutrients/factors included in the medium (dos Santos et al. 2021) and/or the manipulations experienced by the gametes/embryos during the IVP procedures may be playing a role in the programming of abnormal development.

6 Sex Effect in AOS

Another factor that we believe is important to consider is fetal sex. We have previously reported a higher incidence of AOS among female IVP-derived fetuses on day 105 of gestation (61.53% vs. 27.27% in males (Rivera et al. 2022) and 40.0% vs. 17.7% (Chen et al. 2013)). Recently, Hattori et al. 2024 reported 10 out of 13 cases of overgrowth in calves derived from IVP (76.92%); of those, 100% (6/6) of IVF-derived females were overgrown, while in males, this was 57.14% (4/7). Moreover, it was reported that the incidence of AOS in IVP-derived calves was 24% in females vs. 10% in males and that females and males derived from IVP had the same birth weight indicating that females were growing more than males during gestation as it is usually the males which are heavier at birth in AI or MOET (Mullaart et al. 2023). This last finding was previously reported in Bos indicus IVP-derived calves (de Rezende et al. 2009), while (Behboodi et al. 2001) observed that the birth weight of IVP females was heavier than in vivo generated males. Previously we reported that female fetuses derived from IVP have different growth patterns between days 56 and 105 of gestation, resulting in fetal weights 1.32-fold higher than those derived from AI ($p < 0.01$), while no effect was observed in male fetuses in the same period ($p = 0.99$) (Rivera et al. 2022; Nava-Trujillo and Rivera 2023a). This is in agreement with the findings of a previous study in sheep in which female fetuses were on average heavier than males (Carolan et al. 1998). These observations warrant further investigation to test the hypothesis that fetal sex is a factor affecting the incidence of AOS, with females being more susceptible, as seems to be the case in monozygotic twins discordant for BWS (Weksberg et al. 2002; Bliek et al. 2009). This has been proposed to be due, at least in part, to the onus of inactivating the X-chromosome which occurs in females but not in male preimplantation stage embryos.

7 AOS in the Clinic

In clinical practice, we now recognize AOS as a spectrum of intertwining phenotypes. Calves may show some or all the typical clinical phenotypes. Importantly calves do not always have to show absolute oversized to be included in this syndrome. With the advent of assisted reproductive technologies, AOS is regularly associated with calves derived from these techniques (Young et al. 1998). However, AOS occurring spontaneously in non-ART calves is now well recognized (Li et al. 2019a). Differentiating between AOS and other aetiologies of fetal oversize can be challenging when the calf is not IVP. When determining the cause of these conditions, it is best to categorize them by gestational length; prolonged versus appropriate

There is no established definition for post-term or prolonged gestation in cattle. In humans, births more than 2 weeks over the expected delivery date are considered post-term and are independently associated with reduced neonatal outcomes (Linder et al. 2017). Most cattle do not exceed a gestation of 290 days, and therefore a gestation length exceeding 300 days can be considered prolonged (Foote 1981). Gestation length, even when in the normal range, directly correlates to birth weight (i.e., longer gestation results in greater birth weight) (Zaborski et al. 2009). As such calves from prolonged gestation are almost invariably large (rare exceptions reported), and in general have poor survival outcomes. These calves may also have long hair-coats, over-erupted teeth, and overgrown hoof capsules and have been reported to weigh from 59 to 98 kg. Once identified these calves require delivery by caesarean section to salvage the dam.

Prolonged gestation and failure to initiate parturition has been documented in Ayrshire, Holstein-Friesian, Guernsey, Jersey, Swedish red and white, and Belgian Blue breeds (Holm and Short 1962; Graves et al. 1991; Cornillie et al. 2007; Buczinski et al. 2007). Prolonged gestation in these cases is secondary to adenohypohoseal hypoplasia/aplasia, cerebellar hypoplasia, and adrenal hypoplasia of the fetus, collectively referred to as "pituitary giants." Dysfunctional fetal hypothalamic-pituitary-adrenal (HPA) axes result in a failure to initiate parturition, and consequently fetal oversize. A hereditary component has been suspected in many of these cases; however, putative genetic variants have not yet been identified. Infectious and toxic agents such as Akabane virus, Bluetongue virus, Bovine Viral Diarrhea virus, or Veratum californicum may also prolong gestation through impairment of the fetal HPA axis and should also be considered in these cases.

In our practice, we now regularly encounter AOS calves derived from IVP embryos having prolonged gestation. In this context, we have defined prolonged gestation as reaching 290 days gestation with little or no signs of impending parturition. Prolonged gestation appears to be equally distributed between primiparous and multiparous recipient cows. We have detected a sex bias for female AOS calves in prolonged gestation pregnancies. However, as female-sexed semen is routinely used in IVP embryos, this observation may be a product of more female embryos being transferred. The mechanism(s) as to why these pregnancies are prolonged are unclear but may relate to the underlying epigenetic modifications driving AOS phenotypes in calves.

Fetal giants have also been rarely recognized in calves that are born at normal gestation length. No specific descriptions of these calves have been made, with calves over 59 kg (130 lb) at birth regarded as "fetal giants" (Roberts 1971). Using this weight as a threshold is likely to remove the effect of breed on inclusion as a fetal giant, as this weight is very large regardless of breed. In humans, a neonate is considered large for gestational age if their weight exceeds the 97th percentile for a given gestational age (approximately 4.5 kg at term), regardless of race (Ye et al. 2014). What is considered large for gestational age and a fetal giant in cattle is still yet to be fully defined; however, we have defined it as a calf that is 75% bigger than expected (approximately 55 kg for Angus and 68 kg for Holstein). Extremely large calves, born at term with similar epigenetic features to IVP-induced AOS, have been documented by the authors (Li et al. 2019a, 2022b).

Spontaneous AOS calves have been observed in both Holstein, Brown Swiss, and Charolais with calves of both male and female sex. These calves were all extremely large and required delivery by caesarean section. Calves affected AOS may have a variety of clinical features in addition to macrosomia, including macroglossia (enlarged tongue), omphalocele (umbilical hernia), gastroschisis (failure of body

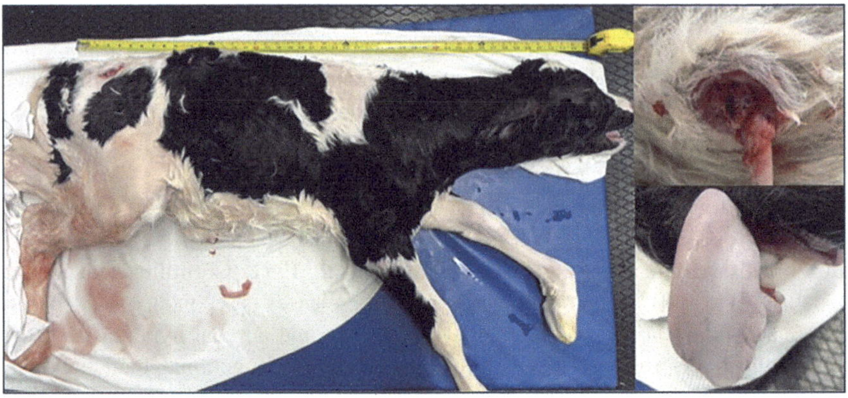

Fig. 2 70 kg AOS Holstein heifer calf delivered by caesarian section. Parturition complicated by a uterine torsion which necessitated the caesarian section. Calf was from an IVP embryo. Calf demonstrates typical associated phenotypes of macroglossia and omphalocele. Calf was deceased prior to delivery and so no cardiovascular data was obtained

wall closure), linear folding of the ears, flexural and angular limb deformities (Fig. 2). A feature that has been recently identified is the persistence of fetal circulation. Many calves retain the anatomic shunts that allow for circulation to bypass the lungs during gestation. Additionally, many of these calves also have defects in the valvular and muscular structures of the heart. The prognosis for survival of AOS calves is guarded. Many die or are euthanized within the first few days of life, especially in cases of persistent fetal circulation.

Estimates for the number of AOS calves born each year are difficult to obtain. Given that ART-associated AOS occurs at a similar rate to BWS in humans, it is estimated that approximately 5–10% of IVP calves born each year in the USA will have AOS (USDA 2018). While the absolute number is unknown, the potential economic impact of these calves is likely to be significant. As these calves invariably result in dystocia requiring resolution by caesarean section it is estimated that AOS costs cattle producers 3–5 million dollars per year (Abdela and Ahmed 2016). This does, however, not account for more mildly affected individuals. Milder case presentation has hampered the estimation of the incidence of similar conditions in humans, as they are rarely investigated, and therefore the true incidence is likely to be much higher. Given the potential for mildly affected individuals to exit in bovine populations, this condition may be more pervasive. There is strong evidence that calf mortality is increasing and this in part may be due to mild AOS. Calf birth weight has a large bearing on neonatal calf outcomes, with increasing birth weight positively associated with an increased risk of calf mortality (Linden et al. 2009). This not only has an economic impact but also has a welfare implication for affected calves. These calves are not only at greater risk of mortality, but they will also be more likely to have a failure of passive transfer of immunity, require more medical interventions, and have indicators of physiologic stress (Uetake 2013). In summary, large for gestational age calves are likely to be encountered by most bovine practitioners. AOS should be considered when evaluating these cases and when counseling clients.

8 Cellular and Molecular and Epigenetic Findings in AOS

8.1 Overview of the Epigenome

We and others have demonstrated that AOS is an epigenetic syndrome, and it is important to provide a short overview of the epigenome to enhance understanding of the subsequent sections. "Epigenetics" is defined as covalent and non-covalent modifications of the DNA and associated histone proteins and the mechanisms by which such modifications influence overall chromatin structure and function. Chromatin is defined as the association of DNA, RNA, and protein. Two types of chromatin exist; open or accessible chromatin referred to as "euchromatin" and closed or inaccessible chromatin which is referred to as "heterochromatin." The epigenetic modification of DNA methylation occurs when the DNA methyltransferases add a methyl (CH3) group donated by S-Adenosyl-L-Methionine onto the fifth carbon of cytosine in a CpG dinucleotide context. This modification is associated with gene silencing when it is found in gene promoters and enhancers. DNA methylation can also prevent the binding of regulatory proteins and may be bound by DNA methyl-biding domain-containing proteins to promote heterochromatin formation and gene silencing. DNA is wrapped around an octamer of proteins known as histones. Histone proteins can be post-translationally modified and these modifications can promote euchromatin or heterochromatin formation. Chromatin is regulated by looping and by compartmentalizing (insulating)sections of DNA by the action of 11-zinc finger protein CCCTC-binding factor (CTCF) together with Cohesin. Beyond chromatin, gene regulation may occur via noncoding RNAs (ncRNAs). These may contain a few dozen nucleotides, for example, micro RNAs (miRNAs) which are regulatory small RNAs that control the cellular level of its target (m)RNA by binding and preventing translation. On the other hand, there are long ncRNAs (lncRNAs) that can act in cis (at the site of transcription) or in trans (elsewhere in the cell or genome) to regulate gene expression and control chromatin configuration. In addition to the locus-specific modifications that control gene expression or chromatin silencing (e.g. transposable elements), two epigenetic mechanisms exist that either control the allele-specific expression of large regions of chromatin (imprinted domains) or silence an entire chromosome in females (X-chromosome inactivation). Imprinted genes occur in clusters through the genome.

These clusters contain both maternally expressed (paternally imprinted) and paternally expressed (maternally imprinted) genes. The correct allelic expression of imprinted genes within a cluster is regulated by an imprinting control region (ICR), a discreet region of differential DNA methylation (DMR), which is acquired during gametogenesis (methylation imprints) and must be maintained after fertilization.

These genes are involved in the growth and development of the fetus and the placenta. Studies in mice have shown that these genes are the reason why parthenogenesis is not possible in mammals (Kawahara et al. 2007, 2008; Li et al. 2018; Wei et al. 2022) and in humans there are several congenital developmental syndromes which are the result of incorrect control of imprinted genes (loss-of-imprinting) such as Beckwith-Wiedemann Syndrome (Online Mendelian Inheritance in Man [OMIM]: 130650), Angelman Syndrome (OMIM:105830), Prader-Willi Syndrome (OMIM:176270, and Silver-Russel Syndrome (OMIM: 180860). The other epigenetic mechanism is X-chromosome inactivation (or "dosage compensation"). In mammals, the female is the homogametic sex and one of the X chromosomes must be randomly silenced to maintain the correct level of gene expression comparable to that of males. X- inactivation occurs during preimplantation development. The silencing of the X-chromosome requires a large cadre of epigenetic modifiers, histone variants and posttranslational modification, small ncRNAs and lncRNAs, chromatin compartmentalization by CTCF and Cohesin, and finally the formation of a heterochromatic knob next to the nuclear membrane known as the Barr body.

8.2 Loss-of-Imprinting in AOS

Beckwith-Wiedemann Syndrome (BWS) is the most common congenital overgrowth syndrome in humans (Brioude et al. 2018). BWS is phenotypically similar to AOS (Li et al. 2019a; Rivera et al. 2021), and its incidence as in cattle, is associated with the use of ART (Halliday et al. 2004; Mussa et al. 2017; Hattori et al. 2019; Hara-Isono et al. 2020). The most commonly recognized epigenetics dysregulations in BWS occur on chromosome 11, which contains two imprinted clusters known as IC2 and IC1. Loss of methylation of IC2 (the ICR of this locus is known as KvDMR1) occurs in more than 50% of BWS patients, while gain of methylation of IC1 (*H19/IGF2* ICR) occurs in around 2.5–10.2% of BWS patients (Carli et al. 2022; Kim et al. 2022). In addition, genetic mutations, which result in secondary epimutations akin loss-of-imprinting in this syndrome, are paternal uniparental disomy of 11p15 in 3.3–27.5% of BWS patients, and mutations of the maternal *CDKN1C* gene, occurring in 2.5–3.3% of BWS patients (Soejima and Higashimoto 2013; Wang et al. 2020; Carli et al. 2022; Kim et al. 2022). The studies about the epigenetic alterations observed in BWS were a point of start to understand the epigenetics dysregulations in AOS, and particular attention has been given to the epigenetics alterations IC2 (KvDMR1) and IC1 (*H19/IGF2*) (Hori et al. 2010; Chen et al. 2013).

In cattle, both KvDMR1 and *H19/IGF2* ICR are located on chromosome 29, and the imprinting patterns seen in humans are conserved with *H19* and *CDKN1C* expressed from the maternal allele while *IGF2* and *KCNQ1OT1* being expressed from the paternally inherited chromosome (Chen et al. 2013). KvDMR1 is the promoter region of the lncRNA *KCNQ1OT1*. KvDMR1 hypomethylation and *KCNQ1OT1* biallelic expression were observed in 1 of 2 AOS IVP cases (Hori et al. 2010). Similarly, biallelic expression of *KCNQ1OT1* resulting from hypomethylation of KvDMR1 was observed by us in 2 of 7 AOS day 105 fetuses (Chen et al. 2013). In that study, we also studied the consequence of biallelic expression of *KCNQ1OT1*, namely the silencing of the clustered maternally expressed imprinted genes *CDKN1C* (a cell cycle regulator) and *PHLDA2* (involved in placenta function) by *KCNQ1OT1*. As is the case for *AIRN*, the lncRNA *KCNQ1OT1* induces heterochromatin formation via Polycomb-induced chromatin contraction (Terranova et al. 2008). To characterize the imprint some more deeply in AOS, we embarked on a sequencing study in which four AOS female fetuses were compared to four AI control female fetuses (Chen et al. 2015). The studies demonstrated that AOS is a multi-locus, tissue-specific loss-of-imprinting condition, with a positive correlation between the number of dysregulated imprinted genes and the fetal weight ($r = 0.90$, $p < 0.05$) (Chen et al. 2015). Further, one or more of the four AOS fetuses in that study had a biallelic expression of *KCNQ1OT1, PEG3, NNAT, NAP1L5, DIRAS3 BEGAIN, MAGEL2, PLAGL1, SNRPN*. A question remains of how early during development is this loss of imprinting programed in the embryo. At least two studies in which the effect of serum supplementation to the culture medium was being characterized identified increased levels of *IGF2* compared with embryos derived from superovulation or AI (Blondin et al. 2000) and decreased DNA methylation of *IGF2R, PEG10*, and *SNRPN* (Lafontaine et al. 2020).

In summary, the most common epimutation identified to date in AOS is loss of methylation at the *IGF2R* ICR on the maternal allele (Young et al. 2001; Chen et al. 2015; Li et al. 2019b), allowing the expression of the normally silent lncRNA *AIRN* which by interacting with histone modifiers (i.e., Polycomb group proteins) silences *IGF2R* on the maternal chromosome. In addition, loss-of methylation at KvDMR1 on the maternal allele and the resulting biallelic expression of *KCNQ1OT1* as well as biallelic expression of other imprinted genes such as *PEG3* is sometimes observed in AOS while gain of methylation at the *H19/IGF2* ICR has never been described in AOS, which perhaps is due to the low number of AOS individuals with a characterized imprintome.

While BWS and AOS have been stamped as loss-of-imprinting syndromes, they also suffer from other epimutations, such as incorrect DNA methylation levels and location as well as global dysregulation of short and long RNAs (Chen et al. 2013, 2015, 2017; Li et al. 2019b, 2022a; Rivera et al. 2022). For example, the expression of genes associated with terms such as "ribosome," "oxidative phosphorylation," "cancer," "extracellular-matrix-receptor interactions," "PI3K-Akt signaling," and "cell-adhesion molecules" pathways are altered AOS fetuses (Chen et al. 2017). Moreover, genes involved in myoblast proliferation, differentiation, migration, and fusion are misregulated in AOS, and their expression is positively correlated with fetal weight (Chen et al. 2017).

microRNAs (miRNAs), transfer RNA (tRNA), and tRNA-derived fragments (tRfs) are small non-coding RNA, which participate in the post-transcriptional regulation of (m)RNA. These small RNAs also have altered expression in AOS, at least in part because of changes in DNA methylation (Li et al. 2019b; Goldkamp et al. 2022a, b). A positive relationship between the weight of AOS fetuses and the number of abnormally expressed miRNAs regulating organ size, cell proliferation, apoptosis, cell survival, tumorigenesis, cell cycle, and cell adhesion was also observed (Li et al. 2019b). The miRNAs *bta-miR-21-5p* and *bta-miR-199a-5p* were found to be upregulated (Li et al. 2019b) and the human analogs of these miRNAs target the cell cycle regulator *CDKN1C*, which has lower levels in AOS fetuses (Hori et al. 2010; Chen et al. 2013) and in BWS (Brioude et al. 2018). In the same study, it was observed that *bta-miR-483* was increased in muscle tissue of AOS fetuses (Li et al. 2019b), and in humans, the levels of the same miRNA are positively correlated with the expression level of *IGF2* (Veronese et al. 2010), which is increased in muscle tissue of AOS fetuses (Chen et al. 2017). Further, several miRNAs belonging to the imprinted cluster *DLK1-DIO3* (Callipyge), a regulator of muscle growth, are downregulated in tongue and muscle tissues of AOS fetuses and individuals with BWS, and this was associated with DNA hypermethylation within the cluster (Li et al. 2019b). In addition, some tRNAs, which play an essential role in the translation of mRNA into proteins, are differentially expressed in AOS compared with AI controls (Goldkamp et al. 2022a). Further, tRFs produced from the processing of tRNA may regulate the etiopathogenesis of AOS (Goldkamp et al. 2022b). Misregulation of tRFs identified in AOS can affect both imprinted and non-imprinted genes, impacting metabolic pathways such as those related to survival, birth weight, growth, muscle physiology, abdominal wall defects, and tumorigenesis (Goldkamp et al. 2022b).

Normal chromosomal architecture includes topologically associating domains (TADs) (Li et al. 2022a). TADs represent a level of spatial organization of the chromosome and are self-interacting genomic regions, which can participate in the regulation of imprinted and non-imprinted gene expression in cis. In addition, chromosomal regions may be regulated by or regulate genes in far regions of the same chromosome (far-cis) or even other chromosomes (trans). We identified changes in chromosome topology associated with AOS (Li et al. 2022a). As mentioned earlier, the imprinted status of the *IGF2R* domain is regulated by an ICR. Normally, the *IGF2R* ICR is methylated on the maternal chromosome and unmethylated on the paternal chromosome. While loss of methylation on the maternal chromosome has long been recognized to play a role in the etiology of AOS (Young et al. 2001), it was not understood how this loss of methylation promotes AOS beyond silencing *IGF2R*. Our recent genome-wide characterization of chromosomal interactions with *IGF2R* ICR and ICR2 in AOS (Li et al. 2022a) revealed that loss of methylation of the *IGF2R* ICR on the maternal chromosome creates a binding site for the architectural protein CTCF, resulting in both gain and loss of contacts within and between chromosomes. More surprisingly, we identified more incorrect contacts and consequently altered gene expression in the normally unmethylated paternal allele, which is thought to be inert in the etiology of LOS (Li et al. 2022a). This suggests that the paternal allele also plays a role in the etiology of AOS. Additionally, we identified conformational changes relating to the dysregulation of genes associated with cell proliferation (*PEG10*), tumorigenesis (*LOC535280, EPHA5, RGS17*), and muscle and bone malformation (*MMP2*), among other processes. Lastly, our most recent work (Li et al. 2023), using HiC, methylome, and transcriptome sequencing technologies in fibroblast-derived from AOS, identified that DMR in AOS showed significant enrichment for CTCF binding sites. Additionally, altered chromosome compartments (i.e., euchromatin and heterochromatin) in AOS were positively correlated with gene expression changes, and the promoters of differentially expressed genes showed significant enrichment for DMRs, and differential TADs. Importantly, altered DNA methylation was identified in several *HOX* gene clusters, transcriptional regulators involved in conferring axial positional information to emerging embryonic tissues from the endoderm, mesoderm, and ectoderm germ layers (Mallo et al. 2010).

8.3 Spontaneous AOS

We were the first to report the spontaneous occurrence of phenotypic AOS (Li et al. 2019a). However, data were absent on whether SAOS was epigenetically similar to IVP-AOS. Therefore, we embarked on a study to deeply characterize the DNA epigenome of calves described as SAOS and compared it to the methylome of IVP-induced AOS (fetuses and calves) and also that of control animals (fetuses and calves). Overall we found congruency in DNA methylation epimutations between types of AOS (Li et al. 2022b).

9 Inherited Epimutations and AOS

The factors promoting the occurrence of these epimutations in SAOS are unknown; however, a partial parental contribution (inherited epimutations) cannot be ruled out.

Our recent case study of a sire, dam, and male full sibling of an SAOS male calf, pinpointed AOS-associated DNA methylation epimutations inherited from both parents, with a higher contribution from the dam (Li et al. 2022b). The fact that the full sibling was of normal size at birth, despite inheriting some of the methylation epimutations, indicates that a threshold of epimutations exists for the development of in utero overgrowth (Li et al. 2022b). Future work is needed to corroborate this hypothesis.

10 The Role of the Placenta in AOS

The placentas of cows carrying pregnancies derived from IVP exhibit phenotypic and functional differences to those of cows impregnated by AI, and extra-embryonic membranes are reportedly shorter during early gestation (days 26–35) in such pregnancies (Pedersen et al. 2020). During the last trimester, studies have observed differences in hydro-allantois (van Wagtendonk-de Leeuw et al. 1998) and in the size, thickness, and number of placentomes. Furthermore, the placentomes of IVP pregnancies have increased weight, decreased surface area, decreased volume density of fetal villi, and increased proportion of blood vessels, and the volume density of the caruncular endometrium is increased (Farin and Farin 1995; Bertolini et al. 2002;

Miles et al. 2004). Whether these differences in placental morphology have a role in the etiopathogenesis of AOS in the context of IVP is controversial (Bertolini and Anderson 2002), and some studies have failed to observe such differences (Constant et al. 2006; Pedersen et al. 2020). For example, Miles et al. 2005 reported culture-system-specific effects on placental weight, efficiency, fluid content, number of placentomes, number of pycnotic cells, and density of blood vessels at day 70, and these placental abnormalities were not correlated with differences in fetal weight. Nonetheless, the Canadian report mentioned above would suggest that an abnormal placental could drive overgrowth in AOS (MacDonald 2011). However, an abnormal placenta alone does not explain other abnormal phenotypes observed in AOS such as body-wall defects, hemi hyperplasia (body asymmetry), and isolated organ overgrowth, which genetically modified mouse models have shown to be the result, at least in part, of improper imprinted gene regulation (Eggenschwiler et al. 1997).

Differences have been observed between the transcriptome of placentas from IVP pregnancies and those from AI pregnancies (Chen et al. 2017). However, there is little information about epigenetic dysregulation in the placenta in the context of IVP-associated AOS. A recent report demonstrated that the abundance of *IGF2* and *IGF2R* transcripts at day 90 was lower in placentas from IVP pregnancies than in embryos flushed after superovulation (Willhelm et al. 2021). In addition, Chen et al. 2013 observed that *H19, IGF2*, and *CDKN1C* were imprinted in the placenta tissue of fetuses derived from AI and IVP, but in two of the seven IVP-AOS fetuses, and *KCNQ1OT1* was biallelically expressed in the placenta. Interestingly, the placenta of the largest control fetus was hypomethylated at KvDMR1, demonstrating the importance of this imprinted region in fetal growth. We reported that in IVP fetuses at day 105 of gestation, there was a negative correlation between fetal weight and the expression of *IGF2R* in cotyledon ($r = -0.52$, $P = 0.0004$) but not in intercotyledon; however, this correlation was only true for male fetuses (-0.54, $P = 0.0021$) (Nava-Trujillo and Rivera 2023b). In addition, day 105 IVP fetuses with AOS had a 29% reduction in cotyledon *IGF2R* when compared to the IVP normal fetuses ($P < 0.008$) (Nava-Trujillo and Rivera 2023b).

11 IGF Signaling and Growth in AOS

The first study trying to establish the mechanism related to the occurrence of AOS was carried out by Maxfield et al. 1997, who observed an increase in the weight of plantaris, gastrocnemius, and tibialis anterior muscles in fetuses derived from in vitro culture and found that that increase was because to hypertrophy of primary fibers and hyperplasia of secondary fibers. Then Reinders et al. 1998, tested the hypothesis that the development of diabetes type II after the transfer of IVP embryo could in part explain the "large calf syndrome," but they didn't find a relation.

In bovine, a positive correlation between fetal weight and the abundance of hepatic insulin-like growth factor 2 (*IGF2*) transcripts has been reported (Willhelm et al. 2021). In 1999, Young et al. observed in sheep that large offspring derived from in vitro culture of in vivo fertilized oocytes had higher expression of IGF2 binding protein (*IGFBP2*) in the liver and a higher level of the corresponding protein in plasma. Subsequently, they observed a negative relation between the level of mRNA for insulin- like growth factor 2 receptor (*IGF2R*; the clearance receptor of IGF2) and the weight of fetuses and that overgrown fetuses had a reduction in the expression of this gene and the corresponding protein in the liver, muscle, and plasma (Young et al. 2001). The alteration in the expression of *IGF2R* was related to a loss of methylation in the differentially methylated region 2 (DMR2; the locus' ICR) of the gene *IGF2R*, which occurred in 9 out of 12 (75%) overgrown fetuses (Young et al. 2001). These findings were the first evidence that an epigenetic alteration (i.e., misregulation of an imprinted gene) is related to AOS, and support the early observations of Willadsen et al. (1991) and Wilson et al. (1995), who using cloned embryos noticed that not all the calves derived from the same source presented overgrowth or malformations. The latter observation demonstrates that while the genome of the embryos was identical, they were dissimilarly affected by the procedures indicating that molecular mechanisms of regulation of the genome, namely "the epigenome," differed between the individuals.

IGF2R is an imprinted gene transcribed only from the maternal allele (Barlow et al. 1991). This gene has two DMRs; the differentially methylated region 1 (DMR1) contains the promoter of the gene, whose difference in methylation between maternal (unmethylated) and paternal (methylated) is not enough to repress the expression of *IGF2R* (Stöger et al. 1993). DMR2 is located in the second intron of *IGF2R* and is methylated on the maternal allele and unmethylated on the paternal allele. This DMR also serves as the promoter of the antisense lncRNA *AIRN* which is only expressed from the unmethylated paternal allele (Stöger et al. 1993; Marcho et al. 2015). In addition to *AIRN*'s ability to attract Polycomb silencing (histone modifiers) to the locus (Braceros et al. 2023), *AIRN* blocks the binding of RNA Pol II to the promoter of *IGF2R* and keeps this gene silenced on the paternal chromosome (Sleutels et al. 2002; Seidl et al. 2006). The expression of *AIRN* RNA is stage-dependent, with expression first detected after day 15 in bovine embryos. By day 70 of gestation, IVP fetuses have a lower liver expression of *AIRN* than in vivo fetuses (Farmer et al. 2013). Whether alterations in the expression of *AIRN* are associated with the occurrence of AOS is an unanswered question.

The identified loss of DNA methylation in a region controlling the expression of *IGF2R* in AOS fetuses by Young (Young et al. 2001) agrees with the results from our laboratory (Chen et al. 2015; Li et al. 2019b, 2022a). However, despite the negative correlation between fetal weight and *IGF2R* transcription ($p < 0.012$) and higher levels of *IGF2* mRNA found in day 105 AOS fetuses (Chen et al. 2017), we did not observe an inverse relationship between the transcripts of *IGF2R* and *IGF2* transcripts in the muscle of bovine fetuses aged 56 or 105 days (Nava-Trujillo and Rivera 2023a). This could well be due to the fact that while imprinted regulation maintains the correct allelic expression at this locus it is the protein product of these genes which are involved in the intricate control of fetal growth. At the protein level, IGF2R functions as a negative regulator of the paternally expressed insulin-like growth factor 2 (IGF2), by mediating its internalization and trafficking to the lysosome for degradation (Oka et al. 1985; El-Magd et al. 2014). In mice, knocking out

IGF2R increased the abundance of IGF2 protein due to reduced degradation and this caused fetal overgrowth (Lau et al. 1994). In addition, a reduction in the ability of IGF2R to orchestrate clearance of IGF2 in AOS cannot be discarded as has been demonstrated to be the case in BWS in which the recycling and trafficking of this protein was impaired (Pileggi et al. 2024).

Beyond its function as a clearance receptor, IGF2R is a signaling receptor. For example, in a study in which Leu27IGF2, an analog of IGF2 that has the ability to bind IGF2R but not insulin-like growth factor 1 receptor or the insulin receptor (Beukers et al. 1991), increased mitogenesis and survival in human placental explants with intact IGF2R expression (Harris et al. 2011). In addition, the same IGF2 analog enhanced the functional development of the placenta and nutrient delivery, promoting fetal growth in mice (Sferruzzi-Perri et al. 2008). The signaling role of IGF2R occurs by activating the extracellular signal-regulated kinases (ERK1/2) pathway, as a reduction in the phosphorylation of these proteins was observed in fetoplacental endothelial cells lacking IGF2R (Sandovici et al. 2022). In addition, the downregulation of IGF2R inhibits the IGF2-stimulated ERK1/2 activation (El-Shewy et al. 2007). Nevertheless, the activation of ERK1/2 by IGF2R is not completely understood as IGF2R lacks tyrosine kinase activity, but it could activate it via the G protein pathway (El-Shewy et al. 2007; Amritraj et al. 2012). Unpublished DNA methylation data from our laboratory suggest that molecules within the IGF2 signaling pathways may also be affected in AOS.

12 Early Identification of AOS

AOS has economic implications for farmers, given the greater probability of abortions, dystocia, the need for cesarean sections, and deaths, and the possible repercussions in the future commercialization of animals for breeding programs (Rivera et al. 2021). Moreover, this can be stressful for farmers and veterinarians, who often opt to induce the parturition to prevent negative outcomes (de Rezende et al. 2009; Aguiar et al. 2019; Lopes et al. 2020). Therefore, an early diagnosis of AOS is crucial to maximize the efficiency of the IVP embryo transfer programs and to make decisions whether to terminate the pregnancy early or prepare for calving-related complications. A pre-transfer epigenetic test may represent an alternative method of identifying embryos with epigenetic marks associated with AOS, especially in culture conditions that increase the risk of AOS, such as serum or BSA supplementation. Embryos could be biopsied and frozen, with transfer only for those exhibiting normal epigenetic patterns. If diagnoses are made before the blastocyst stage, embryos with aberrant epigenomes could be discarded before being cryopreserved. However, as we described before, the extra manipulation related to the biopsy process could increase the birth weight and the incidence of AOS (Agerholm et al. 2023).

Identifying maternal markers would be a useful non-invasive method of recognizing AOS. We recently analyzed the transcriptome of leukocytes and found that LOC783838 and PCDH1 could be potential markers for identifying extreme cases of AOS at both 55 and 105 days of gestation, while the transcript levels of *ACTA2, KDM5A, MAN1A2, MIR2376, PRRC2C, RSBN1, S100A14, SRPK2,* and *TTF1* could be references for the identification of extreme cases of AOS at day 105 of pregnancy (Rivera et al. 2022). Recently, Rabaglino et al. 2023 identified a cluster of genes in the heart (related to oxidative phosphorylation and thermogenesis) of fetuses at day 42 of gestation that were positively correlated with the fetal weight and were co-expressed in the maternal blood. Using a statistical model, trained with the expression of these genes in the fetal heart, it was possible to predict the fetal weight through the expression of those genes in the maternal blood ($R2 = 0.93$). These findings open new diagnostic possibilities to study fetal well-being, and without a doubt, they must be considered to try to identify or predict pathologies like AOS. Ultrasound is routinely used for pregnancy diagnosis, fetal sexing, and embryo- fetal viability. However, the utility of predicting the weight of the fetus is unclear. Burns et al. 2018 observed that the width of the placentome at week 7 and the thickness at week 8 correlated positively with birth weight. Between days 260 and 272 of pregnancy, the diameter of the fetal thoracic aorta was positively correlated to birth weight ($r = 0.62$, $p = 0.01$) in control but not in cloned calves (Buczinski et al. 2011). A tendency for a negative association between amniotic sac volume and amniotic fluid volume, measured at day 42 of gestation, with birth weight has been observed (Rotheneder et al. 2022). Whether ultrasound is an applicable approach to identify AOS is controversial Bertolini et al. 2002. Comparing ultrasound measurements taken on days 37 and 58 of gestation, in vitro-derived fetuses, especially in twin pregnancies, exhibited slow growth and reduced crown-rump and amniotic vesicle length (Bertolini et al. 2002). However, the singleton fetuses that were smaller on days 37 and 58 were heavier calves at birth, although none were categorized as AOS. Contrarily, no differences in the biparietal diameter of cranium or cross-section of the abdomen were observed between days 40 and 110 or between fetuses from multiple ovulation, IVP cultured in SOF, or with somatic cell co-culture (Breukelman et al. 2004). The variability of these results may be explained, at least in part, by the use of different genetic groups and by taking measurements on different days. However, in cloned fetuses, it was suggested that the ultrasonographic diagnosis of AOS must include the evaluation of the fetal vesicle as the biparietal diameter as these were larger at day 62 (Chavatte-Palmer et al. 2006). As part of our effort for the early identification of AOS, we carried out ultrasounds on day 55 of pregnancy (Rivera et al. 2022). We used those ultrasound measurements to determine if it was possible to identify overgrowth on that day of fetuses collected on days 56 and 105 of gestation. We measured abdominal diameter, abdominal height, crown-rump length, head length, thoracic height, thoracic diameter, and biparietal diameter. We found that not one single measurement was useful for this purpose; however, the product of six measurements (abdominal diameter, abdominal height, crown-rump length, head length, thoracic height, and thoracic diameter) on day 56 and day 105 was positively correlated with fetal weight.

Surprisingly, even though biparietal diameter had been suggested as a possible measurement to identify weight, that was not the case in our study. More interestingly, it was possible to predict correctly the two most extreme cases of AOS, whose weights were more than twice the average weight of the controls. Further studies involving large numbers of animals with different categories of recipients (heifer vs. mature cows) are required to improve the accuracy of predicting AOS in a sex and breed-specific manner.

13 Concluding Remarks

After more than 40 years of intensive research and advances and the positioning as a tool of worldwide massive use in livestock, the in vitro production of embryos still has shortfalls and IVP embryos do not yet faithfully recapitulate the natural state. Further work is necessary to eliminate the introduction of epimutations due to culture system conditions and gamete/embryo manipulations. IVP-derived fetuses can develop abnormal phenotypes that compromise the success of the pregnancy, the health and future performance of the offspring and the dam, and the success of the embryo transfer program at the farm level, and this has important economic implications.

AOS is a developmental epigenetic defect promoted by the in vitro manipulation of gametes and/or embryos. Although this pathology is considered rare and a condition that was dealt with when serum was no longer a main additive of the culture medium, recent reports provide evidence of its continued existence in the commercial embryo transfer industry (Mullaart et al. 2023). AOS could pose a challenge to IVP embryo programs, especially as the epigenetic alterations associated may be inherited, as has been previously suggested (Li et al. 2022b). Current genetic programs assume the use of both semen and oocytes from donors derived from IVP and it is possible that epimutations of regions controlling the growth and development of the fetus are accumulating in the cattle population.

While it appears from conversations with those in the embryo transfer industry that the extreme of gigantism (and "large" offspring syndrome) is not observed as often as in the past, there is sufficient evidence that supports, that as a whole, IVP-produced calves are heavier than in vivo produced embryos. This alone is a good motivation to do in depth molecular characterizations that may shed light on possible labile regions of chromatin that when reprogramed by ART promote excessive growth while in utero.

In the mid-1990s, reports alerted of very high birth weight/size in offspring produced by IVP, and the condition was termed "large offspring syndrome." Since it was observed that IVP offspring often presented with a sundry of abnormalities beyond overgrowth, the condition was then termed "abnormal offspring syndrome." In this chapter, we provide ample evidence that this is an epigenetic condition that occurs spontaneously and is promoted by embryo manipulation and culture systems that incorrectly program the embryo to undergo faulty development often resulting in death during the fetal stage (i.e., fetal loss; Fig. 3). Therefore, we propose the use of Abnormal Development Syndrome (ADS), as a term that better encompasses the adverse effects of ART in cattle.

Fig. 3 (continued) Top Panel: The bar graph show a summary of reports documenting late pregnancy loss. FTAI: fixed time artificial insemination, ET: embryo transfer, AI: artificial insemination; IVP: in vitro embryo; MOET: multiple ovulation and embryo transfer; EEM (early embryo mortality) = 30-100 days of gestation. Bottom Figure: Variety of IVP embryos collected on day 56 of gestation. The top left picture is a AI fetus for comparison. Each square of the mat where the fetuses rest = 2.54 cm

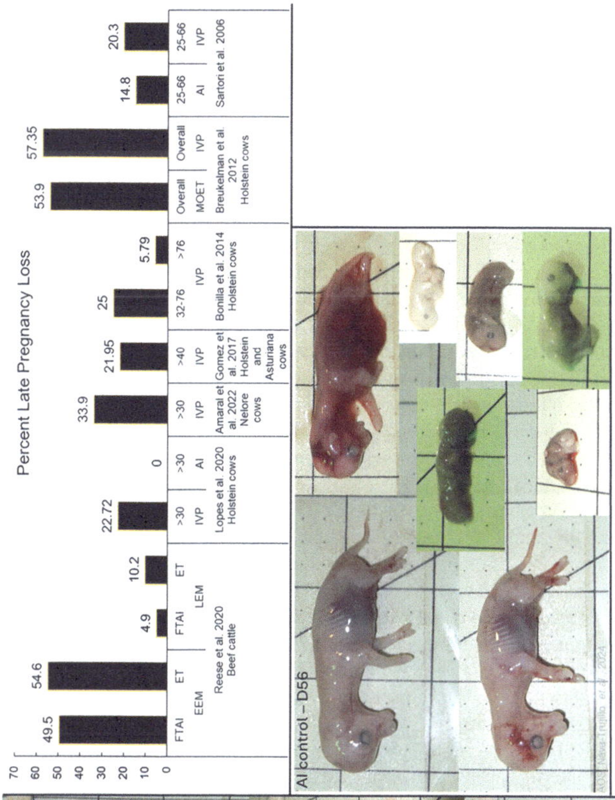

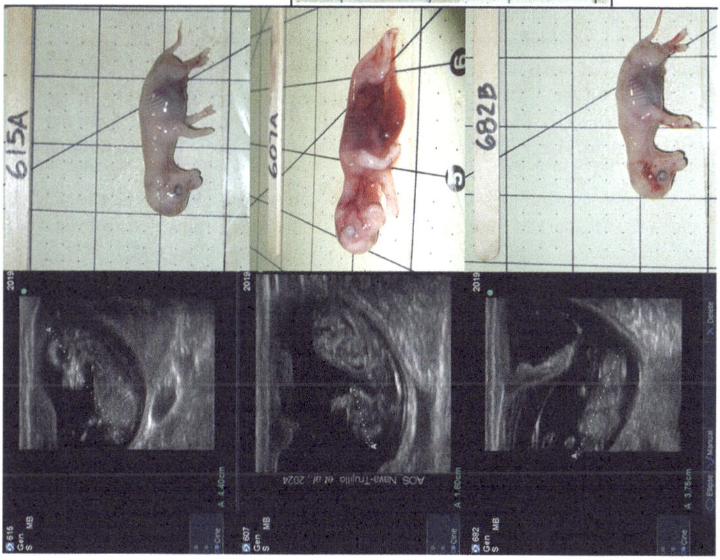

Fig. 3 Examples of fetal loss in IVP embryos. Left panels: Day 55 fetal ultrasounds showing normal appearance of fetuses. The picture on the right is the corresponding fetus which was collected on day 56 of gestation. Fetus 615A is an AI control fetus. IVP fetus 607 had an open body cavity at collection and IVP fetus 682 was hemorrhagic (including the placenta – not shown).

Financial Support Statement This project was supported by the Agriculture and Food Research Initiative competitive grants No. AFRI 2018-67015-27598 and by a University of Missouri Graduate School "Robert E. Waterston Doctoral Fellowship" to HNT.

References

Abdela N, Ahmed W (2016) Risk factors and economic impact of dystocia in dairy cows: a systematic review. J Reprod Infert 7:63–74

Agca Y, Monson RL, Northey DL, Peschel DE, Schaefer DM, Rutledge JJ (1998) Normal calves from transfer of biopsied, sexed and vitrified IVP bovine embryos. Theriogenology 50(1):29–145

Agerholm JS, Madsen SE, Krogh AKH, Najafzadeh V, Secher JB-M (2023) Health assessment of Holstein calves born after in vitro fertilization, biopsy-based genotyping at the blastocyst stage and subsequent embryo transfer. Theriogenology 211:76–83

Aguiar TS, Moreira PSA, Moreti BS, Berber GCM, Pinto-Neto A, Berber RCA (2019) Strategy to improve the parturition ability in crossbred recipients implanted with embryos produced in vitro. Sci Electron Arch 12:81–84

Amaral TF, de Grazia JGV, Martinhao LAG, De Col F, Siqueira LGB, Viana JHM, Hansen PJ (2022) Actions of CSF2 and DKK1 on bovine embryo development and pregnancy outcomes are affected by composition of embryo culture medium. Sci Rep 12:7503

Amritraj A, Posse De Chaves EI, Hawkes C, MacDonald RG, Kar S (2012) Single-transmembrane domain IGF-II/M6P receptor: potential interaction with G protein and its association with cholesterol- rich membrane domains. Endocrinology 153:4784–4798

Barlow DP, Stöger R, Herrmann BG, Saito K, Schweifer N (1991) The mouse insulin-like growth factor type-2 receptor is imprinted and closely linked to the Tme locus. Nature 349(6304):84–87. https://doi.org/10.1038/349084a0

Baruselli PS, Ferreira RM, Vieira LM, Souza AH, Bó GA, Rodrigues CA (2020) Use of embryo transfer to alleviate infertility caused by heat stress. Theriogenology 155:1–11

Bebbere D, Bauersachs S, Fürst RW, Reichenbach H-D, Reichenbach M, Medugorac I, Ulbrich SE, Wolf E, Ledda S, Hiendleder S (2013) Tissue-specific and minor inter-individual variation in imprinting of IGF2R is a common feature of Bos taurus Concepti and not correlated with fetal weight. PLoS One 8:e59564

Behboodi E, Anderson GB, BonDurant RH, Cargill SL, Kreuscher BR, Medrano JF, Murray JD (1995) Birth of large calves that developed from in vitro-derived bovine embryos. Theriogenology 44:227–232

Behboodi E, Groen W, Destrempes MM, Williams JL, Ohlrichs C, Gavin WG, Broek DM, Ziomek CA, Faber DC, Meade HM, Echelard Y (2001) Transgenic production from in vivo-derived embryos: effect on calf birth weight and sex ratio. Mol Reprod Dev 60:27–37

Beilby KH, Kneebone E, Roseboom TJ, van Marrewijk IM, Thompson JG, Norman RJ, Robker RL, Mol BWJ, Wang R (2023) Offspring physiology following the use of IVM, IVF and ICSI: a systematic review and meta-analysis of animal studies. Hum Reprod Update 29:272–290

Bellows DS, Ott SL, Bellows RA (2002) Review: cost of reproductive diseases and conditions in Cattle. Prof Anim Sci 18:26–32

Bertolini M, Anderson GB (2002) The placenta as a contributor to production of large calves. Theriogenology 57:181–187

Bertolini M, Mason JB, Beam SW, Carneiro GF, Sween ML, Kominek DJ, Moyer AL, Famula TR, Sainz RD, Anderson GB (2002) Morphology and morphometry of in vivo- and in vitro-produced bovine concepti from early pregnancy to term and association with high birth weights. Theriogenology 58:973–994

Beukers MW, Oh Y, Zhang H, Ling N, Rosenfeld RG (1991) Insulin-like growth factor ii is highly selective for the type-ii igf receptor in binding, cross-linking and thymidine incorporation experiments. Endocrinology 128:1201–1203

Bliek J, Alders M, Maas SM, Oostra R-J, Mackay DM, van der Lip K, Callaway JL, Brooks A, van't Padje S, Westerveld A, Leschot NJ, Mannens MM (2009) Lessons from BWS twins: complex maternal and paternal hypomethylation and a common source of haematopoietic stem cells. Eur J Hum Genet 17:1625–1634

Blondin P, Farin PW, Crosier AE, Alexander JE, Farin CE (2000) In vitro production of embryos alters levels of insulin-like growth factor-II messenger ribonucleic acid in bovine fetuses 63 days after transfer. Biol Reprod 62:384–389

Bonilla L, Block J, Denicol AC, Hansen PJ (2014) Consequences of transfer of an in vitro-produced embryo for the dam and resultant calf. J Dairy Sci 97:229–239

Braceros AK, Schertzer MD, Omer A, Trotman JB, Davis ES, Dowen JM, Phanstiel DH, Aiden EL, Calabrese JM (2023) Proximity-dependent recruitment of Polycomb repressive complexes by the lncRNA Airn. Cell Rep 42:112803

Brackett BG, Bousquet D, Boice ML, Donawick WJ, Evans JF, Dressel MA (1982) Normal development following in vitro fertilization in the cow. Biol Reprod 27:147–158

Breukelman SP, Reinders JMC, Jonker FH, de Ruigh L, Kaal LMTE, van Wagtendonk-de Leeuw AM, Vos PLAM, Dieleman SJ, Beckers JF, Perényi Z, Taverne MAM (2004) Fetometry and fetal heart rates between Day 35 and 108 in bovine pregnancies resulting from transfer of either MOET, IVP-co-culture or IVP-SOF embryos. Theriogenology 61:867–882

Breukelman SP, Perényi Z, Taverne MAM, Jonker H, van der Weijden GC, Vos PLAM, de Ruigh L, Dieleman SJ, Beckers JF, Szenci O (2012) Characterisation of pregnancy losses after embryo transfer by measuring plasma progesterone and bovine pregnancy-associated glycoprotein-1 concentrations. Vet J 194:71–76

Brioude F, Kalish JM, Mussa A, Foster AC, Bliek J, Ferrero GB, Boonen SE, Cole T, Baker R, Bertoletti M, Cocchi G, Coze C, De Pellegrin M, Hussain K, Ibrahim A, Kilby MD, Krajewska-Walasek M, Kratz CP, Ladusans EJ, Lapunzina P, Le Bouc Y, Maas SM, Macdonald F, Õunap K, Peruzzi L, Rossignol S, Russo S, Shipster C, Skórka A, Tatton-Brown K, Tenorio J, Tortora C, Grønskov K, Netchine I, Hennekam RC, Prawitt D, Tümer Z, Eggermann T, Mackay DJG, Riccio A, Maher ER (2018) Clinical and molecular diagnosis, screening and management of Beckwith–Wiedemann syndrome: an international consensus statement. Nat Rev Endocrinol 14:229–249

Brown BW, Radziewic T (1998) Production of sheep embryos in vitro and development of progeny following single and twin embryo transfers. Theriogenology 49:1525–1536

Buczinski S, Bélanger A-M, Fecteau G, Roy J-P (2007) Prolonged gestation in two Holstein cows: transabdominal ultrasonographic findings in late pregnancy and pathologic findings in the fetuses. J Vet Med A Physiol Pathol Clin Med 54:624–626

Buczinski S, Fecteau G, Lefebvre RC, Smith LC (2011) Assessment of fetal well-being in cattle by ultrasonography in normal, high-risk, and cloned pregnancies. Can Vet J 52:136–141

Burns BM, Hiendleder S, Laing AR, Fordyce G, Herring AD (2018) Ultrasonographic measurements in first trimester concepti identify predictors of birth weight and postnatal development in cattle1. J Anim Sci 96:4186–4194

Camargo LSA, Freitas C, de Sa WF, de Moraes FA, Serapiao RV, Viana JHM (2010) Gestation length, birth weight and offspring gender ratio of in vitro-produced Gyr (Bos indicus) cattle embryos. Anim Reprod Sci 120:10–15

Carli D, Operti M, Russo S, Cocchi G, Milani D, Leoni C, Prada E, Melis D, Falco M, Spina J, Uliana V, Sara O, Sirchia F, Tarani L, Macchiaiolo M, Cerrato F, Sparago A, Pignata L, Tannorella P, Cardaropoli S, Bartuli A, Riccio A, Ferrero GB, Mussa A (2022) Clinical and molecular characterization of patients affected by Beckwith-Wiedemann spectrum conceived through assisted reproduction techniques. Clin Genet 102(4):314–323. https://doi.org/10.1111/cge.14193

Carolan C, McEvoy TG, Staines ME, Sinclair KD, Robinson JJ, Broadbent PJ (1998) Variation in the expression of fetal oversize in day 125 ovine fetuses following transfer of in vitro cultured embryos. Theriogenology 49:229

Chavatte-Palmer P, de Sousa N, Laigre P, Camous S, Ponter AA, Beckers J-F, Heyman Y (2006) Ultrasound fetal measurements and pregnancy associated glycoprotein secretion in early pregnancy in cattle recipients carrying somatic clones. Theriogenology 66:829–840

Chen Z, Robbins KM, Wells KD, Rivera RM (2013) Large offspring syndrome: a bovine model for the human loss-of-imprinting overgrowth syndrome Beckwith-Wiedemann. Epigenetics 8:591–601

Chen Z, Hagen DE, Elsik CG, Ji T, Morris CJ, Moon LE, Rivera RM (2015) Characterization of global loss of imprinting in fetal overgrowth syndrome induced by assisted reproduction. Proc Natl Acad Sci 112:4618–4623

Chen Z, Hagen DE, Ji T, Elsik CG, Rivera RM (2017) Global misregulation of genes largely uncoupled to DNA methylome epimutations characterizes a congenital overgrowth syndrome. Sci Rep 7:12667

Chronopoulou E, Harper JC (2015) IVF culture media: past, present and future. Hum Reprod Update 21:39–55

Constant F, Guillomot M, Heyman Y, Vignon X, Laigre P, Servely JL, Renard JP, Chavatte-Palmer P (2006) Large offspring or large placenta syndrome? Morphometric analysis of late gestation bovine placentomes from somatic nuclear transfer pregnancies complicated by hydrallantois. Biol Reprod 75:122–130

Cornillie P, Van den Broeck W, Simoens P (2007) Prolonged gestation in two Belgian blue cows due to inherited adenohypophyseal hypoplasia in the fetuses. Vet Rec 161:388–391

Crowe AD, Rabaglino MB, Sanchez JM, Moore SG, McDonald M, Lonergan P, Butler ST (2022) 134 Comparison of the effect of artificial insemination and in vitro embryo production on gestation length, calf birthweight, and calving difficulty. Reprod Fertil Dev 35:195–195

Crowe AD, Sánchez JM, Moore SG, McDonald M, Rodrigues R, Morales MF, Orsi de Freitas L, Randi F, Furlong J, Browne JA, Rabaglino MB, Lonergan P, Butler ST (2024) Fertility in seasonal-calving pasture-based lactating dairy cows following timed artificial insemination or timed embryo transfer with fresh or frozen in vitro–produced embryos. J Dairy Sci 107:1788–1804

de Rezende LFC, Zúccari CESN, da Costa e Silva EV, de Rezende CRL, Nasser LFT, de Abreu UGP (2009) Acetato De Triancinolona Como Pré-Indutor Do Parto De Receptoras Deembrião Nelore Produzidos In Vitro. Braz Anim Sci 10:59–67

Dematawena CMB, Berger PJ (1997) Effect of dystocia on yield, fertility, and cow losses and an economic evaluation of dystocia scores for Holsteins. J Dairy Sci 80:754–761

Demetrio DGB, Baumgartner T, Amorim DJ, Oliveira M, Rosales ME, Demetrio CGB, Santos RM (2023) 16 Stillbirth, gestation length, and birth weights from Jersey calves originated from artificial insemination vs in vitro embryo transfer. Reprod Fertil Dev 36:157–157

dos Santos ÉC, da Fonseca Junior AM, de Lima CB, Ispada J, da Silva JVA, Milazzotto MP (2021) Less is more: reduced nutrient concentration during in vitro culture improves embryo production rates and morphophysiology of bovine embryos. Theriogenology 173:37–47

Ecker DJ, Stein P, Xu Z, Williams CJ, Kopf GS, Bilker WB, Abel T, Schultz RM (2004) Long-term effects of culture of preimplantation mouse embryos on behavior. Proc Natl Acad Sci 101:1595–1600

Eggenschwiler J, Ludwig T, Fisher P, Leighton PA, Tilghman SM, Efstratiadis A (1997) Mouse mutant embryos overexpressing IGF-II exhibit phenotypic features of the Beckwith–Wiedemann and Simpson–Golabi–Behmel syndromes. Genes Dev 11:3128–3142

El-Magd MA, Abo-Al-Ela HG, El-Nahas A, Saleh AA, Mansour AA (2014) Effects of a novel SNP of IGF2R gene on growth traits and expression rate of IGF2R and IGF2 genes in gluteus medius muscle of Egyptian buffalo. Gene 540:133–139

El-Shewy HM, Lee M-H, Obeid LM, Jaffa AA, Luttrell LM (2007) The insulin-like growth factor type 1 and insulin-like growth factor type 2/Mannose-6-phosphate receptors independently regulate ERK1/2 activity in HEK293 cells. J Biol Chem 282:26150–26157

Farin PW, Farin CE (1995) Transfer of bovine embryos produced in vivo or in vitro: survival and fetal development. Biol Reprod 52:676–682

Farin PW, Piedrahita JA, Farin CE (2006) Errors in development of fetuses and placentas from in vitro-produced bovine embryos. Theriogenology 65:178–191

Farin CE, Barnwell CV, Farmer WT (2014) Abnormal offspring syndrome. In: Hopper RM (ed) Bovine reproduction. Wiley, Hoboken, NJ, pp 620–638

Farmer WT, Farin PW, Piedrahita JA, Bischoff SR, Farin CE (2013) Expression of antisense of insulin-like growth factor-2 receptor RNA non-coding (AIRN) during early gestation in cattle. Anim Reprod Sci 138:64–73

Foote RH (1981) Factors affecting gestation length in dairy cattle. Theriogenology 15:553–559

Goldkamp AK, Li Y, Rivera RM, Hagen DE (2022a) Characterization of tRNA expression profiles in large offspring syndrome. BMC Genomics 23:273

Goldkamp AK, Li Y, Rivera RM, Hagen DE (2022b) Differentially expressed tRNA-derived fragments in bovine fetuses with assisted reproduction induced congenital overgrowth syndrome. Front Genet 13:1055343

Gómez E, Carrocera S, Uzbekova S, Martín D, Murillo A, Alonso-Guervós M, Goyache F, Muñoz M (2017) Protein in culture and endogenous lipid interact with embryonic stages in vitro to alter calf birthweight after embryo vitrification and warming. Reprod Fertil Dev 29:1932–1943

Goto K, Kajihara Y, Kosaka S, Koba M, Nakanishi Y, Ogawa K (1988) Pregnancies after co-culture of cumulus cells with bovine embryos derived from in vitro fertilization of in vitro matured follicular oocytes. Reproduction 83:753–758

Graves TK, Hansel W, Krook L (1991) Prolonged gestation in a Holstein cow: adenohypophyseal aplasia and skeletal pathology in the offspring. Cornell Vet 81:277–294

Halliday J, Oke K, Breheny S, Algar E, Amor DJ (2004) Beckwith-Wiedemann syndrome and IVF: a case-control study. Am J Hum Genet 75(3):526–528. https://doi.org/10.1086/423902

Hansen PJ (2023) Review: some challenges and unrealized opportunities toward widespread use of the in vitro-produced embryo in cattle production. Anim Int J Anim Biosci 17(Suppl 1):100745

Hara-Isono K, Matsubara K, Mikami M, Arima T, Ogata T, Fukami M, Kagami M (2020) Assisted reproductive technology represents a possible risk factor for development of epimutation-mediated imprinting disorders for mothers aged ≥ 30 years. Clin Epigenetics 12(1):111. https://doi.org/10.1186/s13148-020-00900-x

Harris LK, Crocker IP, Baker PN, Aplin JD, Westwood M (2011) IGF2 actions on trophoblast in human placenta are regulated by the insulin-like growth factor 2 receptor, which can function as both a signaling and clearance receptor1. Biol Reprod 84:440–446

Hasler JF (2000) In vitro production of cattle embryos: problems with pregnancies and parturition. Hum Reprod 15:47–58

Hasler JF, Henderson WB, Hurtgen PJ, Jin ZQ, McCauley AD, Mower SA, Neely B, Shuey LS, Stokes JE, Trimmer SA (1995) Production, freezing and transfer of bovine IVF embryos and subsequent calving results. Theriogenology 43:141–152

Hattori H, Hiura H, Kitamura A, Miyauchi N, Kobayashi N, Takahashi S, Okae H, Kyono K, Kagami M, Ogata T, Arima T (2019) Association of four imprinting disorders and ART. Clin Epigenetics 11(1):21. https://doi.org/10.1186/s13148-019-0623-3

Hattori K, Hoshino Y, Kachi M, Masuda Y, Yamamoto S, Honda S, Minami N, Ikeda S (2024) Analysis of histone H3K4me3 modifications in bovine placenta derived from different calf-production methods. Reprod Domest Anim 59:e14527

Heyman Y, Chavatte-Palmer P, LeBourhis D, Camous S, Vignon X, Renard JP (2002) Frequency and occurrence of late-gestation losses from cattle cloned embryos. Biol Reprod 66(1):6–13. https://doi.org/10.1095/biolreprod66.1.6

Hohnholz T, Volkmann N, Gillandt K, Waßmuth R, Kemper N (2019) Risk factors for dystocia and perinatal mortality in extensively kept Angus Suckler cows in Germany. Agriculture 9:85

Holm LW, Short RV (1962) Progresterone in the peripheral blood of Guernsey and Friesian cows during prolonged gestation. J Reprod Fertil 4:137–141

Holm P, Walker SK, Seamark RF (1996) Embryo viability, duration of gestation and birth weight in sheep after transfer of in vitro matured and in vitro fertilized zygotes cultured in vitro or in vivo. Reproduction 107:175–181

Hori N, Nagai M, Hirayama M, Hirai T, Matsuda K, Hayashi M, Tanaka T, Ozawa T, Horike S (2010) Aberrant CpG methylation of the imprinting control region KvDMR1 detected in assisted reproductive technology-produced calves and pathogenesis of large offspring syndrome. Anim Reprod Sci 122:303–312

Hufana-Duran D, Pedro PB, Venturina HV, Hufana RD, Salazar AL, Duran PG, Cruz LC (2004) Post-warming hatching and birth of live calves following transfer of in vitro-derived vitrified water buffalo (Bubalus bubalis) embryos. Theriogenology 61:1429–1439

Johanson JM, Berger PJ (2003) Birth weight as a predictor of calving ease and perinatal mortality in Holstein Cattle1. J Dairy Sci 86:3745–3755

Kawahara M, Wu Q, Takahashi N, Morita S, Yamada K, Ito M, Ferguson-Smith AC, Kono T (2007) High-frequency generation of viable mice from engineered bi-maternal embryos. Nat Biotechnol 25:1045–1050

Kawahara M, Obata Y, Sotomaru Y, Shimozawa N, Bao S, Tsukadaira T, Fukuda A, Kono T (2008) Protocol for the production of viable bimaternal mouse embryos. Nat Protoc 3:197–209

Keane JA, Ealy AD (2024) An overview of reactive oxygen species damage occurring during in vitro bovine oocyte and embryo development and the efficacy of antioxidant use to limit these adverse effects. Animals (Basel) 14:330

Kim HY, Shin CH, Lee YA, Shin CH, Kim G-H, Ko JM (2022) Deciphering epigenetic backgrounds in a Korean Cohort with Beckwith–Wiedemann syndrome. Ann Lab Med 42:668–677

Kruip TAM, den Daas JHG (1997) In vitro produced and cloned embryos: effects on pregnancy, parturition and offspring. Theriogenology 47:43–52

Lafontaine S, Labrecque R, Palomino JM, Blondin P, Sirard M-A (2020) Specific imprinted genes demethylation in association with oocyte donor's age and culture conditions in bovine embryos assessed at day 7 and 12 post insemination. Theriogenology 158:321–330

Lau MM, Stewart CE, Liu Z, Bhatt H, Rotwein P, Stewart CL (1994) Loss of the imprinted IGF2/cation-independent mannose 6-phosphate receptor results in fetal overgrowth and perinatal lethality. Genes Dev 8:2953–2963

Lazzari G, Wrenzycki C, Herrmann D, Duchi R, Kruip T, Niemann H, Galli C (2002) Cellular and molecular deviations in bovine in vitro-produced embryos are related to the large offspring syndrome. Biol Reprod 67:767–775

Li Z-K, Wang L-Y, Wang L-B, Feng G-H, Yuan X-W, Liu C, Xu K, Li Y-H, Wan H-F, Zhang Y, Li Y-F, Li X, Li W, Zhou Q, Hu B-Y (2018) Generation of Bimaternal and Bipaternal mice from hypomethylated haploid ESCs with imprinting region deletions. Cell Stem Cell 23:665–676.e4

Li Y, Donnelly CG, Rivera RM (2019a) Overgrowth syndrome. The veterinary clinics of North America. Food Anim Pract 35:265–276

Li Y, Hagen DE, Ji T, Bakhtiarizadeh MR, Frederic WM, Traxler EM, Kalish JM, Rivera RM (2019b) Altered microRNA expression profiles in large offspring syndrome and Beckwith-Wiedemann syndrome. Epigenetics 14:850–876

Li Y, Boadu F, Highsmith MR, Hagen DE, Cheng J, Rivera RM (2022a) Allele-specific aberration of imprinted domain chromosome architecture associates with large offspring syndrome. iScience 25:104269

Li Y, Sena Lopes J, Fuster PC, Rivera RM (2022b) Spontaneous and ART-induced large offspring syndrome: similarities and differences in DNA methylome. Epigenetics 17(11):1477–1496

Li Y, Xiao P, Boadu F, Goldkamp AK, Nirgude S, Cheng J, Hagen DE, Kalish JM, Rivera RM (2023) The counterpart congenital overgrowth syndromes Beckwith-Wiedemann Syndrome in human and large offspring syndrome in bovine involve alterations in DNA methylation, transcription, and chromatin configuration. medRxiv. https://doi.org/10.1101/2023.12.14.23299981

Linden TC, Bicalho RC, Nydam DV (2009) Calf birth weight and its association with calf and cow survivability, disease incidence, reproductive performance, and milk production. J Dairy Sci 92:2580–2588

Linder N, Hiersch L, Fridman E, Klinger G, Lubin D, Kouadio F, Melamed N (2017) Post-term pregnancy is an independent risk factor for neonatal morbidity even in low-risk singleton pregnancies. Arch Dis Child Fetal Neonatal Ed 102:F286–F290

Liu X, Huo H, Jin L, Dong Y, Li D, Zhang C, Li S (2022) Genomic imprinting of the IGF2R/AIR locus is conserved between bovines and mice. Theriogenology 180:121–129
Lohuis MM (1995) Potential benefits of bovine embryo-manipulation technologies to genetic improvement programs. Theriogenology 43:51–60
Lopes JS, Alcázar-Triviño E, Soriano-Úbeda C, Hamdi M, Cánovas S, Rizos D, Coy P (2020) Reproductive outcomes and endocrine profile in artificially inseminated versus embryo transferred cows. Animals (Basel) 10:1359
Lucio C, Windeyer MC, Larson K, Pajor EA, Pearson JM (2024) Economic analysis of calving assistance on Western Canadian cow-calf operations. J Agric Appl Econ 56:1–13
MacDonald R (2011) Hydrops in a heifer as a result of in-vitro fertilization. Can Vet J 52:791–793
Mahnani A, Sadeghi-Sefidmazgi A, Keshavarzi H (2018) Performance and financial consequences of stillbirth in Holstein dairy cattle. Animal 12:617–623
Mallo M, Wellik DM, Deschamps J (2010) Hox genes and regional patterning of the vertebrate body plan. Dev Biol 344:7–15
Mangiavacchi PM, Caldas-Bussiere MC, da Silva Mendonça M, Rumpf R, Lemos Júnior PES, Alves CS, da Silva Carneiro W, Dias AJB, Rios ÁFL (2022) Multi-locus DNA methylation analysis of imprinted genes in cattle from somatic cell nuclear transfer. Theriogenology 186:95–107
Marcho C, Bevilacqua A, Tremblay KD, Mager J (2015) Tissue-specific regulation of Igf2r/Airn imprinting during gastrulation. Epigenetics Chromatin 8:10
Martins LT, Neto SG, Tavares KCS, Calderón CEM, Aguiar LH, Lazzarotto CR, Ongaratto FL, Rodrigues VHV, de Sá Carneiro I, Rossetto R, Almeida AP, Fernandes CCL, Rondina D, Dias ACO, Chies JM, Polejaeva IA, Rodrigues JL, Forell F, Bertolini LR, Bertolini M (2016) Developmental outcome and related abnormalities in goats: comparison between somatic cell nuclear transfer- and in vivo-derived concept during pregnancy through term. Cell Reprogram 18:264–279
Maxfield EK, Sinclair KD, Dolman DF, Staines ME, Maltin CA (1997) In vitro culture of sheep embryos increases weight, primary fiber size and secondary to primary fiber ratio in fetal muscle at day 61 of gestation. Theriogenology 47:376
McEvoy TG, Sinclair KD, Broadbent PJ, Goodhand KL, Robinson JJ (1998) Post-natal growth and development of Simmental calves derived from in vivo or in vitro embryos. Reprod Fertil Dev 10(6):459. https://doi.org/10.1071/RD98126
Merton JS, van Wagtendonk-de Leeuw AM, den Daas JHG (1998) Factors affecting birth weight of IVP calves. Theriogenology 49:293
Meyer CL, Berger PJ, Koehler KJ, Thompson JR, Sattler CG (2001) Phenotypic trends in incidence of stillbirth for Holsteins in the United States. J Dairy Sci 84:515–523
Miles JR, Farin CE, Rodriguez KF, Alexander JE, Farin PW (2004) Angiogenesis and morphometry of bovine placentas in late gestation from embryos produced in vivo or in vitro. Biol Reprod 71:1919–1926
Miles JR, Farin CE, Rodriguez KF, Alexander JE, Farin PW (2005) Effects of embryo culture on angiogenesis and morphometry of bovine placentas during early gestation. Biol Reprod 73:663–671
Mullaart E, Geart L, Flapper H, John H, Oback B (2023) Birthweight data of IVP calves. 39th Annual Meeting of the Association of Embryo Technology in Europe (AETE). Animal Reproduction 20(2)
Mussa A, Molinatto C, Cerrato F, Palumbo O, Carella M, Baldassarre G, Carli D, Peris C, Riccio A, Ferrero GB (2017) Assisted reproductive techniques and risk of Beckwith-Wiedemann syndrome. Pediatrics 140(1):e20164311
Nava-Trujillo H, Rivera RM (2023a) Review: large offspring syndrome in ruminants: current status and prediction during pregnancy. Animal 17:100740
Nava-Trujillo H, Rivera RM (2023b) 069 Relationship between placental IGF2, IGF2R, and PHLDA2 and weight of bovine fetuses produced in vivo and in vitro. Anim Sci Proceed 14:473–474

Nava-Trujillo H, D'Ondiz Sánchez A, Soto-Belloso E, Velarde J, Hernandez-Fonseca H, Brackett B (2005) Calves born under tropical conditions after direct transfer of cryopreserved in vitro produced embryos. Revista Cientifica FCV-LUZ 15:429–436
Numabe T, Oikawa T, Kikuchi T, Horiuchi T (2000) Birth weight and birth rate of heavy calves conceived by transfer of in vitro or in vivo produced bovine embryos. Anim Reprod Sci 64:13–20
Oka Y, Rozek LM, Czech MP (1985) Direct demonstration of rapid insulin-like growth factor II Receptor internalization and recycling in rat adipocytes. Insulin stimulates 125I-insulin-like growth factor II degradation by modulating the IGF-II receptor recycling process. J Biol Chem 260:9435–9442
Pedersen HS, Mazzoni G, Pihl M, Kadarmideen HN, Hyttel P, Callesen H (2020) Bovine in vitro produced embryos: development of embryo proper and associated membranes from day 26 to 47 of gestation. Reprod Biol 20:595–599
Pileggi S, Colombo EA, Ancona S, Quadri R, Bernardelli C, Colapietro P, Taiana M, Fontana L, Miozzo M, Lesma E, Sirchia SM (2024) Dysfunction in IGF2R pathway and associated perturbations in autophagy and WNT processes in Beckwith–Wiedemann Syndrome cell lines. Int J Mol Sci 25(7): Article 7
Pimenta-Oliveira A, Oliveira-Filho JP, Dias A, Gonçalves RC (2011) Morbidity-mortality and performance evaluation of Brahman calves from in vitro embryo production. BMC Vet Res 7:79
Pugh A, Hagenson S, Forsyth J, Cockrem K, McMillan V, Walsdorf W, Marsh A, van Wagtendonk AM, Tervit HR (2003) 283 assessment of calves produced by Ivp in a semi-defined medium. Reprod Fertil Dev 16:261–262
Rabaglino MB, Sánchez JM, McDonald M, O'Callaghan E, Lonergan P (2023) Maternal blood transcriptome as a sensor of fetal organ maturation at the end of organogenesis in cattle†. Biol Reprod 109:749–758
Reese ST, Franco GA, Poole RK, Hood R, Fernadez Montero L, Oliveira Filho RV, Cooke RF, Pohler KG (2020) Pregnancy loss in beef cattle: a meta-analysis. Anim Reprod Sci 212:106251
Reinders JMC, Paauw MJG, Hazeleger W, van Wagtendonk-de Leeuw AM, Kemp B (1998) Is the large IVP calf syndrome related to pregnancy diabetes? Theriogenology 49:297
Rérat M, Zbinden Y, Saner R, Hammon H, Blum JW (2005) In vitro embryo production: growth performance, feed efficiency, and hematological, metabolic, and endocrine status in calves. J Dairy Sci 88:2579–2593
Rhon-Calderon EA, Hemphill CN, Vrooman LA, Rosier CL, Lan Y, Ord T, Coutifaris C, Mainigi M, Schultz RM, Bartolomei MS (2024) Trophectoderm biopsy of blastocysts following IVF and embryo culture increases epigenetic dysregulation in a mouse model. Hum Reprod 39:154–176
Rivera RM, Donnelly CG, Patel BN, Li Y, Soto-Moreno EJ (2021) Abnormal offspring syndrome. In: Hopper RM (ed) Bovine reproduction. Wiley, pp 876–895
Rivera RM, Goldkamp AK, Patel BN, Hagen DE, Soto-Moreno EJ, Li Y, Kim CN, Miller C, Williams F, Jannaman E, Xiao Y, Tribulo P, Estrada-Cortés E, Brau-Rodríguez AR, Hansen PJ, Wu Z, Spinka CM, Martin N, Elsik CG (2022) Identification of large offspring syndrome during pregnancy through ultrasonography and maternal blood transcriptome analyses. Sci Rep 12:10540
Rizos D, Ward F, Duffy P, Boland MP, Lonergan P (2002) Consequences of bovine oocyte maturation, fertilization or early embryo development in vitro versus in vivo: implications for blastocyst yield and blastocyst quality. Mol Reprod Dev 61:234–248
Roberts SJ (1971) Veterinary obstetrics and genital diseases: (Theriogenology). GBS Pub. & Distributors, Delhi
Rodríguez-Dorta N, Cognié Y, González F, Poulin N, Guignot F, Touzé J-L, Baril G, Cabrera F, Álamo D, Batista M, Gracia A, Mermillod P (2007) Effect of coculture with oviduct epithelial cells on viability after transfer of vitrified in vitro produced goat embryos. Theriogenology 68:908–913
Rooke JA, McEvoy TG, Ashworth CJ, Robinson JJ, Wilmut I, Young LE, Sinclair KD (2007) Ovine fetal development is more sensitive to perturbation by the presence of serum in embryo culture before rather than after compaction. Theriogenology 67:639–647

Rotheneder S, González-Grajales LA, Beck H, Bootz F, Bollwein H (2022) Variability of bovine conceptus-related volumes in early pregnancy measured with transrectal 3-dimensional ultrasonography. J Dairy Sci 105:4534–4546
Rutigliano HM, Wilhelm A, Hall J, Shi B, Meng Q, Stott R, Bunch TD, White KL, Davies CJ, Polejaeva IA (2017) Cytokine gene expression at the maternal-fetal interface after somatic cell nuclear transfer pregnancies in small ruminants. Reprod Fertil Dev 29:646–657
Saliba WP, Gimenes LU, Drumond RM, Bayão HXS, Di Palo R, Gasparrini B, Rubessa M, Baruselli PS, Sales JNS, Bastianetto E, Leite RC, Alvim MTT (2020) Which factors affect pregnancy until calving and pregnancy loss in Buffalo recipients of in vitro produced embryos? Front Vet Sci 7:577775
Sandovici I, Georgopoulou A, Pérez-García V, Hufnagel A, López-Tello J, Lam BYH, Schiefer SN, Gaudreau C, Santos F, Hoelle K, Yeo GSH, Burling K, Reiterer M, Fowden AL, Burton GJ, Branco CM, Sferruzzi-Perri AN, Constância M (2022) The imprinted Igf2-Igf2r axis is critical for matching placental microvasculature expansion to fetal growth. Dev Cell 57:63-79.e8
Sartori R, Gümen A, Guenther JN, Souza AH, Caraviello DZ, Wiltbank MC (2006) Comparison of artificial insemination versus embryo transfer in lactating dairy cows. Theriogenology 65:1311–1321
Schmidt M, Greve T, Avery B, Beckers JF, Sulon J, Hansen HB (1996) Pregnancies, calves and calf viability after transfer of in vitro produced bovine embryos. Theriogenology 46(3):527–539
Seidl CI, Stricker SH, Barlow DP (2006) The imprinted Air ncRNA is an atypical RNAPII transcript that evades splicing and escapes nuclear export. EMBO J 25:3565–3575
Sferruzzi-Perri AN, Owens JA, Standen P, Roberts CT (2008) Maternal insulin-like growth factor-II promotes placental functional development via the type 2 IGF receptor in the Guinea Pig. Placenta 29:347–355
Silva RR, Demetrio D, Walhof C, Oliveira M, Spricigo J, Santos R (2022) 13 Is pregnancy failure still a major concern for bovine in vitro-produced embryos? Reprod Fertil Dev 35:131–132
Sinclair KD, Broadbent PJ, Dolman DF (1995) In vitro produced embryos as a means of achieving pregnancy and improving productivity in beef cows. Ani Sci 60(1):55–64
Sinclair K, Maxfield E, Robinson J, Maltin C, McEvoy T, Dunne L, Young L, Broadbent P (1997) Culture of sheep zygotes can alter fetal growth and development. Theriogenology 47:380
Sinclair KD, McEvoy TG, Maxfield EK, Maltin CA, Young LE, Wilmut I, Broadbent PJ, Robinson JJ (1999) Aberrant fetal growth and development after in vitro culture of sheep zygotes. Reproduction 116:177–186
Siqueira LG, Tribulo P, Chen Z, Denicol AC, Ortega MS, Negrón-Pérez VM, Kannampuzha-Francis J, Pohler KG, Rivera RM, Hansen PJ (2017) Colony-stimulating factor 2 acts from days 5 to 7 of development to modify programming of the bovine conceptus at day 86 of gestation. Biol Reprod 96:743–757
Sleutels F, Zwart R, Barlow DP (2002) The non-coding Air RNA is required for silencing autosomal imprinted genes. Nature 415:810–813
Soto-Moreno EJ, Balboula A, Spinka C, Rivera RM (2021) Serum supplementation during bovine embryo culture affects their development and proliferation through macroautophagy and endoplasmic reticulum stress regulation. PLoS One 16:e0260123
Stacchezzini S, Fabaro P, Cremonesi F (1997) Field experiences with the transfer of in vitro or in vivo derived piedmontese embryos in holstein recipients. Theriogenology 47:381
Stewart BM, Block J, Morelli P, Navarette AE, Amstalden M, Bonilla L, Hansen PJ, Bilby TR (2011) Efficacy of embryo transfer in lactating dairy cows during summer using fresh or vitrified embryos produced in vitro with sex-sorted semen. J Dairy Sci 94:3437–3445
Stöger R, Kubička P, Liu C-G, Kafri T, Razin A, Cedar H, Barlow DP (1993) Maternal-specific methylation of the imprinted mouse Igf2r locus identifies the expressed locus as carrying the imprinting signal. Cell 73:61–71
Sudano MJ, Paschoal DM, da Silva Rascado T, Magalhães LCO, Crocomo LF, de Lima-Neto JF, da Cruz Landim-Alvarenga F (2011) Lipid content and apoptosis of in vitro-produced bovine embryos as determinants of susceptibility to vitrification. Theriogenology 75:1211–1220

Terranova R, Yokobayashi S, Stadler MB, Otte AP, van Lohuizen M, Orkin SH, Peters AHFM (2008) Polycomb group proteins Ezh2 and Rnf2 direct genomic contraction and imprinted repression in early mouse embryos. Dev Cell 15:668–679

Thibier PM (2001) The animal embryo transfer industry in figures. Embryo Transfer Newsletter

Uetake K (2013) Newborn calf welfare: a review focusing on mortality rates. Anim Sci J 84:101–105

USDA (2018) Retrieved on 26 May 2024, from https://downloads.usda.library.cornell.edu/usda-esmis/files/h702q636h/8p58pg246/sj139446d/Catt-07-20-2018.txt

van Wagtendonk-de Leeuw AM (2006) Ovum pick up and in vitro production in the bovine after use in several generations: a 2005 status. Theriogenology 65:914–925

van Wagtendonk-de Leeuw AM, Aerts BJG, den Daas JHG (1998) Abnormal offspring following in vitro production of bovine preimplantation embryos: a field study. Theriogenology 49:883–894

Veronese A, Lupini L, Consiglio J, Visone R, Ferracin M, Fornari F, Zanesi N, Alder H, D'Elia G, Gramantieri L, Bolondi L, Lanza G, Querzoli P, Angioni A, Croce CM, Negrini M (2010) Oncogenic role of miR-483-3p at the IGF2/483 locus. Cancer Res 70:3140–3149

Viana JHM (2023) 2022 Statistics of embryo production and transfer in domestic farm animals. Embryo Technology Newsletter 41(4):1–14

Walker SK, Heard TM, Seamark RF (1992) In vitro culture of sheep embryos without co-culture: successes and perspectives. Theriogenology 37:111–126

Walker SK, Hartwich KM, Seamark RF (1996) The production of unusually large offspring following embryo manipulation: concepts and challenges. Theriogenology 45:111–120

Wei Y, Yang C-R, Zhao Z-A (2022) Viable offspring derived from single unfertilized mammalian oocytes. Proc Natl Acad Sci 119:e2115248119

Weiller MAA, Schmoeller E, Barbosa AA, de Matos ADC, Brauner CC, Corrêa MN, Pino FABD, Rabassa VR (2024) Metabolic, zootechnical, and health profile of Girolando calves conceived through in vitro fertilization. Ciência Rural 54:e20220615

Weksberg R, Shuman C, Caluseriu O, Smith AC, Fei Y-L, Nishikawa J, Stockley TL, Best L, Chitayat D, Olney A, Ives E, Schneider A, Bestor TH, Li M, Sadowski P, Squire J (2002) Discordant KCNQ1OT1 imprinting in sets of monozygotic twins discordant for Beckwith–Wiedemann syndrome. Hum Mol Genet 11:1317–1325

Willadsen SM, Janzen RE, McAlister RJ, Shea BF, Hamilton G, McDermand D (1991) The viability of late morulae and blastocysts produced by nuclear transplantation in cattle. Theriogenology 35:161–170

Willhelm BR, Ticiani E, Campagnolo K, Rodrigues JL, Roberts AJ, Anderson GB, Bertolini M (2021) Components of the insulin-like growth factor system in in vivo- and in vitro-derived fetuses of cattle, and the association with growth and development. Anim Reprod Sci 234:106856

Wilson JM, Williams JD, Bondioli KR, Looney CR, Westhusin ME, McCalla DF (1995) Comparison of birth weight and growth characteristics of bovine calves produced by nuclear transfer (cloning), embryo transfer and natural mating. Anim Reprod Sci 38:73–83

Ye J, Zhang L, Chen Y, Fang F, Luo Z, Zhang J (2014) Searching for the definition of macrosomia through an outcome-based approach. PLoS One 9:e100192

Young LE, Butterwith SC, Sinclair KD, Maxfield EK, Wilmut I (1997) IGF2 gene expression in ovine fetal development. Theriogenology 47:385

Young LE, Sinclair KD, Wilmut I (1998) Large offspring syndrome in cattle and sheep. Rev Reprod 3:155–163

Young L, Guttierez C, Butterwith S, Robinson J, Broadbent P, McEvoy T, Wilmut I, Sinclair K (1999) Altered IGF binding protein expression is associated with large offspring syndrome in fetal sheep. Theriogenology 51:196

Young LE, Fernandes K, McEvoy TG, Butterwith SC, Gutierrez CG, Carolan C, Broadbent PJ, Robinson JJ, Wilmut I, Sinclair KD (2001) Epigenetic change in IGF2R is associated with fetal overgrowth after sheep embryo culture. Nat Genet 27:153–154

Zaborski D, Grzesiak W, Szatkowska I, Dybus A, Muszynska M, Jedrzejczak M (2009) Factors affecting dystocia in cattle. Reprod Domest Anim 44:540–551

Reproduction and Captive-Breeding of Threatened Wild Birds

Pablo Izquierdo, Alicia Carrero,
and Bárbara Martín-Maldonado

Abstract

The 6th mass extinction is hardly impacting birds, with 159 species extinct in the last 20 years and 12% of known species now threatened. Moreover, bird populations have alarmingly declined worldwide. For example, in Europe, the decline has been by 600 million since 1980. This reduction of biodiversity has far-reaching consequences for ecosystems, exemplified by the endangered scissor-tailed hummingbird's role as a pollinator in Venezuela. Conservation efforts include in-situ and ex-situ strategies. Despite the challenges that it poses, contemporary ex-situ initiatives have successfully rescued species from the brink of extinction. However, captive breeding programs present considerable challenges due to their complexity, time investment, and lack of scientific publications, and success is not guaranteed. Consequently, it becomes essential to prioritize optimizing reproduction to improve the chances of offspring survival. This chapter will comprehensively examine the factors to be considered when formulating an ex-situ program involving captive breeding for endangered bird species, from breeders management and incubation conditions to neonatal care and the release of young birds.

P. Izquierdo · A. Carrero
Wildlife Hospital, GREFA (Grupo de Rehabilitación de la Fauna Autóctona y su Hábitat), C/Monte del Pilar, Majadahonda, Madrid, Spain
e-mail: cria@grefa.org

B. Martín-Maldonado (✉)
Department of Veterinary Medicine, School of Biomedical and Health Sciences, Universidad Europea de Madrid, Villaviciosa de Odón, Spain, Madrid, Spain
e-mail: barbara.martin-maldonado@universidadeuropea.es

J. C. Gardón, K. Satué Ambrojo (eds.), *Assisted Reproductive Technologies in Animals Volume 2*, https://doi.org/10.1007/978-3-031-87198-6_8

Keywords

Wildlife · Biodiversity · Conservation · Ex-situ programs · Captive-breeding · Neonatal cares · Reproductive disorders

1 Introduction: The Need for Captive Breeding in Wild Birds

Amid the 6th mass extinction, birds are taking the worst part of it. According to the International Union for Conservation of Nature (IUCN), 159 bird species have become extinct during the last 20 years, and now 12% of the described bird species are threatened (1354 of 11,197) (IUCN 2023). Specifically in Europe, bird populations have been reduced by 600 million individuals since 1980, which means a decline of 18% in the breeding bird abundance (Burns et al. 2021). Species abundance has been described as a synonym of a healthy ecosystem, where a little change can have catastrophic consequences. The decline of one unique species can lead to drastic biotic changes and contribute to the ecosystem's degradation (Rosenberg et al. 2019). For example, the ecological importance of the scissor-tailed hummingbird (*Hylonnympha macrocerca*), listed as Endangered on the IUCN Red List, lies in its role as a pollinator in Venezuela's forests. The decrease in its population leads to less pollination and therefore loss of vegetal biodiversity and reduction and fragmentation of forests (Leimberger et al. 2022; IUCN 2023). The lack of knowledge and monitoring of some threatened species obstructs their recovery from extinction. Nevertheless, the extinction of some of the most singular species didn't go unnoticed, such as the dodo (*Raphus cucullatus*) in the 1690s, or more recently the kamao (*Myadestes myadestinus*) in the 1980s (Reynolds et al. 1997; Roberts and Solow 2003).

Most of the decline factors have their origin in human activity. Among them, habitat fragmentation, due to intensive agriculture or urbanization, highlights. The loss of vegetal biodiversity or the removal of stopover sites in a migration network has a great impact on bird biodiversity and ecosystem balance (Xu et al. 2019; Galla et al. 2020). Reductions in the abundance of species can be primarily linked to decreases in the structural complexity and resources of forest habitats, along with changes in relationships between predators, prey, and competitors (Shahabuddin et al. 2021). The significant reduction in food availability, due to habitat fragmentation or pollution, has played an important role in biodiversity loss. Pesticides and other chemicals employed in pest control for agricultural intensification for decades have enhanced this phenomenon. Proof of this is the 58% decrease in farmland bird populations since the 1980s in Europe (Garcês et al. 2022). The scarcity of food plus the anthropization of ecosystems has forced several species, such as seagulls, storks, or even vultures, to adapt to human presence and use the resources that cities provide (Mason et al. 2013; Miranda 2017). However, feeding in urban wastes involves several risks for animals, such as the accumulation of heavy metals, drugs, microplastics, or antimicrobial-resistant bacteria (Martín-Maldonado et al. 2022).

Also, climate change and warming global temperatures have affected the phenology of some bird species, which have modified their timing of migration and breeding, the incubation process, and even the nest provisioning (DuRant et al. 2019; Halupka et al. 2023; Wheeler et al. 2023). Although the advanced laying dates could suggest an annual increase in the number of fledged young birds, 56.7% of 104 bird species had shown a declining trend in offspring production with large species producing smaller clutches (Halupka et al. 2023). Moreover, embryo development is susceptible to high temperatures: mortality can occur when the egg is exposed to temperatures higher than 40 °C. Fortunately, most of the species show some plasticity of incubation behaviors to adapt to environmental conditions. However, it has been estimated that by 2100 the annual mean surface air temperature could rise to 5.0 °C, increasing the proportion of nests exposed to lethal temperatures (DuRant et al. 2019). Also, global warming impacts the vegetation and leads to habitat loss and expansion of arthropod populations that could act as vectors for important infectious diseases such as malaria. The repercussions of this situation have a double effect: on the one hand, infectious diseases can kill a large number of individuals; on the other hand, the death of adult individuals (breeders) results in a decrease in the annual number of offspring (Garamszegi 2011).

To slow down biodiversity loss, a multitude of conservation programs have been developed focused mainly on unique key species. Those projects can follow two strategies: in-situ when the actions are carried out in the natural habitat of the species to be conserved, or ex-situ, which implies the collection and keeping in captivity of individuals from a threatened species to serve as captive breeders for reinforcement programs and safeguard populations (Moloney et al. 2023). This is the case of the passenger pigeon (*Ectopistes migratorius*), which was thought to once be the most abundant bird in the northern hemisphere. In the late 1800s, this species declined in such numbers that it became almost extinct in a few decades. Massive exploitation of its meat and large-scale loss of habitat were the main factors to drove the passenger pigeon to the brink (Bucher 1992; Hung et al. 2014). Nevertheless, there was an attempt to revert the situation through ex-situ conservation (Dolman et al. 2015). The last known group of passenger pigeons was kept by Professor Charles Otis Whitman, at the University of Chicago, who attempted to breed them in captivity in the 1890s (Sealy 2012). As part of a behavioral study, Whitman applied cross-fostering techniques by using rock doves as foster parents. This technique, widely used today, was unsuccessful. One famous bird was bred during the process though: Martha. Named after Martha Washington (George Washington's wife), Martha was the last survivor of her species. Martha's story is a sad one as she remained in Cincinnati's Zoo for several years, alone in her cage, until she died in 1914 (Sealy 2012; Avery 2014; Jørgensen and Gladstone 2022).

The last century has witnessed an exponential increase in ex-situ initiatives, bearing in mind that the extinction crisis would lead other species to the same fate. Many zoos, non-governmental organizations (NGOs), and private institutions have championed saving species through ex-situ conservation: blue-throated macaw (*Ara glaucogularis*), pink pigeon (*Nesoenas mayeri*), Mauritius kestrel (*Falco punctatus*), Bonelli's eagle (*Aquila fasciata*), Californian condor (*Gymnogyps californianus*), or Andean condor (*Vultur gryphus*) (Snyder and Snyder 1989; Jones et al. 1992, 1995; Nicoll et al. 2004; Astore et al. 2017; Campos et al. 2021; USFWS 2021; Shopland et al. 2022; Egea-Casas et al. 2023). These species, thanks to intensive captive management and habitat restoration, have made a comeback from the brink of extinction, increasing their numbers from just a few individuals to a few hundred. For example, in 1974 only four Mauritius kestrels remained in the wild, but after the implementation of the captive breeding program, in 2000 the population was estimated to be 500–800 individuals (Jones et al. 1995; Nicoll et al. 2004).

2 First Considerations

It is important to highlight that, once achieved, captive breeding inside an ex-situ program is not the ultimate tool for a species' conservation. It also needed an in-situ program to ensure the survival of released individuals and their reproduction in the wild, reducing the impact of human activities in their habitat (Moloney et al. 2023). However, most of the time, captive breeding can help wild populations reinforce or even re-establishment. IUCN recommended this conservation strategy for 2199 threatened species globally (IUCN 2020).

In captive breeding, genetics is a key factor: inbreeding and outbreeding depression, hybridization, and epigenetics are among the most important genetic challenges, as they can reduce fertility and involve hatching failure, and offspring survival (Lamka et al. 2022; Witzenberger and Hochkirch 2011; Wright et al. 2021; Farquharson et al. 2021; Bussolini et al. 2023). In general terms, genetic diversity loss over time is related to the number of individuals in a population, but also to the genetically effective population size. This way, genetic drift has a larger impact on small populations, as is the case of threatened ones. However, during last-minute efforts to save a species, genetic selection becomes most of the time unachievable, as we must use the breeders and the scientific evidence available (Crates et al. 2022). As the captive reproduction goes on, crossings must be directed to increase the gene flow and reduce the inbreeding depression and genetic drift (Wright et al. 2021). With conscientious management, captive breeding can have positive effects on genetic diversity as Gautschi et al. confirmed in bearded vultures (Gautschi et al. 2003). To this, most captive breeding programs have geneticists as advisors or as staff fully involved in the process.

Even so, side effects of captivity, even when crossings are carefully planned, can take an important toll on the species. Epigenetics, a phenomenon still not well understood, may affect the adaptability of captive-bred individuals for a different range of conditions, both in captivity and in the wild (Farquharson et al. 2021; Holt and Comizzoli 2022; Crates et al. 2022). Dietary requirements, fight-or-flee response, and disease resilience are among the factors considered to be affected by epigenetic changes in captive populations (Holt and Comizzoli 2022), and what about the behavior of the species? Do the captive-bred birds behave and breed the same way as their wild relatives? All the species have some behavioral plasticity that allows them to cope with new threats. This flexibility can concede associative learning, boldness, and anatomical, physiological, and dietary adaptations to animals at an individual level, which are key for the success of a population during Human-Induced Rapid Environmental Change (HIREC). In this sense, a few risk factors determine if a species can successfully face HIREC: shyness, migration, specialist level, and neocortex size and development. On the other hand, the smaller the difference between the previous conditions and the HIREC conditions, the better its ability to adapt to its new environment (Mason et al. 2013). Comparing captive- versus wild-born individuals, a recent meta-analysis determined that captive-born animals have a lower reproductive success than wild-born ones, probably because the last ones are adapted to environmental pressures (Farquharson et al. 2018; Badia-Boher et al. 2021). Modified behavior in captive-born animals may result from artificial environments or lower maternal contact during their early life stages. (Mason et al. 2013). Adaptability and survival of captive-bred birds are compromised, so captive-breeding programs and release methodologies have had to evolve during the last decades to address this issue. For example, new technologies allow the assessment of the breeder's pedigree to avoid inbreeding depression. Also, new methodologies of raising have been developed to avoid imprinting and enhance early learning processes, which make a huge impact on early survival and migratory patterns (Galla et al. 2020; Wright et al. 2021; Crates et al. 2022).

Additionally, captive breeding programs pose challenges in terms of difficulty and time commitment, with no assurance of success. In general terms, the global failure of captive breeding has been estimated between 33% and 89% (Badia-Boher et al. 2021). For example, LIFE+ Cantabrian capercaillie project (LIFE+09 NAT/ES/513) took time between 2010 and 2016 with a global budget of 5.9 million euros to face the decline of this subspecies endemic to the Iberian Peninsula. Among the actions that the program involved, captive breeding for the reinforcement of wild populations was one of the specific aims. Despite all the money and research invested in captive breeding, only five specimens could be released, and four of them died of natural causes shortly after (LIFE+ Urogallo Cantábrico 2016). Prioritizing the optimization of reproduction becomes crucial to enhance offspring survival and reduce stress on the animals (Bussolini et al. 2023).

Overall, there's so much more than just breeding for these ex-situ programs to be a successful tool for species survival in the wild. In this chapter, we will review all the aspects to consider when designing an ex-situ program with captive breeding for endangered bird species.

3 The Breeders

As previously mentioned, the selection of breeders must be carefully planned. In dramatic situations, one must just work with what is available, but most of the situations would enable a program to develop some kind of selection.

Historically many bird propagation programs were initiated with trapped wild individuals. The first captive breeding attempts involving the endangered California condor were initiated when the remaining 22 birds were trapped in the wild. These individuals were proven to be excellent breeders and the first breeding results did not take long (Snyder and Snyder 1989). Nevertheless, other taxa might not be as adaptable as condors and wild birds may take several years to show any interest in breeding after being taken into captivity (if they ever do) (Mason 2010).

Wild trapped birds, when working with large populations, allow us to select different sub-populations and low-related individuals to start a breeding program. By knowing the genetic variability of the wild population, the extraction can be directed to maximize the genetic representation within the captive group (Heinrichs et al. 2019; Crates et al. 2022). In some cases, although birds may come from the wild, they usually are collected from wildlife rescue centers and similar sources (Platt et al. 2007). However, these individuals may not be as suitable as the previous group as they might be too handicapped to be successful breeders. If possible, captive-bred birds should make the most of a breeding stock. This group is adapted to captive conditions and previous husbandry of the birds is known (imprinting, feeding habits, etc.) (Platt et al. 2007; Crates et al. 2022).

4 Aviaries

Aviary design, rather than beautiful or eye-appealing, must be functional. It also should allow staff to take care of many aviaries (if many offspring are needed) with a low effort. In times when resources are limited, captive breeding programs must be economically viable and worth the costs. There are as many aviary designs as groups or taxa in birds (Sokół and Pluta 2022; Alverson et al. 2023). Moreover, within the same group or related species, several designs might be found specifically adapted for the requirements of the species (and sometimes, the individuals). Nevertheless, some designs perform better when housing wild birds for breeding purposes.

When designing an aviary for the propagation of a wild bird species, stress is a major factor that must be considered. Whether physical or psychological, stress inhibits the hormonal chain of events that eventually lead to breeding. Hence, aviaries must be designed to allow some degree of tranquil environment for the birds to carry on with their daily routines (nest-building, courtship, egg-laying, etc.) (Phillips et al. 2018). Birds of prey are proven to be amongst the most stress-sensitive birds when within the captive breeding scheme. When facing this problem, Hurrell (1977) came up with a design called "Skylight and Seclusion" which is widely used nowadays. Skylight and seclusion aviaries are built in a way that their inhabitants are isolated (Fig. 1) (Platt et al. 2007). Peregrine falcons (*Falco peregrinus*), Bonelli's eagles, cinereous vultures (*Aegypius monachus*), and common kestrels (*Falco tinnunculus*) have been bred in skylight and seclusion aviaries for decades (Kirkwood 1980; Stanford 2011; Johnson 2014). Similarly, other stress-sensitive species can surely improve their performance in captivity when housed in skylight and seclusion aviaries.

For some others, open enclosures made of mesh are a better option. Vultures, for example, seem to tolerate human disturbance to a bigger extent than other birds of

Fig. 1 Skylight and seclusion aviary for Bonelli's eagles

prey and many programs around the globe are currently breeding them in open enclosures (Frey and Llopis 2014). This design provides a cooler environment due to improved air circulation, which is useful when working in warmer climates. For bearded vultures (*Gypaetus barbatus*), cape vultures (*Gyps coprotheres*), condors, and other new-world and old-world vultures open enclosures are recommended (Frey and Llopis 2014). However, it is important to highlight that a side effect of the design is found in the behavior of the offspring. In some cases, depending on the facilities, birds might show a higher degree of tameness, probably caused by their early exposure to human presence (pers. obs.).

Finally, for some other groups, aviary design is highly variable. Waterfowl, passerines, and others are bred regularly in different settings, but some common infrastructures (baths, feeding platforms, and nest boxes) can be found when addressing their propagation in captivity.

Aviaries, regardless of the design, must have systems to provide shelter, nesting surfaces (nest boxes, platforms, and burrows), feeding sites, and water. These features should be placed and designed to be low maintenance and easy to check and fix. Nesting material, when needed, should be provided regularly and away from the nesting area to improve their ethological enrichment (Stevens et al. 2021). No matter how shy or sensitive the species is, disturbance must be kept at a minimum during the breeding season (Dickens and Bentley 2014). To this end, providing water and feeding the birds is suggested to be done from the outside of the aviaries. Also, easy access to the nest from the outside (in the case of avian species that use nest boxes and platforms) is highly recommended.

5 Monitoring and Behavior

Daily, record-keeping is mandatory. Once the breeding stock is settled down, keepers must keep a record of any behavior, feeding routine, and incidence. This information is an important decision-making tool to improve the future reproduction of the target species. Some birds show highly social behavior as juveniles. This stage is crucial for learning and mating, and very useful for the future of a breeding program (van Overveld et al. 2020). Many species of birds are in different stages and ages (as juveniles, during migration, colony breeders, etc.) and, by allowing this behavior, we can improve the breeding success.

For proper monitoring, the installation of a closed-circuit television (CCTV) system provides a unique insight into birds' lives. Through a 24/7 recording system, staff can gather information about the breeding stock with a lower time investment. Birds, like many other animals, are active during specific times of the day, regardless of their keeper's schedules and shifts. By going through the recordings, peak activity hours can be checked straightforwardly. CCTV systems should be installed so certain areas of the aviaries are always under surveillance. Nesting areas, preferred perches, and feeding grounds are focal points of birds' activities and should be easy to check.

When CCTV is not possible, peep holes and one-way mirrors are an option, but in these cases, regular monitoring is challenging when other daily tasks must be carried out.

5.1 Ethology

Through close observation and profound knowledge of the species' ethology, one should be able to establish the proper combination of males and females to achieve successful reproduction. Gregarious and social species are usually easier to pair up as they can be safely housed together in large groups. These groups enable the birds to find the right mate and breed successfully (Alverson et al. 2023). For example, in some wild birds, such as the North Island brown kiwi (*Apterix mantelli*), sex ratios could be altered to reduce the number of birds within an aviary, as one male can reproduce with several females, e.g., in some waterfowl species (Undin and Castro 2022).

Territorial and monogamous birds pose a challenge though. These species develop a series of courtship behaviors in the wild that can hardly be replicated in a captive environment to their full extent (Platt et al. 2007; Ota 2020; Boersma et al. 2022). When dealing with such species, many combinations of male and female must be made until the right pair is found. In these cases, CCTV systems are crucial during this process and can save us a lot of time. One thing that captive breeding programs experience is intraspecific aggression. In a confined environment, non-compatible birds don't have the chance of getting away from their conspecifics and this situation can easily cause some casualties. For some species, the difference in size between males and females means that one can prey on the other when the breeding season is over. This is commonly found in species such as sparrowhawks in captivity. For some species, e.g., bearded vulture, even when the pair has been established, some degree of aggression can be expected between male and female even with experienced pairs.

5.2 Physical and Clinical Influences on Wild Birds' Reproduction

Any animals, including wild birds, must be in optimal clinical conditions to fully perform their reproductive function. Any primary or secondary pathologies, whether chronic or acute, will negatively influence the performance of this complex organic process (Crates et al. 2022).

In many cases, wild birds admitted to wildlife rehabilitation centers (WRC) have injuries whose consequences affect their ability to survive in the wild. These animals are identified as unrecoverable for wildlife, but sometimes they can be suitable for captive breeding programs. Of course, their health status should be thoroughly assessed by specialists to determine whether the lesions they had could harm their reproductive capacity. If they can be potential breeders, their introduction to mate with another individual of the same species will be considered. However, even if these animals are correctly accepted as a pair, effective reproduction will not always take place due to several factors that need to be into account.

- *Musculoskeletal injuries*, such as limb amputations or myopathies, as well as those that cause chronic pain, represent a limiting factor for breeding. In this sense, a missing limb may alter the avian behavior (display, courtship, and so forth) to the point that the bird is not able to mate or copulate properly (Crosta et al. 2021). These situations may also lead to dystocia in females during laying, even if the reproductive behavior is correct (Tristan 2010).
- *Sensory organ deficits*, e.g., total or partial blindness, or cognitive alterations as a result of craniocerebral trauma or intoxications that have left chronic sequelae, may limit fertility or inhibit sexual normal activity (Crosta et al. 2003).
- Any debilitating *infectious pathology*, like systemic disorders, can greatly impact reproductive performance. For example, trichomoniasis can cause oral and digestive lesions, leading the host to a starving status and impeding breeding. However, some specific agents, such as *Chlamydophila* and *Salmonella*, can target directly on reproductive organs. Moreover, vertical transmission through the egg has been reported in many bird species for both pathogens (Bailey and Lierz 2017; Malik et al. 2021). *Chlamydophila* is more prevalent in psittacines than raptors, but it remains one of the most important pathogens to survey in breeding programs (Bailey and Lierz 2017). Also, other bacteria, such as *E. coli*, viruses, like paramyxovirus and influenza, and fungi, such as *Aspergillus*, can cause gonad disorders (Bailey and Lierz 2017; Lloyd 2008).

- Health disorders derived from *captivity*:
  - *Metabolic alterations* due to inadequate diets and lack of exercise can also interfere with reproduction. These conditions include obesity, atheromatosis, "fatty liver," and visceral or articular gout, among others, which may induce permanent fertility problems (Facon et al. 2014; Crosta et al. 2003).
  - *Musculoskeletal issues* derived from captivity, such as pododermatitis or bumblefoot. This pathology causes pain and discomfort, that can inhibit sexual activity, and can lead to severe infections with osteomyelitis and bone necrosis (Díez et al. 2020).
- To prevent some of these pathologies, it is important to understand the biology and genetics of different species. For instance, falcons are more susceptible to pododermatitis than other species. Therefore, it is essential to prepare aviaries with appropriate perches and maintain the animals in good body condition to prevent overweight and avoid this condition (Crosta et al. 2003).
- *Age-related pathologies* are also very important to consider, especially when working with geriatric birds. Proper management and feeding routines have led to an increased lifespan in captive avian species. Older individuals develop issues such as osteoarthritis or other degenerative musculoskeletal pathologies, cataracts, chronic renal and/or hepatic disease, cardiomyopathies, or neoplasms, among others, which will directly compromise their health and therefore their fertility (Chitty 2020). In addition, although the effects of age on the reproductive system of wild birds are not yet well established, it is believed that the reproductive potential in geriatric animals, both females and males, may be diminished. Some species of geriatric animals exhibit low sperm counts, decreased sperm volume, decreased sperm motility, and poor sperm morphology. Reproductive failure may occur due to decreased receptivity, which can be caused by a drop in hormone production over time in both males and females. Additionally, egg production declines in females due to the extensive energy expenditure required (Tristan 2010). Nevertheless, in our experience, lesser kestrels (*Falco naumanni*) captive breeding programs include individuals older than 15 years (sometimes in their 20 s), when their wild conspecifics rarely survive longer than 6–7 years (pers. obs.). This increased longevity poses a challenge in this species as the breeding rates of this small falcon are overall lower after 10 years. From a pragmatic point of view, increased lifespan decreases the productivity of captive breeding programs, which still must provide housing and feeding for birds that no longer reproduce. However, these birds can be used as foster parents or housed in aviaries designed for educational purposes as animal ambassadors.
- To determine the reasons for reproductive failure in our captive couples, it is important to consider all the mentioned factors and perform periodic veterinary checks. This allows for early diagnosis and treatment of any pathologies before the next breeding season and helps to avoid reproduction in animals with serious health issues.

6 Feeding

Feeding is one of the most important aspects of bird husbandry and breeding. Food has a great influence on longevity, fitness, behavior, breeding success, and quality of the obtained offspring. When designing feeding routines, proper weight management must be considered as being overweight is a key factor for several problems such as bumblefoot, atherosclerosis, hepatic diseases, and others (Wyss et al. 2013; Facon et al. 2014; Crosta et al. 2003).

In captive breeding programs, weighing the birds cannot be done regularly unless the aviaries include some kind of device that permits weighing the birds without them being caught (on perches or platforms). Otherwise, birds must be weighed during clinical checks, when breeding season is over. Over the years and experience, keepers can develop an idea of the right amount of food for every species and individual. On birds, regardless of the species, behavior usually gives us a clue about both the health and condition of the individual. Record keeping (amount of food provided, behavior observed during feeding, etc.) is also encouraged.

Nevertheless, when keeping many birds, overweight and underweight birds are something we must deal with. Overweight is widely found in avian collections. A lack of physical activity, due to captive conditions such as aviary design, and overfeeding is a common problem and a difficult one to tackle. In these cases, reproduction pays a toll: there is a strong negative relationship between body weight and reproductive efficiency in terms of gonadal form and function (Dickens and Bentley 2014). Also, birds can be underweight for many reasons too. Poor feeding routine is, of course, the main reason for an underweight bird, but other hidden diseases or debilitating conditions may lead to an unfit bird (Paraschiv et al. 2020).

In captive breeding propagation programs, not only the breeding birds are to be kept in perfect condition but also their offspring. Many aspects of the food provided may interfere with the quality of the captive-bred birds and their ability to reproduce themselves or adapt to the wild conditions (when the main goal is a release program) (Moss et al. 2012; Nägeli et al. 2022). Growth rate, weight, feather development, and pigmentation are among the features more influenced by food in young birds. These issues have a strong influence on the ability of individuals to survive and reproduce in the long term.

Some developing research is finding that offspring's sex ratio can be related to food resources and their influence on female fitness before reproduction (Bukaciński et al. 2020). Although its mechanisms are still to be understood, the Trivers-Willard Hypothesis has been found in several animal groups. This hypothesis proposed by Robert Trivers and Dan Willard in 1973 suggests that females adjust the sex ratio of offspring in response to maternal condition, to maximize their reproductive success (Nager et al. 1999). This ability makes sense when the difference in size between males and females may affect their survival in situations where food resources are scarce. It is then not difficult to conceive a captive environment where some females or breeding pairs have access to more food than others within the same breeding facilities or different institutions. A biased sex ratio can compromise a captive propagation program and a reintroduction program, and this effect must be considered.

7 Avian Reproductive Physiology and Endocrinology

The reproductive cycle in birds of many avian species is controlled by the Hypothalamic-Pituitary-Gonadal (HPG) regulatory axis through the action of several hormones (Ubuka and Bentley 2011). Environmental factors stimulate the hypothalamus and pituitary gland to secrete gonadotropins, specifically luteinizing and follicle-stimulating hormones (LH and FSH respectively) (Pollock and Orosz 2002). In males, FSH promotes testicular growth and sperm development, while LH stimulates testosterone production. In females, FSH promotes the development of the ovarian follicles and LH induces ovulation (Saguesse 2015). The number of ovulations in females depends on the species. Most wild bird species are indeterminate layers, so they continue to lay more eggs if eggs are removed from the nest (Johnson 1999). However, some species, such as the common kestrel, are determinate layers and will not lay additional eggs if eggs are removed or destroyed (Kennedy 1991).

Then, the ovary produces estrogens, progesterone, and androgens (Saguesse 2015). Early rise in estrogen levels has been linked to nest-building behavior and to calcium mobilization for eggshell production. Sources of calcium for shell production include intestinal absorption from the diet, renal control of calcium levels, and mobilization of bone calcium deposits (McClelland et al. 2021). Estrogens also stimulate food intake, oviduct growth, vitellogenesis, as well as the synthesis of several oviductal proteins. Moreover, in combination with prostaglandins and progesterone, estrogens may stimulate the release of arginine vasotocin. This hormone, along with prostaglandins promotes oviduct contractions during oviposition, which can vary from seconds to hours, depending on the species (Pollock and Orosz 2002).

Even though the avian reproductive cycle is not fully understood, it has been suggested that many factors such as photoperiod, seasonal changes, food provision, the presence of nesting material, and the presence of a mate can interfere. Among them, photoperiod has been considered the most significant environmental factor for reproductive activity in most avian species (Juss et al. 1995).

Moreover, exposure to environmental toxins is a particular concern for the endocrine system of wild birds and may have a direct impact on their reproductive physiology. Birds that have been exposed to environmental toxins and are now part of captive breeding programs may be susceptible to sublethal accumulative effects (Coppock and Dziwenka 2022). Among these noxious agents, organochlorine and organophosphate pesticides, heavy metals, industrial pollutants, petroleum hydrocarbons, and in a few reports, herbicides and fungicides have been identified as potential sources of endocrine disruption and reproductive impairment in birds (Giesy et al. 1994; Jenssen et al. 2010; Zabala et al. 2020). These chemicals can cause a wide range of effects in adult birds. For example, thyroid function disruption during incubation can cause teratogenesis. In general, reproductive disturbances can affect all stages of reproduction, starting with sublethal stress, reduced fertility, suppression of egg formation, eggshell thinning, impaired incubation, and chick-rearing behaviors (Fry 1995; Zabala et al. 2020; Coppock and Dziwenka 2022).

7.1 Common Avian Reproductive Disorders

Many different reproductive disorders can affect, both directly and indirectly, the reproductive tract in avian species. Related clinical signs usually range from mild symptoms to sudden death, this is why reports of reproductive disorders in wild birds are mostly limited to post-mortem findings (Saguesse 2015). Therefore, our understanding of this system is largely based on the domestic chicken (*Gallus gallus*).

In females, endocrine disorders, infectious diseases, toxic exposure, and trauma are common causes of avian ovarian issues in wild birds. Oophoritis, or inflammation of the ovary, is usually secondary to systemic infections and has the same infectious origin as salpingitis (see Sect. 5.2). Ovarian cysts, like neoplasms, are rare in wild birds, but they can occur in these species (Fig. 2). Both disorders are believed to result from ovarian dysfunction caused by endocrine or anatomical alterations, among other factors (Carrasco and González 2017). Rupture of follicles, follicular cysts, oviduct problems, and other issues can eventually lead to coelomitis, which can severely compromise the animal's life.

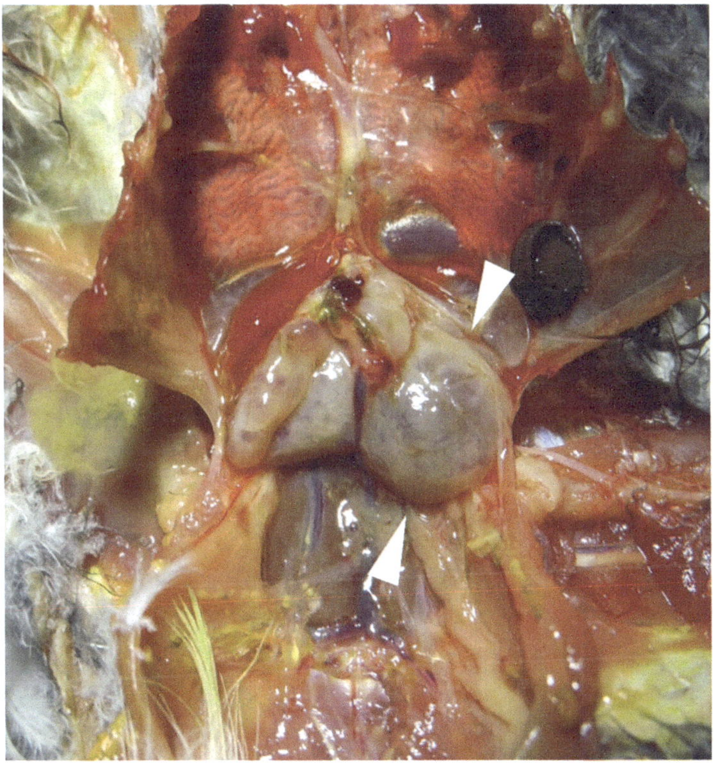

Fig. 2 Ovarian cysts in a lesser kestrel (*Falco naumanni*)

Oviductal diseases such as salpingitis, or the inflammation of the oviduct, have many times an infectious etiology (Fig. 3) (Crosta et al. 2003). Noninfectious causes of salpingitis include trauma, egg or eggshell fragments after its rupture along the oviduct, and inflammation due to oviposition disorders such as egg binding. While obesity has not been described as a primary cause of salpingitis in wild birds, excessive abdominal fat has been associated with salpingitis in domestic fowl and should be considered as a potential risk factor (Bowles 2006).

Females should be monitored closely when the eggs are expected soon. Activity around the nests increases gradually. Whether natural or artificial incubation, a closer look is mandatory. Many females, especially elders, may succumb to one of the many complications. The failure of an egg to pass through the reproductive tract is one of the most diagnosed problems in companion avian species and can also occur in wild birds kept in captivity. Obesity, systemic diseases, age-related disorders (Fig. 4), stress, calcium imbalances, masses along the oviduct or in the coelomic cavity (Fig. 5), malformed eggs, and oviductal disease are all potential causes of egg binding and dystocia in these species (Orosz et al. 1997). Egg binding and dystocia can have serious consequences, including compression of internal organs, abdominal hernias, breakage of unpassed eggs, and oviductal prolapse or rupture with secondary tissue necrosis and coelomitis. These are all life-threatening emergencies (Rosen 2012). Dystocia in wild birds kept in captivity is typically identified post-mortem due to the severity of the situation, rapid symptom progression, and difficulty in early diagnosis without continuous monitoring (Hussain et al. 2022). In contrast, pet birds living at home can be monitored more closely by their owners to prevent such situations (Crosta et al. 2003). To minimize the risk of these situations, it is important to maintain good environmental conditions, provide a balanced diet, and implement preventive clinical management. To avoid health problems such as stress, trauma, obesity, malnutrition, and hypocalcemia, it is important to provide animals with a balanced diet containing adequate calcium and phosphorus, access to sunlight, and a stress-free environment with appropriate temperature and humidity levels. This management may help to prevent some common causes of egg binding and dystocia in wild birds (Crosta et al. 2003).

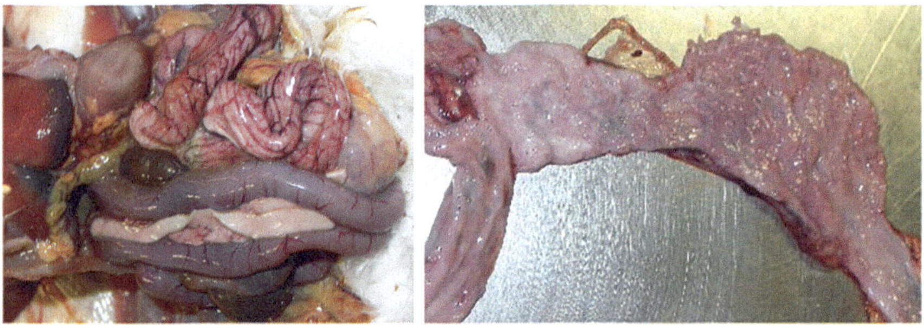

Fig. 3 Infectious salpingitis in a barn owl (*Tyto alba*): in-situ (left) and detail of the oviductal mucosa (right)

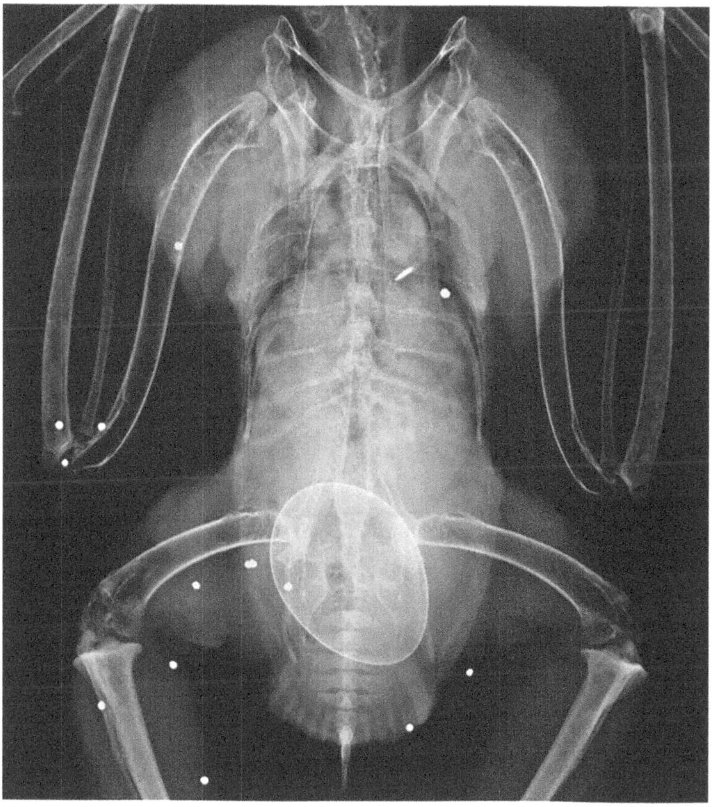

Fig. 4 Radiograph of egg binding in a Bonelli's eagle (*Aquila fasciata*) with cardiac disease

In males, while there may be several diseases that affect the male avian reproductive system, not all of them are currently known. The documented diseases that affect the male reproductive system are largely limited to the testis, specifically to orchitis, testicular neoplasms, and phallus prolapse (Hoggard and Craig 2022; Kimura et al. 2022; Walsh and Shivaprasad 2023).

8 Clutch Management

Any captive breeding program must deal with eggs one way or another. The most suitable approach is not to interfere, but there are many situations when keepers have to deal with deserted eggs, dead incubating females, rescued eggs from poachers, and so on (Platt et al. 2007).

As one of the goals of a captive breeding program is to increase the productivity of breeders, it usually involves removing eggs or full clutches to increase the hatching rate, chick survival, or the number of eggs per female you want to obtain. There are two main procedures for this:

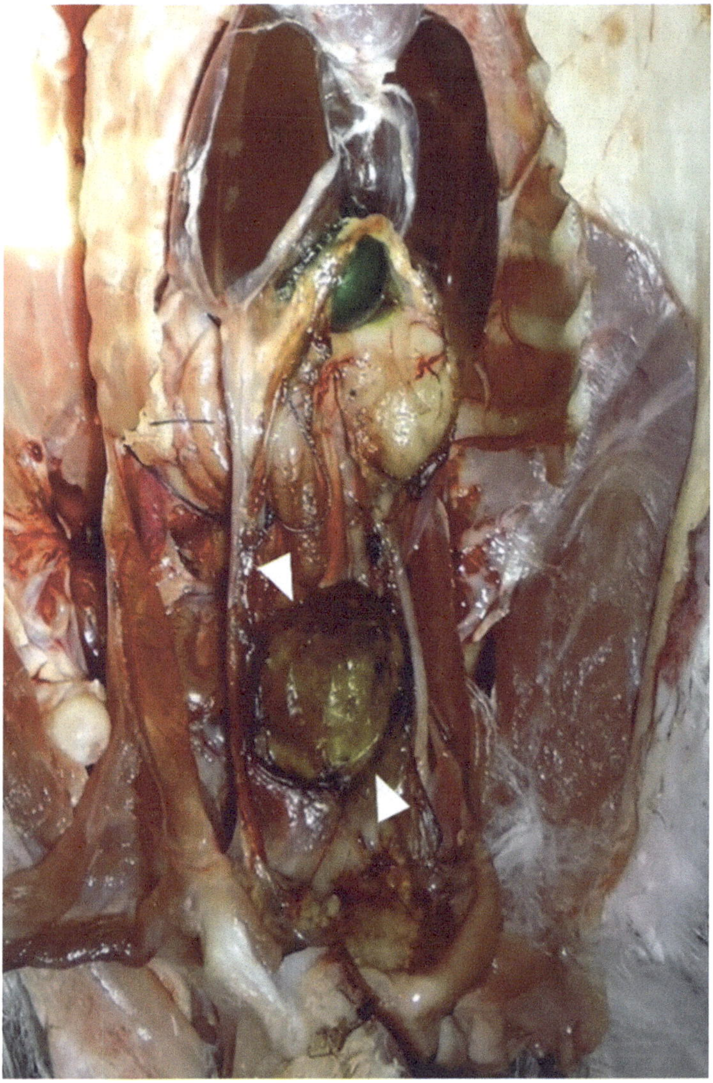

Fig. 5 Oviductal abscess in a Bonelli's eagle (*Aquila fasciata*)

- *Pulling eggs*: many species of birds are prepared to lay clutches of many eggs. For example, most species of waterfowl lay clutches of up to 20 eggs in some cases. It takes a while for them to complete a full clutch as eggs are laid, on average, every 24–48 h. Nevertheless, birds can adjust the number of eggs they lay in response to external stimuli. This adaptive feature enables them to replace broken or stolen eggs and, this way, keep their reproductive success against the odds (Figuerola and Green 2006). In a captive environment, this capacity can be used in our favor by pulling freshly laid eggs from nests and increasing this way, the

productivity of a single female. However, this methodology does not work in situations where females are nest parasites (they lay eggs inside host females' nests), such as the cuckoo (*Cuculus canorus*), or the male carries on with the incubation duties, such as emu (*Dromaius novaehollandiae*), rheas (*Rhea pennata*), and ostriches (*Struthio* spp.) amongst others.
- *Pulling clutches*: a more usual approach is to remove a whole clutch of eggs once we are certain the female has stopped laying. By closely monitoring the laying female, it is easy to calculate when to do so and how many days of incubation under the hen have passed by. A short period of natural incubation is usually recommended as the developing embryo can withstand a higher degree of neglect (Hao and Huiwen 2020). Under suboptimal incubation conditions, a more developed embryo still stands a chance. Waiting for a long period before removing a whole clutch is not recommended as females usually lose their capacity to start laying eggs again.

When handling eggs, changing temperatures can affect hatching rates. Freshly laid eggs are usually hardy and can be stored for a few days in some species (mostly those from precocial birds). Eggs near the hatching date can also withstand some degree of cooling. When handling eggs, they should be moved horizontally. In some cases, vertical storage can be an option, for example, poultry are usually handled and incubated in this position.

9 Artificial Incubation

9.1 Incubator Room and Incubators of Wild Birds

A proper incubator room should be available at least a month earlier than the first egg is expected. When incubating wild avian species, the incubators found in the market might not fit our propagation program. Incubators have been developed over the decades to incubate poultry and, over generations, both the incubators and domestic species of birds, have co-evolved. However, in our experience, when incubating wild avian species, hatching rates are usually lower using these models. This results in incubators that, unfortunately, provide a suboptimal environment for the developing embryo in wild non-domesticated species. There are, nevertheless, companies already designing incubators that fit better in a wild species propagation program and they are worth the investment.

It is highly recommended to start running incubators at least a month before they are expected to be needed. This way, technicians have enough time to get used to the machines and test them. Incubators must be run in the right environment to perform at their best. Environmental temperature and relative humidity requirements are provided by manufacturers.

- *Temperature*: most avian embryos develop properly at similar temperatures and there is plenty of research on temperature requirements for high hatching success. When comparing different groups of birds, these temperatures don't differ much and are within the range of 36–38 °C (Platt et al. 2007). The only exception is found in the *Megapodiidae* family. These birds bury their eggs in compost molds or volcanic substrates and don't incubate their eggs. They rely on the substrate's temperature to do the job and their contribution to the process is restricted to adding layers of substrate or removing them to maintain the proper environment for the developing embryo (Heij 1997). In these birds, the incubation temperature is rather low compared to other avian species and highly variable.

- *Humidity*: as embryos develop, the weight of the egg decreases, mainly through evaporation through the pores of the shell (Vieco-Galvez et al. 2021). The amount of weight that an egg loses through the whole incubation period is crucial to successful hatching. Eggs that lose much weight make it harder for the embryo to hatch and the same is true for the ones that don't lose enough (Kavtarashvili and Buyarov 2021). Regarding wild birds, egg porosity and thickness are usually adapted to the environment in which those eggs are laid, and some research needs to be done to find the right conditions. In this context, a common mistake is assuming that waterfowl eggs need a high relative humidity during development when some of the species might build the nests away from the water bodies or up in trees where humidity is usually lower.
- *Turning*: eggs require regular turning for proper development (Fig. 6). On average, most eggs are turned once an hour for mid-size species. Passerines and small birds turn their eggs more often and this might be crucial because of their short incubation period (Platt et al. 2007). Turning should always occur along the long axis. Lack of turning usually decreases hatching success, so properly designed incubators usually include turning systems (moving trays or rollers). Hand turning is also an option, highly recommended indeed.

Fig. 6 Bonelli's eagle (*Aquila fasciata*) eggs in the incubator

9.2 Monitoring Egg Development: Some Considerations

There are a few techniques to follow up embryo development. Pin-pointing problems that might arise is only possible if a closer look is kept on incubators and eggs. When working with poultry, average conditions are kept in the incubators, and the eggs are checked once or twice during the whole period. As previously mentioned, incubation conditions for poultry are well established and no changes are necessary during the incubation to adapt the incubators to the eggs. Wildlife eggs are different: every egg is usually precious.

Two monitoring techniques are highly recommended during wildlife egg incubation:

Regular weight: weighting eggs can give us a clue about eggshell quality. The eggshell quality can be compromised for many reasons, such as old females, poor diet, and illness of breeders, and this is an issue that must be addressed (Castilla et al. 2009). In the meantime, odd-shell eggs can still be detected and incubated successfully. As guidance, most of the eggs must lose 17% of their fresh weight from the beginning to the end of incubation (Parry-Jones and Ferguson 2003; Platt et al. 2007). Nevertheless, they are tolerant to some degree of deviation. A strong and healthy embryo might still hatch in any scenario within the 15–20% weight loss range (Parry-Jones and Ferguson 2003). In wild birds, the eggs must be weighed every couple of days to estimate the final weight loss under the established conditions. This enables us to change the environment in the short term and adjust the weight loss by modifying the relative humidity in the incubator. The higher the humidity, the slower the weight loss.

Candling: one of the essential tools is the candler. A powerful light placed at the blunt end of the egg is usually enough to provide an insight into the egg. Depending on the eggshell characteristics (thickness and pigmentation), the air sac and the first stages of embryo development can be observed. Candling is very useful for detecting embryo mortality and in decision-making when the hatching is around the corner (Platt et al. 2007). Candling must be done by carefully turning the egg and trying to identify the different structures within the egg.

9.3 Problematic Eggs

Several issues can be encountered when incubating wild birds' eggs. Due to factors such as improper diet, age, or genetics, the quality of the clutch might be compromised (Castilla et al. 2009; Zabala et al. 2020; Coppock and Dziwenka 2022). The typical scenario is eggs that lose way too much weight during incubation (Duvall 2023). The reason for that can be a high density of pores per square centimeter, a thinner eggshell, a compromised cuticula, or an abnormal size of the pores. These eggs usually lose weight in the 20–24% range (pers. obs.). In some cases, a high-humidity environment cannot solve the problem and other measures can be taken. At GREFA's WRC, we use adhesive clear bandages to partially cover the egg to reduce evapotranspiration. It usually works out well and we should see a considerable reduction in weight loss.

On the other hand, some eggs might show the opposite situation. A thicker eggshell or fewer pores per square centimeter as usual might reduce evapotranspiration and, at average relative humidities, these eggs lose way less than 15% of their weight during incubation (pers. obs.). This scenario is by far, more dangerous. The embryos usually don't make it to the end of incubation and, when examined, edema is common (Duvall 2023). The first option is to run the incubators completely dry, although this might not reduce the relative humidity lower than the environmental humidity in the facilities. Another option is to drill a small hole in the eggshell by the air sac (avoiding the outer membranes of the embryo) to enhance evapotranspiration. Although it might seem very invasive, this procedure is widely used to deal with this problem and there is no risk if the membranes are left untouched. By monitoring the weight loss, we can decide whether one hole is enough, or a second hole is recommended.

9.4 Hatching

Hatching is one of the most delicate parts of the captive breeding program. Due to dietary issues, genetics, breeders' age, and other factors, a poor hatching rate can be expected (Assersohn et al. 2021). Unlike poultry, the length of the incubation period within the wide range of avian species can vary much: 22–38 days for owls, 30–35 for falcons, and up to 61 for eagles (Parry-Jones and Ferguson 2003; Platt et al. 2007). Furthermore, a slightly different incubation period can be found within the same species depending on the eggs. For example, cinereous vultures can hatch between 53 and 57 days after the onset of incubation.

Through regular candling, it is usually easy to spot subtle differences in the air sac profile which indicate that the embryo is turning inside the egg. These movements are crucial to ensure that the embryo is capable of hatching. As a rule of thumb, the embryo must place its head under its left wing facing, with the beak pointing to the air sac. Not only does this position enable the embryo to reach the air sac in search of oxygen, but it is also the only way that the embryo can break the eggshell by using the structure called the egg tooth (Parry-Jones and Ferguson 2003).

Timing is crucial: internal pip, external pip, and hatching take different times depending on the species.

Keeping a record of every single species is recommended and is very useful for detecting any abnormality during these last steps. A fair amount of embryo mortality happens before hatching. As it was mentioned before, low or high weight losses might impede hatching. Nevertheless, for many different reasons, a perfectly developed embryo might find difficulties due to malposition. This issue is widely found in birds although its frequency tends to be very low (Assersohn et al. 2021; Duvall 2023). As artificial incubation has a long way to go to fit the wild avian embryo, malposition is a common occurrence in captive breeding programs. Detecting this issue on time requires skills and equipment that not always are available. When suspecting an embryo is struggling to hatch, X-rays can be helpful (Fig. 7). By placing the egg on the X-ray detector panel and letting it rotate to find its balance, the image should show us the head and beak in the middle facing towards the air sac and the back of the embryo on the left side of the egg. To hatch, the embryo must have moved the head below the right wing enabling it to reach the air sac and swift to pulmonary breathing (Duvall 2023).

At this point, several malpositions have been described: some embryos find themselves upside down, some might move their head down the wrong wing or one of the feet might be placed over the head (Duvall 2023). If a malposition is detected and we can find where the head is placed by any means, a small hole can be made to allow fresh air into the egg. The chorioallantoic membrane is highly vascularized by this point and pinching any vessel can be lethal for the embryo. Nevertheless, when internal pip is about to happen, the area of the chorioallantoic membrane around the head and beak is usually deprived of active vessels to allow a safe internal pip for the embryo. When assisting the embryo is usually safer to make a small hole in the eggshell near the beak if the time is right. By moisturizing the membrane

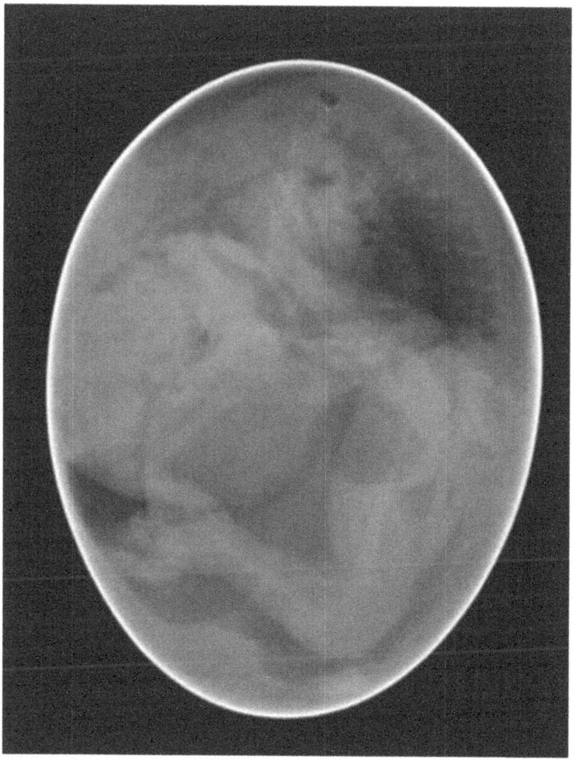

Fig. 7 Left-wing overhead malposition detected by X-ray image

with sterile physiological saline, small vessels are easy to spot, but it could be necessary electrocautery (Duvall 2023).

Finally, removing small pieces of eggshell and membrane is a slow process, depending on the species. The interval from the first pip to the hatch varies from 24 to 72 h, so, we might assist the chick for up to 3 days (Platt et al. 2007). By the end of the process, the umbilical cord that connects the chick to the chorioallantoic membrane should have collapsed and the yolk sac should be inside the celomic cavity of the chick.

10 Neonatal Cares

10.1 Hand-Rearing and Other Considerations

Although some facilities rely on adult birds to rear their offspring from day 1, there is usually a peak mortality during the first week after hatching. These casualties might be related to a lack of experience within the pair, extreme weather conditions, and, of course, accidents (Parry-Jones and Ferguson 2003; Duvall 2023). When designing a captive propagation program, hand-rearing must be taken into account if every single chick is precious.

Some factors are more important than others during these very first hours (and days). Temperature is the most important factor, and many casualties can be avoided if it is carefully monitored. During the first 24 h, the hatchlings must be kept at the same temperature as during hatching. 36.5 °C should be sufficient for these very first hours during which the hatchlings are not fed. Then, the temperature should be dropped daily by 1 °C until they are adapted to room temperature (10–13 days) (Platt et al. 2007). Brooders can be acquired in the market and if the temperature can be modified and kept steady all of them can be used for even the most sensitive species. For altricial birds, brooders must be easy to clean, as they will be used for quite a long time. If different ages of birds are expected to be hand-reared, several brooders will be needed. A common practice among avian breeders is to count on several brooders set at decreasing temperatures so the birds can be moved as they age and become hardier (Parry-Jones and Ferguson 2003). On the other hand, precocial birds are less of a challenge. During the first 24–48 h, most precocial birds will be capable of moving, drinking, and eating by themselves (Cones and Crowley 2020). This early independence allows them to tolerate a more variable environment. With these birds, it is common practice to provide infrared bulbs at different highs depending on the age of the young birds (Fig. 8). They should be able to get closer to the infrared bulb as needed and, if different environments are provided within the brooding cage, temperature is not a problem.

Once the temperature is controlled, the diet of nestlings is another factor to consider and there is plenty of research on this topic for any species. Nevertheless, there are several aspects in common. A proper weight record must be kept for the most delicate species. These records are useful when troubleshooting. For altricial birds, it is wise to weigh the birds before and after every feed to make sure that the weight

Fig. 8 Three Bonelli's eagle (*Aquila fasciata*) nestlings under infrared bulbs

is increasing as it is supposed to. The more birds of a single species are hand-reared, the more information can be obtained from the data. Precocial birds should be provided with available food all the time and they should spend most of the time foraging within their enclosures. Weighting the birds once a day, if possible, will help us to spot the weak ones that might need more attention. As they still have plenty of reserves in the yolk sac, altricial nestlings should not be fed until they are 8–12 h old (Platt et al. 2007). During the first stages of development, hatchlings might eat around 8–10% of their weight in every feed and some weight must be gained between feeds (pers. obs.). The wrong food or frequency can be solved early on by looking at the growth rate and comparing it with previous data or research on the species. Later, some birds might consume around 20% of their weight every day and altricial will need less attention and a lower feeding frequency (pers. obs.).

One thing to keep in mind when hand-rearing birds is imprinting. Imprinting is, in its narrow definition, a form of early learning that defines the future reaction of the individual to certain stimuli (Bailey and Lierz 2017; Jones and Heidenreich 2021). In aviculture, the most common form of imprinting is related to the behavior that hand-reared birds show towards humans. When dealing with wild birds, wrong practices may cause fully or partially imprinted birds that show abnormal social and breeding behaviors towards their kind (Jones and Heidenreich 2021). Regarding the imprinting process, we find differences between wild altricial and precocial species. In altricial species, due to their slow development, imprinting takes place at later stages and is usually a relatively slow process. Precocial birds, on the other hand, get imprinted on their parents and their kind within the very first hours after hatching (Bolhuis 2018). Once the timing is known, imprinting can be even manipulated to obtain certain behaviors that can be of interest. If the birds are going to be included in captive breeding programs by natural means, imprinting on humans should be avoided so the birds always recognize their species. Nevertheless, wild birds are easier to handle in captivity when they have been exposed to human presence early on. Birds also learn during their early days to differentiate a threat and processes usually included within the imprinting definition. For birds that are going to be included in a release program, imprinting should not be altered at all as it might compromise their survival.

In some cases, the most sensitive species are reproduced in captivity by artificial insemination, a technique widely used for the hybridization of birds of prey, that has been applied in conservation for many decades (Bailey and Lierz 2017; Fausto et al. 2023). Species such as the houbara bustard (*Chlamydotis undulata*) are reproduced by hundreds by artificial insemination: in this species, the future breeders are hand-reared and fully imprinted on humans to ensure they become semen donors or receptive females (Sorci et al. 2023).

Overall, most of the programs aim to get birds as wild as possible, and, sooner or later, the nestlings are introduced with adults to continue the task. When working with birds, the process of adoption or fostering is very straightforward (Slagsvold 2019). The young birds, and sometimes the eggs just before hatching, are introduced into the nests of the adults. These adults should be incubating eggs (their own or dummy eggs) to be capable of adopting (Bailey and Lierz 2017). If the timing is right, birds can't count or tell different sizes apart. Adults can foster more nestlings along the rearing process if the food supply is enough. Different nestlings can be exchanged along the way too.

10.2 Neonatology

The health of chicks depends on many factors, including the health of the parents and breeding history, the condition of any siblings, and any problems encountered during incubation and hatching (Bailey and Magno 2016). A full veterinary examination is essential to assess the condition of the chick after hatching.

The first 7–10 days are crucial, so close attention should be paid to its development to identify any problems early on. Physical indicators of correct development include balanced weight gain, proper hydration, normal mucous membranes, and skin color, correct head, beak, and limb position, normal feces, urine, and urate production and appearance, proper digestion time (crop emptying), and regular feather development (Duvall 2023). It is also important to monitor the nestling's behavior and its feeding response, as any changes or abnormalities may indicate a developing problem that needs to be monitored.

Because young nestlings do not hide their symptoms, identifying sick nestlings is easier than identifying sick adult birds. If they are seriously ill, young birds can die very quickly. Therefore, it is crucial to begin symptomatic treatment as soon as a sick nestling is detected (Crosta 2009).

10.2.1 Acquired Problems

- *Incomplete absorption of the yolk sac* is a common abnormality found in neonatal wild birds. Normally, before hatching, the yolk sac is retracted into the abdominal cavity and the nestling's navel closes over it. If the yolk sac is not fully absorbed, it can lead to trauma and infection. The prognosis for these individuals depends on the size of the unretracted yolk sac (Duvall 2023).
- Some dipteran insects such as black flies are *blood-sucking ectoparasites* and have been reported to cause the death of juvenile raptors and nestlings both in the wild and in captivity. Biting flies can cause direct trauma and anemia, as well as secondary infections in wounds (Franke et al. 2016). Moreover, nestlings may jump from the nest prematurely due to irritation of black flies (Dykstra et al. 2012; Duvall 2023). Indoor enclosures may be a temporary solution to prevent injury to affected birds.
- Neonatal disorders such as fibrous *osteodystrophy, rickets, and osteomalacia* are significant concerns due to unbalanced diets combined with rapid growth. The lack of calcium, phosphorus, or vitamin D3, as well as limited exposure to sunlight, interfere with bone development and lead to metabolic bone diseases in nestlings. Long bones and joints usually are the most affected leading to rolled toes, chondrodystrophy, splayed legs, or angel wings. Moreover, these conditions can be aggravated with unsuitable substrates. To mitigate these problems, it is crucial to provide balanced diets, adequate access to UV light, and warm, non-slip surfaces, particularly during the initial 7–10 days of life (Bailey and Magno 2016).
- *Omphalitis and yolk sac infection* are also common causes of morbidity and mortality in wild nestlings (Fig. 9). Among the bacteria reported, *E. coli* stands up (Malik et al. 2021). Both vertically and horizontally transmitted infections have been reported. If the infection ascends from the navel, it may result in fatal septicemia, which can occur within a few hours (Jones 2008) (Table 1).
- *Infections of the gastrointestinal tract* are also a leading cause of death in nestlings and can be influenced by various factors, such as intestinal flora, nutrition, environmental conditions, and pathogens (Table 1). Newly hatched birds have a completely sterile digestive tract, which quickly becomes colonized by microorganisms from the nest or incubator (Bailey and Magno 2016).
- *Other infectious disorders* (Table 1).

To prevent infections, it is crucial to maintain strict hygiene in the facilities, incubators, hatchers, and brooders. Proper disinfection of all equipment that comes into direct or indirect contact with the nestlings is also necessary. It is essential to properly disinfect the nestling's navel after hatching and until it is fully closed to prevent the entry of pathogens.

10.2.2 Congenital Problems

Wild avian nestlings born in captivity have been reported to occasionally exhibit congenital disorders. The most frequent congenital malformations described in chicks concern ophthalmic problems or musculoskeletal disorders (Barreiro et al.

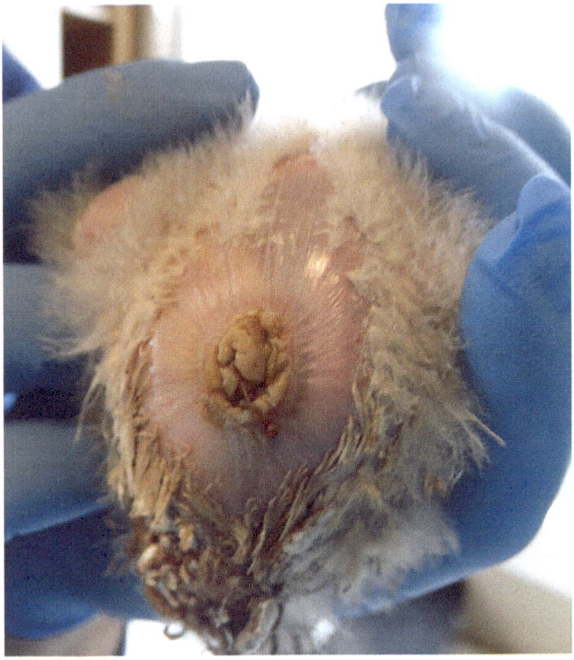

Fig. 9 Omphalitis in a 5-day-old cinereous vulture (*Aegypius monachus*)

2003). Congenital ocular abnormalities include eye abnormalities such as microphthalmia, microphakia, retinal dysplasia, malformation of the ciliary body, choroid and pecten, colobomas (Fig. 10), and congenital cataracts (Rexford 1956; Buyukmihci et al. 1988). Some of these problems may not be life-threatening. However, they can affect their ability to survive in the wild and prevent them from being released.

11 Pre-Release Considerations

The main objective of a reinforcement project is that individuals released lead to positive population growth (Harris et al. 2023). However, despite the clear potential, reintroductions often fail to establish self-sustaining wild populations and prove to be resource-intensive, raising doubts about their reliance on effectively applied science for understanding when and how ex-situ management can be a viable conservation measure (Fischer and Lindenmayer 2000; McGowan et al. 2017; Gross et al. 2023).

On one hand, it is important to enhance the survival of captive-bred individuals in the wild, through a good training program during nestlings and fledglings' development. The estimated rates of mortality of captive-born individuals in the wild are high when compared to the rates of wild-born conspecifics (Harris et al. 2023).

Table 1 Most relevant infectious disorders reported in nestlings (Leger 2012; Duvall 2023)

	Clinical signs
Bacteria	
Escherichia coli	Omphalitis, failure of yolk sac absorption
Klebsiella pneumonia	Omphalitis
Proteus vulgaris	Omphalitis
Clostridium spp.	Omphalitis
Streptococcus faecalis	Omphalitis
Salmonella spp.	Lethargy, anorexia, acute respiratory signs, omphalitis, and sudden death
Staphylococcus aureus	Omphalitis, yolk sac infection, and septicemia
Bordetella avium	"Lockjaw syndrome": Inability to articulate temporomandibular joint and open the beal. Also, nasal discharge and a mildly swollen head
Ornithobacterium rhinotracheale	Respiratory distress with airsacculitis
Pseudomonas aeruginosa	Necrosuppurative omphalitis, sudden death
Mycoplasma spp.	Sudden death
Viruses	
Adenovirus	Diarrhea, lethargy, anorexia, sudden death
Polyomavirus	Abnormal feather development, skin discoloration, ascites, hemorrhages and sudden death
Circovirus	Abnormal feather development, feather necrosis, immunosuppression, depression, anorexia, and diarrhea
Avipoxvirus	Benign cutaneous proliferation, fatal systemic diphtheritic infection of respiratory or digestive systems
Herpesvirus	Depression, regurgitation, lethargy
West Nile virus	Nystagmus, torticollis, ataxia, seizures, and neurologic signs
Avian influenza	Neurologic signs
Fungi	
Candida albicans	White punctiform plaques at the oropharynx, reduced growth, weight loss, crop stasis
Aspergillus spp.	Tracheitis, airsacculitis, pneumonia
Parasites	
Trichomonas spp.	Yellow caseous plaques at the oropharynx, reduced growth, weight loss, respiratory distress, and anorexia
Cryptosporidium spp.	Upper respiratory disease with conjunctivitis, tracheitis, and respiratory distress
Eimeria spp.	Weakness, emaciation, greenish diarrhea, dyspnea, and sudden death
Caryospora spp.	Diarrhea, depression, and weight loss
Protocalliphora larvae	Anemia and weakness
Dermanyssus spp.	Anemia, weakness, and sudden death
Simuliid flies	Anemia and secondary infections

Most of the time, the death of captive-born individuals in the wild is due to the phenotypic cost of captivity (inbreeding, adaptation to captivity, non-genetic changes in behavior, etc.), but also the release method and timing are key to success. For example, the employ of an acclimation cage has been described as a useful tool for

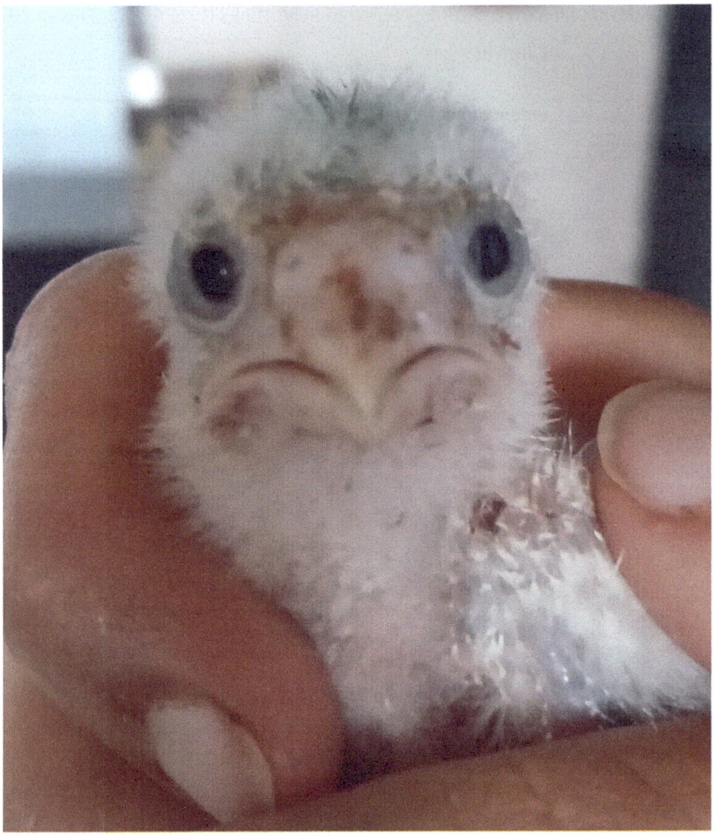

Fig. 10 Palpebral coloboma in a lesser kestrel's nestling (*Falco naumanni*)

releasing captive-born birds of prey, such as the red kite (*Milvus milvus*), while for crested ibis (*Nipponia nippon*) the release during the non-breeding season with immediately food supplementation had a positive effect on post-release survival (Li et al. 2022; Iglesias-Lebrija et al. 2023). Moreover, the timing of release may make a great difference in both the short-term and long-term of captive-bred individuals' survival (Harris et al. 2023).

On the other hand, we must be sure that released individuals are not asymptomatic carriers of pathogens or antimicrobial-resistant bacteria that can alter the balance of the wild populations.

11.1 Wildlife Disease Risk Analysis

Diseases represent a major risk to released and wild-born individuals, even being the cause of failure of several ex-situ programs (Warne and Chaber 2023). Animals can become more susceptible to some pathogens when they are immunocompromised or malnourished, or when they have been unexposed to that agent, so they have not developed their immunity (Louten 2016). The lack of native pathogens contacts harms released captive-bred individuals diminishing herd immunity and increasing susceptibility to infections in the wild (Raidal and Peters 2017). Active surveillance in the release area may help to prevent this situation through vaccination if possible. In contrast, pathogens prevalent in captive facilities often differ from those present in wild habitats (Crates et al. 2022). A clear example was the fatal toxoplasmosis reported in free-living "Alala" (*Corvus hawaiiensis*), the most endangered corvid in the world (Work et al. 2000). Parasites, like coccidia, have been the pathogen class most involved in translocation-significant disease incursions in birds, but also viruses, fungi, and bacteria have been described (Viggers et al. 1993; Spalding et al. 2008; Low 2010; Perrott and Armstrong 2011).

To enhance the health of all animals involved in the project, captive-born and wild-born, it is imperative to perform a disease risk analysis (DRA). This assessment should be done before, during, and after releasing captive-born individuals into the wild, and should include a complete clinical examination with coprological, hematological, and serological analysis, biochemistry, and necropsies, where appropriate. Since 2014, the IUCN published useful guidelines for risk identification and management; however almost none of the projects described in the literature reports the DRA performance (Jakob-Hoff et al. 2014). Depending on the region to which the captive-born birds will be reintroduced, it is recommended to include the detection of avian influenza, mycoplasmosis, poxvirus and salmonellosis, botulism, Newcastle and West Nile virus, and trichomonosis, among others (Giacinti et al. 2021).

11.2 Phenotypic Costs of Captivity

Multiple studies suggest that ex-situ conservation may lead to individuals with genetic and performance deficits that may permanently impact the performance of captive-bred animals post-release, particularly in programs with multiple captive generations (Farquharson et al. 2021; Harris et al. 2023; Gross et al. 2023). The capture of breeders in these programs acts as a bottleneck, reducing genetic diversity despite strict captive breeding. Inherent factors like small population size, inbreeding, and relaxed selection increase the risk of undesirable genetic traits (Furlan et al. 2020). Releasing captive-born individuals into the wild with such traits can harm genetic diversity and reduce effective population sizes (Crates et al. 2022; Gross et al. 2023). Furthermore, the captive phenotype often diverges from the wild type, and the release of individuals with this altered phenotype can affect fitness costs in the wild, so conservation actions may be counterproductive (Stojanovic 2023). When talking about fitness, we refer not only to body size, body condition, and movement but also to any phenotypic adaptation to survive and reproduce successfully (Gross et al. 2023). Although captive breeding aims to preserve genetic variation, it may inadvertently maintain phenotypes performing poorly in the wild.

Phenotypic modifications can involve vocalization (e.g., the loss of territorial broadcast calls), morphology (e.g., wing shape and limb length), physiology (e.g., endocrine disruption in response to stress), or behavior (e.g., timing in migration depart) (Domínguez et al. 2010; Burnside et al. 2017; Love et al. 2017; Tanimoto et al. 2017; Mellor et al. 2018; Stojanovic 2023).

Without domestication selection, phenotypic plasticity during captive development can lead to deviations from the wild type. However, non-natural conditions in captivity can influence phenotypes, impacting post-release performance compared to wild conspecifics. A recent meta-analysis demonstrated that captive-born birds underperform relative to their wild counterparts ($p < 0.05$), especially in survival and reproduction success. This performance was significantly better in captive-born individuals exposed to a good pre-release enrichment than in captive-born unenriched individuals (Gross et al. 2023). Some authors reported that lower survival rates in captive-bred birds may result from artificial rearing conditions, limiting the development of essential survival behaviors (Harris et al. 2023).

Then, using enrichment measures and behavior-based management becomes crucial to address these concerns. Enrichment includes fostering pre-release organisms with experienced conspecifics, training, implementing soft-release phases with supplemental feeding or shelter, and reducing the captive phase duration. Some examples are simulating natural environments, antipredator training, fostering pre-release organisms with experienced conspecifics, implementing soft-release phases with supplemental feeding or shelter, and reducing the captive phase duration (Lumsden and Drever 2002; Brown et al. 2006; Tetzlaff et al. 2019; Zhu et al. 2023; Gross et al. 2023). Behavior-based management (BBM) involves active training and can help captive-bred individuals learn survival skills, like avoiding predators or finding food resources (White et al. 2005; Roberts and Luther 2023). Some of the BMM techniques with a higher impact on survival are forage training, migration training, conspecific presence + mentor-reared young, or acclimation cage (Roberts and Luther 2023).

Furthermore, practitioners should employ adaptive management to identify detrimental phenotypic changes in captive animals by establishing comprehensive monitoring protocols for all aspects of a captive breeding program. Nevertheless, this monitoring process demands substantial resources to devise and execute approaches that generate ample data, making more expensive those programs (Crates et al. 2022).

12 Release Methodologies

During their development, nestlings need good enrichment and training that prepare them for their release in the wild. Several release methods have been described, and the selection depends on the species and ecosystem that we are working on (Crates et al. 2022). It is important to highlight the need for performing modeling approaches to balance competing demands of different techniques with recovery goals for wild populations and to carefully select the release locations with plenty of resources (Heinsohn et al. 2022; Crates et al. 2022).

Overall, two main strategies are followed regarding the release of captive-bred individuals:

- *Soft or adaptative release*: this strategy allows animals to adapt in-situ before release, and after release, they are supported for a while (Resende et al. 2021). These releases promote better reintroduction outcomes with a lower dispersal (Knox et al. 2017). To improve the success ratio, anti-predator training, and environmental enrichment can be introduced during fledgling development (Resende et al. 2021). There are many different techniques, but two of them stand up:
  - As established by the Peregrine Fund by the end of the XX century, birds of prey are widely released by *hacking* (Dzialak et al. 2006). This methodology involves releasing the young birds before fledgling in an artificial platform or nest situated in the release area. The birds are fed and monitored during the last stages of their development. After fledging, the individuals usually remain in the area and come back for food as needed (Badia-Boher et al. 2021). This methodology is based on philopatric behavior that is developed during these early stages.
  - Other birds are not easily released by this methodology and the *acclimatation cage* is a better choice. The release group is kept in the cage for a long period. In some species, such as vultures, the process might last for 6–12 months. During this time, the individuals get used to the environment. When released, the bird shows a lower dispersal behavior. This methodology must take into account migration patterns, breeding periods, and so on when a release date has to be decided. Many gregarious birds benefit from this methodology such as vultures or corvids such as choughs (Burgess et al. 2011; Mihoub et al. 2014).
- Finally, *hard releases*, where no post-release support is provided, are usually less successful (Resende et al. 2021). Predation, early dispersive behavior, and starvation are the usual outcomes when releasing captive-bred birds through this methodology.

Acknowledgments The authors want to thank all the staff and volunteers of GREFA's Wildlife Rescue Center for their support.

Competing Interests The authors have no conflicts of interest to declare that are relevant to the content of this chapter.

References

Alverson DP, Martin MS, Hebebrand CT, Greggor AL, Masuda B, Swaisgood RR (2023) Designing a mate choice program: tactics trialed and lessons learned with the critically endangered honeycreeper, 'akikiki (*Oreomystis bairdi*). Conserv Sci Pract 5(4):e12923. https://doi.org/10.1111/csp2.12923

Assersohn K, Marshall AF, Morland F, Brekke P, Hemmings N (2021) Why do eggs fail? Causes of hatching failure in threatened populations and consequences for conservation. Anim Conserv 24(4):540–551. https://doi.org/10.1111/acv.12674

Astore V, Estrada R, Jacome NL (2017) Reintroduction strategy for the Andean condor conservation program, Argentina. Int Zoo Yearb 51(1):124–136. https://doi.org/10.1111/izy.12140

Avery M (2014) A message from Martha: the extinction of the passenger pigeon and its relevance today. Bloomsbury Nature Writing, A&C Black, London. ISBN: 978–14–7290-626-7

Badia-Boher JA, Hernández-Matías A, Viada C, Real J (2021) Raptor reintroductions: cost-effective alternatives to captive breeding. Anim Conserv 25(2):170–181. https://doi.org/10.1111/acv.12729

Bailey TA, Lierz M (2017) Veterinary aspects of bird of prey reproduction. Vet Clin North Am Exot Anim Pract 20(2):455–483. https://doi.org/10.1016/j.cvex.2016.11.008

Bailey T, Magno MN (2016) Neonatology. In: Samour J (ed) Avian medicine, 3rd edn. Elsevier Health Sciences, St. Louis, MO

Barreiro A, de Trocóniz PF, Vila M, López-Beceiro AM, Pereira JL (2003) Congenital skeletal abnormalities in a tawny owl chick (*Strix aluco*). Avian Dis 47(3):774–776. https://doi.org/10.1637/6097

Boersma J, Jones JA, Enbody ED, Welklin JF, Ketaloya S, Nason D, Karubian J, Schwabl H (2022) Male White-shouldered Fairywrens (*Malurus alboscapulatus*) elevate androgens greater when courting females than during territorial challenges. Horm Behav 142:105158. https://doi.org/10.1016/j.yhbeh.2022.105158

Bolhuis JJ (2018) Development of perceptual mechanisms in birds: predispositions and imprinting. In: Moss C (ed) Neuroethological studies of cognitive and perceptual processes. Chapter 8. Routledge, New York, pp 158–184. ISBN: 978-04-2949-870-1

Bowles HL (2006) Evaluating and treating the reproductive system. Clin Avian Med 2:519–539

Brown GE, Rive AC, Ferrari MC, Chivers DP (2006) The dynamic nature of antipredator behavior: prey fish integrate threat-sensitive antipredator responses within background levels of predation risk. Behav Ecol Sociobiol 61:9–16. https://doi.org/10.1007/s00265-006-0232-y

Bucher EH (1992) The causes of extinction of the passenger pigeon. In: Curr Ornithol. Springer US, Boston, MA, pp 1–36. https://doi.org/10.1007/978-1-4757-9921-7_1

Bukaciński D, Bukacińska M, Chylarecki P (2020) Effect of food availability on offspring sex ratios in replacement clutches of mew gulls (*Larus canus*) and black-headed gulls (*Chroicocephalus ridibundus*) in the highly unstable environment of the Vistula River. J Ornithol 161:829–847. https://doi.org/10.1007/s10336-020-01761-0

Burgess M, Woolcock D, Hales R, Hales A (2011) A pilot release of captive-bred red-billed choughs into Cornwall, UK. In: Global reintroduction perspectives. IUCN, Gland

Burns F, Eaton MA, Burfield IJ, Klvaňová A, Šilarová E, Staneva A, Gregory RD (2021) Abundance decline in the avifauna of the European Union reveals cross-continental similarities in biodiversity change. Ecol Evol 11(23):16647–16660. https://doi.org/10.1002/ece3.8282

Burnside RJ, Collar NJ, Dolman PM (2017) Comparative migration strategies of wild and captive-bred Asian Houbara *Chlamydotis macqueenii*. Ibis 159(2):374–389. https://doi.org/10.1111/ibi.12462

Bussolini LT, Crates R, Magrath MJ, Stojanovic D (2023) Identifying factors affecting captive breeding success in a critically endangered species. Emu 123:1–9. https://doi.org/10.1080/01584197.2023.2194541

Buyukmihci NC, Murphy CJ, Schulz T (1988) Developmental ocular disease of raptors. J Wildl Dis 24(2):207–213. https://doi.org/10.7589/0090-3558-24.2.207

Campos CI, Martinez MA, Acosta D, Diaz-Luque JA, Berkunsky I, Lamberski NL, Cruz-Nieto J, Russello MA, Wright TF (2021) Genetic diversity and population structure of two endangered neotropical parrots inform in-situ and ex-situ conservation strategies. Diversity 13(8):386. https://doi.org/10.3390/d13080386

Carrasco DC, González MS (2017) Reproductive disorders in commonly kept fowl. Vet Clin North Am Exot Anim Pract 20(2):509–538. https://doi.org/10.1016/j.cvex.2016.11.009

Castilla AM, Herrel A, Van Dongen S, Furio N, Negro JJ (2009) Determinants of eggshell strength in endangered raptors. J Exp Zool A Ecol Genet Physiol 311(4):303–311. https://doi.org/10.1002/jez.532

Chitty J (2020) Care of the geriatric raptor. Vet Clin North Am Exot Anim Pract 23(3):503–523. https://doi.org/10.1016/j.cvex.2020.05.003

Cones AG, Crowley PH (2020) Optimal maternal incubation strategies for altricial and precocial birds. Ecol Model 436:109290. https://doi.org/10.1016/j.ecolmodel.2020.109290

Coppock RW, Dziwenka MM (2022) Reproductive and developmental toxicity in avian species. In: Reproductive and developmental toxicology, pp 1461–1486. https://doi.org/10.1016/B978-0-323-89773-0.00072-2

Crates R, Stojanovic D, Heinsohn R (2022) The phenotypic costs of captivity. Biol Rev 98(2):434–449. https://doi.org/10.1111/brv.12913

Crosta L (2009) Avian neonatology & pediatrics. In: 34th World Small Animal Veterinary Association World Congress Proceedings, Sao Paulo, Brazil

Crosta L, Gerlach H, Bürkle M, Timossi L (2003) Physiology, diagnosis, and diseases of the avian reproductive tract. Vet Clin North Am Exot Anim Pract 6(1):57–83

Crosta L, Petrini D, Sawmy S (2021) Reproduction management of herds/flocks of exotic animals: investigating breeding failures in birds, reptiles, and small mammals. Vet Clin North Am Exot Anim Pract 24(3):661–695. https://doi.org/10.1016/j.cvex.2021.06.001

Dickens MJ, Bentley GE (2014) Stress, captivity, and reproduction in a wild bird species. Horm Behav 66(4):685–693. https://doi.org/10.1016/j.yhbeh.2014.09.011

Díez CR, González F, López I, Suárez L, Moraleda V, Rodríguez C (2020) Pododermatitis in raptors admitted in a wildlife rehabilitation Centre in Central Spain. Prev Vet Med 175:104875. https://doi.org/10.1016/j.prevetmed.2019.104875

Dolman PM, Collar NJ, Scotland KM, Burnside RJ (2015) Ark or park: the need to predict relative effectiveness of ex-situ and in situ conservation before attempting captive breeding. J Appl Ecol 52(4):841–850. https://doi.org/10.1111/1365-2664.12449

Domínguez J, Vidal M, Tapia L (2010) Morphological changes in European goldfinches (*Carduelis carduelis*) released by bird trappers. Anim Welf 19(4):385–389. https://doi.org/10.1017/S096272860000186X

DuRant SE, Willson JD, Carroll RB (2019) Parental effects and climate change: will avian incubation behavior shield embryos from increasing environmental temperatures? Integr Comp Biol 59(4):1068–1080. https://doi.org/10.1093/icb/icz083

Duvall A (2023) Raptor pediatrics. Vet Clin North Am Exot Anim Pract 27:359–378. https://doi.org/10.1016/j.cvex.2023.11.012

Dykstra CR, Hays JL, Simon MM, Wegman AR (2012) *Protocalliphora* (Diptera: Calliphoridae) infestations of nestling red-shouldered hawks in Southern Ohio. Wilson J Ornithol 124(4):783–787. https://doi.org/10.1676/1559-4491-124.4.783

Dzialak MR, Lacki MJ, Carter KM, Huie K, Cox JJ (2006) From the field: an assessment of raptor hacking during a reintroduction. Wildl Soc Bull 34(2):542–547. https://doi.org/10.2193/0091-7648(2006)34[542:AAORHD]2.0.CO;2

Egea-Casas O, López-López P, Álvarez E, Cortone G, Galán M, Iglesias-Lebrija JJ, Lo Valvo M, Martínez J, Merlino S, Viada C, Di Vittorio M (2023) Similar post-fledging behavior observed in wild and reintroduced juvenile Bonelli's eagle in two Mediterranean islands. Ornithol Appl 125(4):duad037. https://doi.org/10.1093/ornithapp/duad037

Facon C, Beaufrere H, Gaborit C, Albaric O, Plassiart G, Ammersbach M, Liegeois JL (2014) Cluster of atherosclerosis in a captive population of black kites (*Milvus migrans* subsp.) in France and effect of nutrition on the plasma lipid profile. Avian Dis 58(1):176–182. https://doi.org/10.1637/10555-041513-Case.1

Farquharson KA, Hogg CJ, Grueber CE (2018) A meta-analysis of birth-origin effects on reproduction in diverse captive environments. Nat Commun 9(1):1055. https://doi.org/10.1038/s41467-018-03500-9

Farquharson KA, Hogg CJ, Grueber CE (2021) Offspring survival changes over generations of captive breeding. Nat Commun 12(1):3045. https://doi.org/10.1038/s41467-021-22631-0

Fausto AM, Taddei AR, Batocco F, Belardinelli MC, Carcupino M, Schiavone A, Saia S, Castillo A, Marzoni M (2023) Raptors bred in captivity: semen characteristics and assisted reproduction outcome in goshawk (*Accipiter gentilis*). PeerJ 11:e15094. https://doi.org/10.7717/peerj.15094
Figuerola J, Green AJ (2006) A comparative study of egg mass and clutch size in the Anseriformes. J Ornithol 147:57–68. https://doi.org/10.1007/s10336-005-0017-5
Fischer J, Lindenmayer DB (2000) An assessment of the published results of animal relocations. Biol Conserv 96(1):1–11. https://doi.org/10.1016/S0006-3207(00)00048-3
Franke A, Lamarre V, Hedlin E (2016) Rapid nestling mortality in Arctic peregrine falcons due to the biting effects of black flies. Arctic 69:281–285. https://doi.org/10.14430/arctic4580
Frey H, Llopis A (2014) Bearded vulture European endangered species Programme (EEP): guidelines for housing bearded vultures in captivity. Vulture Conservation Foundation, Zürich. https://4vultures.org/our-work/captive-breeding/bearded-vulture-eep/
Fry DM (1995) Reproductive effects in birds exposed to pesticides and industrial chemicals. Environ Health Perspect 103(suppl 7):165–171
Furlan EM, Gruber B, Attard CR, Wager RN, Kerezsy A, Faulks LK, Beheregaray LB, Unmack PJ (2020) Assessing the benefits and risks of translocations in depauperate species: a theoretical framework with an empirical validation. J Appl Ecol 57(4):831–841. https://doi.org/10.1111/1365-2664.13581
Galla SJ, Moraga R, Brown L, Cleland S, Hoeppner MP, Maloney RF, Richardson A, Slater L, Santure AW, Steeves TE (2020) A comparison of pedigree, genetic and genomic estimates of relatedness for informing pairing decisions in two critically endangered birds: implications for conservation breeding programmes worldwide. Evol Appl 13(5):991–1008. https://doi.org/10.1111/eva.12916
Garamszegi LZ (2011) Climate change increases the risk of malaria in birds. Glob Chang Biol 17(5):1751–1759. https://doi.org/10.1111/j.1365-2486.2010.02346.x
Garcês A, Silva F, Pires I (2022) Silent Europe: the collapse of common Bird species. J Adv Vet Res 12(5):645–647. ISSN: 2090-6277/2090-6269
Gautschi B, Müller JP, Schmid B, Shykoff JA (2003) Effective number of breeders and maintenance of genetic diversity in the captive bearded vulture population. Heredity 91(1):9–16. https://doi.org/10.1038/sj.hdy.6800278
Giacinti J, McKenzie C, McRuer DD, Sinclair D, Stevens B, Bowman J, Gomer T, Heydon C, Nituch L, Jardine C (2021) Disease risks associated with translocation of wildlife–information for veterinarians and wildlife rehabilitators in Ontario. Department of Pathobiology, University of Guelph, CWHC Ontario/Nunavut Region. https://www.cwhc-rcsf.ca/publications.php#manuals-and-guides
Giesy JP, Ludwig JP, Tillitt DE (1994) Deformities in birds of the Great Lakes region. Environ Sci Technol 28(3):128A–135A. https://doi.org/10.1021/es00052a720
Gross IP, Wilson AE, Wolak ME (2023) The fitness consequences of wildlife conservation translocations: a meta-analysis. Biol Rev 99:348. https://doi.org/10.1111/brv.13025
Halupka L, Arlt D, Tolvanen J, Millon A, Bize P, Adamík P et al (2023) The effect of climate change on avian offspring production: a global meta-analysis. Proc Natl Acad Sci 120(19):e2208389120. https://doi.org/10.1073/pnas.2208389120
Hao L, Huiwen H (2020) Artificial incubation of Tetrao parvirostris. Anim Husb Feed Sci 12(1):41–42. https://doi.org/10.19578/j.cnki.ahfs.2020.01.009
Harris SM, Robinson OJ, Hingrat Y, Le Nuz E, Ruiz-Gutierrez V (2023) Understanding the demography of a reinforced population: long-term survival of captive-bred and wild-born houbara bustards in Morocco. Biol Conserv 284:110185. https://doi.org/10.1016/j.biocon.2023.110185
Heij CJ (1997) The behaviour of the Moluccan Megapode, *Eulipoa wallacei* (caves: Megapodidae) In nesting grounds. Treubia 31(2):169–176. https://doi.org/10.14203/trb.v31i2.619
Heinrichs JA, McKinnon DT, Aldridge CL, Moehrenschlager A (2019) Optimizing the use of endangered species in multi-population collection, captive breeding and release programs. Glob Ecol Conserv 17:e00558. https://doi.org/10.1016/j.gecco.2019.e00558
Heinsohn R, Lacy R, Elphinstone A, Ingwersen D, Pitcher BJ, Roderick M, Schmelitschek E, Van Sluys M, Stojanovic D, Tripovich J, Crates R (2022) Population viability in data deficient

nomadic species: what it will take to save regent honeyeaters from extinction. Biol Conserv 266:109430. https://doi.org/10.1016/j.biocon.2021.109430

Hoggard NK, Craig LE (2022) Medullary bone in male budgerigars (*Melopsittacus undulatus*) with testicular neoplasms. Vet Pathol 59(2):333–339. https://doi.org/10.1177/03009858211069126

Holt WV, Comizzoli P (2022) Conservation biology and reproduction in a time of developmental plasticity. Biomol Ther 12(9):1297. https://doi.org/10.3390/biom12091297

Hung CM, Shaner PJL, Zink RM, Liu WC, Chu TC, Huang WS, Li SH (2014) Drastic population fluctuations explain the rapid extinction of the passenger pigeon. Proc Natl Acad Sci 111(29):10636–10641. https://doi.org/10.1073/pnas.1401526111

Hurrell LH (1977) Breeding in skylight and seclusion facilities. In: Papers on the veterinary medicine and domestic breeding of diurnal birds of prey, vol 328. British Falconers' Club, Oxford, pp 215–226

Hussain Z, Ali Z, Ahmad R (2022) Causes of morbidity and mortality in wild animals and birds at captive breeding facilities of Punjab. Pak J Zool 54(5):2337–2351. https://doi.org/10.17582/journal.pjz/20210805120803

Iglesias-Lebrija JJ, Hernández R, Álvarez E, Fajardo Í, Benítez JR, Díaz JP, Galdón A, Marco M, Carrasco M, De la Fuente S, Galán M, Moraleda V, Sanz A (2023) Juvenile dispersal of red kites *Milvus milvus* in North Africa. Contribution from GPS monitoring of individuals released in Andalusia. Go-South Bull 20:103–107

IUCN (2020) IUCN red list captive breeding recommendations. Conservation planning Specialist Group. https://www.cpsg.org/iucn-red-list-captive-breeding-recommendations. Accessed 6 Apr 2024

IUCN (2023) The IUCN red list of threatened species. Version 2023-1. https://www.iucnredlist.org/. Accessed 16 Feb 2024

Jakob-Hoff RM, MacDiarmid SC, Lees C, Miller PS, Travis D, Kock R (2014) Manual of procedures for wildlife disease risk analysis. World Organisation for Animal Health, Paris, p 149. ISBN: 978-92-9044-957-7

Jenssen BM, Aarnes JB, Murvoll KM, Herzke D, Nygård T (2010) Fluctuating wing asymmetry and hepatic concentrations of persistent organic pollutants are associated in European shag (*Phalacrocorax aristotelis*) chicks. Sci Total Environ 408(3):578–585. https://doi.org/10.1016/j.scitotenv.2009.10.036

Johnson AL (1999) Reproduction in the female. In: Causey Whittow G (ed) Sturkie's avian physiology, 5th edn. Academic Press, San Diego, pp 569–596. ISBN: 978-00-8054-208-9

Johnson D (2014) The peregrine falcon in inner London and factors affecting its existence in urban London. In: London Birds Report no. 74 of 2014. London Natural History Society, London. ISBN: 0-901009-42-3

Jones MP (2008) Raptors: paediatrics and behavioural development and disorders. In: BSAVA manual of raptors, pigeons and passerine birds. BSAVA Library, Gloucester, pp 250–259. https://doi.org/10.22233/9781910443101.22

Jones MP, Heidenreich B (2021) Behavior of birds of prey in managed care. Vet Clin North Am Exot Anim Pract 24(1):153–174. https://doi.org/10.1016/j.cvex.2020.09.007

Jones CG, Swinnerton KJ, Taylor CJ, Mungroo Y (1992) The release of captive-bred pink pigeons. Dodo J Wildl Preserv Trusts 28:92

Jones CG, Heck W, Lewis RE, Mungroo Y, Slade G, Cade TOM (1995) The restoration of the Mauritius kestrel *Falco punctatus* population. Ibis 137:S173–S180. https://doi.org/10.1111/j.1474-919X.1995.tb08439.x

Jørgensen D, Gladstone I (2022) The passenger pigeon's past on display for the future. Environ Hist 27(2):347–353

Juss TS, King VM, Kumar V, Follett BK (1995) Does unusual entrainment of the circadian system under T36h photocycles reduce the critical daylength for photoperiodic induction in Japanese quail? J Biol Rhythms 10(1):17–32. https://doi.org/10.1177/074873049501000102

Kavtarashvili A, Buyarov V (2021) The reasons for the deterioration of chicken eggshell quality at high temperatures: a review. In: International Conference on efficient production and processing (ICEPP-2021). February, 2021. Kazan, Russia. E3S web of conferences, vol 247. https://doi.org/10.1051/e3sconf/202124701015

Kennedy ED (1991) Determinate and indeterminate egg-laying patterns: a review. Condor 93(1):106–124. https://doi.org/10.2307/1368612

Kimura TH, Rocha GA, da Rocha Neto HJ, Cavalcante Filho MF, Rodrigues MC, Catenacci LS (2022) Prolapse and amputation of phallus in a greater rhea (*Rhea americana*) kept in captivity: case report. Res Soc Dev 11(3):e41411326015. https://doi.org/10.33448/rsd-v11i3.26015

Kirkwood JK (1980) Management of a colony of common kestrels (Falco tinnunculus) in captivity. Lab Anim 14(4):313–316. https://doi.org/10.1258/002367780781071148

Knox CD, Jarvie S, Easton LJ, Monks JM (2017) Soft-release, but not cool winter temperatures, reduces post-translocation dispersal of jewelled geckos. J Herpetol 51(4):490–496. https://doi.org/10.1670/16-078

Lamka GF, Harder AM, Sundaram M, Schwartz TS, Christie MR, DeWoody JA, Willoughby JR (2022) Epigenetics in ecology, evolution, and conservation. Front Ecol Evol 10:871791. https://doi.org/10.3389/fevo.2022.871791

Leger JS (2012) Nondomestic avian pediatric pathology. Vet Clin North Am Exot Anim Pract 15(2):233–250. https://doi.org/10.1016/j.cvex.2012.03.005

Leimberger KG, Dalsgaard B, Tobias JA, Wolf C, Betts MG (2022) The evolution, ecology, and conservation of hummingbirds and their interactions with flowering plants. Biol Rev 97(3):923–959. https://doi.org/10.1111/brv.12828

Li M, Tuohetahong Y, Lin F, Dong R, Wang H, Wu X, Ye X, Yu X (2022) Factors affecting post-release survival and dispersal of reintroduced crested ibis (*Nipponia nippon*) in Tongchuan City, China. Avian Res 13:100054. https://doi.org/10.1016/j.avrs.2022.100054

LIFE Urogallo Cantábrico (2016) Urgent action programme for the conservation of the Cantabrian capercaillie and its habitat. LIFE+ Cantabrian Capercaillie Project. *Fundación Biodiversidad.* Ministerio de Agricultura y pesca, alimentación y medio ambiente, España. https://lifeurogallo.es/sites/default/files/resumen-divulgativo-lifeurogallo-layman-ingles.pdf. Accessed 23 Feb 2024

Lloyd C (2008) Raptors: gastrointestinal tract disease. In: BSAVA manual of raptors, pigeons and passerine birds. BSAVA Library, Gloucester, pp 260–269. https://doi.org/10.22233/9781910443101.23

Louten J (2016) Virus transmission and epidemiology. In: Essential human virology, pp 71–92. https://doi.org/10.1016/B978-0-12-800947-5.00005-3

Love AC, Lovern MB, DuRant SE (2017) Captivity influences immune responses, stress endocrinology, and organ size in house sparrows (*Passer domesticus*). Gen Comp Endocrinol 252:18–26. https://doi.org/10.1016/j.ygcen.2017.07.014

Low M (2010) Which factors limited Stitchbird population growth on Mokoia Island? N Z J Ecol 34:269–271

Lumsden HG, Drever MC (2002) Overview of the trumpeter swan reintroduction program in Ontario, 1982-2000. Waterbirds 25:301–312

Malik YS, Arun Prince Milton A, Ghatak S, Ghosh S (2021) Avian Salmonellosis. In: Malik YS, Prince Milton AA, Ghatak S, Ghosh S (eds) Role of birds in transmitting zoonotic pathogens. Livestock diseases and management book series. Springer, Singapore. https://doi.org/10.1007/978-981-16-4554-9_15

Martín-Maldonado B, Rodríguez-Alcázar P, Fernández-Novo A, González F, Pastor N, López I, Suárez L, Moraleda V, Aranaz A (2022) Urban birds as antimicrobial resistance sentinels: White storks showed higher multidrug-resistant *Escherichia coli* levels than seagulls in Central Spain. Animals 12(19):2714. https://doi.org/10.3390/ani12192714

Mason GJ (2010) Species differences in responses to captivity: stress, welfare and the comparative method. Trends Ecol Evol 25(12):713–721. https://doi.org/10.1016/j.tree.2010.08.011

Mason G, Burn CC, Dallaire JA, Kroshko J, Kinkaid HM, Jeschke JM (2013) Plastic animals in cages: behavioural flexibility and responses to captivity. Anim Behav 85(5):1113–1126. https://doi.org/10.1016/j.anbehav.2013.02.002

McClelland SC, Cassey P, Maurer G, Hauber ME, Portugal SJ (2021) How much calcium to shell out? Eggshell calcium carbonate content is greater in birds with thinner shells, larger clutches and longer lifespans. J R Soc Interface 18(182):20210502. https://doi.org/10.1098/rsif.2021.0502

McGowan PJ, Traylor-Holzer K, Leus K (2017) IUCN guidelines for determining when and how ex-situ management should be used in species conservation. Conserv Lett 10(3):361–366. https://doi.org/10.1111/conl.12285

Mellor E, Brilot B, Collins S (2018) Abnormal repetitive behaviours in captive birds: A Tinbergian review. Appl Anim Behav Sci 198:109–120. https://doi.org/10.1016/j.applanim.2017.09.011

Mihoub JB, Princé K, Duriez O, Lécuyer P, Eliotout B, Sarrazin F (2014) Comparing the effects of release methods on survival of the Eurasian black vulture *Aegypius monachus* reintroduced in France. Oryx 48(1):106–115. https://doi.org/10.1017/S0030605312000981

Miranda AC (2017) Mechanisms of behavioural change in urban animals: the role of microevolution and phenotypic plasticity. In: Murgui E, Hedblom M (eds) Ecology and conservation of birds in urban environments. Springer, Cham, pp 113–132. https://doi.org/10.1007/978-3-319-43314-1_7

Moloney DJ, Collins C, Holloway P, O'Riordan R (2023) The Conservationist's toolkit: a critical review of the need for a conceptual framework of both in-situ and ex-situ conservation strategies to ensure the success of restoration ecology. Biol Conserv 287:110345. https://doi.org/10.1016/j.biocon.2023.110345

Moss EH, Hipkiss T, Oskarsson I, Häger A, Eriksson T, Nilsson LE, Halling S, Nilson PO, Hörnfeldt B (2012) Long-term study of reproductive performance in Golden eagles in relation to food supply in boreal Sweden. J Raptor Res 46(3):248–257. https://doi.org/10.3356/JRR-11-48.1

Nägeli M, Scherler P, Witczak S, Catitti B, Aebischer A, van Bergen V, Kormann U, Grüebler MU (2022) Weather and food availability additively affect reproductive output in an expanding raptor population. Oecologia 198:125. https://doi.org/10.1007/s00442-021-05076-6

Nager RG, Monaghan P, Griffiths R, Houston DC, Dawson R (1999) Experimental demonstration that offspring sex ratio varies with maternal condition. Proc Natl Acad Sci 96(2):570–573. https://doi.org/10.1073/pnas.96.2.570

Nicoll MAC, Jones CG, Norris K (2004) Comparison of survival rates of captive-reared and wild-bred Mauritius kestrels (*Falco punctatus*) in a re-introduced population. Biol Conserv 118(4):539–548. https://doi.org/10.1016/j.biocon.2003.09.028

Orosz SE, Dorrestein GM, Speer BL (1997) Urogenital disorders. In: Altman RB, Clubb SL, Dorrestein GM, Quesenberry K (eds) Avian medicine and surgery. W.B. Saunders, Philadelphia, pp 614–644. ISBN: 978-07-2165-446-1

Ota N (2020) Tap dancers in the wild: field observations of multimodal courtship displays in socially monogamous songbirds. Sci Nat 107(4):30. https://doi.org/10.1007/s00114-020-01686-x

Paraschiv IA, Rizac RI, Soare T, Ciobotaru-Pîrvu E, Militaru M (2020) Particularities of necropsy in cases of birds kept in captivity. Sci Work Ser C Vet Med 66(2):70. ISSN 2065-1295

Parry-Jones J, Ferguson A (2003) Strigiformes (owls). In: Management guidelines for the welfare of zoo animals. The Federation of Zoological Gardens of Great Britain and Ireland, London. ISSN: 0963-1712

Perrott JK, Armstrong DP (2011) *Aspergillus fumigatus* densities in relation to forest succession and edge effects: implications for wildlife health in modified environments. Ecohealth 8:290–300. https://doi.org/10.1007/s10393-011-0716-8

Phillips CJ, Farrugia C, Lin CH, Mancera K, Doneley B (2018) The effect providing space in excess of standards on the behaviour of budgerigars in aviaries. Appl Anim Behav Sci 199:89–93. https://doi.org/10.1016/j.applanim.2017.10.015

Platt JB, Bird DM, Bardo L (2007) Captive breeding 21. In: Bird DM, Bildstein KL (eds) Raptor research and management techniques, Chapter 21. Hancock House Publishers, Surrey, BC, pp 383–400. ISBN: 978-08-8839-639-6

Pollock CG, Orosz SE (2002) Avian reproductive anatomy, physiology and endocrinology. Vet Clin North Am Exot Anim Pract 5(3):441–474. https://doi.org/10.1016/S1094-9194(02)00010-5

Raidal SR, Peters A (2017) Psittacine beak and feather disease: ecology and implications for conservation. Emu 118(1):80–93. https://doi.org/10.1080/01584197.2017.1387029

Resende PS, Viana-Junior AB, Young RJ, Azevedo CS (2021) What is better for animal conservation translocation programmes: soft-or hard-release? A phylogenetic meta-analytical approach. J Appl Ecol 58(6):1122–1132. https://doi.org/10.1111/1365-2664.13873

Rexford DL (1956) An anomalous condition in the eye of some hawks. Auk 73(3):457–457. https://doi.org/10.1093/auk/73.3.457

Reynolds MH, Snetsinger TJ, Herrmann CM (1997) Kauai's endangered solitaires: Update on population status and distribution. Trans Western Sect Wildl Soc 33:49–55. https://pubs.usgs.gov/publication/70179305

Roberts JL, Luther D (2023) An exploratory analysis of behavior-based and other management techniques to improve avian conservation translocations. Biol Conserv 279:109941. https://doi.org/10.1016/j.biocon.2023.109941

Roberts DL, Solow AR (2003) When did the dodo become extinct? Nature 426:245–245. https://doi.org/10.1038/426245a

Rosen LB (2012) Avian reproductive disorders. J Exot Pet Med 21(2):124–131

Rosenberg KV, Dokter AM, Blancher PJ, Sauer JR, Smith AC, Smith PA, Stanton JC, Panjabi A, Helft L, Parr M, Marra PP (2019) Decline of the north American avifauna. Science 366(6461):120–124. https://doi.org/10.1126/science.aaw1313

Saguesse MD (2015) Disorders of the endocrine system. In: Samour J (ed) Avian medicine, 3rd edn. Elsevier, St. louis, MO. ISBN: 978-0-7234-3832-8

Sealy SG (2012) Hour of egg laying of the mourning dove in Manitoba, and a look at an early dataset on laying times of captive mourning doves and passenger pigeons. Blue Jay 70(1):30–36

Shahabuddin G, Goswami R, Krishnadas M, Menon T (2021) Decline in forest bird species and guilds due to land use change in the Western Himalaya. Glob Ecol Conserv 25:e01447. https://doi.org/10.1016/j.gecco.2020.e01447

Shopland S, Barbon AR, Greenwood A, Tatayah V, Barrows M (2022) Translocation of european captive pink pigeons Nesoenas mayeri to Mauritius: disease risk analysis for production of a pre-export quarantine protocol. J Zoo Aqua Res 10(4):171–182. https://doi.org/10.19227/jzar.v10i4.551

Slagsvold T (2019) Social learning in birds studied by cross-fostering in the wild. Integr J Vet Biosci 3(2):1–4. http://urn.nb.no/URN:NBN:no-80191

Snyder NF, Snyder HA (1989) Biology and conservation of the California condor. In: Power DM (ed) Current ornithology, vol 6. Springer, Boston, MA, pp 175–267. https://doi.org/10.1007/978-1-4757-9918-7_5

Sokół R, Pluta P (2022) Endoparasites in western capercaillies (*Tetrao urogallus*) and black grouse (tetrao tetrix) kept in various types of aviaries. J Wildl Dis 58(1):114–121. https://doi.org/10.7589/JWD-D-21-00017

Sorci G, Hussein HA, Levêque G, Saint Jalme M, Lacroix F, Hingrat Y, Lesobre L (2023) Ranking parameters driving siring success during sperm competition in the north African houbara bustard. Commun Biol 6(1):305. https://doi.org/10.1038/s42003-023-04698-1

Spalding MG, Carpenter JW, Novilla MN (2008) Disseminated visceral coccidiosis in cranes. In: Atkinson CT, Thomas NJ, Hunter DB (eds) Parasitic diseases of wild birds. Wiley Online Books, New York, pp 181–194. https://doi.org/10.1002/9780813804620.ch9

Stanford M (2011) Introduction to veterinary management of raptors. In: British Veterinary Zoological Society Proceedings, Port Lumpne, United Kingdom. Accesed 18 Apr 2024. https://doi.org/10.5555/20113303703

Stevens A, Doneley R, Cogny A, Phillips CJ (2021) The effects of environmental enrichment on the behaviour of cockatiels (*Nymphicus hollandicus*) in aviaries. Appl Anim Behav Sci 235:105154. https://doi.org/10.1016/j.applanim.2020.105154

Stojanovic D (2023) Altered wing phenotypes of captive-bred migratory birds lower post-release fitness. Ecol Lett 26(5):789–796. https://doi.org/10.1111/ele.14200

Tanimoto AM, Hart PJ, Pack AA, Switzer R, Banko PC, Ball DL, Sebastián-González E, Komarczyk L, Warrington MH (2017) Changes in vocal repertoire of the Hawaiian crow, *Corvus hawaiiensis*, from past wild to current captive populations. Anim Behav 123:427–432. https://doi.org/10.1016/j.anbehav.2016.11.017

Tetzlaff SJ, Sperry JH, DeGregorio BA (2019) Effects of antipredator training, environmental enrichment, and soft release on wildlife translocations: a review and meta-analysis. Biol Conserv 236:324–331. https://doi.org/10.1016/j.biocon.2019.05.054
Tristan T (2010) The aging raptor. Vet Clin North Am Exot Anim Pract 13(1):51–84. https://doi.org/10.1016/j.cvex.2009.10.001
Ubuka T, Bentley GE (2011) Neuroendocrine control of reproduction in birds. In: Norris DO, Lopez KH (eds) Hormones and reproduction of vertebrates. Academic Press, New York, pp 1–25. https://doi.org/10.1016/B978-0-12-374929-1.10001-0
Undin M, Castro I (2022) Predicting breeding systems to guide conservation strategies: a kiwi example. Ethology 128(7):538–549. https://doi.org/10.1111/eth.13286
USFWS (2021) California condor recovery program. 2020 annual population status. Ventura, CA; 2018. https://nrm.dfg.ca.gov/FileHandler.ashx?DocumentID=170382&inline. Accessed 24 Feb 2024
van Overveld T, Blanco G, Moleón M, Margalida A, Sánchez-Zapata JA, de la Riva M, Donázar JA (2020) Integrating vulture social behavior into conservation practice. Condor 122(4):duaa035. https://doi.org/10.1093/condor/duaa035
Vieco-Galvez D, Castro I, Morel PC, Chua WH, Loh M (2021) The eggshell structure in apteryx; form, function, and adaptation. Ecol Evol 11(7):3184–3202. https://doi.org/10.1002/ece3.7266
Viggers KL, Lindenmayer DB, Spratt DM (1993) The importance of disease in reintroduction programmes. Wildl Res 20:687–698. https://doi.org/10.1071/WR9930687
Walsh AL, Shivaprasad HL (2023) Unusual cases of chlamydiosis in psittacine birds. Vet Pathol 60(3):360–368. https://doi.org/10.1177/03009858231162905
Warne RK, Chaber AL (2023) Assessing disease risks in wildlife translocation projects: a comprehensive review of disease incidents. Animals 13(21):3379. https://doi.org/10.3390/ani13213379
Wheeler M, Ton R, Brandl HB, Schuett W, Griffith SC (2023) Reduced parental brood visit rate in wild zebra finches *Taeniopygia castanotis* is correlated with high maximum daily ambient temperature. J Avian Biol 2025:e03152. https://doi.org/10.1111/jav.03152
White TH Jr, Collazo JA, Vilella FJ (2005) Survival of captive-reared Puerto Rican parrots released in the Caribbean National Forest. Condor 107(2):424–432. https://doi.org/10.1093/condor/107.2.424
Witzenberger KA, Hochkirch A (2011) Ex situ conservation genetics: a review of molecular studies on the genetic consequences of captive breeding programmes for endangered animal species. Biodivers Conserv 20:1843–1861. https://doi.org/10.1007/s10531-011-0074-4
Work TM, Massey JG, Rideout BA, Gardiner CH, Ledig DB, Kwok OCH, Dubey JP (2000) Fatal toxoplasmosis in free-ranging endangered 'Alala from Hawaii. J Wildl Dis 36(2):205–212. https://doi.org/10.7589/0090-3558-36.2.205
Wright BR, Hogg CJ, McLennan EA, Belov K, Grueber CE (2021) Assessing evolutionary processes over time in a conservation breeding program: a combined approach using molecular data, simulations and pedigree analysis. Biodivers Conserv 30(4):1011–1029. https://doi.org/10.1007/s10531-021-02128-4
Wyss F, Wenker C, Hoby S, Gardelli B, Studer-Thiersch A, Von Houwald F, Schumacher V, Clauss M, Doherr MG, Häfeli W, Furrer S, Béchet A, Robert N (2013) Factors influencing the onset and progression of pododermatitis in captive flamingos (*Phoenicopteridae*). Schweiz Arch Tierheilkd 155(9):497–503. https://doi.org/10.1024/0036-7281/a000499
Xu Y, Si Y, Wang Y, Zhang Y, Prins HH, Cao L, de Boer WF (2019) Loss of functional connectivity in migration networks induces population decline in migratory birds. Ecol Appl 29(7):e01960. https://doi.org/10.1002/eap.1960
Zabala J, Trexler JC, Jayasena N, Frederick P (2020) Early breeding failure in birds due to environmental toxins: a potentially powerful but hidden effect of contamination. Environ Sci Technol 54(21):13786–13796. https://doi.org/10.1021/acs.est.0c04098
Zhu J, Gosnell JS, Akallal L, Goltsman M (2023) Fear changes traits and increases survival: a meta-analysis evaluating the efficacy of antipredator training in captive-rearing programs. Restor Ecol 31(3):e13674. https://doi.org/10.1111/rec.13674

Artificial Intelligence and Its Impact on Predicting Embryo and Sperm Characteristics

Marcelo Fábio Gouveia Nogueira, Alexandre Mendes de Almeida Júnior, Matheus Búbola Duarte, and José Celso Rocha

Abstract

Currently, artificial intelligence (AI) is pervasive in several tasks, applications, and support services. In the medical field, AI already is used for auxiliary image diagnostics and to predict some disease signals. However, little is done in the veterinary field with domestic animals. This chapter intends to introduce the basic concepts of AI (and its specific ramifications as artificial neural networks and genetic algorithms) and a state-of-art of its application with mouse and cattle embryos and/or sperm. Bovine embryo has, by far, been the main issue of AI studies. The development of AI-based software—to automatically evaluate the morphological quality of in vitro-produced bovine embryos—was an early step. Intuitively, the more daring goal should be the predictive application towards the pregnancy success of those embryos. Also, and briefly, we will be commenting on some comparative details of the AI use in human embryos and the prospective application to other species (such as equine embryos).

Keywords

Artificial neural network · Genetic algorithm · Spermatozoa · Blastocyst · Cattle

M. F. G. Nogueira · A. M. de Almeida Júnior · M. B. Duarte · J. C. Rocha (✉)
Department of Biological Sciences, School of Sciences, Humanities and Languages, São Paulo State University (UNESP), Assis, Brazil
e-mail: marcelo.fabio@unesp.br; matheus.bubola@unesp.br; jose.celso@unesp.br

J. C. Gardón, K. Satué Ambrojo (eds.), *Assisted Reproductive Technologies in Animals Volume 2*, https://doi.org/10.1007/978-3-031-87198-6_9

1 Introduction

Technological advancement has reached a new level with the introduction of artificial intelligence (AI), a technology that is having a disruptive impact across various sectors of society. AI utilizes a wide range of techniques and models, such as computer vision, deep learning, machine learning, and symbolic AI, enabling machines to perceive, analyze, and learn from data. In addition, it enables decision-making and the generation of rapid responses. These capabilities allow for rapid decision-making and response generation, driving extensive adoption across diverse economic and social sectors worldwide. Currently, it is practically impossible to discuss advances in education, transportation, healthcare, agriculture, and manufacturing without considering the automation of these processes through AI models (Rawas 2024).

In healthcare, AI is emerging as a powerful tool, offering significant advances in personalized medical care, diagnostic precision, and the improvement of clinical trials. More sophisticated AI models are pushing the boundaries beyond human capability by processing and interpreting vast amounts of data, providing fast and reliable responses for clinical use. The implementation of these techniques in medical devices promises immediate information delivery to healthcare professionals, potentially aiding in decision-making processes. Moreover, AI can connect complex patient data—such as age, medical history, and lifestyle habits—to their pathological condition, identifying patterns that would be impossible for doctors to recognize with the same speed and accuracy. Although still in development, it is expected that AI will eventually provide reliable information in a democratic and transparent manner, reaching even the most underprivileged social sectors (Zhang et al. 2024).

More advanced AI models, such as large language models (LLMs), are particularly relevant in medical applications In fact, the programming language has been shown to be capable of answering complex questions assertively, even achieving high scores in medical tests, such as the Med-PaLM2 model, which achieved an accuracy of 86.5% in the United States Medical Licensing Examination (USMLE) (Singhal et al. 2023). Another advantage is the integration of AI with retrieval-augmented generative models (RAG), capable of generating responses based on relevant documents and databases and retrieved responses, avoiding the generation of false or biased results. The automated operation of these techniques is crucial for obtaining reliable and fast responses, as it eliminates errors from mechanical and repetitive tasks (Zhang et al. 2024). Rodrigues et al. (2024) propose the application of deep learning (DL) networks beyond the morphological analysis of magnetic resonance images, aiming at the prognosis of several neural diseases, such as Alzheimer's and Huntington's disease. Therefore, AI was used to automate the segmentation process of the hypothalamus and subregions for volumetric and structural evaluation, considering different modalities and image resolution, which presented satisfactory results (Rodrigues et al. 2024). Although advances in AI present promising results, its applicability in medicine still faces resistance, often based on its own technological limitations and clashes with ethical and regulatory issues.

Some constraints on the broader use of AI relate to the need to overcome technical limitations, such as performance variability, unpredictability, and fluctuations due to the lack of real intelligence. One of the most relevant challenges is to solve the occurrence of "hallucinations" generated by language models—i.e., incorrect information that appears plausible. Hallucinations in AI models can seriously compromise the accuracy of diagnoses and treatment decisions, especially when these systems are used autonomously. In parallel, ethical issues surrounding the acquisition of confidential patient data also pose significant restrictions to the progress of algorithm testing and analysis (Zhang et al. 2024).

The profound impact of AI on health, both human and animal, is undeniable, including in more specific areas such as assisted reproductive technologies (ART). In humans, ART has emerged as a remarkable biomedical achievement, being able to significantly improve pregnancy success in cases of infertility (Amjad et al. 2024). Conversely, ART in animal reproduction is often associated with the enhancement and control of genetic inheritance. In cattle, in vitro embryo production (IVP) has gained scientific and technological prominence, driven by the growing economic relevance of livestock farming in some countries and in the global economy. In both contexts, AI can play a central role in the evolution and success of ART by optimizing and standardizing each stage of the process—from selecting high-quality gametes, performing in vitro fertilization (IVF), to culturing, selecting, and assessing the quality of embryos, and finally, to implanting and cryopreserving gametes and embryos. New technologies based on complex data analysis and highlight automation present important results in overcoming suboptimal ART conditions, where models such as machine learning (ML) stand out as sophisticated techniques with extreme application potential (Amjad et al. 2024).

In cattle, suboptimal conditions in IVP stages still represent a significant challenge, resulting in low blastocyst production rates per oocyte, reduced developmental competence, lower pregnancy probabilities, and poor embryo cryotolerance compared to their in vivo-derived counterparts. These limitations are related both to inherent differences between in vitro and in vivo conditions and to the lack of consistent and reproducible criteria and methods for objectively evaluating quality and success at each stage. In this context, AI models based on ML and DL expand the possibilities for analyzing large experimental datasets, generating precise and rapid responses superior to those obtained by conventional manual methods. ML and DL techniques enable the prediction of outcomes, pattern discovery, and more comprehensive assessments, particularly in the evaluation of gamete and embryo quality. For example, neural networks have been shown to be an efficient tool for analyzing sperm images, evaluating morphological characteristics, sperm subpopulations, and plasma and nuclear integrity, enabling the identification of factors associated with fertility (Campanholi et al. 2023). Additionally, these models would also be able to evaluate the cumulus-oocyte complex, providing insightful data regarding oocyte quality (Cavusoglu et al. 2023). The collection of such information is crucial for improving in vitro maturation and IVF success rates, which are critical for obtaining high-quality bovine embryos.

Another possible AI application would be the automated selection of high-quality embryos. Using detailed morphological data, such as trophectoderm area, zona pellucida thickness, and inner cell mass morphology, ML algorithms have demonstrated the ability to identify embryos with higher success potential, reducing intersubjective variability in assessments conducted by embryologists. In cattle, the standard evaluation method involves stereomicroscopy based on three scores recommended by the International Embryo Technology Society. Although widely used, the predictive value of this method directly depends on the embryologist's experience and intrinsic factors related to the professional's well-being, such as fatigue and emotional state. Therefore, AI would offer benefits in terms of time efficiency and cost reduction associated with manual labor, as well as greater precision in measuring the results of different in vitro production systems. This would include classifying systems that produce high-quality embryos, predicting pregnancy success rates (i.e., the percentage of recipients' pregnancies after embryo transfer), and recommending additional techniques that can be safely employed, such as cryopreservation (Rocha et al. 2017b).

Beyond embryo morphology analysis, AI enables the correlation of specific information from recipient females, as proposed in human ART, by analyzing physiological parameters and estrous cycles, promoting more accurate predictions about the success of the technique or the number of cycles needed for a successful pregnancy. Despite the clear benefits of ART, implementing these technologies also faces significant challenges. The quality, quantity, and homogeneity of the data used to train ML models are essential for improving algorithm performance. Small inconsistencies or insufficient information can compromise result accuracy, and for this reason, AI models must undergo continuous testing and data refinement stages (Amjad et al. 2024).

2 Artificial Intelligence

With the rapid technological advance of computer hardware and the advent of ChatGPT, AI is becoming extremely widespread. Its growing use is mainly due to its robust applicability to problems that cannot be easily solved by humans or traditional computing structures (Haenlein and Kaplan 2019). AI tools can process and analyze large amounts of data in a short space of time (Tobin et al. 2019). However, the history of AI dates back to the 1930s with Alan Turing and the creation of the term artificial intelligence in 1956. Figure 1 illustrates the timeline of AI to the present day.

2.1 Artificial Neural Network

The artificial neural network (ANN) is an artificial intelligence technique that tries to mimic the human nervous system, i.e., it is based on the functioning of human neurons. However, it uses mathematical and computational parameters to find

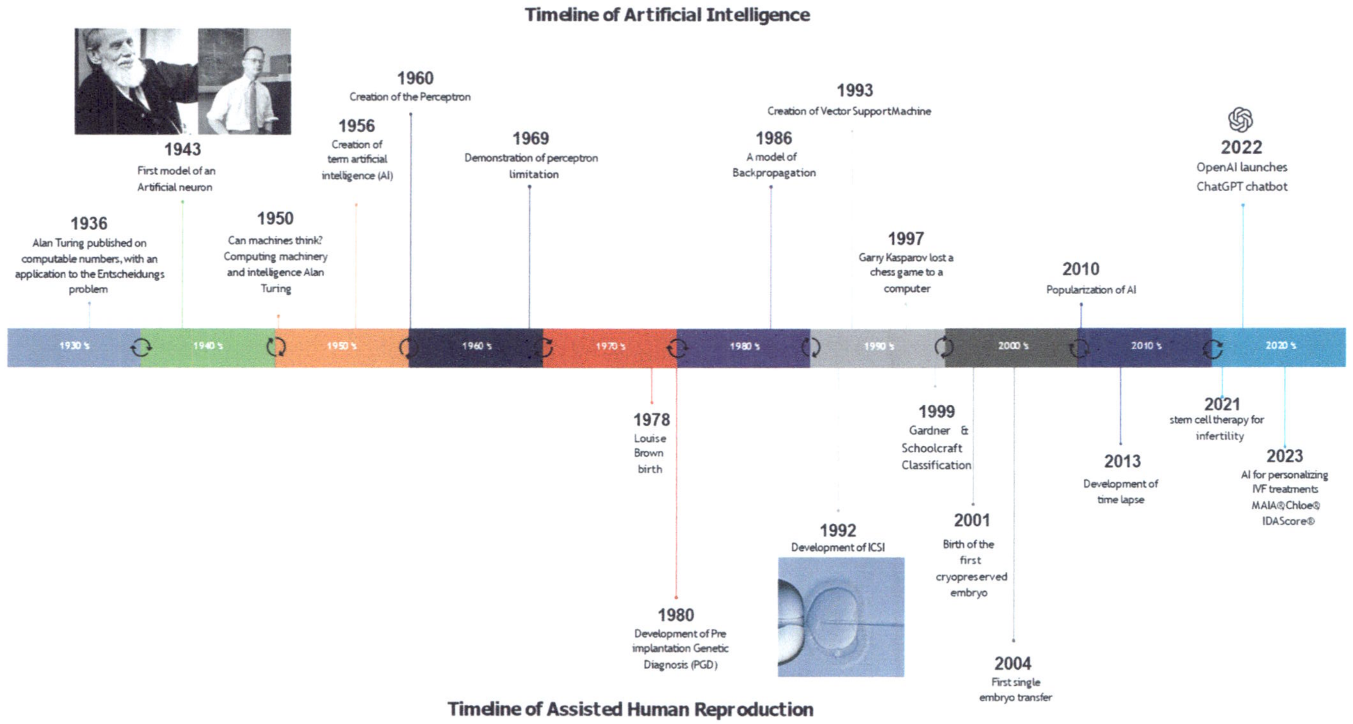

Fig. 1 AI timeline. (Adapted from Fernandez et al. 2020)

patterns and solve problems. Currently, one of the most widely used ANN configurations are multilayer perceptron artificial neural networks (MLP), which have an architecture made up of three levels of layers (input, hidden, and output; Fig. 2). The input layer is where the variables of the problem studied are presented to the ANN and the output layer is where the results are obtained. The intermediate (hidden) layers are responsible for processing the information, performing the weighted combination of the inputs, applying the activation functions, and adjusting the weights of the connections between the neurons, thus enabling the ANN to extract relevant characteristics of the problem studied. With this, it is expected that the ANN will learn to perform classifications, regressions, or other data processing tasks pertinent to the object of study (Fogel et al. 1997; Basheer and Hajmeer 2000; Desai and Shah 2021).

For the ANN to be trained, it needs a learning algorithm—which consists of adjusting the weights of the processing connections for the highest possible predictive accuracy. This can be of two types: the supervised algorithm—in which the network is provided with sets of input/output data examples, adjusting its weights based on the output template, i.e., supervising the result and correcting it—and the unsupervised algorithm, in which the learning process does not require sets of examples (template) for the network to learn (Silva et al. 2010; Braga et al. 2000). Among the supervised learning algorithms is Backpropagation. This algorithm is divided into two stages: feedforward and backforward. In the feedforward stage, the data provided in the input layer is propagated in a single direction in the ANN to the output layer. The responses produced by the output are then compared to the values of the standard responses (template) and, if they are greater than a previously established error in relation to the expected responses, the back forward stage begins. At this point, the errors are backpropagated to adjust the network weights, culminating in the input layer, where the feedforward stage begins again. In this way, the action

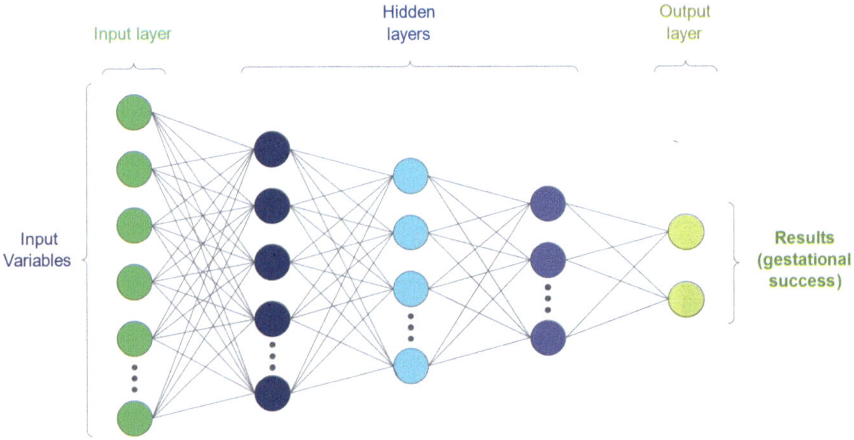

Fig. 2 Architecture of an Artificial Neural Network Multilayer Perceptron (MLP)

of this learning algorithm allows for a gradual reduction in the errors produced by the ANN's responses, compared to those desired (Braga et al. 2000).

Three stages are required to evaluate the performance of an ANN model: training, validation, and testing. During training, the model is fed a set of data to learn about the problem, adjusting the weights as necessary. In the validation stage, a separate set of data (never fed to the ANN) is used to evaluate the model's performance and adjust the hyperparameters, which control the learning process. Finally, in the test stage, an independent data set (also never made available to the ANN) is used to evaluate the model's final performance, generating predictions for the output values (Abiodun et al. 2018; Santos et al. 2005; Braga et al. 2000).

In the case of MLP, the inputs need to be a set of data related to the desired output, usually transformed into numerical values. In this sense, it is necessary for an image—for example of a murine, bovine, or human embryo—to be represented by mathematical values. To do this, it is necessary to apply digital image processing techniques, using filters such as the Hough transform (Atherton and Kerbyson 1999) to determine the circularity of the embryo, and texture analysis (Haralick et al. 1973; Tuceryan and Jain 1993; Soille 2013), or the Watershed transform (Beucher 1992)—which proposes a morphological approach to the problem by interpreting surfaces—can mathematically represent the embryo. Figure 3 illustrates an example of complete segmentation of a bovine embryo.

Rocha et al. (2017a) used the digital processing technique and obtained 36 mathematical variables derived from a digital image of a bovine embryo. These variables, defined to represent the main characteristics of the blastocyst, were reduced,

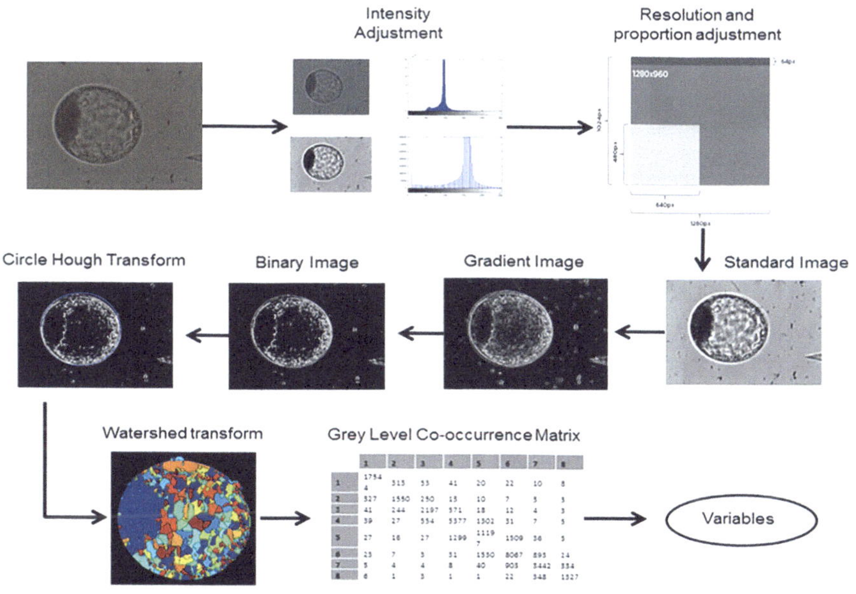

Fig. 3 Flowchart illustrating the segmentation of a bovine embryo image produced in vitro. (From Rocha et al. 2017a)

after collinearity analysis, to 24 and used as input for an ANN that obtained 84% accuracy for blind test data.

Using the same digital image processing technique, but on human embryos, Chéles et al. (2022) determined 33 mathematical variables that characterize the blastocyst (Fig. 4).

In research with murine embryos, Matos et al. (2014) used 12 mathematical variables, extracted from two-dimensional images of blastocysts, as input for an ANN and obtained an accurate prediction of 75% for the blind test data.

Another technique widely used today is deep learning (DL). Deep learning (Alzubaidi et al. 2021) is a sub-area of machine learning. DL uses deep neural networks in its architecture. These networks are so called because their architecture is made up of multiple layers of artificial neurons, to find non-linear relationships more efficiently than traditional neural networks, which use fewer layers and neurons (Dong et al. 2021).

In particular, within the area of DL, there is an AI technique called convolutional neural networks (CNNs; Yao et al. 2019) which are suitable for image manipulation. CNNs are particularly suitable for computer vision tasks, including pattern recognition, image classification, object detection, and image segmentation.

Thus, if we consider an image in the form of a matrix of pixels and each pixel is represented by numerical values, this image can be a CNN input. When we apply this technique to the image, it scans the image in small regions, capturing local features (such as edges and patterns), a process called convolution. It then produces a

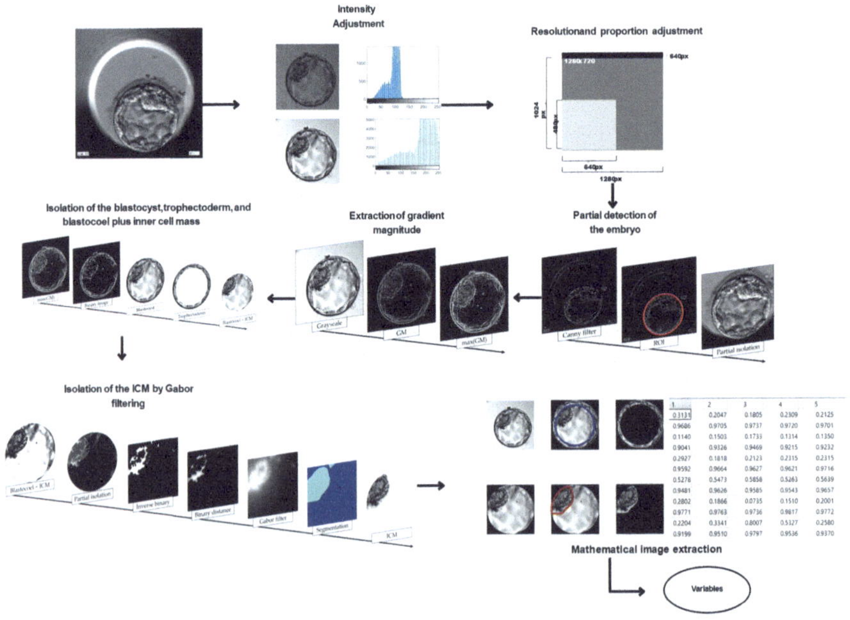

Fig. 4 Illustration of the segmentation process of a human blastocyst. (Adapted from Chéles et al. 2022)

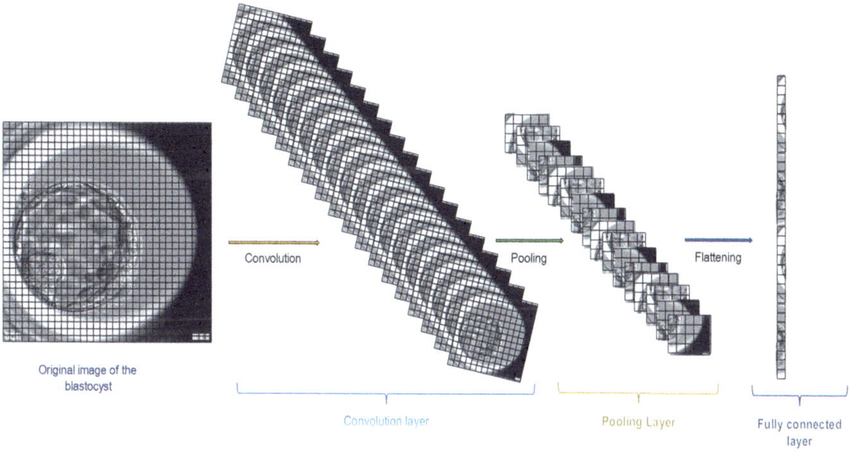

Fig. 5 Illustration of the extraction of a CNN's feature map

new matrix, now with the data of interest filtered out. This process is repeated for several layers and at the end a feature map is extracted from the initial image (Ajit et al. 2020; Subeesh and Mehta 2021; Li et al. 2021). This feature map (Fig. 5) can be used, for example, as input for an ANN MLP in embryonic classification.

However, using deep learning, and CNNs in particular, requires a great deal of computational effort (Alzubaidi et al. 2021), largely due to the need for a large amount of data for training and the complexity of their architecture.

2.2 Genetic Algorithm

Due to the computational effort required to find a suitable AI for a proposed problem and the complexity of its architecture, it is necessary to use techniques to optimize the AI learning process. In this sense, the genetic algorithm technique is one of the most suitable.

The genetic algorithm (GA) is an AI technique used to optimize processes. GA simulate the biological evolution of species, as proposed by Darwin and Wallace (Eiben and Smith 2015). The technique begins with the random creation of a population of candidate solutions, for example, a population of ANNs. Each of these ANNs has its architecture and functions represented in a chromosome, made up of the parameters to be optimized. The population is then evaluated for its suitability or quality in relation to the problem in question. Based on the suitability of the solutions, the best ones are selected for the next-generation population. In this way, the fittest ANNs (i.e., the most accurate at solving the problem) have a higher probability of being selected. From the second population onwards, recombination (or crossing over, the production of recombination between the best individuals in the population), mutation (changes to the chromosomes of individuals), and migration (the addition of random, unknown individuals) are carried out. The process of selection, recombination, mutation, and migration is repeated for several generations until a stopping criterion is reached. This criterion can be a maximum number of generations, an acceptable solution, or a predefined condition. In this way, the application of the GA technique (Fig. 6) allows for sufficient variation to identify optimal solutions to the problem in question (Tanomaru 1995; de Lacerda and De Carvalho 1999; Carr 2014; Katoch et al. 2021).

Recently, computer software has made it possible to use a combination of numerical data (representative of the problem and question) and images of the problem as input for AI. In this way, it is possible to use the deep learning technique in an even more comprehensive way, achieving very expressive accuracy results which are, however, challenging for computer processing (Fig. 7).

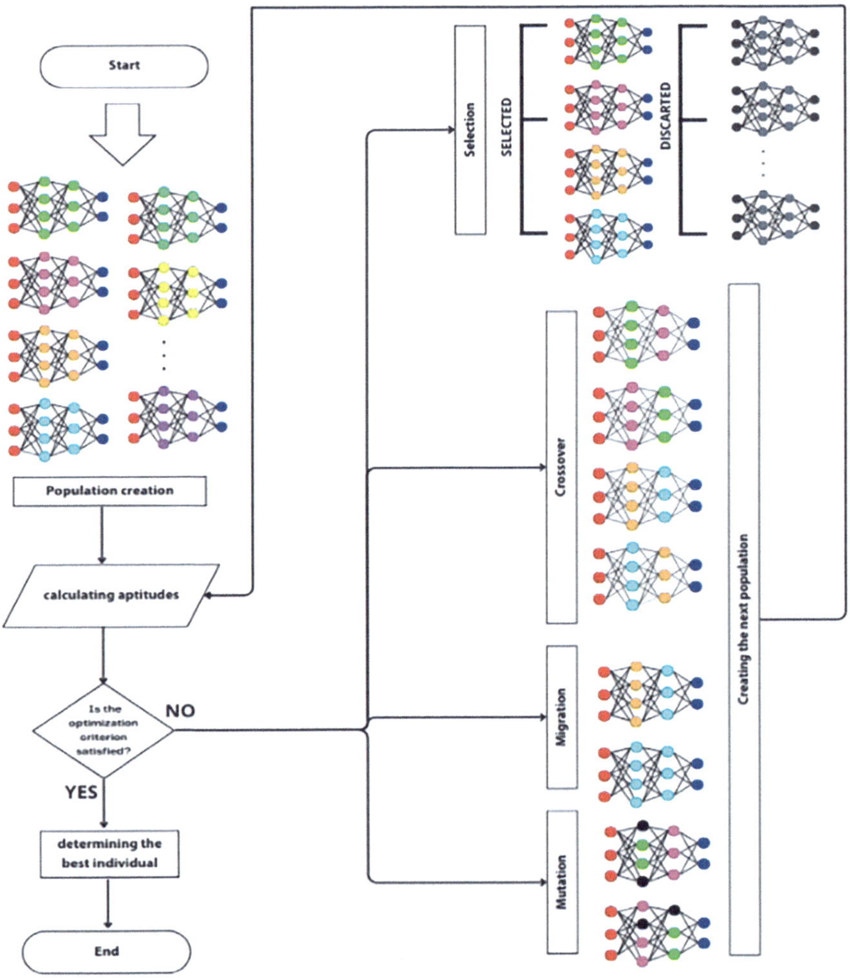

Fig. 6 Flowchart illustrating the application of the GA technique

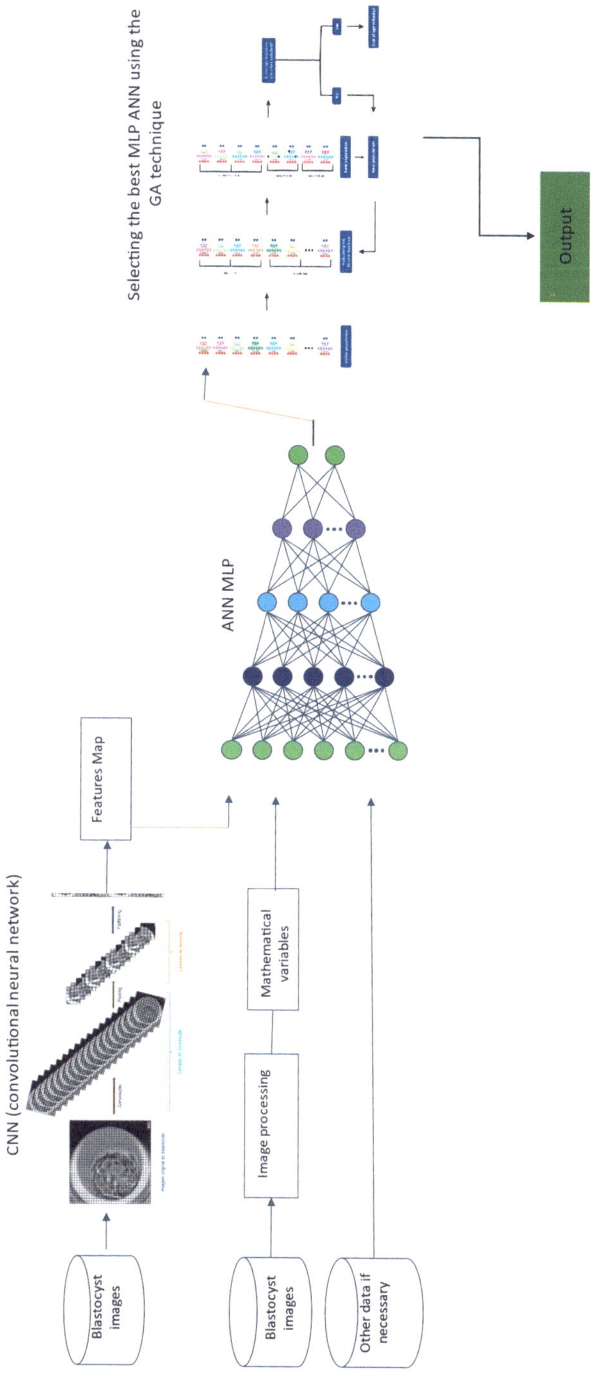

Fig. 7 Flowchart illustrating the application of the AI technique using DL with hybrid inputs—numbers and images

3 Assisted Reproduction Techniques

The use of AI in assisted reproductive technologies (ART) of domestic animals is yet in the beginning. Contrary to the ART in human beings—where in the last 5 years there was a rise of multiple approaches, several published articles and private companies offering commercial devices or software—domestic animals, as bovine species, are in primordial understanding of AI's utilization, value and operation in farm routine. Some of the few examples of AI's use, retrieved from cattle, are on the evaluation of the morphological quality of the embryo (in this case, the blastocyst stage) and on the prediction of PIVE from sperm variables. Translating the experience with human embryos could be a strong basis to jump the species barrier and go on the other apart cattle (e.g., equine species).

3.1 Embryonic Classification

One of the earliest publications on an artificial neural network model for classifying the morphological quality of mammalian embryos (Matos et al. 2014) used in vivo-derived mouse blastocysts as a model (Fig. 8). In this study, 12 numerical variables were extracted manually from two-dimensional images of murine blastocysts. The ANN architecture consisted of an input layer (18 numerical variables), an intermediate layer (containing 13 artificial neurons), and the output layer (i.e., the four grades of embryonic morphological quality: excellent, good, fair, and poor). The ANN was developed using the Backpropagation supervised learning algorithm and the prediction error was calculated using the confusion matrix. In this example, the result obtained was 87.9% correct (training data), 70.0% correct (validation data), and 75.0% correct (test data), which showed that it was a good, proposed model with great potential for application to embryos of other species.

More research has been published on the bovine species. A fully automated system was proposed for the extraction of numerical variables (via digital image processing) followed by ANN analysis for the morphological classification of bovine blastocysts produced in vitro (Rocha et al. 2017a, b). In these studies, the three best ANNs (chosen by the genetic algorithm) provided small variations in predictive accuracy, as observed among experienced embryologists (who analyze the same embryos with different perceptions and results). This observation—that using more than just one ANN for prediction—was important and followed the choice (in these studies) of choosing 3 experienced embryologists to analyze the embryo images. The use of three embryologists to analyze the same digital image of the embryo makes it possible to exclude outlier information (by using the mode of the three analyses). Thus, the supervised training of ANNs, using the mode value, introduced less bias and increased the robustness of the artificial neural network in contrast to when the training was based on a single evaluator (as done in the work by Matos et al. 2014). The result for predicting embryo morphological quality (in 3 quality grades: excellent/good, fair, and poor) was 76.4% with the test data. More importantly, the Kappa index (which measures the agreement among different evaluators

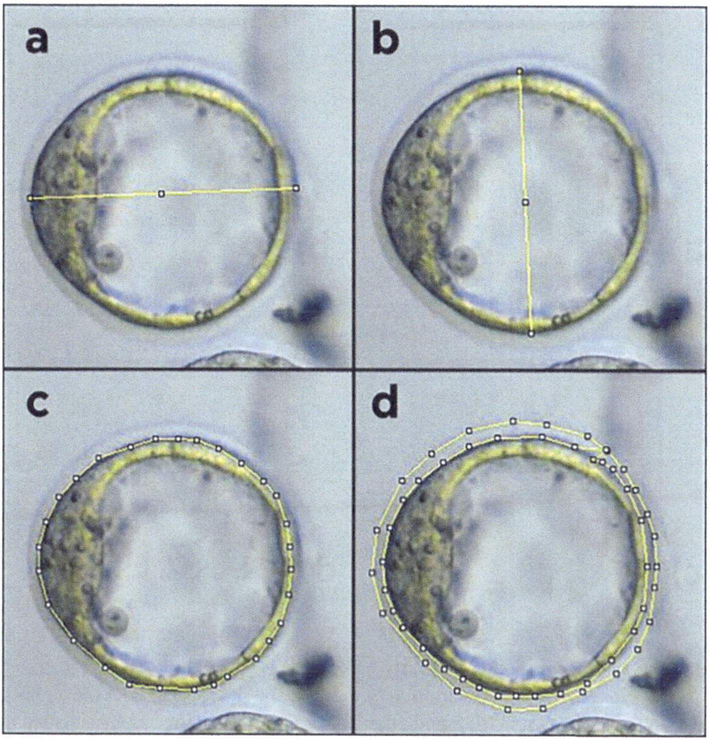

Fig. 8 Example of image processing from a mouse blastocyst to retrieve numerical variables and to use in an ANN. (From Matos et al. 2014)

on the same object of analysis) between the three best ANNs (0.616) was better than the Kappa index among the three experienced embryologists used to validate the mode (i.e., the training template), which was 0.571 (Rocha et al. 2017a). This showed that the result, using the mode between three ANNs, is more constant and reproducible than the mode between three humans (with experience in evaluating embryos). This work also proposed a user-friendly graphic interface so that the lay user could receive the information after processing the image and using the ANN (Fig. 9).

A new layer of automation was proposed with the use of the smartphone to capture the image of the bovine blastocyst, instead of using the inverted microscope as in the previously mentioned works (Rocha et al. 2017a, b). By using the smartphone camera to capture the digital image of the bovine embryo—juxtaposing it with the eyepiece of the stereoscopic microscope—it would be easier, in terms of equipment, to access the use of the previously trained and published ANN (Ciniciato et al. 2017; Nogueira et al. 2018). Although this approach proved to be feasible—that is, the smartphone would only be the tool for capturing the digital image and sending it wirelessly to a server computer, which would perform the AI analysis using the previously trained ANN software—the loss of resolution of the primary digital

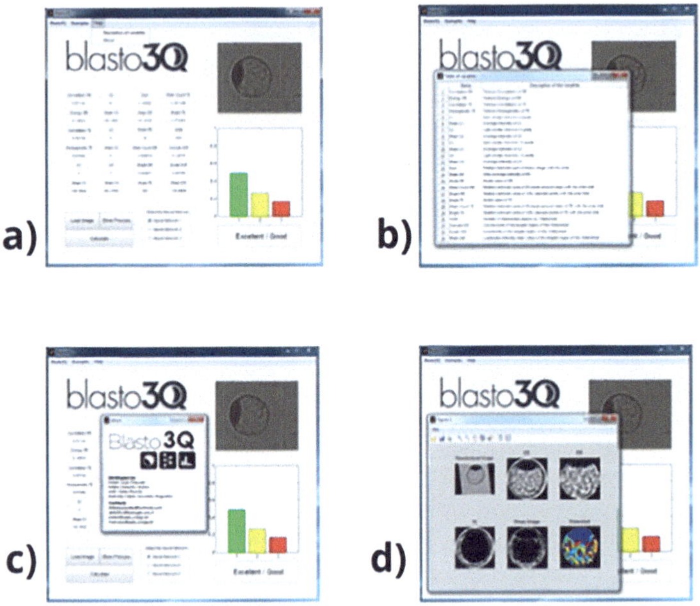

Fig. 9 Tabs and buttons of the user-friendly interface. (**a**) tab with the variables and about the authors; (**b**) tab with the nominal description of the 24 variables; (**c**) tab with the authors' details and e-mail correspondences; and (**d**) Standard, ER, RR, TE, Intersection, and Watershed images of the selected embryo. (From Rocha et al. 2017a)

image (in relation to that obtained with the inverted microscope by Rocha et al. 2017a; Fig. 10) negatively impacted the system's accuracy. In general, the images captured by the smartphone were correctly segmented and processed in 77.8%, and 66.7% were analyzed with results in agreement with the control group, in which the digital image was obtained via an inverted microscope (Nogueira et al. 2018).

3.2 Sperm Assessment

Few if any work has been made on the use of AI to select sperm cells from domestic animals. A recent article focusing on the bovine species tried to correlate parameters from subpopulations of spermatic cells to predict in vitro embryo production (Campanholi et al. 2023). In this work, the authors used 18 input variables from sperm to predict the output of embryos from PIVE in the Senepol breed. Of the three best ANNs obtained, one of them achieved an accuracy of 90.7% (test)—although it dropped to 72.4% when a new data set was tested (Fig. 11). It is a potential mark that AI can be used on the male gamete to predict the overall result (i.e., embryo production following in vitro production). Clear is that the use of only sperm cell variables (to predict embryo production) is far less reliable than adding

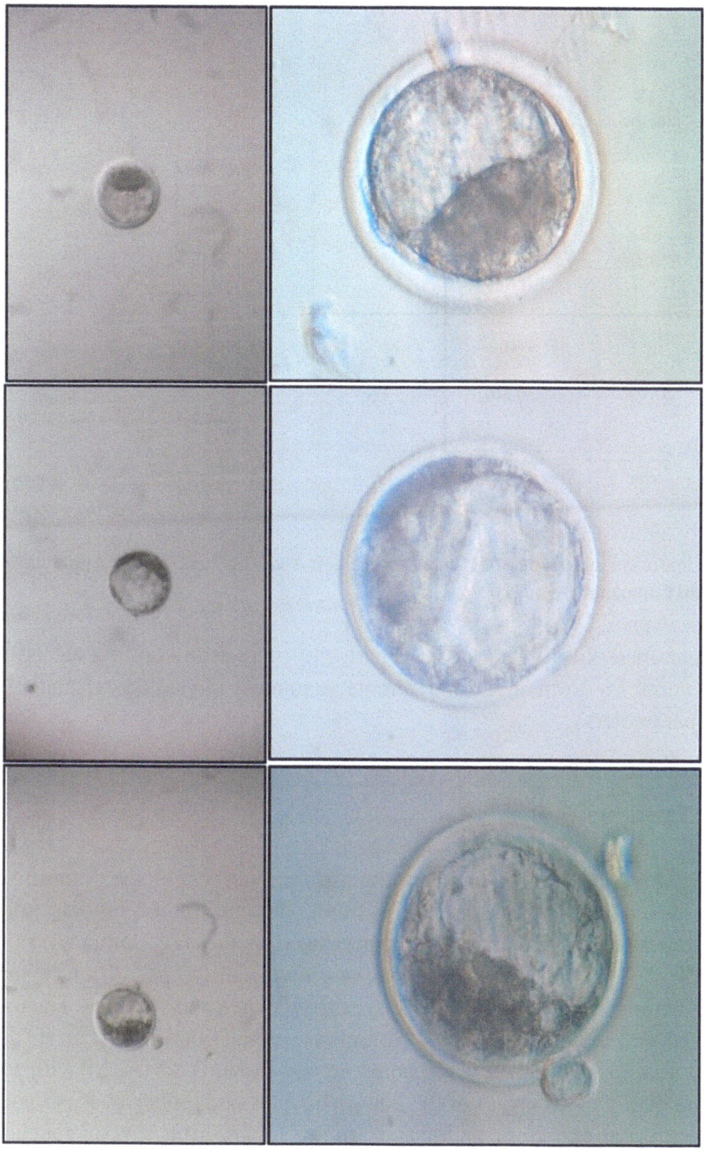

Fig. 10 Illustrative images captured by the smartphone without a macro lens (left column) and with an inverted microscope (right column), from three different embryos (upper, middle, and bottom lines; from Nogueira et al. 2018)

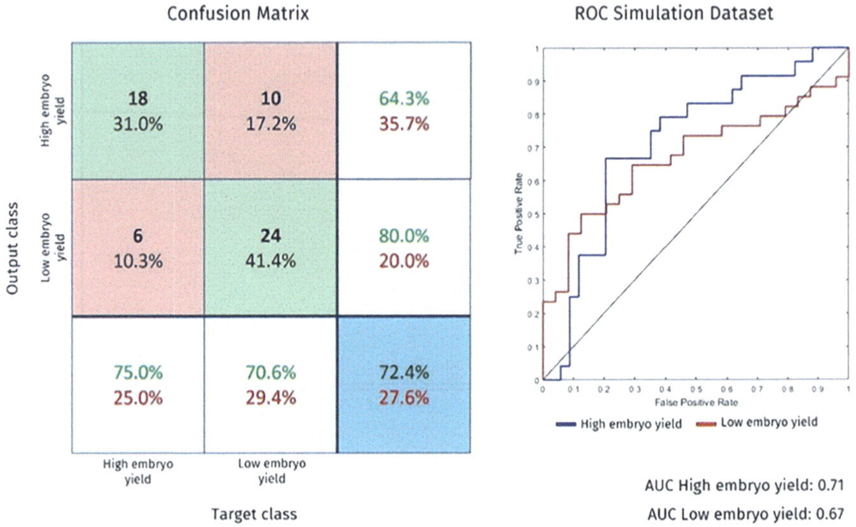

Fig. 11 Confusion matrix and ROC curve for the best ANN architecture, when applying blind test data. (From Campanholi et al. 2023)

female parameters too. The ideal scenario, to predict any embryonic issue, is to retrieve variables from male and female (gametes, physiological and biodemographic parameters).

4 Conclusions

The use of AI to assist reproductive techniques and process on the domestic animal-assisted reproduction is feasible and a trend. Beyond animal models (mice) and cattle, AI experience in a much more experienced species (i.e., humans) could bring insights to the equine species. This one is growing in the use of intracytoplasmic sperm injection (ICSI) to produce embryos from mares and stallions. The use of the human species' experience with ICSI embryo production (and AI from the time-lapse frame's movie) could bring some easier attempts to try AI with variables (numerical and from images) of the equine oocytes and embryos. It is logical that, from the murine and cattle published articles (apart from several recent human papers), the following species should be equine, ovine, caprine, or canine. The discrepancy of embryonic morphology and other subtle variables (among different species) should be in mind to avoid a non-sustainable expectancy of the AI's accuracy when new species of domestic animals are attempted.

Acknowledgments We would like to thank the funding agency São Paulo Research Foundation (FAPESP) for the grants #2006/06491-2, #2011/06179-7, #2012/20110-2, #2012/50533-2, #2017/19323-5, #2018/19053-0, #2018/24252-2, #2019/26749-4, #2020/07634-9, #2021/11747-6 and #2024/12714-2.

References

Abiodun OI, Jantan A, Omolara AE, Dada KV, Mohamed NA, Arshad H (2018) State-of-the-art in artificial neural network applications: a survey. Heliyon 4(11):e00938. https://doi.org/10.1016/j.heliyon.2018.e00938

Ajit A, Acharya K, Samanta A (2020) A review of convolutional neural network. In: 2020 International Conference on Emerging Trends in Information Technology and Engineering (ic-ETITE). IEEE, Piscataway, NJ, pp 1–5. https://doi.org/10.1109/ic-ETITE47903.2020.049

Alzubaidi L, Zhang J, Humaidi AJ, Al-Dujaili A, Duan Y, Al-Shamma O, Farhan L (2021) Review of deep learning: concepts, CNN architectures, challenges, applications, future directions. J Big Data 8:1–74. https://doi.org/10.1186/s40537-021-00444-8

Amjad F, Hajari MA, Dalman A, Baharvand H, Montazeri L (2024) Bioengineering: a promising approach for standardization and automation of assisted reproductive technology. F S Rev 5(1):1–21. https://doi.org/10.1016/j.xfnr.2024.01.002

Atherton TJ, Kerbyson DJ (1999) Size invariant circle detection. Image Vis Comput 17(11):795–803. https://doi.org/10.1016/S0262-8856(98)00160-7

Basheer IA, Hajmeer M (2000) Artificial neural networks: fundamentals, computing, design, and application. J Microbiol Methods 43(1):3–31. https://doi.org/10.1016/S0167-7012(00)00201-3

Beucher S (1992) The watershed transformation applied to image segmentation. Scanning Microsc 1992(6):28. https://digitalcommons.usu.edu/microscopy/vol1992/iss6/28

Braga ADP, Carvalho ACPLF, Ludermir TB (2000) Redes neurais artificiais: teoria e prática. Editora LTC. ISBN: 9788521615644

Campanholi SP, Garcia Neto S, Pinheiro GM, Nogueira MFG, Rocha JC, Losano JDA, Siqueira AFP, Nichi M, Assumpção MEOD, Basso AC, Monteiro FM, Gimenes LU (2023) Can in vitro embryo production be estimated from semen variables in senepol breed by using artificial intelligence? Front Vet Sci 10:1254940. https://doi.org/10.3389/fvets.2023.1254940

Carr J (2014) An introduction to genetic algorithms. Whitman College Mathematics Department, Walla Walla, WA, pp 1–40

Cavusoglu T, Gokhan A, Sirin C, Tomruk C, Kilic KD, Olmez E, Er O, Gullu K (2023) Classification of bovine cumulus-oocyte complexes with convolutional neural networks. Med Rec 5(3):489–495. https://doi.org/10.37990/medr.1292782

Chéles DS, Ferreira AS, de Jesus IS, Fernandez EI, Pinheiro GM, Dal Molin EA, Nogueira MFG (2022) An image processing protocol to extract variables predictive of human embryo fitness for assisted reproduction. Appl Sci 12(7):3531. https://doi.org/10.3390/app12073531

Ciniciato DS, Takahashi MB, Nogueira MFG, Rocha JC (2017) Potential use of smartphone as a tool to capture embryo digital images from stereomicroscope and to evaluate them by an Artificial Neural Network. In: International Conference on Computer-Human Interaction Research and Applications, pp 185–189. https://doi.org/10.5220/0006518501850189

de Lacerda EG, De Carvalho ACPLF (1999) Introdução aos algoritmos genéticos. Sistemas inteligentes: aplicaçoes a recursos hıdricos e ciências ambientais 1:99–148

Desai M, Shah M (2021) An anatomization on breast cancer detection and diagnosis employing multi-layer perceptron neural network (MLP) and convolutional neural network (CNN). Clin eHealth 4:1–11. https://doi.org/10.1016/j.ceh.2020.11.002

Dong S, Wang P, Abbas K (2021) A survey on deep learning and its applications. Comput Sci Rev 40:100379. https://doi.org/10.1016/j.cosrev.2021.100379

Eiben AE, Smith JE (2015) Introduction to evolutionary computing. Springer, Berlin, Heidelberg. https://doi.org/10.1007/978-3-662-44874-8

Fernandez EI, Ferreira AS, Cecílio MHM, Chéles DS, de Souza RCM, Nogueira MFG, Rocha JC (2020) Artificial intelligence in the IVF laboratory: overview through the application of different types of algorithms for the classification of reproductive data. J Assist Reprod Genet 37(10):2359–2376. https://doi.org/10.1007/s10815-020-01881-9

Fogel DB, Wasson EC, Boughton EM, Porto VW (1997) A step toward computer-assisted mammography using evolutionary programming and neural networks. Cancer Lett 119(1):93–97. https://doi.org/10.1016/S0304-3835(97)00259-0

Haenlein M, Kaplan A (2019) A brief history of artificial intelligence: on the past, present, and future of artificial intelligence. Calif Manage Rev 61(4):5–14. https://doi.org/10.1177/0008125619864925

Haralick RM, Shanmugam K, Dinstein IH (1973) Textural features for image classification. IEEE Trans Syst Man Cybern 6:610–621. https://doi.org/10.1109/TSMC.1973.4309314

Katoch S, Chauhan SS, Kumar V (2021) A review on genetic algorithm: past, present, and future. Multimed Tools Appl 80:8091–8126. https://doi.org/10.1007/s11042-020-10139-6

Li Z, Liu F, Yang W, Peng S, Zhou J (2021) A survey of convolutional neural networks: analysis, applications, and prospects. IEEE Trans Neural Netw Learn Syst 33(12):6999–7019. https://doi.org/10.1109/TNNLS.2021.3084827

Matos FD, Rocha JC, Nogueira MFG (2014) A method using artificial neural networks to morphologically assess mouse blastocyst quality. J Anim Sci Technol 56:1–10. https://doi.org/10.1186/2055-0391-56-15

Nogueira MFG, Bertogna Guilherme V, Pronunciate M, Dos Santos PH, Bezerra L, da Silva D, Rocha JC (2018) Artificial intelligence-based grading quality of bovine blastocyst digital images: direct capture with juxtaposed lenses of smartphone camera and stereomicroscope ocular lens. Sensors 18(12):4440. https://doi.org/10.3390/s18124440

Rawas S (2024) AI: the future of humanity. Discov Artif Intell 4(1):1–14

Rocha JC, Passalia FJ, Matos FD, Takahashi MB, Ciniciato DDS, Maserati MP, Nogueira MFG (2017a) A method based on artificial intelligence to fully automatize the evaluation of bovine blastocyst images. Sci Rep 7(1):7659. https://doi.org/10.1038/s41598-017-08104-9

Rocha JC, Passalia FJ, Matos FD, Takahashi MB, Maserati MP Jr, Alves MF, Nogueira MFG (2017b) Automatized image processing of bovine blastocysts produced in vitro for quantitative variable determination. Sci Data 4(1):1–9. https://doi.org/10.1038/sdata.2017.192

Rodrigues L, Bocchetta M, Puonti O, Greve D, Londe AC, França M, Appenzeller S, Rittner L, Iglesias JE (2024) High-resolution segmentations of the hypothalamus and its subregions for training of segmentation models. Sci Data 11(1):1–10. https://doi.org/10.1038/s41597-024-03775-2

Santos AMD, Seixas JMD, Pereira BDB, Medronho RDA (2005) Usando redes neurais artificiais e regressão logística na predição da hepatite A. Rev Bras Epidemiol 8:117–126. https://doi.org/10.1590/S1415-790X2005000200004

Silva IND, Spatti DH, Flauzino RA (2010) Redes neurais artificiais para engenharia e ciências aplicadas. Artliber Editora, São Paulo

Singhal K, Tu T, Gottweis J, Sayres R, Wulczyn E, Hou L, Clark K, Pfohl S, Cole-Lewis H, Neal D, Schaekermann M, Wang A, Amin M, Lachgar S, Mansfield P, Prakash S, Green B, Dominowska E, Arcas BA, Tomasev N, Liu Y, Wong R, Semturs C, Mahdavi SS, Barral J, Webster D, Corrado GS, Matias Y, Azizi S, Karthikesalingam A, Natarajan V (2023) Towards expert-level medical question answering with large language models. arXiv preprint. pp 1–30. https://doi.org/10.48550/arXiv.2305.09617

Soille P (2013) Morphological image analysis, principles and applications. Springer Science & Business Media, Berlin. https://doi.org/10.1007/978-3-662-05088-0

Subeesh A, Mehta CR (2021) Automation and digitization of agriculture using artificial intelligence and internet of things. Artif Intell Agric 5:278–291. https://doi.org/10.1016/j.aiia.2021.11.004

Tanomaru J (1995) Staff scheduling by a genetic algorithm with heuristic operators. In: 1995 IEEE International Conference on systems, man and cybernetics. Intelligent systems for the 21st century, pp 1951–1956

Tobin S, Jayabalasingham B, Huggett S, de Kleijn M (2019) A brief historical overview of artificial intelligence research. Inf Serv Use 39(4):291–296. https://doi.org/10.3233/ISU-190060

Tuceryan M, Jain AK (1993) Texture analysis. In: Handbook of pattern recognition and computer vision, pp 235–276. https://doi.org/10.1142/9789814343138_0010

Yao G, Lei T, Zhong J (2019) A review of convolutional-neural-network-based action recognition. Pattern Recogn Lett 118:14–22. https://doi.org/10.1016/j.patrec.2018.05.018

Zhang P, Shi J, Kamel Boulos MN (2024) Generative AI in medicine and healthcare: moving beyond the 'peak of inflated expectations'. Fut Internet 16:462. https://doi.org/10.3390/fi16120462

The manufacturer's authorised representative in the EU is Springer Nature Customer Service Centre GmbH, Europaplatz 3, 69115 Heidelberg, Germany. If you have any concerns regarding our products, please contact ProductSafety@springernature.com

Printed and bound by CPI Group (UK) Ltd, Croydon, CR0 4YY
01/10/2026
02225929-0007